Gene Amplification in Mammalian Cells

Gene Amplification in Mammalian Cells

A Comprehensive Guide

edited by

Rodney E. Kellems

Baylor College of Medicine
Houston, Texas

CRC Press
Taylor & Francis Group
Boca Raton London New York

CRC Press is an imprint of the
Taylor & Francis Group, an **informa** business

CRC Press
Taylor & Francis Group
6000 Broken Sound Parkway NW, Suite 300
Boca Raton, FL 33487-2742

First issued in paperback 2019

ISBN-13: 978-0-8247-8756-1 (hbk)
ISBN-13: 978-0-367-40266-2 (pbk)

Visit the Taylor & Francis Web site at
http://www.taylorandfrancis.com

and the CRC Press Web site at
http://www.crcpress.com

Library of Congress Cataloging-in-Publication Data

Gene amplification in mammalian cells: a comprehensive guide / edited by Rodney E. Kellems.
p. cm.
Includes bibliographical references and index.
ISBN 0-8247-8756-0
1. Gene amplification. 2. Gene amplification--Research--Methodology. I. Kellems, Rodney E.
QH450.3.G46 1992
599'.087'322--dc20 92-26048
CIP

Foreword

The experimental study of variable gene copy number has a venerable history, starting with Sturtevant's studies on the *bar* locus in *Drosophila* some 65 years ago [1]. Gene duplications are common in bacteria under appropriate selection conditions (see Ref. 2). The demonstration in 1968 of developmentally regulated amplification of genes, most dramatically shown by Brown and Dawid with *Xenopus* ribosomal genes [3], raised the question of whether high-level differentiation specific gene expression was the result of selective gene amplification in higher eukaryotes. With the advent of molecular biology in the mid-1970s, this question was answered with an essentially uniform "no."

Gene amplification remained a "dormant" subject as applied to mammalian cells until 1978–1979 when documentation of gene amplification as a mechanism for acquired resistance in cultured mammalian cells employing molecular biology techniques by the Schimke [4] and the Stark [5] groups opened a new era for the study of gene amplification. It should be noted, however, that their definitive studies were presaged by the cytogenetic studies of Beidler and Spengler with HSRs (homogeneously staining chromosome regions) in methotrexate-resistant Chinese hamster lung cells [6] and of Levan et al. with minute chromosomes of Sewa cells [7]. The finding of gene amplification was initially surprising, inasmuch as both research groups had originally assumed that enzyme overproduction was going to be the result of an "up/promoter" mutation. More recently, a large number of cases of gene amplification in cultured mammalian cells have been found, examples of which are described in this volume. One gets the impression that with appropriate selection protocols, virtually any gene can be amplified, unless it or a closely linked gene is otherwise detrimental to cell growth or survival. For instance, we have reported that the trileucyl peptide calpain inhibitor ALLN, which is hydrophobic, is cytotoxic and resistance can be readily selected in CHO cells that results from amplification of the MDR1 gene [8]. Recently, we repeated the selection procedure with the addition of verapamil to prevent function of the P-170 glycoprotein exit pump and found that resistance develops, which is associated with overproduction (and gene amplification) of an aldo-ketoreductase, an enzyme that inactivates the active form of ALLN (R. Sharma et al., in preparation). Surely the list of amplified genes will continue to grow.

The status of experimental gene amplification in mammalian cells as of 10 years ago was represented in *Gene Amplification*, the outcome of a conference held at the Banbury Conference Center of the Cold Spring Harbor Laboratories in 1981 [9]. This current volume, edited by Rodney Kellems, is a welcome updating of the status of the field. Many of the current contributors were authors in 1982, many are scientific "offspring" of those earlier contributors, and still others are new, vigorous contributors.

What follows are a few brief thoughts about where gene amplification has gone from those early days and where it might be going.

I. AMPLIFICATION IN INDUSTRY

It is surprising to me how rapidly the phenomenon of gene amplification was pursued within a commercial context and what a major role it has played as a means to produce certain proteins for medical usages where bacteria or yeast do not make biologically active proteins and where mammalian cells do. The fact that DHFR amplification (methotrexate resistance) in Chinese hamster ovary cells is the most commonly employed technique commercially is in some part a historical accident of the time (the early 1980s) because scientists who joined various companies were familiar with this system. Clearly, this is an example of extremely rapid technology transfer from the laboratory to industry. It is also important to note that scientists associated with the industry are making important and basic contributions to the field.

II. AMPLIFICATION IN CANCER BIOLOGY

Prehaps the major area in which gene amplification is observed in the "real world" is in cancer biology. Interestingly, although there are examples of gene amplification in clinical resistance (DHFR and methotrexate), it does not appear that gene amplification is a common mechanism for clinical cancer resistance, as opposed to its common occurrence as studied in the laboratory. Why this difference exists is clearly worthy of understanding in the future. On the other hand, oncogene amplification is extremely common. From the time of initial discovery of gene amplification in mammalian cells, various predictions of a role for amplification in carcinogenesis were made [10–13]. Starting in the mid-1980s, reports of oncogene amplification commenced, and now amplification of a variety of oncogenes has become a common, but certainly not universal, theme, with respect to either the cancer type, the oncogene involved, or the percentage of a specific cancer type with a specific oncogene amplification.

Perhaps there are two extreme views of oncogene amplification and cancer. One assertion would be that amplification of an oncogene is a necessary and/or a sufficient event(s) to result in a cancer cell. Ancillary to this view might be a relationship between the copy number of an amplified gene and tumor behavior. In the instances of experimental gene amplification and drug resistance, the correlation between gene copy number and the biological response of cells, i.e., the degree of resistance, is fairly good. However, in the case of oncogene amplification, such simplistic relationships do not appear to hold. The other extreme view would be that oncogene amplification has no role in the malignant process and is a neutral manifestation of an underlying state of a malignant cell that generates a high degree of genetic instability, of which gene amplification is one, but by no means the only, type of genomic alteration found.

Probably there is no simple and single answer, and the understanding will lie somewhere in the middle of these extremes, depending on the oncogene and the stem-cell type involved, including the unique logic of control of proliferation and differentiation of that cell type. Most thinking about gene amplification assumes that overproduction of the oncogene product is the casual end-point. However, an alternative consideration for oncogene amplification might involve binding to amplified DNA segments of regulatory proteins present in limited amounts in cells, thus reducing their availability for regulation of other genes and thereby altering cell regulatory behavior unrelated to expression of an oncogene per se. This type of mechanism may be involved in instances where the degree of oncogene amplification vastly exceeds any copy number that may seem reasonable, based on drug resistance amplifications at clinically reasonable drug levels (say 5- to 10-fold at most).

III. GENE COPY DIFFERENCES AMONG INDIVIDUAL HUMANS

Another area where variable gene copy may play a more significant role than currently understood is in individual responses to various environmental agents. It is most often assumed that all individuals have one copy of a gene per haploid genome. Unless an investigator is highly sensitive to looking for copy number differences, such differences will be overlooked in standard use of restriction fragment analyses. However, there are some examples where gene copy number differences have been reported. For instance, Ohlsson et al. [14] reported a family with an additional copy of a beta-interferon gene that was inherited as a Mendelian trait. What might constitute a selection for this amplification is not clear. Prody et al. [15; also Soreq et al., this volume] reported a father and son with complex amplification of a "true" cholinesterase gene and suggested that chronic exposure to the insecticide parathion may have played a role in selection of this amplification event(s). More recently, Ingelman-Sundberg and his colleagues [16] have reported in preliminary form a family with 10- to 15-fold amplification of the CYP2D gene (a member of the P-450 drug-metabolizing gene family), detected by virtue of lack of response to certain pharmacological agents. Interestingly, the P-450 gene family constitutes an evolutionary amplification process (a member of a multigene family), at least one member of which appears to undergo amplification in germ cell lineages in historical time as well. Thus, the question arises as to how often differences among individuals, perhaps most readily observed in relation to responses to drugs, may result from genetically inherited differences in gene copy number. Perhaps even more interesting are the questions of what exposures, what selection pressures, and what mechanism(s) have led to the fixation of multiple gene copies in germ cells in individuals of a species. The realization that variations in gene copy number occur may lead to new understanding of individual differences among members of a species.

IV. AMPLIFICATION MECHANISMS

A major effort in the field of gene amplification in mammalian cells in recent years has involved attempts to understand the mechanism(s) whereby genes are amplified. Early studies on the definition of amplified DNA segments concluded that a variable-length chromosome was amplified [17,18]. These studies employed cell lines subjected to multiple step selections over long time periods and concluded that multiple recombination events had occurred, within the gene as well as in flanking regions, and indicated a high degree of genomic "flux." More recent studies, described in this volume, have focused on the study of cell populations as soon as possible after the initial amplification event(s) have occurred and have employed studies of the structure of amplified sequences as well as their localization by use of newly available and highly sensitive in situ hybridization techniques. Certain of these types of studies favor some form of sister chromatid exchange phenomenon as a primary event (see Stark, this volume). An interesting and rather common structure of amplified arrays of genes is their existence as inverted (head to head) structures (see Fried et al., this volume). Such structures are most readily accounted for by aberrant replication and recombination models, as discussed in this volume (see Fried et al. and Wahl et al. for extensive discussions). I suspect that the details of how amplified genes are generated and exactly what segment of a chromosome is amplified will be variable from one cell population to another, and that a single mechanism will not account for all instances of amplification, whether they are a part of the initial, i.e., primary, event, or as cells are either maintained under constant selection or subjected to additional selection pressure.

V. AMPLIFICATION OF GENES AS A RESPONSE TO PERTURBATION OF CELL CYCLE PROGRESSION

I have been struck over the years with the frequency with which cell variants with amplified genes bear a number of cytogenetic abnormalities in addition to the cytogenetic consequences of specific gene amplification. These include differing degrees of aneuploidy (polyploidy) and presence of deletions and translocations, including extreme cases of the existence of so-called "marker chromosomes," which basically constitute breakage and recombination of chromosome fragments to constitute unrecognizable new chromosomes (see Fougere-Deschartrette et al. [19] for some examples). Such chromosomal changes, including aneuploidy, deletions, translocations, and amplifications, are, of course, common cytogenetic abnormalities of cancer cells. Thus, an understanding of mechanisms of gene amplification may provide insight into types of genomic alterations (including rearrangement events) that are so very common in cancer and are not explicable within the context of the understanding of point mutational events.

Such "global" cytogenetic changes in genome structure suggest that cell variants selected for a specific amplified gene may have been subjected to events that variously alter overall chromosome integrity. Interestingly, all the selection protocols employed to generate amplification events involve extensive cell killing and the use of critically narrow ranges of drug concentrations to detect amplification events. Any number of studies in my laboratory have shown that pretreatment with various agents, including inhibitors of DNA synthesis, hypoxia, UV light, carcinogens, and gamma radiation, can increase the frequency of gene amplification in rodent cell lines (CHO and mouse) (see Ref. 20 for references). Characteristically, the degree of cell killing, the time of exposure, and the time following removal of the perturbing agent when selection pressure is added are all critical variables. Such treatments markedly increase the frequency of sister chromatid exchanges [21] and result in metaphase chromosomes with varying degrees of chromosome breaks, as well as metaphase spreads with extensive extrachromosomal DNA [21,22]. We have interpreted our studies to indicate that such pretreatments alter normal cell cycle progression events and dissociate DNA synthesis and chromosome condensation from the events of mitosis [22]. Such treatments facilitate sister chromatid exchange phenomena, are likely to alter replication progression, and result in chromosome breakage. A major consequence of such events is cell death. However, recombinational resolution of stalled/altered replication forks or recombinational repair of broken chromosomes can form the substratum for selection of specific alterations, including amplification events (and the potentially reciprocal deletion event). This hypothesis suggests that amplification events are "induced" by any variety of agents or growth conditions that affect cell cycle progression, including, perhaps, the selecting agent itself. Several groups [23,24], employing the Luria-Delbrück fluctuation analysis, have concluded that CAD gene amplification occurs spontaneously and is not induced by the selecting agent (PALA). Although it is, indeed, the case that these workers have obtained subpopulations with differing frequencies (and rates) of CAD gene amplification, it is not clear, given the inherent heterogeneity of such mammalian cell populations (for instance, see Ref. 25), that what was segregated was a preexisting amplification event or a subpopulation of cells with properties (amplifiability or inherent variable response to the drug) that result in differing response to a single selection condition (drug concentration) that is based on results with the entire (potentially heterogeneous) cell population.

VI. AMPLIFICATION PROPENSITY–AMPLIFICATION PHENOTYPE

One of the interesting questions currently being asked is why certain types of cells or cell lines amplify genes readily whereas others do not (see Tlsty and Giulotto and Perry, this volume, for discussions). Various investigators have reported that primary cell cultures or diploid fibroblast cell lines do not amplify genes at any reasonably detectable rate [26–28]. Results from our laboratory indicate that, whether studied by stepwise or single-step protocols, rodent cell lines amplify genes (DHFR, CAD, MDR1) far more readily than do most cell lines derived from human sources (R. Sharma, in preparation). We have recently provided data consistent with the hypothesis that rodent cell lines differ from most human cell lines in the manner in which they enter and exit mitosis when treated with inhibitors of DNA synthesis (aphidicolin, hydroxyurea), disruption of the mitotic spindle apparatus (colcemid), and gamma radiation [29,30]. Most human cell lines, whether transformed or not, display a "stringent" mitotic checkpoint control in that when cell cycle events are perturbed, they "sense" the perturbation and prevent entry into and/or exit from mitosis. This "stringent" mitotic checkpoint control is highly analogous to the behavior of yeast. In contrast, rodent cell lines display a "relaxed" mitotic checkpoint control, and rapidly progress into a second cell cycle and dissociate mitosis from the cell cycle. Interestingly, primary (embryonic) rodent and hamster cells display the "stringent" mitotic checkpoint control [29,30]. In our studies (R. Sharma et al., in preparation), the "relaxed" control property is highly correlated with a high propensity for gene amplification. Such findings are consistent with the concept that disruption of normal cell cycle progression is a prior event, and that gene amplification is one among a number of possible outcomes, and is detected following appropriate selection conditions.

VII. AMPLIFICATION IS A FIELD IN ITS EARLY GROWTH PHASE

The advances in gene amplification have been quite remarkable since the first discoveries some 14 years ago. Given the interest in the subject, its importance, the many forms it takes throughout biology, and the increasing number of excellent investigators in the field, I shall eagerly look forward to the next compendium and the new discoveries and new concepts that will surely develop in the interim.

ACKNOWLEDGMENTS

I thank the many members of my laboratory who have contributed to the field of gene amplification over the past 14 years, many of whom have continued to be active investigators in the field.

REFERENCES

1. Sturtevant, A. M. The effects of unequal crossing over at the *bar* locus in *Drosophila*. *Genetics*, *10*:117–147 (1925).
2. Anderson, R. P., and Roth, J. R. Tandem genetic duplications in phage and bacteria. *Annu. Rev. Microbiol*. *31*:473–505 (1975).
3. Brown, D. P., and Dawid, I. Specific gene amplification in oocytes. *Science*, *160*:272–280 (1968).
4. Alt, F. W., Kellems, R. E., Bertino, J. R., and Schimke, R. T. Selective multiplication of dihydrofolate reductase genes in methotrexate-resistant variants of cultured murine cells. *J. Biol. Chem.* *253*:1357–1370 (1978).
5. Wahl, G. M., Padgett, R. A., and Stark, G. R. Gene amplification causes overproduction of the first three enzymes of UMP synthesis in *N*-(phosphoacetyl 1-aspartate) resistant hamster cells. *J. Biol. Chem.* *254*:8579–8689 (1979).
6. Beidler, J. L., and Spengler, B. A. Metaphase chromosome anomaly: Association with drug resistance and cell-specific products. *Science* *191*:185–187 (1976).
7. Levan, A., Levan, G., and Mitelman, F. Chromosomes and cancer. *Hereditas*, *86*:15–30 (1977).
8. Sharma, R. C., Inoue, S., Roitelman, J., Schimke, R. T., and Simoni, R. D. Peptide transport by the multidrug resistance pump. *J. Biol. Chem.* *267*:5731–5734 (1992).
9. Schimke, R. T., ed. *Gene Amplification*, Cold Spring Harbor Laboratory Press, Cold Spring Harbor, NY, 1982.
10. Paul, M. L. Gene amplification model of carcinogenesis. *Proc. Natl. Acad. Sci. USA*, *78*:2465–2468 (1981).
11. Varshavsky, A. On the possibility of metabolic control of replicon "misfiring" relationship to emergence of malignant phenotypes in mammalian cell lineages. *Proc. Natl. Acad. Sci. USA*, *78*:3673–3677 (1981).
12. Schimke, R. T. Gene amplification, drug resistance, and cancer. *Cancer Res.* *44*: 1735–1742 (1984).

13. Stark, G. R., and Wahl, G. M. Gene amplification. *Annu. Rev. Biochem. 53*:447–491 (1984).
14. Ohlsson, M., Feder, J., Cavalli-Sforza, L. L., and von Gabain, A. Proc. Close linkage of alpha and beta interferons and infrequent duplication of beta interferon in humans. *Proc. Natl. Acad. Sci. USA*, *82*:4473–4476 (1985).
15. Prody, C. A., Dreyfus, P., Zamir, R., Zakut, H., and Soreq, H. De novo amplification within a "silent" human cholinesterase gene in a family subjected to prolonged exposure to organophosphate insecticides. *Proc. Natl. Acad. Sci. USA*, *86*:690–694 (1989).
16. Johansson, I., Dahl, M., Bertilsson, L., Sjoqvist, F., and Ingelman-Sundberg, M. Gene amplification of an active CYP2D gene as a cause of extremely rapid metabolism of debrisoquine. *J. Basic Clin. Physiol. Pharm. 3*:242 (1992).
17. Federspeil, N. A., Beverley, S. M., Schilling, J. W., and Schimke, R. T. Novel DNA rearrangements are associated with dihydrofolate reductase gene amplification. *J. Biol. Chem. 259*:9127–9140 (1984).
18. Giulotto, E., Saito, I., and Stark, G. R. Structure of DNA formed in the first step of CAD gene amplification. *EMBO J. 5*:2115–2121 (1986).
19. Fougere-Deschartrette, C., Schimke, R. T., Weil, D., and Weiss, M. C. A study of chromosomal changes associated with amplified dihydrofolate reductase genes in rat hepatoma cells and their dedifferentiated variants. *J. Cell Biol. 99*:497–502 (1984).
20. Schimke, R. T. Gene amplification in cultured cells. *J. Biol. Chem. 263*:5989–5993 (1988).
21. Hill, A. B., and Schimke, R. T. Increased gene amplification in L5178Y mouse lymphoma cells with hydroxyurea-induced chromosomal aberrations. *Cancer Res. 45*:5050–5057 (1985).
22. Schimke, R. T., Hoy, C., Rice, G., Sherwood, S. W., and Schumacher, S. I. Enhancement of gene amplification by perturbation of DNA synthesis in cultured mammalian cells. *Cancer Cells*, *6*:317–326 (1989).
23. Kemp, T. D., Swyryd, E. A., Bruist, M., and Stark, G. R. Stable mutations of mammalian cells that overproduce the first three enzymes of pyrimidine nucleotide biosynthesis. *Cell*, *9*:541–550 (1976).
24. Tlsty, T. D., Hargolin, B. M., and Lum, K. Differences in the rates of gene amplification in nontumorigenic and tumorigenic cell lines as measured by Luria-Delbrück fluctuation analysis. *Proc. Natl. Acad. Sci. USA*, *86*:9441–9445 (1989).
25. Brown, P. C., Tlsty, T. D., and Schimke, R. T. Enhancement of methotrexate resistance and dihydrofolate reductase gene amplification by treatment of mouse 3T6 cells with hydroxyurea. *Mol. Cell. Biol. 3*:1097–1107 (1981).
26. Wright, J. A., Smith, H. S., Watt, F. M., Hancock, M. C., Hudson, D. L., and Stark, G. R. DNA amplification is rare in normal human cells. *Proc. Natl. Acad. Sci. USA*, *87*:1791–1795 (1990).
27. Tlsty, T. D. Normal diploid human and rodent cells lack a detectable frequency of gene amplification. *Proc. Natl. Acad. Sci. USA*, *87*:3132–3136 (1990).
28. Serrano, E. E., and Schimke, R. T. Flow cytometric analysis of mammalian glial cultures treated with methotrexate. *Glia*, *3*:539–549 (1990).
29. Kung, A. L., Sherwood, S. W., and Schimke, R. T. Cell line-specific differences in

the control of cell cycle progression in the absence of mitosis. *Proc. Natl. Acad. Sci. USA*, *87*:9553–9556 (1991).

30. Schimke, R. T., Kung, A. L., Rush, D. F., and Sherwood, S. W. Differences in mitotic control among mammalian cells. *C.S.H.S.Q.B.* *56*:417–425 (1992).

Robert T. Schimke
Department of Biological Sciences
Stanford University, Stanford, California

Preface

During the past decade it has become apparent that the process of gene amplification in mammalian cells is of substantial importance to molecular geneticists, genetic engineers, and medical scientists. For the molecular geneticist the ability to selectively amplify specific genes allows the isolation of mammalian cell lines from which the amplified genes can be easily cloned and subsequently studied. For the genetic engineer the use of amplifiable genetic markers allows the cotransfer and coamplification of nonselectable genes of academic or commercial value in virtually any mammalian cell. For the medical scientist gene amplification is important because of its consequences regarding drug resistance and cancer. As a result of such widespread interest, information concerning gene amplification has been published in a large variety of journals. In this volume the leading contributors to our understanding of mammalian gene amplification summarize their work and in doing so provide a convenient and authoritative source of information on this important topic. No comparable volume has appeared since the proceedings of a Banbury Conference on gene amplification were published in 1982.

An Introduction provides a historical background and biological perspective to gene amplification. The remainder of the book consists of 40 chapters divided into five parts. Part I includes chapters dealing with the molecular and cytogenetic organization of amplified DNA. Sensitive methods allowing the detection and molecular cloning of amplified genes are presented, and the relationship between gene amplification and DNA replication is considered. Part II, entitled Gene Amplification and Drug Resistance, contains a large number of relatively brief chapters describing selection protocols that have been useful for the isolation of mammalian cell lines with amplified copies of specific genes. The resulting cell lines have been useful as genetically enriched sources of the amplified genes and have allowed these genes and their cDNAs to be cloned without difficulty. In Part III experts in mammalian gene expression illustrate the use of amplifiable genetic markers to achieve high levels of protein production in mammalian cells. The availability of a collection of amplifiable genetic markers has added a significant new dimension to the genetic manipulation of mammalian cells. In Part IV leading medical oncologists and molecular geneticists present evidence for the presence of amplified copies of specific cellular oncogenes in tumor cell lines and primary tumor tissue. The relationship between gene amplification and tumor progression is considered. Part V includes chapters from a number of laboratories that have made considerable contributions to our understanding of the mechanism of gene amplification in mammalian cells. Genetic and environmental factors affecting the frequency and mechanism of gene amplification are presented.

I am grateful to the authors for accepting my invitation to contribute a chapter to this work. Their enthusiastic response indicated their belief that this volume is timely and important. We hope the rest of our colleagues agree with this view and find the book a useful source of ideas and information. On behalf of all contributors I express our sincere appreciation to Lisa Honski and Joseph Stubenrauch of Marcel Dekker, Inc. Their talented professional assistance was critical to the successful completion of this project. On a personal note I wish to extend special thanks to my wife, Hope Northrup, for her enthusiasm, encouragement, and understanding.

Rodney E. Kellems

Contents

Contributors

Mirit Aladjem Department of Cell Research and Immunology, George S. Wise Faculty of Life Sciences, Tel Aviv University, Tel Aviv, Israel

Kari Alitalo Cancer Biology Laboratory, Departments of Virology and Pathology, University of Helsinki, Helsinki, Finland

Irene L. Andrulis Samuel Lunenfeld Research Institute, Mount Sinai Hospital and Departments of Molecular and Medical Genetics and Pathology, University of Toronto, Toronto, Ontario, Canada

Stuart M. Arfin Department of Biological Chemistry, University of California, Irvine, California

John F. Ash Department of Anatomy, University of Utah, Salt Lake City, Utah

Giuseppe Attardi Division of Biology, California Institute of Technology, Pasadena, California

J. Carl Barrett Laboratory of Molecular Carcinogenesis, National Institute of Environmental Health Sciences, National Institutes of Health, Research Triangle Park, North Carolina

Michael T. Barrett Samuel Lunenfeld Research Institute, Mount Sinai Hospital, and Department of Molecular and Medical Genetics, University of Toronto, Toronto, Ontario, Canada

Lenore K. Beitel* Department of Biochemistry and McGill Cancer Centre, McGill University, Montreal, Quebec, Canada

Revital Ben-Aziz-Aloya Department of Biological Chemistry, The Life Sciences Institute, The Hebrew University of Jerusalem, Jerusalem, Israel

June L. Biedler Laboratory of Cellular and Biochemical Genetics, Memorial Sloan-Kettering Cancer Center, New York, New York

Garrett M. Brodeur Department of Pediatrics, Washington University School of Medicine, St. Louis, Missouri

Gérard Buttin Department of Immunology, Institut Pasteur, Paris, France

Susan M. Carroll Department of Pathology, University of California San Diego School of Medicine, La Jolla, California

Chuck C.-K. Chao Department of Biochemistry, Chang Gung Medical College, Taoyuan, Taiwan, Republic of China

A. Craig Chinault Department of Biochemistry and The Institute for Molecular Genetics, Baylor College of Medicine, Houston, Texas

Philip Coffino Departments of Microbiology and Immunology and Medicine, University of California, San Francisco, California

Frank R. Collart Biology and Medical Research Division, Argonne National Laboratory, Argonne, Illinois

Steven J. Collins Molecular Medicine Program, Fred Hutchinson Cancer Research Center and Department of Medicine, University of Washington Medical School, Seattle, Washington

Gray F. Crouse Department of Biology, Emory University, Atlanta, Georgia

Matt E. Cruzen Department of Biological Chemistry, University of California, Irvine, California

Michelle Debatisse Department of Immunology, Institut Pasteur, Paris, France

Bruno Robert De Saint Vincent Department of Immunology, Institut Pasteur, Paris, France

Gal Ehrlich Department of Biological Chemistry, The Life Sciences Institute, The Hebrew University of Jerusalem, Jerusalem, Israel

---

**Current affiliation*: Lady Davis Institute for Medical Research, Sir Mortimer B. Davis Jewish General Hospital, Montreal, Quebec, Canada.

Salvatore Feo* Eukaryotic Gene Organization and Expression Laboratory, Imperial Cancer Research Fund, London, England

Mike Fried Eukaryotic Gene Organization and Expression Laboratory, Imperial Cancer Research Fund, London, England

Steven C. Gerken† Department of Biological Chemistry, University of California, Irvine, California

Dalia Ginzberg Department of Biological Chemistry, The Life Sciences Institute, The Hebrew University of Jerusalem, Jerusalem, Israel

Elena Giulotto Department of Genetics and Microbiology, University of Pavia, Pavia, Italy

Averell Gnatt Department of Biological Chemistry, The Life Sciences Institute, The Hebrew University of Jerusalem, Jerusalem, Israel

Deborah L. Grady Genomics and Structural Biology Group, Los Alamos National Laboratory, Los Alamos, New Mexico

Mark T. Groudine Division of Basic Sciences, Fred Hutchinson Cancer Research Center and Department of Radiation Oncology, University of Washington Medical School, Seattle, Washington

Joyce L. Hamlin Department of Biochemistry, University of Virginia School of Medicine, Charlottesville, Virginia

David Hassin Department of Cell Research and Immunology, George S. Wise Faculty of Life Sciences, Tel Aviv University, Tel Aviv, Israel

Edith Heard‡ Eukaryotic Gene Organization and Expression Laboratory, Imperial Cancer Research Fund, London, England

C. Edgar Hildebrand Genomics and Structural Biology Group, Los Alamos National Laboratory, Los Alamos, New Mexico

Harri Hirvonen Department of Medical Biochemistry, University of Turku, Turku, Finland

Eliezer Huberman Biology and Medical Research Division, Argonne National Laboratory, Argonne, Illinois

Current affiliations:
*Institute of Developmental Biology of the Italian National Research Council, Palermo, Italy.
†Department of Human Genetics, University of Utah, Salt Lake City, Utah.
‡Unité de Genetique Moléculaire Moraine, Institut Pasteur, Paris, France.

Lee F. Johnson Department of Molecular Genetics, The Ohio State University, Columbus, Ohio

Edward M. Johnstone Central Nervous System Molecular Biology Research, Lilly Research Laboratories, Indianapolis, Indiana

Shulamit Karby Department of Cell Research and Immunology, George S. Wise Faculty of Life Sciences, Tel Aviv University, Tel Aviv, Israel

Randal J. Kaufman Department of Molecular and Cellular Genetics, Genetics Institute, Cambridge, Massachusetts

Rodney E. Kellems Department of Biochemistry and The Institute for Molecular Genetics, Baylor College of Medicine, Houston, Texas

Kris J. Kontis* Department of Biological Chemistry, University of California, Irvine, California

Päivi J. Koskinen Cancer Biology Laboratory, Departments of Virology and Pathology, University of Helsinki, Helsinki, Finland

Sharon S. Krag Department of Biochemistry, The Johns Hopkins University School of Hygiene and Public Health, Baltimore, Maryland

Yaron Lapidot-Lifson† Department of Obstetrics and Gynecology, The Edith Wolfson Medical Center, Tel Aviv University, Tel Aviv, Israel

Sara Lavi Department of Cell Research and Immunology, George S. Wise Faculty of Life Sciences, Tel Aviv University, Tel Aviv, Israel

Te-Chang Lee Institute of Biomedical Sciences, Academia Sinica, Taipei, Taiwan, Republic of China

Rueyling Lin Department of Biochemistry and The Institute for Molecular Genetics, Baylor College of Medicine, Houston, Texas

Victor Ling Division of Molecular and Structural Biology, The Ontario Cancer Institute and Department of Medical Biophysics, University of Toronto, Toronto, Ontario, Canada

Sheila P. Little Central Nervous System Molecular Biology Research, Lilly Research Laboratories, Indianapolis, Indiana

---

**Current affiliation*: Department of Microbiology and Molecular Genetics, University of California, Irvine, California.

†Also affiliated with the Department of Biological Chemistry, The Life Sciences Institute, The Hebrew University of Jerusalem, Jerusalem, Israel.

Tomi P. Mäkelä Cancer Biology Laboratory, Departments of Virology and Pathology, University of Helsinki, Helsinki, Finland

Barry J. Maurer* Division of Biology, California Institute of Technology, Pasadena, California

James G. McArthur† Department of Biochemistry and McGill Cancer Centre, McGill University, Montreal, Quebec, Canada

Eckart U. Meese‡ Departments of Radiation Oncology and Human Genetics, The University of Michigan, Ann Arbor, Michigan

David W. Melton Institute of Cell and Molecular Biology, Edinburgh University, Edinburgh, Scotland

Paul S. Meltzer Departments of Radiation Oncology and Pediatrics, The University of Michigan, Ann Arbor, Michigan

Lewis Neville Department of Biological Chemistry, The Life Sciences Institute, The Hebrew University of Jerusalem, Jerusalem, Israel

Zlil Ophir Department of Cell Research and Immunology, George S. Wise Faculty of Life Sciences, Tel Aviv University, Tel Aviv, Israel

Maria G. Pallavicini Department of Laboratory Medicine, University of California, San Francisco, California

Giovanni Pauletti§ Division of Biology, California Institute of Technology, Pasadena, California

Peter G. Pauw Department of Biology, Gonzaga University, Spokane, Washington

Mary Ellen Perry Molecular Biology Laboratory, Imperial Cancer Research Fund, London, England

David S. Pfarr SmithKline Beecham Pharmaceuticals, King of Prussia, Pennsylvania

Mazin B. Qumsiyeh Cytogenetics Laboratory, T. C. Thompson Children's Hospital, Chattanooga, Tennessee

---

Current affiliations:
*Department of Pediatrics, Children's Hospital of Michigan, Detroit, Michigan.
†Department of Molecular and Cellular Biology, University of California at Berkeley, Berkeley, California.
‡Department of Human Genetics, University of Saar, Hamburg, Germany.
§Department of Medicine, UCLA School of Medicine, Los Angeles, California.

Mitchell E. Reff Department of Gene Expression, IDEC Pharmaceuticals Corporation, La Jolla, California

Igor B. Roninson Department of Genetics, University of Illinois at Chicago, Chicago, Illinois

Kalle Saksela Cancer Biology Laboratory, Departments of Virology and Pathology, University of Helsinki, Helsinki, Finland

Edward A. Sausville Laboratory of Biological Chemistry, Division of Cancer Treatment, National Cancer Institute, National Institutes of Health, Bethesda, Maryland

Jane R. Scocca Department of Biochemistry, The Johns Hopkins University School of Hygiene and Public Health, Baltimore, Maryland

Shlomo Seidman Department of Biological Chemistry, The Life Sciences Institute, The Hebrew University of Jerusalem, Jerusalem, Israel

Louis Siminovitch Samuel Lunenfeld Research Institute, Mount Sinai Hospital, Toronto, Ontario, Canada

Marilyn L. Slovak Department of Cytogenetics, City of Hope National Medical Center, Duarte, California

Hermona Soreq Department of Biological Chemistry, The Life Sciences Institute, The Hebrew University of Jerusalem, Jerusalem, Israel

Barbara A. Spengler Laboratory of Cellular and Biochemical Genetics, Memorial Sloan-Kettering Cancer Center, New York, New York

Raymond L. Stallings Genomics and Structural Biology Group, Los Alamos National Laboratory, Los Alamos, New Mexico

Clifford P. Stanners Department of Biochemistry and McGill Cancer Centre, McGill University, Montreal, Quebec, Canada

George R. Stark Department of Molecular Biology, Imperial Cancer Research Fund, London, England

D. Parker Suttle Department of Biochemical and Clinical Pharmacology, St. Jude Children's Research Hospital, Memphis, Tennessee

Lars Thelander Department of Medical Biochemistry and Biophysics, University of Umeå, Umeå, Sweden

Floyd Thompson Arizona Cancer Center, University of Arizona, Tucson, Arizona

Thea D. Tlsty Department of Pathology, Lineberger Comprehensive Cancer Center, University of North Carolina, Chapel Hill, North Carolina

Franck Toledo Department of Immunology, Institut Pasteur, Paris, France

Jeffrey M. Trent Departments of Radiation Oncology and Human Genetics, The University of Michigan, Ann Arbor, Michigan

Florence W. L. Tsui Department of Rheumatology/Immunology, The Toronto Hospital, Western Division, Toronto, Ontario, Canada

Lela Veinot Department of Anatomy, University of Toronto, Toronto, Ontario, Canada

Geoffrey M. Wahl Gene Expression Laboratory, The Salk Institute for Biological Studies, La Jolla, California

Karen Ward* Central Nervous System Molecular Biology Research, Lilly Research Laboratories, Indianapolis, Indiana

Richard H. Wilson Department of Genetics, University of Glasgow, Glasgow, Scotland

Bradford E. Windle Department of Molecular Biology, The Cancer Therapy and Research Center, San Antonio, Texas

Florian M. Wurm Cell Culture and Fermentation Research and Development, Genentech, Inc., South San Francisco, California

Haim Zakut Department of Obstetrics and Gynecology, The Edith Wolfson Medical Center, Tel Aviv University, Tel Aviv, Israel

*No longer affiliated with Lilly Research Laboratories.

Gene Amplification in Mammalian Cells

Introduction to Gene Amplification: Historical Background and Biological Perspective

Rodney E. Kellems *Baylor College of Medicine, Houston, Texas*

Much of the early work concerning gene amplification was covered in a volume edited by Robert T. Schimke and published in 1982 [1]. During the subsequent 10 years many new examples of gene amplification were reported, and the relevance of gene amplification to biotechnology and medicine became firmly established. This book deals specifically with the significance of gene amplification in mammalian cells and provides extensive coverage concerning the role of gene amplification in drug resistance, cancer, and biotechnology. This introductory chapter briefly reviews the examples of developmentally regulated gene amplification and provides an abbreviated historical background on spontaneously occurring gene amplification in bacteria, yeast, *Drosophila*, and other insects. To set the stage for material specifically covered in this book, the research leading up to the discovery of gene amplification in mammalian cells, is also presented.

It is important to distinguish two fundamentally different categories of gene amplification events: (1) developmentally regulated gene amplification and (2) spontaneously occurring gene amplification. Developmentally regulated gene amplification is part of a genetically orchestrated program to synthesize enormous amounts of a particular gene product (RNA or protein) within a short period of time [2]. There are very few examples of this type of gene amplification, and none is known for mammalian organisms. Spontaneously occurring gene amplification is a widespread genetic phenomenon and presumably occurs throughout the entire phylogenetic spectrum. I will begin with a consideration of developmentally regulated gene amplification.

I. DEVELOPMENTALLY REGULATED GENE AMPLIFICATION

A. Amplification of rRNA Genes During Amphibian Oogenesis

The first reported example of developmentally regulated gene amplification followed the discovery of ribosomal RNA (rRNA) gene amplification in *Xenopus laevis* oocytes [3,4]. The frog oocyte is the largest known cell and contains approximately 10^{12} ribosomes. These ribosomes are intended to serve the protein synthetic needs of the early embryo, which does not begin the synthesis of its own ribosomes until gastrulation. The synthesis of massive quantities of rRNA is facilitated by rRNA gene amplification (Fig. 1). The amplified rDNA is diluted out during early embryogenesis. In most eukaryotes the genes encoding rRNA consist of a large multigene family with several hundred germ line copies. Because of the large size of the *Xenopus* oocyte and the large number of ribosomes required for early embryogenesis, the 450 germ line copies are insufficient to meet the enormous production requirements of amphibian oogenesis. Initially a 10–40-fold amplification of rRNA genes occurs in the premeiotic germ cells of both males and females [5]. The mechanism of this gene amplification is unknown. The amplified rDNA, present in the form of extrachromosomal circles, is lost during spermatogenesis but is further amplified in the developing oocyte. During a 3 week period of oogenesis encompassing early meiotic prophase, additional rRNA gene amplification results in a 2500-fold increase in rRNA genes. At the end of the gene amplification process more than 70% of nuclear DNA encodes rRNA and exists as extrachromosomal circles ranging in size from 12 to 210 kilobases (kb) [6,7]. This corresponds to 1 to 20 rRNA genes per extrachromosomal circle, each encoding the 45S rRNA precursor. The amplification of the extrachromosomal rRNA genes is believed to result primarily from a rolling circle DNA replication mechanism. Mechanisms generating the original extrachromosomal circles have not yet been determined but may involve site-specific excision or homologous recombination between tandemly arranged copies of the chromosomal rRNA genes. The analysis of amplified rDNA from individual oocytes indicates that only a subset of the original 450 rRNA genes is amplified. The subset is not always the same from oocyte to oocyte. Thus, no one rRNA gene is invariably recruited for amplification, nor is the entire germ line complement of rRNA genes amplified in individual oocytes. The amplified rDNA is diluted out during early embryogenesis. The amplification of rRNA genes during amphibian oogenesis is the only known example of developmentally regulated gene amplification in vertebrate animals.

B. Developmentally Regulated Gene Amplification in Insects

Developmentally regulated gene amplification has been observed in several circumstances of insects. In each case the amplification is associated with polytene

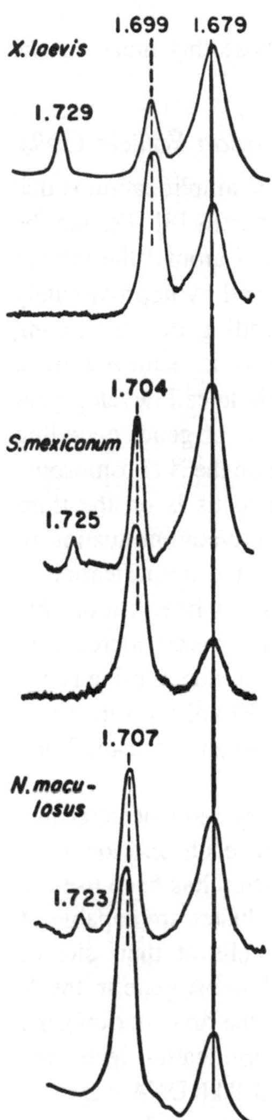

Figure 1 Ribosomal RNA gene amplification in amphibian oocytes. Tracings of germinal vesicle and somatic cell DNA centrifuged to equilibrium in CsCl in the analytical centrifuge. For each species the germinal vesicle DNA is traced on the top, and the somatic DNA is traced below. The band at the density of 1.679 is the deoxyadenylate–deoxythymidylate copolymer (dAT) carrier added at the beginning of the isolation. The density of the DNA in the main band is the same in germinal vesicle and somatic cell DNA in all cases; in addition to this band, the germinal vesicle preparations show a high density component that contains sequences complementary to rRNA. (Reprinted from Ref. 3 with permission.)

chromosomes found in cells of specific types, such as those that make up the salivary gland and the ovarian follicular epithelium.

1. Chorion Gene Amplification in Drosophila Ovarian Follicle Cells

The best studied example of developmentally regulated gene amplification is that of chorion gene amplification in *Drosophila* ovarian follicle cells [8]. During the last 5 hr of oogenesis, an eggshell, or chorion, is deposited around the mature oocyte. Chorion proteins are synthesized within a 2 hr period by approximately 1000 cells that make up the follicular epithelium surrounding the developing oocyte. The enormous synthetic capacity of follicle cells is achieved by a combination of polytenization (45C) that occurs during follicle cell development and subsequently by a selective amplification of two clusters of genes encoding chorion proteins [9]. One cluster of chorion genes is located on the X chromosome and is amplified 20-fold during choriogenesis. The other locus is on the third chromosome and is amplified approximately 80-fold. Selective amplification of chorion genes occurs just before choriogenesis. Note that the amplification is beyond the 45-fold increase in DNA content resulting from polytenization that occurs during the development of follicle cells. Thus the overall increase in chorion gene dosage is up to 3600 relative to a nonpolytenized, nonamplified cell. Amplification extends over chromosomal domains of 50–100 kb, with maximal copy levels occurring within the chorion gene clusters in the center of the amplified domain (Fig. 2).

Chorion gene amplification is controlled by cis-regulatory sequences termed amplification control elements (ACEs), which map within each chorion gene cluster [10–12]. Transposon-mediated germ line transformation has been used to show that specific DNA sequences from the chorion gene cluster are capable of inducing a normal program of amplification in follicle cells at their site of chromosomal insertion. ACE1 is located upstream of the 3′-most gene in the X chromosome chorion gene cluster and ACE3 is situated 5′ to the first chorion gene in the cluster on chromosome 3. Because of its higher amplification level, the chromosome 3 cluster has been more intensively studied. A 3.8 kb DNA fragment from this region is capable of directing tissue-specific amplification and expression of the two chorion genes that it contains. Similar findings pertain to a 4.7 kb fragment of DNA encompassing the X chromosome chorion gene cluster. Specific internal deletions of 510 and 467 base pairs (bp) within these fragments drastically reduce their ability to induce follicle cell specific amplification. These deletions serve to define the locations of ACEs. Additional sequences, termed amplification enhancer regions (AERs) have also been identified within the chorion gene cluster on chromosome 3 [13,14]. These sequences are not sufficient to promote amplification but do have a modulating effect on the function of ACE3. It is possible that ACEs and AERs cooperate in effecting follicle cell specific amplification of the chorion gene clusters.

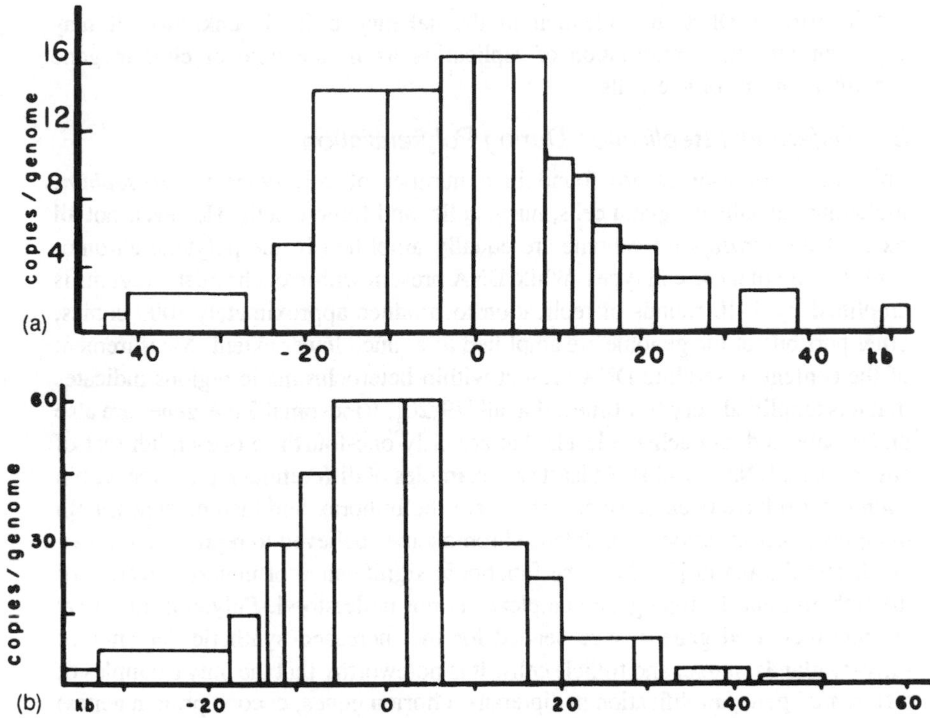

Figure 2 Chorion gene amplification in *Drosophila* follicle cells. The extent of DNA amplification at the chorion gene loci on chromosomes 3 and X are shown. The width and position of each bar corresponds to the size and location of a genomic *Eco*RI fragment at either the third chromosome (a) or the X chromosome (b) chorion gene locus. The extent of amplification is indicated on the vertical axis. For each chorion locus the gradient of amplification is symmetrically distributed around the center of the chorion gene family. (Reprinted from Ref. 9 with permission.)

2. *Amplification of Cocoon Protein Genes in Rhynchosciara*

Approximately 20 DNA puffs appear in the salivary gland polytene chromosomes of the insect *Rhynchosciara* at the time of cocoon formation [15,16]. Genes encoding abundant cocoon proteins are present in at least some of these puffs. Genomic and complementary DNA (cDNA) clones encoding cocoon proteins have been isolated and used to demonstrate that a 16-fold DNA amplification precedes the high level of transcriptional activity that is associated with these puffs while the cocoon is being formed [17,18]. The requirement for rapid synthesis of large quantities of cocoon proteins presumably explains the need for gene amplification. Like chorion gene amplification discussed above, the cocoon protein gene amplification occurs over a preexisting background of polytenization. Although the mechanism of DNA amplification in the salivary puffs is unknown, it may represent repetitive reinitiation of replication as in the case of chorion gene amplification in follicle cells.

3. Differential Replication During Polytenization

Polytene chromosomes are found in a number of cell types in *Drosophila*, including the salivary gland cells, nurse cells, and follicle cells. However, not all parts of the *Drosophila* genome are equally amplified in the polytene chromosomes found in these cell types. While DNA present within euchromatic regions is amplified by 9–10 rounds of replication to produce approximately 1000 copies, other portions of the genome are amplified to a much lower extent. Measurement of the content of satellite DNA present within heterochromatic regions indicates that it is amplified very few times, if at all [19,20]. Ribosomal RNA genes are also underreplicated and achieve levels that are only one-fourth to one-eighth that of euchromatic DNA [21–23]. At least two examples of differential replication within euchromatin have been detected. These are the bithorax and histone genes, both mapping to constrictions in polytene chromosomes believed to represent intercalary heterochromatin [24,25]. The functional significance of underreplication of the bithorax and histone gene complexes is not understood. Polytenization may provide increased gene dosage needed for the increased synthetic demands of salivary glands, nurse, and follicle cells. It is noteworthy that the only examples of differential gene amplification in dipterans (chorion genes, cocoon protein genes) are limited to cells with polytene chromosomes.

B. Gene Amplification and Selective Elimination in Ciliated Protozoa

The ciliated protozoa (e.g., *Tetrahymena*, *Oxytricha*) contain two nuclei: a micronucleus that contains a diploid complement of DNA and a macronucleus that contains highly amplified copies of selected regions of micronuclear DNA [26]. The macronucleus is transcriptionally active and contains only that DNA needed for vegetative growth, whereas the micronucleus serves as the germ line repository of sexually transmitted genetic information. At each conjugal event the old macronucleus is eliminated and a new macronucleus is regenerated from micronuclear DNA. During regeneration of new macronuclei, repeated rounds of DNA replication occur without nuclear division, resulting in polyploidization of the entire haploid DNA content. Fragmentation of the polyploid chromosomal DNA occurs, followed by the selective elimination of approximately 95% of the DNA. The remaining DNA is further amplified to achieve approximately 45 copies per macronucleus in the case of *Tetrahymena*. Thus selective enrichment of specific DNA sequences in the ciliates results from an initial nonselective amplification followed by a selective elimination of sequences not required for vegetative growth. Polyploidization is presumably required to provide a large concentration of transcription templates to enable ciliates to produce enough gene products to achieve their very large size.

The gene amplification process in *Tetrahymena* has been best characterized for rRNA genes [27,28]. During regeneration of the macronucleus, the single copy of micronuclear rDNA is excised as a 10.5 kb linear molecule. This is transformed into a 21 kb linear molecule having tandemly repeated CCCCAA sequences at both termini and two rRNA genes in palindromic arrangement with the 5′ ends in the center. The resulting palindromic rDNA, like other linear DNA molecules in the macronucleus, has an origin of DNA replication and telomeres. The origin is located very near the center and the telemeres consist of 40–70 copies of the C_4A_2 repeat at each end of the 21 kb palindrome. The germ line micronucleus contains a single copy of the rRNA gene per haploid genome, whereas in the macronucleus the rRNA genes are amplified nearly 10,000-fold and comprise 1–2% of the total macronuclear DNA [27,28]. Other genes are amplified by similar mechanisms but in general not to the same extent.

II. SPONTANEOUSLY OCCURRING GENE AMPLIFICATION

Whereas developmentally regulated gene amplification is a rarely used gene regulatory strategy, spontaneously occurring gene amplification is a common process that generates cells and organisms with new genetic potential. The resulting phenotypes may provide the organism with a selective growth advantage under certain environmental conditions. Spontaneously occurring gene amplification has been observed in bacteria, yeast, insects, and mammalian organisms.

A. Gene Duplication and Amplification in Bacteria

In 1977, in an extensive review, Anderson and Roth [29] concluded that tandem duplications were frequently occurring genetic events in bacteria. Subsequent publications have served to reinforce this view. Gene amplification in bacteria can occur as a chromosomal or extrachromosomal event. Many aspects of gene amplification in bacteria show considerable analogy with gene amplification in mammalian cells (see Section Two, Chapters 6–24).

1. *Chromosomal Events*

Novick and colleagues [30,31] provided one of the earliest examples of gene amplification in *Escherichia coli*. Following long-term growth in a chemostat containing limiting amounts of lactose as the sole carbon source, these investigators routinely recovered bacterial strains that produced a fourfold elevation of β-galactosidase. The following phenotype was associated with such cells: (1) the enzyme produced from the cultures selected for growth on limiting lactose was indistinguishable from wild-type enzyme, (2) the hypersynthesis of β-galactosidase was an unstable property of the cells, (3) the hypersynthesis phenotype was genetically linked to the *lac* chromosomal locus, and (4) diploidy for the *lac* locus was demonstrated genetically. They concluded that the phenotype resulted from spontaneously occurring duplications of the *lac* gene that conferred a selective advantage to the resulting cells. Nearby markers were haploid, indicating that the size of the duplicated region was relatively small. A spontaneous duplication frequency of 10^{-3} was estimated.

Cells expressing increased β-galactosidase can also be isolated by selection for growth on sodium lactobionate, a poorly utilized substrate [32]. In this case unstable, hypersynthesizing lines (presumably analogous to those isolated following chemostat growth on limiting lactose) were selected at a frequency of 10^{-4}.

A more recent study [33] of gene amplification in the *lac* region of *E. coli* involved the construction of a strain having a "leaky" lac^- mutation in which the synthesis of β-galactosidase is reduced to approximately 8% that of wild type. Lac^+ revertants can be isolated by growth on lactose minimal plates and appear at a frequency of 10^{-7}–10^{-6}. The low frequency of revertants in comparison with the studies referred to above is presumably due to the high stringency of the selection used in this case. The revertants fell into two categories: stable (40%) and unstable (60%). Analysis of more than 100 of the unstable revertants indicated that the *lac* region was amplified from 40- to 200-fold and that the size of the amplified region ranged from 7 to 37 kb. The revertant phenotype and the amplified genes were lost when cells were grown under nonselective conditions. In none of the stable revertants was the *lac* region amplified. The genetic basis of the stable revertant phenotype was not determined. A role for homologous recombination in the amplification and loss of the *lac* region was indicated by the following observations. It was not possible to isolate *lac* amplification mutations when using cells that were $recA^-$. When *lac*-amplified DNA was introduced into $recA^-$ cells, the amplified DNA was not lost in the absence of selection. These findings suggest that, as with most cases of gene amplification in bacteria, a functional homologous recombination system is required for the acquisition or loss of amplified DNA.

A variety of selection schemes have been used to isolate bacterial mutants containing amplified, or duplicated, copies of specific genes [34–38]. Examples of these selection schemes are presented in Table 1 and illustrate the use of enzyme inhibitors, growth on limiting or unnatural carbon sources, or the use of strains with mutant loci that require increased gene dosage for survival. In all cases selected cell lines appear at a relatively high frequency and possess an unstable phenotype characterized by increased levels of a specific enzyme that permits growth under selective conditions. Available evidence indicates that gene amplification accounts for the increased enzyme phenotypes and that the amplified genes are unstable unless introduced into a $recA^-$ background. With sufficient selection pressure, the copy number of amplified genes can reach several hundred. Most of the published work has related to amplification of genomic DNA sequences and has demonstrated that the size of the duplicated regions may range

Table 1 Selective Overproduction of Enzymes in Bacteria

Selection	Enzyme overproduced (fold)	Ref.
Growth on limiting lactose or lactobionate	β-Galactosidase (200×)	30–33
Chloramphenicol resistance	Chloramphenicol acetyltransferase (25×)	43–45
Ampicillin resistance (penicillin)	β-lactamase (35×) (penicillinase)	41
Aminotriazole resistance (a histidine analogue)	Imidazoleglycerol–phosphate dehydratase (10×)	29
Growth on xylitol	Ribitol dehydrogenase (20×)	36
Growth on limiting glycine	Glycyl–tRNA synthetase (4×)	34, 35
Growth on limiting arabinose, malate, sorbitol, or melibiose	The permease system specific for the limiting carbon source	39
In vivo passage through rabbit small intestine	Cholera toxin	37

from a few kilobases to as much as 25% of the bacterial genome [29,35]. The importance of homologous recombination in the gene amplification process has been demonstrated in most studies.

The use of specific selection schemes limits the number of genetic loci that can be examined to those that provide a selectable phenotype. Anderson and Roth devised a more generalized transduction scheme to identify and quantitate gene duplication events at discrete genetic loci throughout the *Salmonella* genome [39]. With their method they have measured the frequency of spontaneous tandem duplications at 38 chromosomal sites and the amount of material included in 199 independent duplications. The results of their studies indicate that gene duplications occur throughout the bacterial genome but that duplications are much more frequent at certain loci. A very high duplication frequency was observed near the cluster of ribosomal RNA genes, termed *rrn* loci, especially near the *purD* locus, which was duplicated in 3% of the population. Duplication in other regions of the genome occurred with frequencies in the range of 10^{-4}–10^{-3}. Further analysis showed that 114 of 115 frequently occurring duplications in the neighborhood of the *rrn* loci had *rrn* genes at the end points of the duplication. Similar results and conclusions had been reported previously. Hill and colleagues, who analyzed the duplication of the *glyT* locus in *E. coli*, found that *glyT* duplications arise by recombination between the four *rrn* loci located within this region of the genome. Duplications occur in three size classes (40, 163, and 250 kb), defined by the positions of the *rrn* loci. The smallest size class is the most prevalent. These findings suggest that unequal recombination between the homologous *rrn* sequences occurs frequently and causes a high duplication frequency in this region of the *E. coli* and *Salmonella* genomes. In all cases the duplications are unstable unless placed in a *recA*$^-$ background, again providing evidence for homologous recombination in the process of gene duplication. The analysis of gene duplications in other regions of the genome have also indicated the importance of homologous recombination between tandemly duplicated sequences [41].

2. Extrachromosomal Events

A very striking example of the role of gene amplification in bacterial adaption was provided by the analysis of drug resistance factors (R-factors), extrachromosomal genetic elements that confer drug resistance [42]. R-factors consist of two portions, the resistance transfer factor (RTF) and the R-determinant. The former encodes replicative functions and the latter encodes enzymes that confer resistance to specific drugs. Rownd and his colleagues [43–45] investigated the R-factor NR1, which reversibly dissociates into RTF and R-determinants in *Proteus mirabilis*. They and others have presented evidence that when cells are grown in the presence of drugs to which the R-determinants confer resistance (e.g., chloramphenicol) the number of R-determinants per cell is amplified [43–47]. Under these conditions R-factors harboring multiple tandem copies of R-determinants are formed. Also present are autonomous poly-R-determinants. This form of gene amplification was initially observed as an increase in the quantity and density of plasmid DNA carried by bacteria following selection in chloramphenicol (Fig. 3). The ratio of plasmid DNA to genomic DNA increases by as much as 10-fold. A model to account for these events is provided in Figure 4 [48]. Homologous recombination between direct repeats (IS elements) is believed to play a major role in the generation of a multitude of plasmid molecules, some of which provide increased selective advantage.

More recently Gutterson and Koshland [49] have shown that when recombinant plasmids carrying segments of the bacterial genome are introduced into mutant *E. coli* strains defective in plasmid replication (*pol A* mutants), recombination occurs between the plasmid insert sequences and the homologous segment of the bacterial genome. Such transformants can be selected on the basis of drug resistance markers carried by the plasmid. Plasmid sequences integrated into the genome were amplified up to 30-fold by selection for resistance to increasing concentrations of ampicillin or tetracycline. This strategy of gene transfer and amplification is similar to ones commonly used with cultured mammalian cells (see Section Three, Chapters 25–27).

B. Gene Amplification in Yeast

Cells selected for copper resistance contain increased copies of a gene, termed *CUP1*, that encodes the copper binding protein chelatin [50,51]. The genes are arranged in tandem arrays consisting of a repeat unit 1.5–2.0 kb long. In some

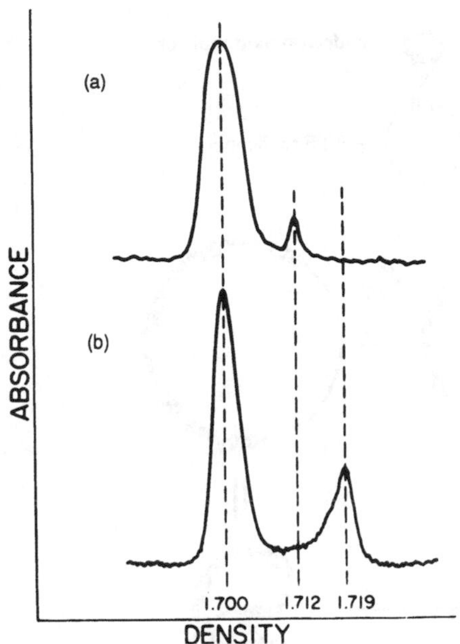

Figure 3 Extrachromosomal gene amplification in bacteria. Density profiles of DNA extracted from cultures of *Proteus mirabilis* grown in the absence (a) or presence (b) of chloramphenicol or streptomycin. The DNA appearing at 1.7000 is chromosomal DNA. The DNA appearing at 1.712 or 1.719 is R-factor DNA carrying genes encoding drug resistance. Following selection for increasing levels of drug resistance, the density and quantity of R-factor DNA increases and can account for more than 20% that of chromosomal DNA. (Reprinted from Ref. 43 with permission.)

strains up to 12 tandemly arranged copies have been observed. Short repeats flanking the *CUP1* gene are suspected of playing a role in the gene duplication process by facilitating unequal exchange events. Strains of yeast used in wine making have tandemly arranged copies of the chelatin gene, suggesting that the wine making process provides selection for increased levels of chelatin. In recent studies Fogel and his colleagues [52] have demonstrated that unequal crossing over at the amplified *CUP1* locus leads recombination products with increased and decreased numbers of *CUP1* genes. Gene conversion (i.e., nonreciprocal recombination) products are observed approximately 11% of the time and are believed to arise from mispairing among identical repeat units, leading to the formation of unpaired loops containing an integral number of repeat units. Conversions involving the loops lead to nonreciprocal changes in the gene copy numbers present in the tandem arrays. Fogel et al. [52] conclude that known properties of gene conversion account for all segregations they observe.

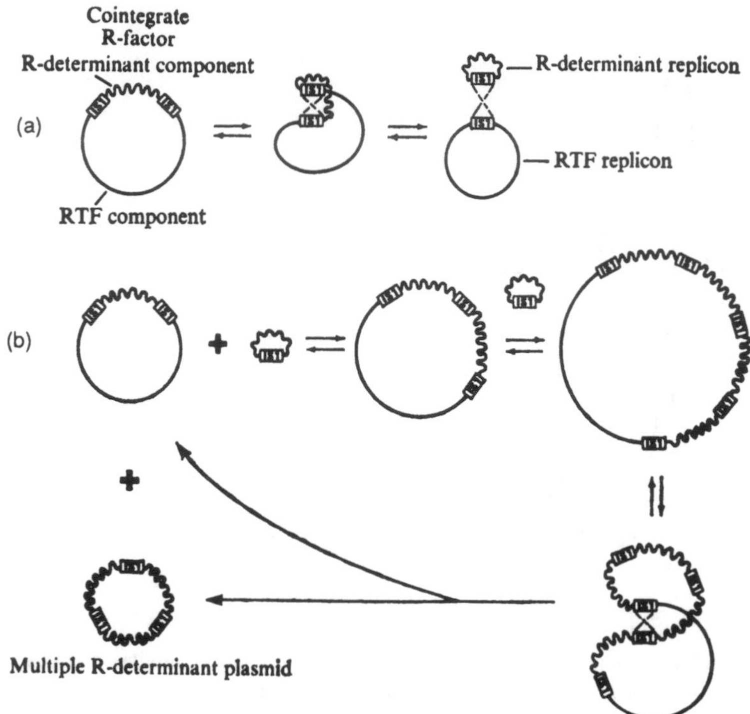

Figure 4 Model to account for changes in structures of R-factors following selection for drug resistance. The model depicted provides for (a) the formation of independent RTF and R-determinant replicons and (b) the production of polygenic R-determinant units. (Reprinted from Ref. 48 with permission.)

C. Spontaneously Occurring Gene Amplification in Insects

1. *Gene Duplications Account for the Bar Phenotype in Drosophila*

Bar is the name given to a sex-linked eye mutation that was first described in 1914 [53]. The mutation is dominant as judged from the fact that females carrying one copy of the *bar* locus have eyes distinctly smaller than the wild-type round eye but not as small as bar. Homozygous bar females have the complete bar (i.e., reduced eye) phenotype (Fig. 5). The bar mutation created a considerable amount of interest among *Drosophila* geneticists at the early part of this century [53–57] because of several interesting and unusual features. The mutation shows a relatively high frequency of reversion from bar to the rounded wild-type eye. Furthermore, some bar mutants undergo further mutation, giving rise to a more reduced eye phenotype termed double-bar by Sturtevant [55]. Double-bar has the interesting feature of reverting to bar or wild type with relatively high frequency.

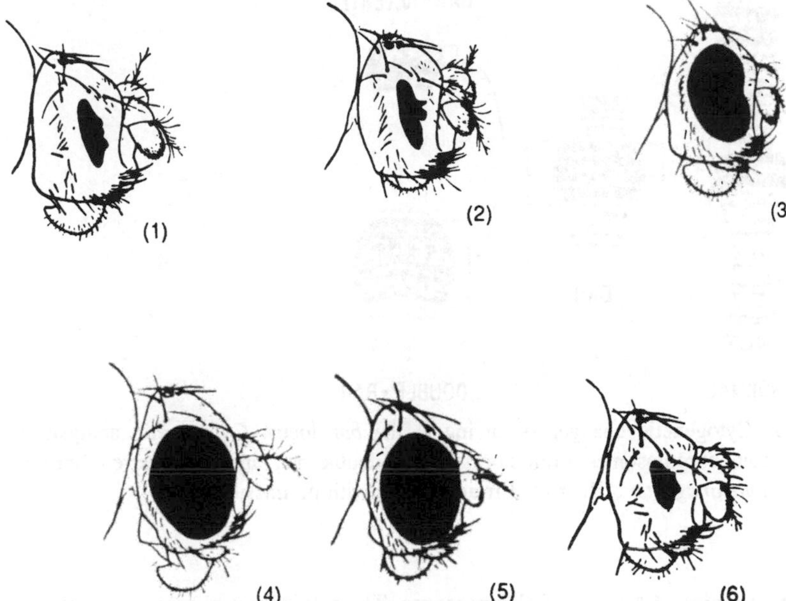

Figure 5 Eye phenotypes associated with genetic changes at the *bar* locus in *Drosophila*. Bar is the name given to an X-linked mutation that results in reduced eye size. Wild-type eyes are termed round. Eye phenotypes associated with genetic changes at the *bar* locus are illustrated as follows: (1) homozygous bar female, (2) bar male, (3) bar-over-round female, (4) female homozygous for round, obtained by reversion, (5) male that carries round, obtained by reversion, and (6) double-bar male. (Reprinted from Ref. 55 with permission.)

Initial studies suggested that these mutational changes occurred predominantly in females and were generally associated with nonreciprocal recombination between homologous chromosomes [55,56]. It has been shown that intrachromosomal recombination, possibly involving sister chromatid exchange, can also occur and presumably accounts for rare reversions in males and some reversion in females [57]. A significant advance in our understanding of the bar mutation was made possible by cytogenetic analysis of salivary gland chromosomes of wild-type, bar, and double-bar strains of *Drosophila*. In 1936 Bridges reported that the bar mutation is a tandem duplication of seven bands in the 16A region of the X chromosome [58]. Bar-reverted corresponds to a loss of the extra segment, whereas double-bar is a triplication of the 16A region (Fig. 6). Chromosomes having at least seven tandem copies of this region have been reported [59,60]. Thus, the various eye phenotypes associated with mutational changes at the *bar* locus can be accounted for by nonreciprocal recombination between repeated

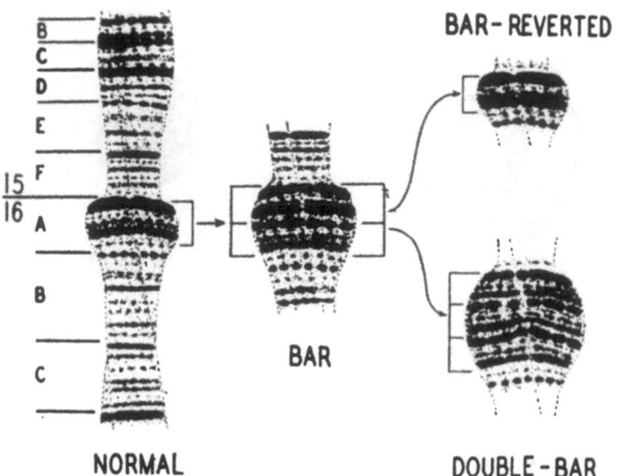

Figure 6 Cytogenetic changes occurring at the *bar* locus. Cytogenetic analysis of salivary gland chromosomes from normal, bar, double-bar, and bar-reverted lines of *Drosophila melanogaster*. (Reprinted from Ref. 58 with permission.)

segments at region 16A of the X chromosome. The role of homologous recombination in the events at the *bar* locus are reminiscent of similar processes described for gene amplifications in bacteria.

2. Gene Amplification and Insecticide Resistance

The gene duplications occurring at the *bar* locus were discovered because early drosophila geneticists showed considerable interest in inherited eye phenotypes. Gene amplification at other genetic loci in insects was more recently discovered because of interest in understanding insecticide resistance, a concern of substantial economic importance. A number of studies have indicated that insecticide resistance is often accompanied with increased levels of certain enzymes that have the common feature of modifying insecticides to increase their water solubility [61]. For example, cytochrome P450 mediated, mixed function oxidases catalyze the hydroxylation of aromatic hydrocarbons; glutathionine *S*-transferases catalyze the addition of glutathione to electrophiles; and nonspecific hydrolases cleave esters and amides to create products with increased charge. One of the best characterized examples is the massive overproduction of esterase B associated with resistance to organophosphate insecticides. In aphids and mosquitoes the enhanced enzyme production is known to be the result of gene amplification [62,63]. The amplification of esterase B genes in mosquitoes is believed to have been selected originally in the 1960s when the intensive use of organophosphate insecticides began. A recent report indicates that esterase B gene amplification probably arose as a single initial event in the 1960s and that the worldwide presence of esterase B amplification today is due to migration rather than newly arising gene amplifications [64]. Another documented case of gene amplification in insects is the association of metallothionein gene amplification with metal tolerance in natural populations of *Drosophila melanogaster* [65].

III. GENE AMPLIFICATION IN MAMMALIAN CELLS

A. Isolation and Biochemical Analysis of Methotrexate-Resistant Cells

The discovery of gene amplification in mammalian cells stems primarily from efforts of Maire Hakala and her colleagues at Roswell Park Memorial Institute and Glenn Fischer at Yale University School of Medicine. In the late 1950s they initiated efforts to understand the biochemical basis of amethopterin (methotrexate) resistance in mammalian cells [66–68,69a,b]. Their experimental approach was to isolate highly drug-resistant tumor cell lines and then determine the biochemical basis for the drug-resistant phenotype. Highly methotrexate-resistant cells were isolated by selecting cells for growth in the presence of increasing concentrations of the drug. Depending on selective conditions, the resulting cells had up to 155 times the original level of folic acid reductase activity (now more commonly called dihydrofolate reductase, DHFR) (Fig. 7). In 1961 Hakala and her colleagues presented convincing biochemical evidence that only the quantity of the enzyme had changed and that the Michaelis constant and turnover number had remained the same [68]. They concluded:

> Resistance to amethopterin in this case results from the large excess of folic acid reductase which acts as a specific physiological inactivator by binding the antagonist. Enough of the enzyme remains free to carry on its normal functions.

Hakala also found that for murine Sarcoma 180 cells the methotrexate-resistant phenotype was unstable and that "the loss of resistance closely parallels the decrease of folic acid reductase." Thus, the seminal report in 1961 [68] clearly describes the phenotypic characteristics associated with spontaneously occurring gene amplification and drug resistance: stepwise selection, increased enzyme production, and phenotypic instability. Hakala's critical contributions were to illustrate that a stepwise selection procedure is required to obtain highly methotrexate-resistant cells and to provide a biochemical explanation for the drug-resistant phenotype. The methotrexate-resistant Sarcoma 180 cells isolated and characterized by Hakala were the ones to be used many years later in Schimke's laboratory at Stanford to determine the genetic basis for the drug-resistant phenotype (see below). Many other reports concerning methotrexate resistance appeared throughout the 1960s and early 1970s [69–79]. In some cases drug resistance and the associated overproduction of dihydrofolate reductase were

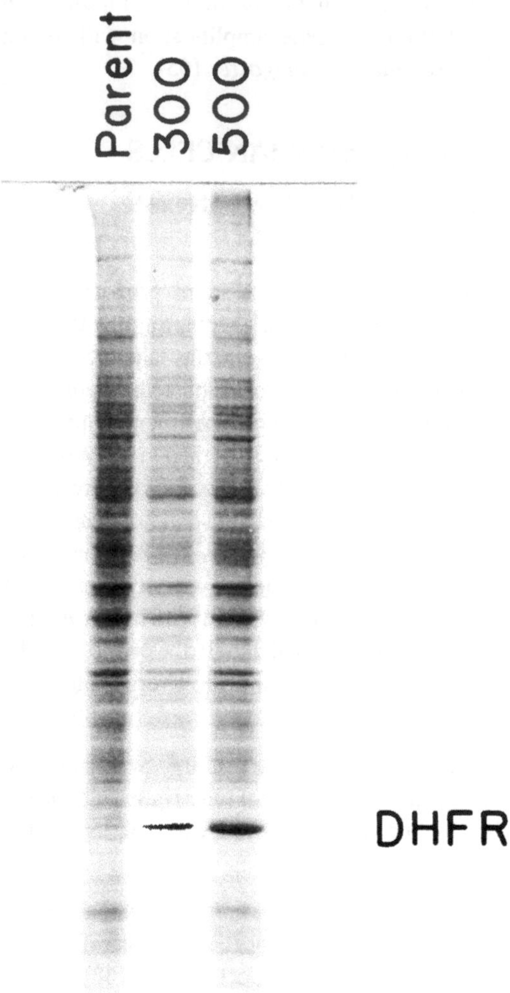

Figure 7 Comparison of proteins made by parental and methotrexate-resistant Hela cells. Protein extracts were prepared from a parental population of Hela cells and those selected for the ability to grow in 300 or 500 μM methotrexate. Approximately 30 μg of protein was electrophoretically separated by polyacrylamide gel electrophoresis and visualized by staining with Coomassie Blue. Dihydrofolate reductase (DHFR) is a major protein in the methotrexate-resistant cells.

stable phenotypes that were maintained for extended periods of time in the absence of selection. It did not become clear until later why the drug-resistant phenotype was stable in some cells and not in others.

B. Analysis of Dihydrofolate Reductase Gene Expression in Methotrexate-Resistant Cells

Although initial interest in methotrexate resistance was concentrated among oncologists and pharmacologists, the phenotype of increased dihydrofolate reductase production stimulated the interest of biochemists and geneticists. A number of biochemical and genetic techniques were developed during the 1970s that made it possible to carry the phenotypic analysis of methotrexate-resistant cells several steps further and to eventually determine the genetic basis of increased DHFR levels. Reports from the laboratories of John Littlefield [80] and Robert Schimke [81] used immunochemical approaches to show that increased levels of DHFR protein resulted from increased synthesis and that changes in protein stability did not contribute to the increased levels found in the drug-resistant cells. Additional studies [82,83] used in vitro translation assays as well as solution hybridization with sequence-specific probes to show that methotrexate-resistant cells possessed increased levels of *DHFR* messenger RNA (mRNA).

C. Amplification of Dihydrofolate Reductase Genes in Methotrexate-Resistant Cells

In 1976 Biedler and Spengler of the Sloan-Kettering Cancer Center published two reports showing that a very striking cytogenetic feature was associated with highly methotrexate-resistant cells (Figs. 8 and 9). In one article [84] they wrote: "Large, homogeneously staining chromosome regions [HSRs] which lack the longitudinal differentiation ordinarily revealed by cytogenetic 'banding' methods have been found in antifolate-resistant Chinese hamster cells and also in human neuroblastoma cells established in vitro." In another paper published the same year [85] Biedler and Spengler "conclude that meaningful association is evident between preferential synthesis of a protein (dihydrofolate reductase) and a newly arising, unique chromosome abnormality (HSR)." These findings culminated years of effort by Biedler to identify cytogenetic abnormalities associated with methotrexate-resistant cells that overproduce DHFR [86,87]. Her critical contributions were to relate HSRs to enzyme overproduction in drug-resistant cells and to point to the existence of HSRs in some human neuroblastomas. The latter observation forms the basis of Section Four (Chapters 28–33) in this volume.

Biedler's "meaningful association" between HSRs and the overproduction of DHFR was clarified by a number of reports from Schimke's laboratory in 1978 and 1979. In 1978 Alt et al. used subtractive hybridization procedures to purify cDNA sequences complementary to *DHFR* mRNA and used this probe to quantitate

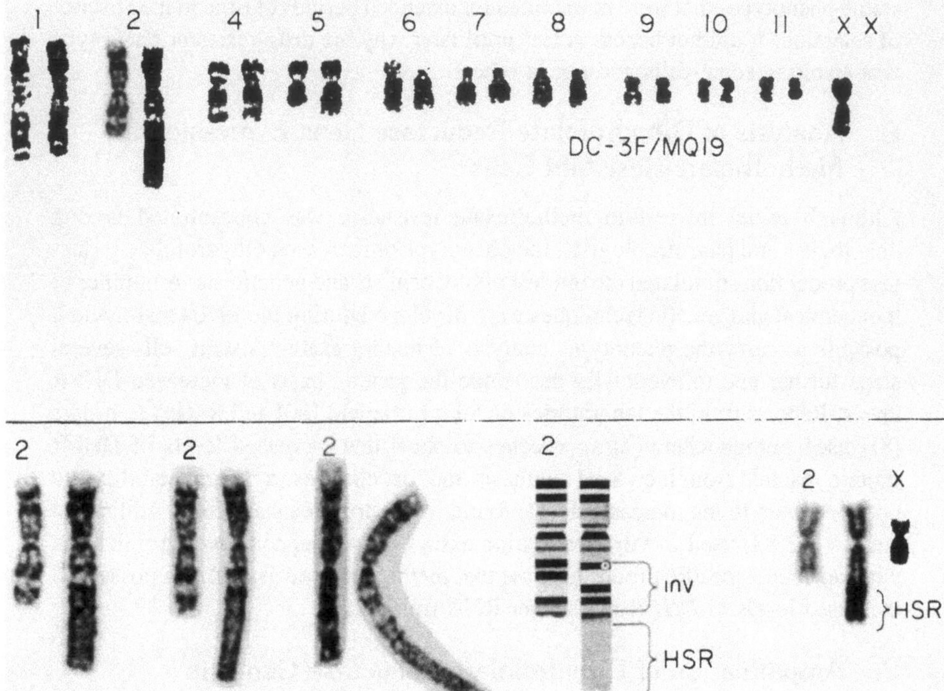

Figure 8 The cytogenetic findings associated with methotrexate resistance. Top: G-banded karyotype of a Chinese hamster cell selected for resistance to the antifolate methasquin. Bottom: Three examples of chromosome 2 and its homologue with an HSR, a diagrammatic representation of the pair, and a set including an X chromosome from a C-banded cell. (Reprinted from Ref. 85 with permission.)

DHFR mRNA and gene copies in methotrexate-resistant murine sarcoma and leukemia cell lines [88]. They found that gene amplification fully accounts for the overproduction of DHFR in stable and unstable lines of methotrexate-resistant cells. This was the first report to provide a genetic basis for the methotrexate-resistant phenotype. Later that year, Nunberg et al. [89] showed by in situ hybridization (Fig. 10) that the amplified *DHFR* genes localized to the HSRs present in a line of highly methotrexate-resistant Chinese hamster ovary (CHO) cells. Another report [90] linking amplified DHFR genes with an HSR present in methotrexate-resistant murine leukemia cells followed in 1979. Also in 1979 Kaufman et al. [91] showed that amplified *DHFR* genes in unstable methotrexate-resistant cells are associated with double minute chromosomes. An example of

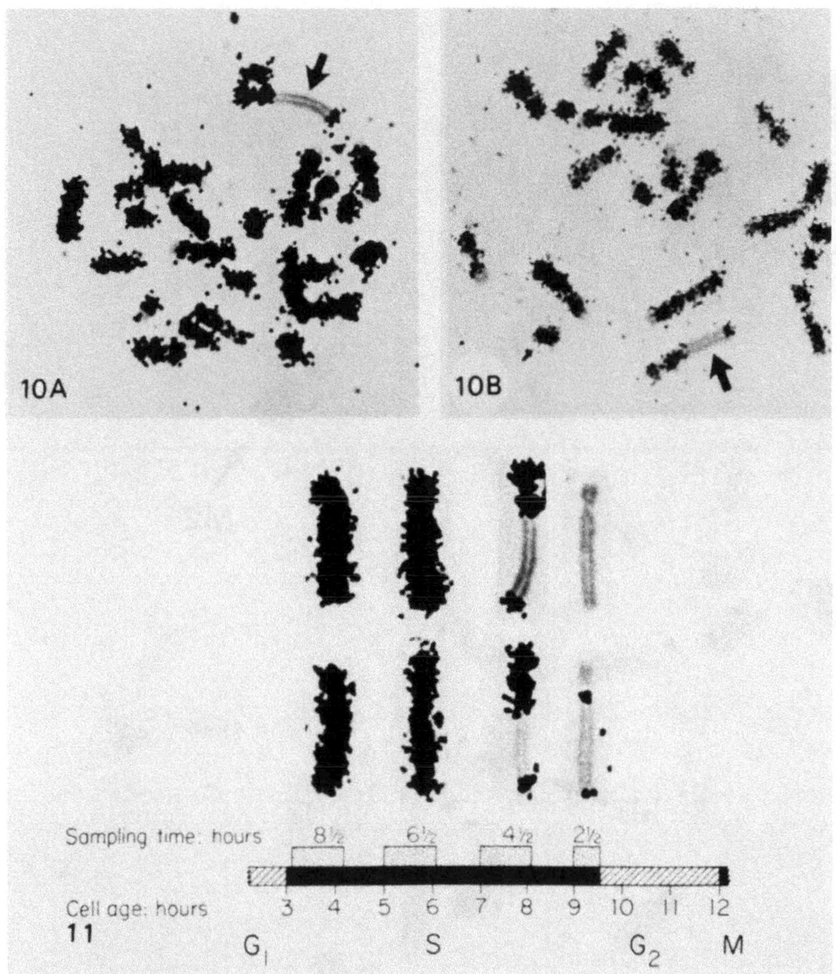

Figure 9 DNA replication timing of HSR: the HSR behaves as a unit during DNA replication. Top: Autoradiograph of metaphase chromosomes of an antifolate-resistant hamster cell sampled 4.5 hr after pulse labeling with tritiated thymidine. In each cell, the long segment of the marker chromosome corresponding to the HSR of G-banded cells is indicated by an arrow. Note the absence of silver grains. Bottom: Labeling patterns of marker chromosomes from eight individual cells sampled at four different times, as indicated, after tritiated thymidine pulse. Estimated cell cycle phase is diagrammed. DNA synthesis in long chromosome segments corresponding to HSR appears completed by mid-S phase, as indicated by absence of silver grains at 2.5 and 4.5 hr postlabeling. (Reprinted from Ref. 85 with permission.)

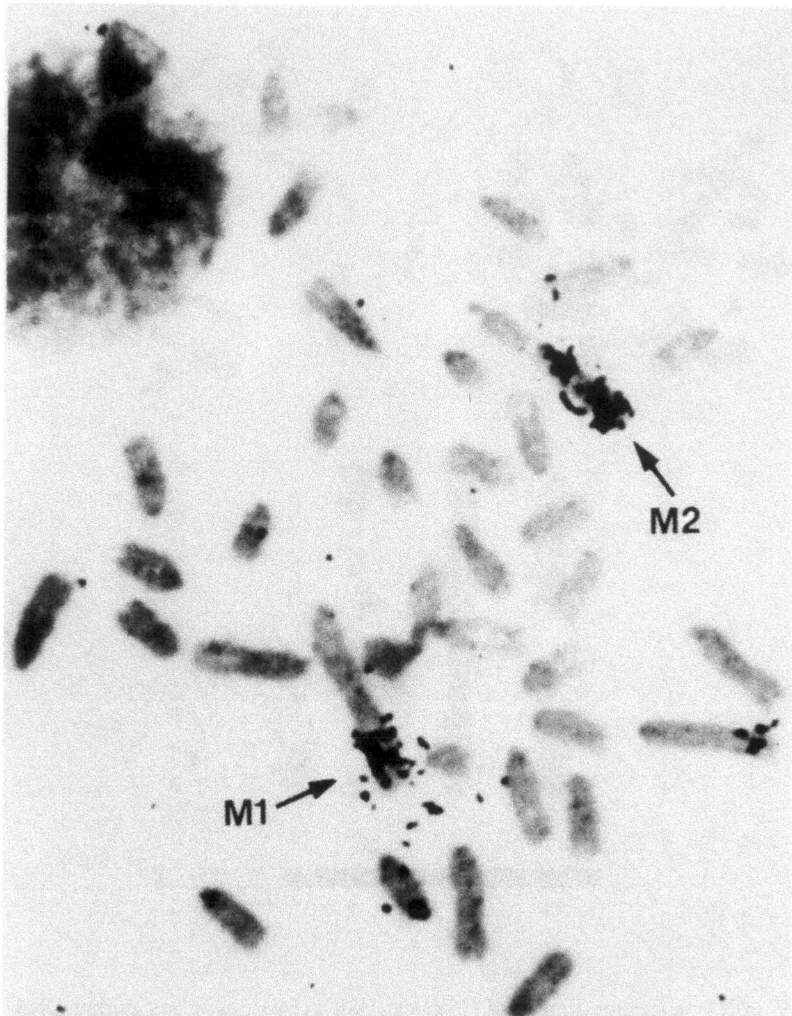

Figure 10 Methotrexate-resistant L5178YR metaphase cell hybridized with tritium-labeled *DHFR* cDNA sequences. After exposure to an autoradiographic emulsion for 5 days, silver grains are located almost exclusively on HSRs of M1 and M2, localizing the amplified *DHFR* genes to these regions. Some labeling is also observed at the distal end of one normal-sized chromosome.

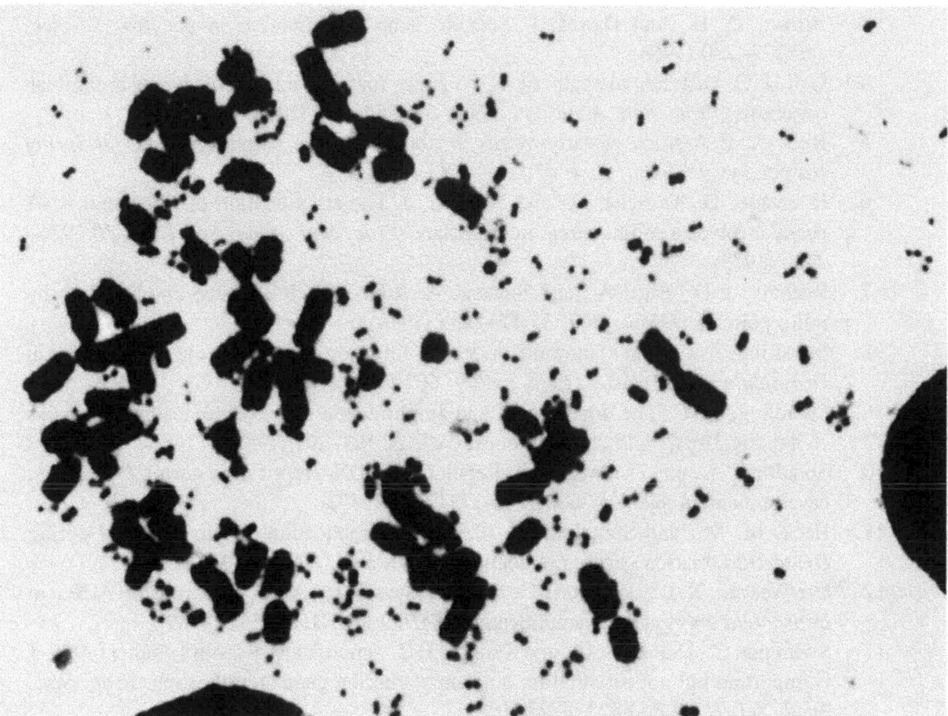

Figure 11 Double minute chromosomes. Double minute chromosomes are easily visualized in a methotrexate-resistant line of Sarcoma 180 cells originally isolated by Hakala [68]. This cell has been heavily stained with Giemsa; the paired nature of the minute chromosomes and their variable size are apparent.

double minute chromosomes in methotrexate-resistant Sarcoma 180 cells is shown in Figure 11. This series of manuscripts from Schimke's laboratory capped more than 20 years of efforts by oncologists, pharmacologists, biochemists, and geneticists to understand the molecular basis of methotrexate resistance. This work served as a catalyst to stimulate an enormous amount of research concerning gene amplification in mammalian cells. The major findings from this research activity are assembled in this book.

REFERENCES

1. Schimke, R. T. *Gene Amplification*, Cold Spring Harbor Laboratory, Cold Spring Harbor, New York, 1982.
2. Kafatos, F. C., Orr, W., and Delidakis, C. Developmentally regulated gene amplification. *Trends Genet. 1*:301–306 (1985).

3. Brown, D. D., and Dawid, I. Specific gene amplification in oocytes. *Science*, *160*:272–280 (1968).
4. Gall, J. G. Differential synthesis of the genes for ribosomal RNA during amphibian oogenesis. *Proc. Natl. Acad. Sci. USA*, *60*: 553–560 (1968).
5. Bird, A. P. A study of early events in ribosomal gene amplification. *Cold Spring Harbor Symp. Quant. Biol. 42*:1179–1183 (1978).
6. Hourcade, D., Dressler, D., and Wolfson, J. The amplification of ribosomal RNA genes involves a rolling circle intermediate. *Proc. Natl. Acad. Sci. USA*, *70*:2926–2930 (1973).
7. Rochaix, J.-D., Bird, A., and Bakken, A. Ribosomal RNA gene amplification by rolling circles. *J. Mol. Biol. 87*:473–487 (1974).
8. Spradling, A. C., and Mahowald, A. P. Amplification of genes for chorion proteins in *Drosophila melanogaster*. *Cell*, *16*:599–607 (1980).
9. Spradling, A. C. The organization and amplification of two chromosomal domains containing *Drosophila* chorion genes. *Cell*, *27*:193–201 (1981).
10. Spradling, A., and Orr-Weaver, T. Regulation of DNA replication during *Drosophila* development. *Annu. Rev. Genet. 21*:373–403 (1987).
11. Heck, M. M., and Spradling, A. C. Multiple replication origins are used during *Drosophila* chorion gene amplification. *J. Cell Biol. 110*:903–914 (1990).
12. Orr-Weaver, T. L., Johnston, C. G., and Spradling, A. C. The role of ACE3 in *Drosophila* chorion gene amplification. *EMBO J. 8*:4153–4162 (1989).
13. Swimmer, C., Delidakis, C., and Kafatos, F. C. Amplification-control element ACE-3 is important but not essential for autosomal chorion gene amplification. *Proc. Natl. Acad. Sci. USA*, *86*:8823–8827 (1989).
14. Delidakis, C., and Kafatos, F. C. Amplification enhancers and replication origins in the autosomal chorion gene cluster of *Drosophila*. *EMBO J. 8*:891–901 (1989).
15. Bonaldo, M. F., Santelli, R. V., and Lara, F. J. S. The transcript from a DNA puff of *Rhynchosciara* and its migration to the cytoplasm. *Cell*, *17*:827–833 (1979).
16. Winter, C. E., de Bianchi, A. G., Terra, W. R., and Lara, F. J. S. Protein synthesis in the salivary glands of *Rhynchosciara americana*. *Dev. Biol. 75*:1–12 (1980).
17. Glover, D. M., Zaha, A., Stocker, A. J., Santelli, R. V., Pueyo, M. T., De-Toledo, S. M., and Lara, F. J. S. Gene amplification in *Rhynchosciara* salivary gland chromosomes. *Proc. Natl. Acad. Sci. USA*, *79*:2947–2951 (1982).
18. Millar, S., Hayward, D. C., Read, C. A., Browne, J. J., Dantelli, R. V., Vallejo, F. G., Pueyo, M. T., Zaha, A., Glover, D. M., and Lara, F. J. S. Segments of chromosomal DNA from *Rhynchosciara americana* that undergo additional rounds of DNA replication in the salivary gland DNA puffs have only weak ARS activity in yeast. *Gene*, *34*:81–86 (1985).
19. Hammond, M. P., and Laird, C. D. Control of DNA replication and spatial distribution of defined DNA sequences in salivary gland cells of *Drosophila melanogaster*. *Chromosoma*, *91*:267–278 (1985).
20. Hammond, M. P., and Laird, C. D. Chromosome structure and DNA replication in nurse and follicle cells of *Drosophila melanogaster*. *Chromosoma*, *91*:279–286 (1985).
21. Hennig, W., and Meer, B. Reduced polyteny of ribosomal RNA cistrons in giant chromosomes of *Drosophila hydei*. *Nature New Biol. 233*:70–72 (1971).

22. Spear, B. B., and Gall, J. G. Independent control of ribosomal gene replication in polytene chromosomes of *Drosophila melanogaster*. *Proc. Natl. Acad. Sci. USA*, *70*: 1359–1363 (1973).
23. Spear, B. B. The genes for ribosomal RNA in diploid and polytene chromosomes of *Drosophila melanogaster*. *Chromosoma*, *48*:159–179 (1974).
24. Spierer, A., and Spierer, P. Similar level of polyteny in bands and interbands of *Drosophila* giant chromosomes. *Nature*, *307*:176–178 (1984).
25. Lifschytz, E. Sequence replication and banding organization in the polytene chromosomes of *Drosophila melanogaster*. *J. Mol. Biol.* *164*:17–34 (1983).
26. Blackburn, E. H., and Karrer, K. M. Genomic reorganizations in ciliated protozoans. *Annu. Rev. Genet.* *20*:501–521 (1986).
27. Pearlman, R. E., Anderson, P., Engberg, J., and Nilsson, J. R. Synthesis of ribosomal DNA in conjugating *Tetrahymena*. *Exp. Cell Res.* *123*:147–155 (1979).
28. Yao, M.-C., Zhu, S.-G., and Yao, C.-H. Gene amplification in *Tetrahymena thermophila*: Formation of extrachromosomal palindromic genes coding for rRNA. *Mol. Cell. Biol.* *5*:1260–1267 (1985).
29. Anderson, R. P., and Roth, J. R. Tandem genetic duplications in phages and bacteria. *Annu. Rev. Microbiol.* *31*:473–505 (1977).
30. Novick, A., and Horiuchi, T. Hyperproduction of β-galactosidase by *Escherichia coli* bacteria. *Cold Spring Harbor Symp. Quant. Biol.* *26*:234–245 (1961).
31. Horiuchi, T., Horiuchi, S., and Novick, A. The genetic basis of hypersynthesis of β-galactosidase. *Genetics*, *48*:157–169 (1964).
32. Langridge, J. Mutations conferring quantitative and qualitative increases in β-galactosidase activity in *Escherichia coli*. *Mol. Gen. Genet.* *105*:74–83 (1969).
33. Tlsty, T. D., Albertini, A. M., and Miller, J. H. Gene amplification in the *lac* region of *E. coli*. *Cell*, *37*:217–224 (1984).
34. Folk, W. R., and Berg, P. Isolation and partial characterization of *Escherichia coli* mutants with altered glycyl transfer ribonucleic acid synthetase. *J. Bacteriol.* *102*:193–203 (1971).
35. Straus, D. S., and Straus, L. D. Large overlapping tandem genetic duplications in *Salmonella typhimurium*. *J. Mol. Biol.* *103*:143–153 (1976).
36. Rigby, P. W. J., Burleigh, B. D., and Hartley, B. S. Gene duplication in experimental enzyme evolution. *Nature*, *251*:200–204 (1974).
37. Mekalanos, J. J. Duplication and amplification of toxin genes in *Vibrio cholerae*. *Cell*, *35*:253–263 (1983).
38. Sonti, R.V., and Roth, J. R. Role of gene duplication in the adaptation of *Salmonella typhimurium* to growth on limiting carbon sources. *Genetics*, *123*:19–28 (1989).
39. Anderson, P., and Roth, J. Spontaneous tandem genetic duplications in *Salmonella typhimurium* arise by unequal recombination between rRNA (rrn) cistrons. *Proc. Natl. Acad. Sci. USA*, *78*:3113–3117 (1981).
40. Hill, C. W., Grafstrom, R. H., Harnish, B. W., and Hillman, B. S. Tandem duplications resulting from recombination between ribosomal RNA genes in *Escherichia coli*. *J. Mol. Biol.* *116*:407–428 (1977).
41. Edlund, T., and Normark, S. Recombination between short DNA homologies causes tandem duplication. *Nature*, *292*:269–271 (1981).

42. Clowes, R. C. Molecular structure of bacterial plasmids. *Bacteriol Rev. 36*:361–405 (1972).
43. Hashimoto, H., and Rownd, R. H. Transition of the R factor NR1 in *Proteus mirabilis*: Level of drug resistance of nontransitioned and transitioned cells. *J. Bacteriol. 123*:56–68 (1975).
44. Perlman, D., and Rownd, R. H. Transition of R factor NR1 in *Proteus mirabilis*: Molecular structure and replication of NR1 deoxyribonucleic acid. *J. Bacteriol. 123*:1013–1034 (1975).
45. Perlman, D., Twose, T. M., Holland, M. J., and Rownd, R. H. Denaturation mapping of R factor deoxyribonucleic acid. *J. Bacteriol. 123*:1035–1042 (1975).
46. Mattes, R., Burkardt, H. J., and Schmitt, R. Repetition of tetracycline resistance determinant genes on R plasmid pRSD1 in *Escherichia coli*. *Mol. Gen. Genet. 168*:173–184 (1979).
47. Schmitt, R., Bernhard, E., and Mattes, R. Characterization of Tn1721, a new transposon containing tetracycline resistance genes capable of amplification. *Mol. Gen. Genet. 172*:53–65 (1979).
48. Cohen, S. N. Transposable genetic elements and plasmid evolution. *Nature*, *263*:731–738 (1976).
49. Gutterson, N. I., and Koshland, D. E. Replacement and amplification of bacterial genes with sequences altered in vitro. *Proc. Natl. Acad. Sci. USA*, *80*:4894–4898 (1983).
50. Fogel, S., and Welch, J. W. Tandem gene amplification mediates copper resistance in yeast. *Proc. Natl. Acad. Sci. USA*, *79*:5342–5346 (1982).
51. Welch, J. W., Fogel, S., Cathala, G., and Karin, M. Industrial yeasts display tandem gene iteration at the *Cupl* region. *Mol. Cell. Biol. 3*:1353–1361 (1983).
52. Welch, J. W., Maloney, D. H., and Fogel, S. Unequal crossing-over and gene conversion at the amplified *CUPl* locus of yeast. *Mol. Gen. Genet. 222*:304–310 (1991).
53. Tice, S. C. A new sex-linked character in *Drosophila*. *Biol. Bull. 26*:221–230 (1914).
54. Zeleny, C. The direction and frequency of mutation in the bar-eye series of multiple allelomorphs of *Drosophila*. *J. Exp. Zool. 34*:203–233 (1921).
55. Sturtevant, A. H. The effects of unequal crossing over at the *bar* locus in *Drosophila*. *Genetics, 10*:117–147 (1925).
56. Morgan, L. V. Proof that bar changes to not-bar by unequal crossing-over. *Proc. Natl. Acad. Sci. USA*, *17*:270–272 (1931).
57. Peterson, H. M., and Laughnan, J. R. Intrachromosomal exchange at the *bar* locus in *Drosophila*. *Proc. Natl. Acad. Sci. USA, 50*:126–133 (1963).
58. Bridges, C. B. The bar "gene" a duplication. *Science*, *83*:210–211 (1936).
59. Ashburner, M. *Drosophila: A Laboratory Handbook*, Cold Spring Laboratory Press, Cold Spring Harbor, New York, 1989.
60. Rapoport, I. A. Multiple linear repetitions of sections of chromosomes and their evolutionary significance. *Zh. Obsch. Biol.* (Moscow) *1*:235–270. (In Russian, English summary, translation from E. B. Lewis.)
61. Devonshire, A. L., and Field, L. M. Gene amplification and insecticide resistance. *Annu. Rev. Entomol. 36*:1–23 (1991).

62. Field, L. M., Devonshire, A. L., and Forbe, B. G. Molecular evidence that insecticide resistance in peach-potato aphids (*Myzus persicae*) results form amplification of an esterase gene. *Biochem. J. 251*:309–312 (1988).
63. Mouches, C., Pasteur, N., Berge, J. B., Hyrein, O., Raymond, M., et al. Amplification of an esterase gene is responsible for insecticide resistance in a California *Culex* mosquito. *Science*, *233*:778–780 (1986).
64. Raymond, M., Callaghan, A., Fort, P., and Pasteur, N. Worldwide migration of amplified insecticide resistance genes in mosquitos. *Nature*, *350*:151–153 (1991).
65. Maroni, G., Wise, J., Young, J. E., and Otto, E. Metallothionein gene duplications and metal tolerance in natural populations of *Drosophila melanogaster*. *Genetics*, *117*:739–744 (1987).
66. Hakala, M. T. Prevention of toxicity of amethopterin for Sarcoma-180 cells in tissue culture. *Science*, *126*:255 (1957).
67. Hakala, M. T., and Taylor, E. The ability of purine and thymine derivatives and of glycine to support the growth of mammalian cells in culture. *J. Biol. Chem. 234*:126 (1959).
68. Hakala, M. T., Zakrzewski, S. F., and Nichol, C. A. Relation of folic acid reductase to amethopterin resistance in cultured mammalian cells. *J. Biol. Chem. 236*:952–958 (1961).
69a. Fischer, G. A. Nutritional and amethopterin-resistant characteristics of leukemic clones. *Cancer Res. 19*:372–376.
69b. Fischer, G. A. Increased levels of folic acid reductase as a mechanism of resistance to amethopterin in leukemic cells. *Biochem. Pharm. 7*:75–77 (1961).
70. Friedkin, M., Crawford, E. S., Humphreys, S. R., and Golding, H. The association of increased dihydrofolate reductase with amethopterin resistance in mouse leukemia. *Cancer Res. 22*:600–606 (1962).
71. Kashet, E. R., Crawford, E. J., Friedkin, M., Humphreys, S. R., and Golding A. *Biochemistry*, *3*:1928–1931 (1964).
72. Sartorelli, A. C., Both, B. A., and Bertino, J. R. Folate metabolism in methotrexate-sensitive and -resistant Ehrlich ascites cells. *Arch. Biochem. Biophys. 108*:53–59 (1964).
73. Raunio, R. P., and Hakala, M. T. Comparison of folate reductases of Sarcoma 180 cells, sensitive and resistant to amethopterin. *Mol. Pharmacol. 3*:279–283 (1967).
74. Hillcoat, B. L., Perkins, J. P., and Bertino, J. R. Dihydrofolate reductase from L1210R murine lymphoma: Further studies on the binding of substrates and inhibitors to the enzyme. *J. Biol. Chem. 242*:4777–4781 (1967).
75. Perkins, J. P., Hillcoat, B. L., and Bertino, J. R. Dihydrofolate reductase from a resistant subline of the L1210 lymphoma: Purification and properties. *J. Biol. Chem. 242*:4771–4776 (1967).
76. Littlefield, J. W. Hybridization of hamster cells with high and low folate reductase activity. *Proc. Natl. Acad. Sci. USA*, *62*:88–95 (1969).
77. Nakamura, H., and Littlefield, J. W. Purification, properties, and synthesis of dihydrofolate reductase from wild-type and methotrexate-resistant hamster cells. *J. Biol. Chem. 247*:179–187 (1972).
78. Courtney, V. D., and Robins, A. M. Loss of resistance to methotrexate in L5178Y mouse leukemia grown in vitro. *J. Natl. Cancer Inst. 49*:45–53 (1972).

79. Jackson, R. C., and Huennekens, F. M. Turnover of dihydrofolate reductase in rapidly dividing cells. *Arch. Biochem. Biophys. 154*:192–198 (1973).
80. Hanggi, U. J., and Littlefield, J. W. Altered regulation of the rate of synthesis of dihydrofolate reductase in methotrexate-resistant hamster cells. *J. Biol. Chem. 251*:3075–3080 (1976).
81. Alt, F. W., Kellems, R. E., and Schimke, R. T. Synthesis and degradation of folate reductase in sensitive and methotrexate-resistant lines of S-180 cells. *J. Biol. Chem. 251*:3036–3074 (1976).
82. Chang, S. E., and Littlefield, J. W. Elevated dihydrofolate reductase messenger RNA levels in methotrexate-resistant BHK cells. *Cell, 7*:391–396 (1976).
83. Kellems, R. E., Alt, F. W., and Schimke, R. T. Regulation of folate reductase synthesis in sensitive and methotrexate-resistant Sarcoma 180 cells. *J. Biol. Chem. 251*:6987–6993 (1976).
84. Biedler, J. L., and Spengler, B. A. Metaphase chromosome anomaly: Association with drug resistance and cell-specific products. *Science, 191*:185–187 (1976).
85. Biedler, J. L., and Spengler, B. A. A novel chromosome abnormality in human neuroblastoma and antifolate-resistant Chinese hamster cell lines in culture. *J. Natl. Cancer Inst. 57*:683–695 (1976).
86. Biedler, J. L., Albrecht, A. M., Hutchison, D. J., et al. Drug resistance, dihydrofolate reductase, and cytogenetics of amethopterin-resistant Chinese hamster cells in vitro. *Cancer Res. 32*:153–161 (1972).
87. Biedler, J. L., Albrecht, A. M., and Spengler, B. A. Non-banding (homogeneous) chromosome regions in cells with very high dihydrofolate reductase levels. *Genetics, 77*:s4–s5 (1974).
88. Alt, F. W., Kellems, R. E., Bertino, J. R., and Schimke, R. T. Selective multiplication of dihydrofolate reductase genes in methotrexate-resistant variants of cultured murine cells. *J. Biol. Chem. 253*:1357–1370 (1978).
89. Nunberg, J. H., Kaufman, R. J., Schimke, R. T., Urlaub, G., and Chasin, L. A. Amplified dihydrofolate reductase genes are localized to a homogeneously staining region of a single chromosome in a methotrexate-resistant Chinese hamster ovary cell line. *Proc. Natl. Acad. Sci. USA, 75*:5553–5556 (1978).
90. Dolnick, B. J., Berenson, R. J., Bertino, J. R., Kaufman, R. J., Nunberg, J. H., and Schimke, R. T. Correlation of dihydrofolate reductase elevation with gene amplification in a homogeneously staining chromosomal region in L5178Y cells. *J. Cell Biol. 83*:394–402 (1979).
91. Kaufman, R. J., Brown, P. C., and Schimke, R. T. Amplified dihydrofolate reductase genes in unstable methotrexate-resistant cells are associated with double minute chromosomes. *Proc. Natl. Acad. Sci. USA, 76*:5669–5673 (1979).

I

Organization and Structure of Amplified Genes

1

Cytogenetic Consequences of Gene Amplification: HSRs and DMs in Antifolate-Resistant and Multidrug-Resistant Chinese Hamster and Mouse Cells and in Human Neuroblastoma Cells with Amplified N-*myc* Proto-Oncogenes

June L. Biedler and Barbara A. Spengler *Memorial Sloan-Kettering Cancer Center, New York, New York*

I. INTRODUCTION

There are two striking types of chromosomal structure that are the consequence of preferential amplification of genes in somatic mammalian cells: (1) chromosomal segments containing integrated, amplified genes, exemplified by homogeneously staining regions (HSRs), and (2) extrachromosomal amplified DNA sequences often microscopically visualizable as double minute chromosomes (DMs). HSRs, first described in 1976 [1], were initially found in drug-resistant cells and also in human neuroblastoma cells. The small, paired chromatin bodies known as DMs were first observed in 1962 in a lung carcinoma [2] and subsequently reported for other human tumors, including neuroblastoma. However, in these early studies, the significance of DMs was not understood.

In the 1950s and 1960s, we and others attempted to elucidate the genetic basis of drug resistance in mammalian cells by karyotype analysis, in the hope of detecting new structural or numerical chromosome abnormalities that would indicate that the variant phenotype was the result of an altered genotype. Many such studies (which preceded the discovery and use of chromosome banding procedures for chromosome identification) were of aminopterin- or methotrexate (MTX)-selected cells, because antifolates were among the earliest of the cancer chemotherapeutic agents to be clinically effective and because development of resistance was soon recognized as a treatment obstacle. In our laboratory this cytogenetic approach was used first to examine MTX-resistant sublines of the transplantable mouse leukemia L1210 (see Section II) and subsequently Chinese hamster sublines selected with MTX or the quinazoline antifolate, methasquin (MQ). All these cells were characterized by substantial elevation in dihydrofolate reductase (DHFR) activity, eventually known to result from substantial increases in amount of the protein. These studies culminated in the detection of long, nonbanding HSRs in the high enzyme overproducer hamster cells, which could only be interpreted as signifying an amplification process (see Section II). Thus, the first indication that preferential amplification of genes could and did occur in somatic cells was this novel and specific chromosomal anomaly, the HSR.

II. ANTIFOLATE RESISTANCE: GENETIC BASIS OF DIHYDROFOLATE REDUCTASE OVERPRODUCTION

A. Mouse Leukemia L1210

The first strong correlation between a specific structural chromosomal alteration and a specific biochemical mechanism of resistance was described in 1965 in our MTX-resistant mouse leukemia L1210 cells [3]. In a large series of cell lines initiated in vivo from a very small number of cells and serially passaged in MTX-treated mice, we noted that the short arm of a subtelocentric marker chromosome tended to become elongated to different extents (Fig. 1). This elongation coincided exactly with elevation of target enzyme DHFR activity and development of

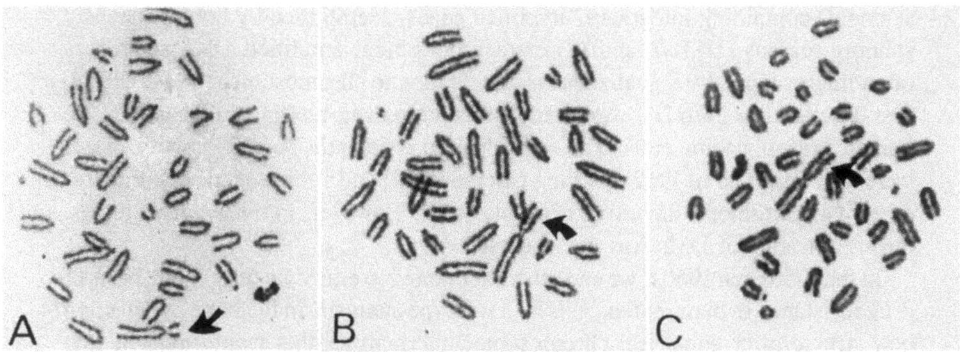

Figure 1 Mouse leukemia L1210 cell with a subtelocentric marker chromosome (A) and MTX-resistant cells with elongations of the short arm (B,C).

resistance, and it was stated that "the chromosome appears to be involved in the genetic regulation of dihydrofolate reductase" [3]. These experiments indicated clearly that the structurally altered chromosomes occurred as a consequence of drug exposure and were associated with a specific biochemical change. However, it was difficult to conceptualize what sort of genetic alteration could lead to chromosome elongation.

B. Chinese Hamster Cells

To better define chromosomal alteration associated with antifolate resistance and increased DHFR activity, we developed in vitro a large series of independently derived Chinese hamster lung cell sublines, which were obtained by stepwise selection in MTX or MQ (Table 1). Although there was extensive chromosomal rearrangement, there was again discernible specificity, with unusually long (pre-

Table 1 Cytogenetic Abnormalities Associated with Gene Amplification in Chinese Hamster Sublines with Elevated Dihydrofolate Reductase Activity and Gene Copy Number[a]

Cell line	Relative DHFR activity	Relative DHFR gene copy number	Abnormal chromosome region		
			Type	Location	Code (Fig. 3)
DC-3F	1	1	None		
DC-3F/MQ10	2.9	3	ABR	2p	a
DC-3F/MQ31	3.9	4	ABR	2p	b
DC-3F8/A55	4.6	5	ABR	2p	c
DC-3F/A1	21		ABR	2p	d
DC-3F8/A40	35		ABR, ABR	4q,9p	
DC-3F/MQ20	49		ABR	2p	e
DC-3F/MQ21	86	20	ABR, HSR	9p,2p	f
DC-3F/A XXVIII	95		ABR, HSR	9q,Xp	
DC-3F8/A75	121		HSR	2p	g
DC-3F/MQ29	122		HSR	4q	
DC-3F/MQ25	129		HSR	2p	h
DC-3F/MQ8	144		HSR	2p	i
DC-3F/MQ19	151	120	HSR	2p	j
DC-3F/A3	170	53	HSR	2p	k
DC-3F/Ab-17	281		HSR	2p	l
DC-3F/A3-P-U	18	7	None		

[a]Abbreviations used: A, methotrexate; MQ, methasquin; ABR, abnormally banding region; HSR, homogeneously staining region.

sumably translocated) segments on a single chromosome 2 especially and on several other chromosomes as well [4]. When trypsin-Giemsa banding methods became available, we reexamined these sublines. The mystifying finding was that the long segments did not band; that is, they did not display the cross-striational arrays of darkly to lightly stained bands typifying normal hamster chromosomes (Fig. 2). The segments stained homogeneously, with intermediate intensity, and were termed, simply, homogeneously staining regions or HSRs. Since HSRs comprised 2–5% of the total chromosome complement and were present in the resistant sublines with the highest levels of DHFR activity (Table 1), we hypothesized that the regions were functionally involved in excessive production of the enzyme and were chromosomal manifestations of an amplification process [1]. This hypothesis was confirmed by Schimke and coworkers [5], who showed that their MTX-resistant mouse cells could have as many as 200 copies of *dhfr* genes. Subsequently, we and others demonstrated by DNA in situ hybridization techniques that HSRs were indeed sites of amplified *dhfr* genes (Fig. 2B). The specific

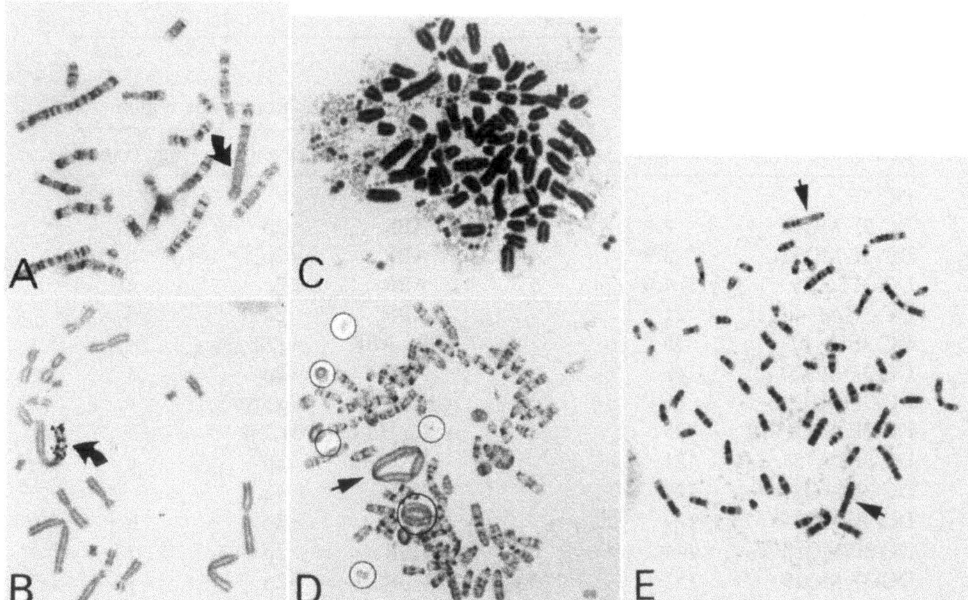

Figure 2 Antifolate-resistant Chinese hamster cell (A) with prototype HSR and (B) labeled in situ with a *dhfr*-specific probe. (C) Highly resistant MAZ/A/MQ mouse cell with large numbers of DMs in several size classes. Trypsin-Giemsa banded mouse cells showing DMs and homogeneously staining ring chromosomes (D), with examples circled or indicated by an arrow, and presence of both DMs and HSRs (arrows) in the same cell (E).

karyotypic alterations that we had previously observed in our MTX-resistant mouse leukemia L1210 cells likely comprised similar structures [3].

In many antifolate-resistant cell lines the amplified *dhfr* genes are not intrachromosomal, but rather are contained in DMs. Starting in the late 1970s, DMs were observed in human tumor cells not under overt selection pressure as well as in cells selected with different cytotoxic drugs. It was deduced or demonstrated that DMs and HSRs are alternative chromosomal manifestations of gene amplification and, as extensively reviewed [6,7] and illustrated in Figure 2, that amplification may give rise to a variety of abnormal chromosome structures such as HSRs and DMs, homogeneously staining chromosomes including ring chromosomes, and abnormally banding regions (ABRs). We initially detected ABRs in the antifolate-resistant Chinese hamster sublines with low DHFR levels (Table 1). In sublines with only low level *dhfr* gene amplification, the ABRs were sometimes as long or longer than the HSRs of high enzyme sublines [8]. Molecular cytogenetic studies suggested that such ABRs arose as a consequence of multiple insertions of amplified *dhfr* genes and other, accompanying DNA sequences as well [9].

C. Extrachromosomal Gene Amplification Model

A conclusion drawn from our cytogenetic studies of antifolate-resistant Chinese hamster cells is that the chromosomal location of HSRs or ABRs is not a reliable indicator of the permanent location of the *dhfr* gene. Of 18 HSRs and ABRs (Table 1), 6 were present on chromosomes other than chromosome 2, and 12 were on chromosome 2p, known to harbor the *dhfr* gene. Of those 12, however, only 5 were found at or near the 2p23–25 native gene locus (Fig. 3). In HSR-containing sublines, the normal, distal portion of the 2p arm was translocated to another chromosome, and there was exact correspondence (as determined microscopically) between the break point of the translocated segment and the (proximal) HSR site [10]. The simplest explanation for this series of observations is that

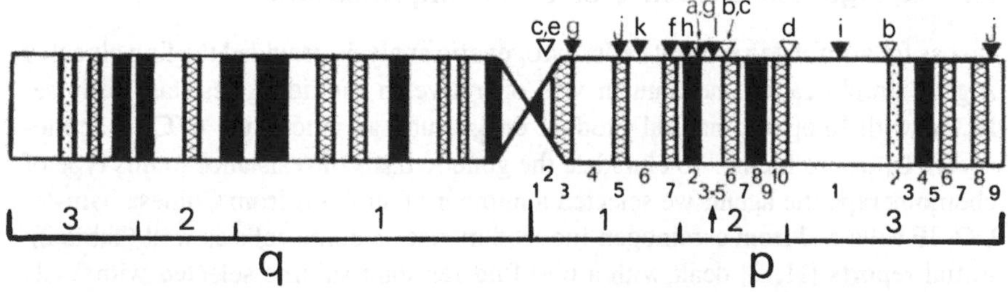

Figure 3 Diagram of Chinese hamster chromosome 2 with sites of HSRs (solid triangles) and ABRs (open triangles). Letters above symbols identify cell lines as listed in Table 1. (Adapted from Ref. 10.)

amplification of the gene and accompanying DNA sequences did not (usually) occur in loco. Rather, the initially amplified genes were transiently extrachromosomal and became reintegrated at sites more or less near but only occasionally at the exact site of the native *dhfr* gene [6,10]. Determining factors for reinsertion sites could be (1) the duration of the extrachromosomal phase—a brief time period (rapid reinsertion) would best account for the observed preferential location of ABRs and HSRs on the 2p arm, and (2) the location and prevalence of drug-induced DNA strand breaks, logical sites of reintegration. That chromosome breakage did occur is attested to by the extent of structural chromosomal rearrangement, including translocation of the chromosome 2p segment from the homologue carrying the amplified genes.

These cytogenetic findings do not appreciably illuminate the mechanism of HSR formation, but one inference can be drawn. Since ABRs are present in the sublines with the lowest DHFR levels (Table 1), it follows that they represent early steps in the formation of HSRs. If so, then ABRs (which are consistent in band pattern for any one subline) may comprise stretches of reintegrated nucleotide sequences, including *dhfr* genes and accompanying DNA sequences initially episomal or DM-like in nature. Subsequent selection pressure from increased drug concentration could be accompanied by excision, amplification, and reintegration of units containing *dhfr* genes, and by excision and loss of non-*dhfr* gene sequences, resulting in tandem arrays of amplicons manifested as HSRs.

We do not know whether the hypothetical extrachromosomally amplified *dhfr* gene is the consequence of excision of the gene or of a gene copy mechanism. However, our observation of the gradual decline in HSR length in the revertant DC-3F/A3-P-U line (Table 1) and for an interstitial HSR in a vincristine-selected subline (see Section Three) grown in absence of drug indicates that excision can and does occur.

III. MULTIDRUG RESISTANCE: GENETIC BASIS

A. Cytogenetic Evidence of Gene Amplification

Just as for antifolate-resistant cells, cytogenetic analysis provided the first clue that a gene amplification mechanism was operative in multidrug-resistant cells selected with lipophilic natural product drugs such as vincristine (VCR), actinomycin D, or colchicine. To elucidate the genetic basis of resistance to this type of chemotherapeutic agent, we selected a number of sublines from Chinese hamster DC-3F cells and from carcinogen-induced mouse sarcoma cells as well (Table 2). Initial reports [11,12] dealt with a 650-fold resistant subline selected with VCR from the same parental Chinese hamster lung cell line, DC-3F, that was used for derivation of sublines resistant to antifolates (Table 1). A somewhat darkly staining HSR was located interstitially on a rearranged chromosome consisting of part of

Table 2 Cytogenetic Abnormalities Associated with Gene Amplification in Multidrug-Resistant Chinese Hamster and Mouse Cells with Amplified P-Glycoprotein-Encoding Genes

				Chromosome abnormality	
Cell line[a]	Relative resistance	Relative amplification	Relative mRNA	Type[b]	Location
DC-3F	1	1	1		
DC-3F/VCRd-5L	2,740	30	16	HSR	8p19/Xq21
DC-3F/VCRm	560	25		ABRs	6q13,10p21
DC-3F/AD X	10,000	3	16	ABR	1q26
DC-3F/AD XC	70,000	5	97	ABR	1q26
DC-3F/DM XX	1,530	20	22	HSR	1q26
MAZ	1	1			
MAZ/VCR	3,940	~50		DMs (183)	
QUA	1				
QUA/ADj	3,370			DMs (106)	

[a]Abbreviations: VCR, vincristine; AD, actinomycin D; DM, daunorubicin.
[b]Numbers in parenthesis indicate mean number of DMs per cell.

the q arm of an X chromosome and a chromosome 8 (Fig. 4). In the next several years, before molecular evidence of gene amplification was obtained, there were numerous reports of sublines in the multidrug resistance category that displayed HSRs, ABRs, or DMs (reviewed in Ref. 13). However, the identity of the encoded product was uncertain. In 1984 studies in other laboratories demonstrated the presence of amplified DNA sequences in multidrug-resistant Chinese hamster cells and in 1985 P-glycoprotein was identified as the primary product of the

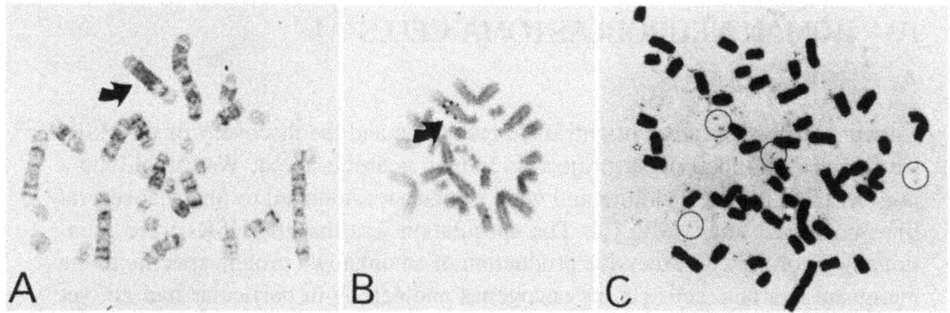

Figure 4 Multidrug-resistant Chinese hamster cells (A) with a typical darkly staining HSR and (B) labeled in situ with a P-glycoprotein-specific probe. Mouse MAZ/VCR cell (C) with numerous very small DMs; examples are circled.

amplified genes (reviewed in Ref. 13 and detailed elsewhere in this volume). Principal molecular and cytogenetic characteristics of the multidrug-resistant hamster, mouse, and human cells studied in our laboratory are described in a recent review [14] and will be highlighted here.

Comparison of the antifolate-resistant and multidrug-resistant cell systems reveals similar types of amplification-associated aberrant chromosome structures. However, HSRs in the latter cells routinely stain more darkly and somewhat less homogeneously, perhaps reflecting a greater complexity of the amplicon or of the amplification process itself. Multidrug-resistant cells, even when highly resistant, sometimes had only small numbers of genes coding for P-glycoprotein (Table 2), in marked contrast to the very high levels of *dhfr* gene amplification (> 1000 genes) attainable [7]. The amplified genes for both types of resistance tended to be chromosomally integrated in Chinese hamster cells and to reside on DMs in mouse cells, indicative of a species-related phenomenon. The resistance phenotype was unstable in the absence of drug. Whereas decrease in resistance and DM number in DM-containing resistant cells was rapid, resistance and HSR length declined gradually; the HSRs appeared to shorten in small decrements [14]. Our studies thus indicated that examination of phenotypic stability could distinguish between gene amplification and some other mechanism of resistance development.

B. Chromosomal Location of Amplified Genes

Of the four independently derived DC-3F sublines listed in Table 2 (DC-3F/AD XC was selected from DC-3F/AD X), two had HSRs or ABRs positioned on chromosome 1q26, the apparent site of the native P-glycoprotein gene(s), and two had the amplified genes located on some other chromosome. These findings are not inconsistent with an extrachromosomal gene amplification model and, for the former two sublines, possibly reflect rapid reintegration and/or homologous recombination processes.

IV. HUMAN NEUROBLASTOMA CELLS

A. HSRs and DMs

Concurrently with studies of antifolate resistance and the discovery of the HSR, we initiated an independent project on human neuroblastoma. We established a number of cell lines in culture and were somewhat astounded to find that several lines contained long HSRs [1]. The speculation was that the HSRs were functionally involved with excessive production of an unknown protein specific to the malignant neuronal cells [1], as oncogenes and *MYCN* in particular had not yet been discovered. It is now well known that many neuroblastoma cell lines [15] and a subset of tumors in the patient amplify and overexpress the N-*myc* proto-oncogene and that the genes are manifested cytogenetically as DMs or HSRs

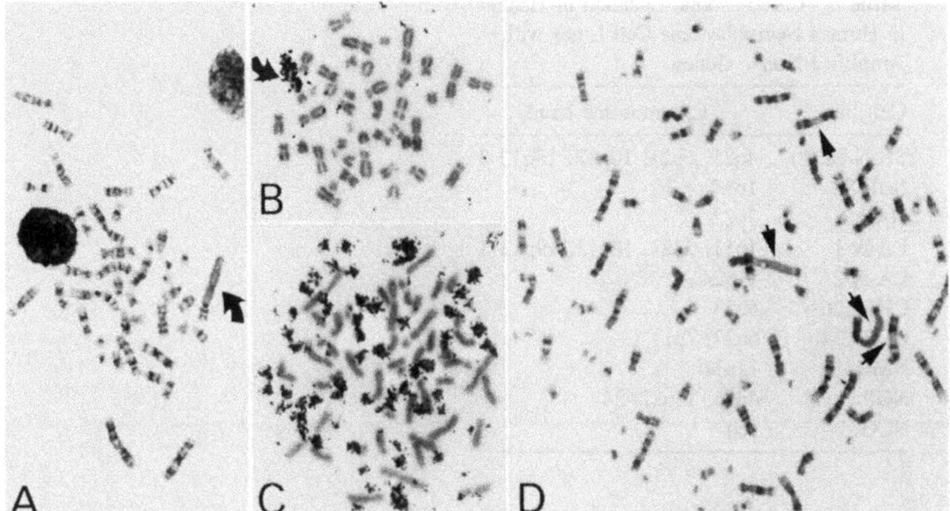

Figure 5 Human neuroblastoma SK-N-BE(2) cell (A) with prototype HSR and (B) labeled in situ with an *MYCN*-specific probe. (C) Hybridization of CHP-234 cells with the same probe, directly demonstrating presence of *MYCN* genes in the rather large DMs. Actinomycin D-selected BE(2)-C/ACT cell with four typical HSRs (arrows) comprising amplified *MYCN* genes and with multiple DMs containing newly acquired amplified genes encoding P-glycoprotein (D).

(Fig. 5). One cell line, NAP, lost DMs with time in culture but gained first one then multiple HSRs, suggesting that HSRs conferred a selective advantage under in vitro conditions. In general, HSR-containing cell populations grow faster than DM-containing cells, indicating that the N-*myc* gene product encoded by stably integrated genes provides proliferative advantage.

B. Chromosomal Location of HSRs

As shown in Table 3, our data and data reported by others demonstrate that *MYCN*-containing HSRs have no preferential chromosome site [6,15,16]. The N-*myc* gene maps to chromosome 2p24. In only one instance, in a small proportion of LA-N-1 cells, did we find an HSR close to or at that site (Table 3). The variability in HSR location is somewhat unexpected, since there was far less structural chromosome rearrangement than in the drug-resistant Chinese hamster cells. A notable feature of many of the cell lines established in vitro from neuroblastoma is that they are karyotypically near-diploid, have relatively few structurally rearranged chromosomes, and tend to remain karyotypically quite stable over long cultivation periods [15,16]. The simplest interpretation of the cytogenetic findings for the

Table 3 Chromosomal Location of HSRs in Human Neuroblastoma Cell Lines with Amplified N-*myc* Genes

Cell line	Chromosome band
SK-N-BE(2)	4q25, 6p21, 10p13, 19q13.3
IMR-32	1p34 or 36
NAP	7p22
LA-N-1	1p11, 2p23, 16q13, 19p13.3
LA-N-2	11q24
CHP-126	5q33
CHP-134	6q27, 7p11
NMB	13p11
NGP	4p16, 13q22–24
NCG	16q

neuroblastoma cells is that amplification of *MYCN* did not occur in loco, but, rather, was extrachromosomal. The initially amplified gene or genes either remained extrachromosomal as episomes (as discussed elsewhere in this volume) or DMs or became reintegrated, most often at a site unrelated to that of the native N-*myc* gene.

V. SUMMARY AND OVERVIEW

In this review of the cytogenetic studies carried out in our laboratory over the past 30 years, several themes emerge.

Cytogenetic analysis of drug-resistant cells for the purpose of determining the genetic basis of resistance development, although naive in retrospect, was in actuality highly successful. Fortuitous, indeed, was the choice of methotrexate-resistant cell systems, since one major resistance mechanism involves amplification of *dhfr* genes frequently associated with the presence of readily detectable aberrant chromosome structures. Another fortuitous aspect was that in the L1210 system the chromosome undergoing alteration was about the only identifiable (marker) chromosome in the nonbanded mouse karyotype. Had similar changes occurred only to a normal chromosome 13 homologue, to which the native *dhfr* gene maps, they would have gone undetected.

Our cytogenetic studies were aided by additional serendipitous coincidences. Resistance developing in sublines selected with natural product cancer chemotherapeutic drugs also involved a gene amplification mechanism, revealed by the presence of an HSR in vincristine-resistant cells [11]. In this series of cell lines, ABRs in sublines with low to moderate levels of amplification selected with actinomycin D or daunorubicin were not revealed until we employed in situ hybridization techniques. The major reasons were the inexplicably poor quality of the metaphase chromosome preparations of multidrug-resistant Chinese hamster cells, in contrast to those of antifolate-resistant cells selected from the same parental DC-3F cell line, and the high incidence of structural rearrangement in the karyotype. Finally, in a totally unrelated project, human neuroblastoma tumor cells under systematic study at that time turned out to have spectacular HSRs like those in MTX-resistant hamster cells [1], resulting from high level *MYCN* amplification. This unusual constellation of findings of similar gene amplification-associated chromosomal abnormalities in the three major projects under study suggested that HSRs, in particular, were widespread, common features of cultured cells—a simplistic generalization, as it turned out. However, their discovery and the speculation that they were the consequence of an amplification mechanism [1] speeded up the eventual molecular confirmation and led to the search for the oncogene involved in human neuroblastoma.

Microscopic detection of HSRs and DMs has over the years become a simple way of recognizing the presence of a gene amplification mechanism even when the identity of the amplified gene is not known. High level amplification usually results in readily detectable, distinctive HSRs or DMs. Chromosomal structures characterizing low level amplification are less obvious. In our work, knowledge of the preferential location of HSRs in antifolate-resistant hamster cells enabled detection of ABRs in sublines with low gene copy numbers. In general, identification of a region as an ABR requires an expert cytogeneticist and confirmation of integrated amplified genes by in situ hybridization methods. The development of molecular methods for detecting amplified sequences has further advanced the study of gene amplification.

Subjects that continue to be of great interest, as discussed elsewhere in this volume, are how gene amplification occurs and how such aberrant chromosomal structures as HSRs and DMs are generated. We have speculated that, since ABRs characterize antifolate-resistant cells with low level resistance whereas HSRs are found in sublines with high level resistance, ABRs are direct precursors of HSRs. This scenario, however, is by no means a certainty. This and other possibilities, including sister chromatid exchange and various models of replication, and the existence of multiple and/or different mechanisms in different cells or at different times within one cell population, remain to be tested. What does seem certain, however, from our series of observations, is that amplification (of chromosomally integrated genes) rarely (never?) occurs in loco. Of the approximately 40 HSRs or ABRs that we have localized in mainly near-diploid Chinese hamster and human neuroblastoma cells, only 6 at most were found at or near the site of the native gene. The simplest explanation is that the amplified sequences are initially extrachromosomal and are integrated into the chromosome in one or multiple steps. The probability of and rapidity of chromosomal reintegration may be species dependent. Continued study of the mechanisms involved in initiation, replication, and insertion of amplified sequences should lead to a greater understanding of cell growth and regulation as well as of the role of genomic instability in cancer.

REFERENCES

1. Biedler, J. L., and Spengler, B. A. Metaphase chromosome anomaly: Association with drug resistance and cell-specific products. *Science, 191*:185–187 (1976).
2. Spriggs, A. I., Boddington, M. M., and Clark, C. M. Chromosomes of human cancer cells. *Br. Med. J. 2*:1431–1435 (1962).
3. Biedler, J. L., Albrecht, A. M., and Hutchinson, D. J. Cytogenetics of mouse leukemia L1210. I. Association of a specific chromosome with dihydrofolate reductase activity in amethopterin-treated sublines. *Cancer Res. 25*:246–257 (1965).
4. Biedler, J. L., Albrecht, A. M., Hutchison, D. J., and Spengler, B. A. Drug response, dihydrofolate reductase, and cytogenetics of amethopterin-resistant Chinese hamster cells in vitro. *Cancer Res. 32*:153–161 (1972).
5. Alt, F., Kellems, R. E., Bertino, J. R., and Schimke, R. T. Selective multiplication of dihydrofolate reductase genes in methotrexate-resistant variants of cultured murine cells. *J. Biol. Chem. 253*:1357–1370 (1978).
6. Biedler, J. L., Melera, P. W., and Spengler, B. A. Chromosome abnormalities and gene amplification: Comparison of antifolate-resistant and human neuroblastoma cell systems, in *Chromosomes and Cancer: From Molecules to Man* (J. D. Rowley and J. E. Ultmann, eds.), Bristol-Myers Cancer Symposia, Vol. 5, Academic Press, New York, 1983, pp. 117–138.
7. Albrecht, A. M., and Biedler, J. L. Acquired resistance of tumor cells to folate antagonists, in *Folate Antagonists as Therapeutic Agents*, Vol. 1 (F. M. Sirotnak, J. J. Burchall, W. B. Ensminger, and J. A. Montgomery, eds.), Academic Press, New York, 1984, pp. 317–353.
8. Biedler, J. L., Melera, P. W., and Spengler, B. A. Specifically altered metaphase chromosomes in antifolate-resistant Chinese hamster cells that overproduce dihydrofolate reductase. *Cancer Genet. Cytogenet. 2*:47–60 (1980).
9. Lewis, J. A., Biedler, J. L., and Melera, P. W. Gene amplification accompanies low level increases in the activity of dihydrofolate reductase in antifolate-resistant Chinese hamster lung cells containing abnormally banding regions. *J. Cell Biol. 94*:418–424 (1982).
10. Biedler, J. L. Evidence for transient or prolonged extrachromosomal existence of amplified DNA sequences in antifolate-resistant, vincristine-resistant, and human neuroblastoma cells, in *Gene Amplification* (R. T. Schimke, ed.), Cold Spring Harbor Laboratory, Cold Spring Harbor, New York, 1982, pp. 39–45.
11. Biedler, J. L., Meyers, M. B., Peterson, R. H. F., and Spengler, B. A. Marker chromosome with a homogeneously staining region (HSR) in vincristine-resistant cells. *Proc. Am. Assoc. Cancer Res. 21*:292 (1980).
12. Biedler, J. L., and Peterson, R. H. F. Altered plasma membrane glycoconjugates of Chinese hamster cells with acquired resistance to actinomycin D, daunorubicin, and

vincristine, in *Molecular Actions and Targets for Cancer Chemotherapeutic Agents* (A. C. Sartorelli, J. S. Lazo, and J. R. Bertino, eds.), Bristol-Myers Cancer Symposia, Vol. 2, Academic Press, New York, 1981, pp. 453–482.

13. Biedler, J. L., and Meyers, M. B. Multidrug resistance (*Vinca* alkaloids, actinomycin D, and anthracycline antibiotics), in *Drug Resistance in Mammalian Cells*, Vol. II: *Anticancer and Other Drugs* (R. S. Gupta, ed.), CRC Press, Boca Raton, Florida, 1989, pp. 57–88.
14. Melera, P. W., and Biedler, J. L. Molecular and cytogenetic analysis of multidrug resistance-associated gene amplification in Chinese hamster, mouse sarcoma, and human neuroblastoma cells, in *Molecular and Cellular Biology of Multidrug Resistance in Tumor Cells* (I. Roninson, ed.), Plenum Press, New York, 1991, pp. 117–145.
15. Biedler, J. L., Meyers, M. B., and Spengler, B. A. Homogeneously staining regions and double minute chromosomes, prevalent cytogenetic abnormalities of human neuroblastoma cells, in *Advances in Cellular Neurobiology*, Vol. 4 (S. Federoff and L. Hertz, eds.), Academic Press, New York, 1983, pp. 267–307.
16. Biedler, J. L., Ross, R. A., Shanske, S., and Spengler, B. A. Human neuroblastoma cytogenetics: Search for significance of homogeneously staining regions and double minute chromosomes, in *Advances in Neuroblastoma Research* (A. E. Evans, ed.), Vol. 12, *Progress in Cancer Research and Therapy*, Raven Press, New York, 1980, pp. 81–96.

2

Organization and Structure of Dihydrofolate Reductase Amplicons

Joyce L. Hamlin *University of Virginia School of Medicine, Charlottesville, Virginia*

I. INTRODUCTION

It has generally been assumed that knowledge of the sequence rearrangements that occur during amplification processes will provide insight into the responsible mechanisms. In addition to these interests, we have been concerned in our laboratory with relationships between functional elements in the genome and the proposed domain structure of chromatin, and with the possibility that amplicons might correspond to chromosomal domains. Therefore, we embarked several years ago on a project to clone an entire dihydrofolate reductase (*DHFR*) amplicon equivalent from a methotrexate (MTX)-resistant Chinese hamster ovary (CHO) cell line.

Because the homogeneously staining chromosome regions (HSRs) that contain amplified *DHFR* genes are elongated rather than thickened, it seemed likely that amplicons would be organized into end-to-end arrays in the chromosomal DNA fiber. In the absence of evidence to the contrary, it was also easiest to assume at the outset that amplicons within a single cell line would be relatively uniform in structure. However, it was appreciated that cloning the equivalent of a whole amplicon might be a demanding task, since estimates based on *DHFR* gene copy number and the size of HSRs suggested that individual repeating units might be several hundred kilobases long [1–3].

In fact, until recently, it has not been possible to clone a whole amplicon equivalent from any mammalian source, primarily because most of the cell lines used to generate recombinant genomic libraries contained rather heterogeneous collections of very large amplicons (see, e.g., Refs. 4,5). As a consequence, multiple branch points were encountered as chromosomal walking steps were extended outward from initial clones, and the 5′ end of a contiguous sequence could not be connected to the 3′ end of the same sequence, as it ultimately should be in any tandem array of homogeneous units [4]. In some MTX-resistant murine cell variants, it even appeared that DNA sequences from distant, unrelated loci were joined to the *DHFR* locus in scrambled arrays during the amplification process [4].

In contrast, we had obtained evidence that the *DHFR* amplicons in MTX-resistant Chinese hamster cells might be less heterogeneous, both within and among cell lines, and possibly somewhat smaller than in other systems. These features have made it possible to clone and characterize the equivalent of three distinct *DHFR* amplicon types from two different MTX-resistant Chinese hamster cell lines.

II. MOLECULAR CLONING OF CHINESE HAMSTER *DHFR* AMPLICONS

Several years ago, we developed a MTX-resistant CHO variant (CHOC 400) that harbors ~1000 copies of the *DHFR* gene distributed among three stable chromosomal HSRs [6]. The high copy number made it possible to visualize restriction fragments arising from the amplicons in digests separated on ethidium bromide-stained agarose gels. By comparison to internal standards, the sizes of individual fragments, hence the total length of the core repeating units, could be roughly estimated. By this criterion, the consensus amplicons in CHOC 400 cells had to be at least 135 kb long [6]. Furthermore, in a side-by-side comparison with DNA from three independently isolated MTX-resistant Chinese hamster cell lines, the pattern of amplified restriction fragments seemed to be quite similar (but not identical) [7]. The same conclusion was reached in experiments in which genomic DNA digests from the three different MTX-resistant Chinese hamster cell lines were compared by an in-gel renaturation method [8], which affords a much higher resolution picture of amplified bands owing to removal of single copy background sequences from the agarose gel [9].

Initial recombinant clones from the amplicons in CHOC 400 cells were isolated from a cosmid library prepared in the vector pHC79. Cosmids containing parts of the *DHFR* gene were detected by a standard colony hybridization protocol, utilizing a partial cDNA from murine cells [10] as a specific probe [11]. Clones illuminated specifically by the probe were then pooled and tested for their ability to rescue a *DHFR*-deficient CHO cell line after transfection. In this way, a cosmid (cH1) was identified that contained the entire 26 kb *DHFR* gene except for a small part of the 3′ untranslated region [11].

In concurrent studies, we had also identified a small number of early-labeled restriction fragments (ELFs) from the CHOC 400 amplicon that were preferentially labeled with [^{14}C]thymidine in the beginning of the S period, one of which was therefore likely to contain an origin of replication [12]. These fragments were subsequently isolated in the cosmid cS21 by excising the amplified ELFs from an agarose gel, labeling in vitro with [^{32}P]dCTP, and using them to probe the cosmid library [13].

By connecting cH1 and cS21 together, and by walking outward in both directions from this locus, we were eventually able to construct a circularly permuted map by connecting the 5′ and 3′ ends of this contiguous sequence together in overlapping cosmids [7,14]. Thus, at least one of the amplicon types in the CHOC 400 genome (the type I sequence) appeared to be organized into simple head-to-tail arrays. By summing the lengths of individual fragments in the cosmid set, this sequence was estimated to be at least 260 kb long (Fig. 1a). The type I amplicon was much larger than the 135 kb that was estimated from the observable restriction pattern of genomic DNA [6], undoubtedly because of the presence of many doublets and triplets in the digests. Because the position of the junction between the "head" and "tail" of the type I sequence was unknown at the time, the correct permutation of the circular map could not be drawn.

In quantitative hybridization studies, the type I sequence was shown to represent only 5–10% of the amplicons in CHOC 400 cells [14]. The majority of amplicons were smaller and were shown in hybridization studies to be missing sequences mapping between cosmids cNQ7 and cHDZ23 (Fig. 1a). The easiest rearrangement to imagine was one in which the prevalent amplicon species (the type II) represented a smaller version of the type I sequence that had been trimmed at its ends.

However, in the course of isolating appropriate hybridization probes from cNQ7 to be used in subsequent chromosomal walking steps outward from this position, a cross-reaction was accidentally detected between a prospective end probe from cNQ7 and a second fragment in the same cosmid. A series of rather complicated mapping and hybridization experiments then showed unequivocally that both cNQ7 and cHDZ23 (as well as the corresponding sequences in CHOC 400 genomic DNA) represented palindromes [14]. Moreover, the two palindromic restriction fragments in cNQ7 and cHDZ23 appeared to be present in equal copy numbers in the genome, suggesting that they might correspond to the two ends of the same amplicon type.

The results of the mapping experiments with cosmid clones therefore evoked a picture in which the predominant amplicon type in CHOC 400 cells was ~230 kb long, with the multiple copies organized into alternating head-to-head and tail-to-tail arrays in the genome (where cNQ7 and cHDZ23 represent the "head" and "tail," respectively). This result, although unexpected, became much easier to accept when an independent study using a snapback hybridization procedure showed

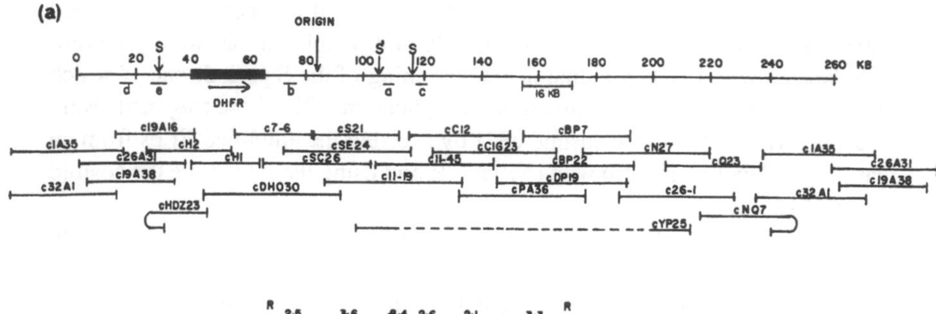

Figure 1 Initial ordering of overlapping recombinant cosmids from the CHOC 400 *DHFR* amplicons. (a) A linear version of the type I amplicon is represented by a circularly permuted map that has been broken arbitrarily in cosmid c32A1, with the 5′ end of cosmid c26A31 placed at zero (a large cosmid collection used to construct this map is shown below). The predicted 230 kb type II amplicon maps between the two inverted sequences depicted in cosmids cHDZ23 and cNQ7. cYP25 is a cosmid that joins sequences in cosmids cSE24 and cN27 and represents a minor amplicon variant (the type III) that contains a 90 kb internal deletion (represented by dashed line). The *DHFR* gene is shown on the linear scale by a closed box, and the position of the *ori*-β/*ori*-γ initiation locus is also shown (origin). *Sfi* I sites, indicated with S above the scale, were determined by digesting cosmids. The positions of probes a–g, which were used in mapping the junction in the type I sequence, are shown below the scale, as is the 16 kb *Eco*RI junction fragment from cosmid cPA36 that contains the junction. A *Hin*dIII map of this fragment is shown at the bottom; the junction actually occurs within the 800 bp fragment.

that some of the amplicons in Syrian hamster cells resistant to *N*-(phosphonacetyl)- L-aspartate (PALA) and in human tumor cells are organized in similar palindromic arrays [15].

III. ANALYZING AMPLIFIED DNA USING CLONED SEQUENCES

The ultimate aims of this extended cloning and mapping project were to determine the sequence arrangement of the parental *DHFR* locus, to study the rearrangements that occurred in this locus during the multiple amplification steps that resulted in the CHOC 400 cell line, and to determine whether *DHFR* amplicons were different in different MTX-resistant Chinese hamster cell variants.

A first priority was to establish the correct permutation of the circularly permuted 260 kb type I map. To do this, it was necessary to identify the two parental restriction fragments (D and K below) that were joined to form the junction (fragment K′D′) between two type I amplicons. This novel junction fragment should hybridize with each of the two parental fragments D and K, but

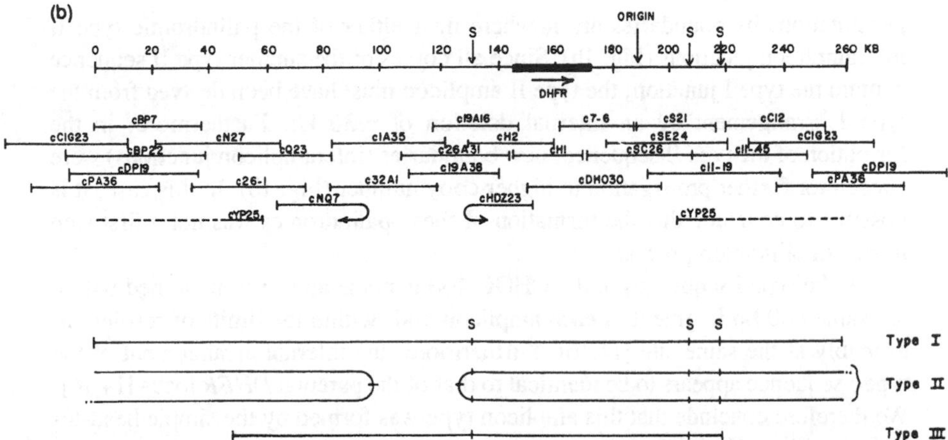

Figure 1 (b) The correct order of recombinant cosmids is shown as their sequences would occur in the CHO genome. The linear scale begins and ends at zero, which represents the type I junction in cPA36. Note that in later studies in which the sizes of *Sfi* I sites were actually determined in the genome, the size estimate for the type I and II amplicons increased to 273 and 240 kb, respectively. A simple diagram of the type I, II, and III amplicons is shown at the bottom. Note that one end of the type II amplicon is joined to the other end of the type II to form giant palindromes in the genome.

the junction fragment itself should differ in size from the parental fragments in a restriction digest.

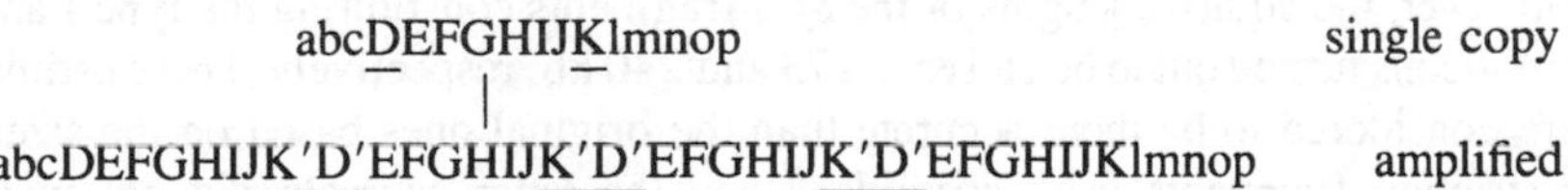

To identify such a junction-containing fragment in a 260 kb amplicon by virtue of its novel size and cross-hybridization with parental fragments is a formidable problem, since there are so many individual fragments generated by restriction enzymes that recognize six-membered sequences (some fragments being so small as to be virtually undetectable). We therefore utilized the infrequent-cutting enzyme, *Sfi* I, to digest the genomic DNA of both CHO and CHOC 400 cells, and analyzed the resulting digests by pulsed-field gradient (PFG) gel electrophoresis. After Southern transfer, the digests were hybridized with appropriate probes, and the type I interamplicon junction was localized relatively easily among the small number of large fragments generated by *Sfi* I [16]. In a series of finer mapping studies, the junction was shown to reside in a single 800 bp *Hin*dIII fragment [16].

The identification of the type I junction allowed us to establish the correct permutation of the circularly permuted type I map vis à vis the parental *DHFR* locus. Interestingly, when the map of the type I amplicon is redrawn in the correct permutation, its boundaries are nowhere near either of the palindromic type II interamplicon junctions (Fig. 1b). Since all copies of the smaller type II sequence contain the type I junction, the type II amplicon must have been derived from the type I arrangement by an internal deletion of ~30 kb. Furthermore, in the formation of the type II sequence, new boundaries (interamplicon junctions) were chosen for further propagation to higher copy number (Fig. 1b). In this case, it is possible to conclude that the formation of these palindromes was not a first step in the amplification process.

All the type I sequences in the CHOC 400 genome appear to be joined within the same 800 bp fragment in each amplicon and, within the limits of resolution, probably at the same site [14,16]. Furthermore, the internal arrangement of the type I sequence appears to be identical to that of the parental *DHFR* locus [14,16]. We therefore conclude that this amplicon type was formed by the simple head-to-tail repetition of an otherwise nonrearranged parental sequence. The two palindromic joints that define the ends of the type II sequence are similarly homogeneous in the CHOC 400 genome [14,16], but the deletion and rearrangement that formed the type II sequence were obviously very complex events.

Analysis of large *Sfi* I fragments separated on PFG gels also allowed us to confirm the proposed structures of the type I and II amplicons, which we had pieced together from rather complicated mapping studies on individual cosmid sequences. By determining the location of *Sfi* I sites in the cosmids, the sizes of the corresponding *Sfi* I fragments in the genome could be predicted (see Fig. 1). Almost without exception, all the predicted fragments from the type I and II amplicons were observed in hybridization studies with appropriate probes [16]. However, the additive lengths of the *Sfi* I fragments constituting the type I and II amplicons turned out to be closer to 273 and 240 kb, respectively. These estimates are considered to be more accurate than the original ones based on the sizes of restriction fragments from cosmids, since the latter estimate did not include fragments less than 1 kb long. Note that a minor amplicon species (the type III, Fig. 1) was predicted from studies on cosmids but could not be identified in *Sfi* I digests.

At this juncture, we had isolated recombinant cosmids representing ~273 kb of contiguous DNA sequence from the *DHFR* locus and had deduced the structure of two major amplicon types from the CHOC 400 genome. When these clones were used as specific hybridization probes to analyze the amplicons in five other independently isolated MTX-resistant Chinese hamster cell lines, the results indicated that all but one of these cell lines had amplified at least the 273 kb sequence represented by the type I amplicon [8]. Coupled with in-gel renaturation analyses showing similar restriction patterns among three of the most resistant of these cell lines [8], it seemed possible at this point that the *DHFR* amplicons in MTX-resistant Chinese hamster cells might often be similar in size and sequence arrangement.

However, application of the in-gel renaturation method to large *Sfi* I fragments separated by PFG gel electrophoresis showed that this was not the case [8]. Because *Sfi* I cuts so rarely, the resulting restriction patterns of the amplicons are very simple, and are largely free of doublets and triplets (which led to underestimates of amplicon size when the same method was used to analyze digests by six-cutters; see Ref. 6). Hence, it was shown that the prevalent amplicons in the DC3F/A3 and MK42 cell lines are greater than 450 and 650 kb long, respectively [8]. Within these core sequences, the amplicons appeared to be very uniform in structure (Fig. 2).

The large size and apparent uniformity of the amplicons in these two cell lines suggested that it might be possible to isolate another replication initiation locus from the additional DNA sequences extending beyond the 273 kb shared with CHOC 400 cells. Utilizing the in-gel renaturation technique to analyze DNA labeled in the early S period in synchronized DC3F/A3 cells, we were, in fact, able to identify two early-labeled fragments in addition to those that characterize the initiation region in the CHOC 400 amplicon [17].

These ELFs were excised from a gel, labeled in vitro with [^{32}P]dCTP, and used

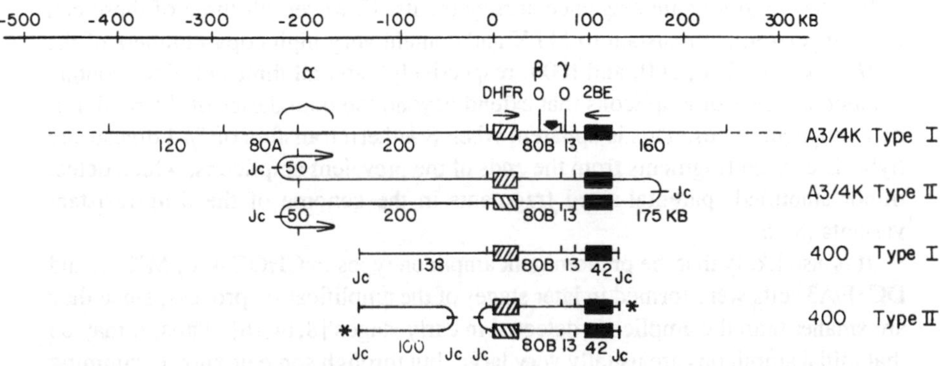

Figure 2 Organization of the *DHFR* amplicons in A3/4K and CHOC 400 cells. Below the linear scale are shown the arrangements of the type I and type II amplicons from A3/4K and CHOC 400 cells. The hairpins shown in the map of the A3/4K type II amplicon correspond to the head-to-head and tail-to-tail palindromic junctions between amplicons. The two ends of the CHOC 400 type II amplicon that are indicated with an asterisk are actually connected together in the genome to form palindromic arrays (see text). The hatched and black boxes represent the *DHFR* and *2BE2121* genes, respectively, and the arrows above each box indicate the direction of transcription. *Ori*-β and *ori*-γ are each indicated with an **O** above the map of the A3/4K type I amplicon, while the general location of *ori*-α is shown with a curly bracket. A matrix attachment site located between *ori*-β and *ori*-γ is indicated with an arrowhead above the map. Small vertical hatch marks above the maps of the three amplicon types indicate the position of *Sfi* I sites, and the sizes (kb) are indicated below each map.

to probe a new cosmid library of DC3F/A3 genomic DNA prepared in the vector pWE-16 [18]. A cosmid (KD23) was isolated that contains both of the additional ELFs and, by chromosomal walking, this new initiation locus was connected to the 5′ end of the 273 kb sequence from CHOC 400 cells [17]. Mapping studies showed that the new locus (termed *ori*-α) is situated ~250 kb upstream from the initiation locus lying close to the *DHFR* gene (recent data indicates that the latter initiation locus may actually contain two origins; hence, it is named *ori*-β/*ori*-γ) (Fig. 2).

Outward walking steps from the ends of this core have now extended the cloned region of the *DHFR* locus to more than 500 kb of contiguous DNA sequence. Within this 500 kb region were found two palindromic junctions that define the two ends of the 450 kb type II amplicon from the DC3F/A3 cell line, which represents the third amplicon to be cloned in its entirety from Chinese hamster cells (C. Ma and J. L. Hamlin, unpublished observations) (Fig. 2). Another major amplicon type in the DC3F/A3 genome is probably ~650 kb long (calculated by summing the length of *Sfi* I fragments in a PFG gel), and this larger amplicon is close to being cloned in its entirety as well (Fig. 2).

These studies demonstrate that the amplicons in the DC3F/A3, MK42, and CHOC 400 cell lines are not infinitely variable, but rather, are characterized by only a few predominant sequence arrangements. However, all three of these cell lines are extremely resistant to MTX and contain very high copy numbers of the *DHFR* locus (~700, 300, and 1000, respectively), and all three cell lines contain smaller numbers of amplicons that extend beyond the boundaries of the predominant amplicon types. This is apparent when Southern transfers of *Sfi* I digests are hybridized with fragments from the ends of the prevalent amplicons, which detect minor amplified, parental-sized fragments in the genome of the drug-resistant variants [8,16].

It is also likely that the predominant amplicon types in CHOC 400, MK42, and DC3F/A3 cells were formed in later stages of the amplification process, since they are smaller than the amplicons detected in early stages [8,14,16]. Thus, it may be that initial amplicons are usually very large, but through some unknown trimming and/or selection process, smaller ones are chosen to be further amplified to high copy number in highly resistant cell lines. This model has been suggested to explain the very large apparent size of amplicons in the earliest stages of *CAD* gene amplification in PALA-resistant Syrian hamster cells [19].

IV. EARLY AMPLIFICATION EVENTS VIEWED AT THE CHROMOSOMAL LEVEL

Recent in situ hybridization studies on the relatively early stages of *DHFR* gene amplification in CHO cells also suggest that initial amplicons can be extremely large [20]. Advantage was taken of the very high sensitivity and resolving power afforded by fluorescent detection methods coupled with the use of multiple cosmid probes. These variations allow even single copy loci to be detected in virtually every cell in a metaphase spread [20]. When applied to CHO cells that had been selected in only two steps with MTX (~40 cell doublings in the presence of drug), some extremely interesting observations were made [20].

The earliest detectable *DHFR* amplicons were almost invariably located on the same chromosome arm as one of the parental, single copy loci (either the long arm of chromosome 2 or its rearranged homologue, Z2). Furthermore, both single copy loci remained intact in their usual positions, and the amplicon clusters were usually positioned more than 50 Mb away (often at the telomere). Even in the relatively early stages of amplification that were examined in this study (two selection steps), most chromosomes already contained 10–25 copies of the *DHFR* amplicon. Finally, there was evidence for a high degree of instability in the chromosomal regions containing amplicon clusters, which were often involved in breakage, fusion, and translocation events.

V. SUMMARY AND CONCLUSIONS

What has been learned about the process of amplification in Chinese hamster cells from the large amount of DNA sequence that we have isolated and characterized, and from recent in situ hybridization studies? Unfortunately, at present, none of our data allows us to discriminate absolutely among the several suggested mechanisms that could be responsible for DNA sequence amplification. However, some solid information has been obtained that will have to be accommodated by any proposed model.

It is clear from the studies on highly resistant Chinese hamster cells that there are preferred amplicon types that can be relatively small (240–650 kb long) and very uniform in structure within a cell line [8,14,16]. In some cases, more than 80% of the amplicons in a cell can be of only one type [14,16]. However, less resistant variants seem to contain a more varied spectrum of much larger amplicons, which could be as large as 10 Mb, based on the frequency of occurrence of novel junction fragments [5] and on in situ hybridization data [20]. Thus, the mechanism(s) giving rise to the early duplicated locus may be different from the mechanism(s) that trim and/or select a smaller version for subsequent amplification to high copy number. We also know that amplicons are usually larger than a single replicon, since two initiation loci have been identified in the relatively large amplicons of DC3F/A3 cells [17].

Results from in situ hybridization studies have provided a wealth of information that must also be accommodated by any viable model for DNA sequence amplification. Whatever the mechanism responsible for the first amplification event, there is a propensity for early amplicons to be very large, for initial events to be intrachromosomal, and for the parental loci to remain intact at their usual locations in the chromosome. On the surface, these findings are most compatible with either unequal sister chromatid exchange or conservative transposition models for amplification (see Ref. 20 for discussion of models).

The in situ hybridization data would appear to rule out a model in which amplification occurs by an onionskin overreplication mechanism, followed by integration of the extra duplexes near the site of the original locus. Our findings are also difficult to reconcile with a model in which the primordial amplicon is deleted from its original locus and then amplifies extrachromosomally (by, e.g., rolling circle replication or unequal sister chromatid exchange).

In future, it will be important to learn in more detail the exact rearrangements that occur during the first amplification steps to produce the very large, terminally located amplicons that we have observed [20]. In our laboratory, we plan to isolate genomic sequences positioned at regular intervals along the length of chromosome 2 for use as probes in fluorescent in situ hybridization studies on the earliest detectable amplification events. In this way, it may be possible to map the internal composition and orientation of the large amount of DNA sequence lying between the single copy loci and the first amplicons. The rearrangements that we define in these studies may help to eliminate certain obvious models for initial amplification events.

ACKNOWLEDGMENTS

I thank the following members of my laboratory, past and present, for their intellectual and experimental contributions to these studies: Jeffrey D. Milbrandt, Nicholas H. Heintz, Jane C. Azizkhan, W. Carlton White, Kay Greisen, Martin Montoya-Zavala, James E. Looney, Chi Ma, Tzeng-Horng Leu, James P. Vaughn, Pieter A. Dijkwel, and, last but not least, Kevin Cox.

REFERENCES

1. Nunberg, J. H., Kaufman, R. J., Schimke, R. T., Urlaub, G., and Chasin, L. A. Amplified dihydrofolate reductase genes are localized to a homogeneously staining region of a single chromosome in a methotrexate-resistant Chinese hamster ovary cell line. *Proc. Natl. Acad. Sci. USA*, *75*:5553–5556 (1978).
2. Dolnick, B.J., Berenson, R. J., Bertino, J. R., Kaufman, R. J., Nunberg, J. H., and Schimke, R. T. Correlation of dihydrofolate reductase elevation with gene amplification in a homogeneously staining chromosomal region in L5178Y cells. *J. Cell Biol.* *83*:394–402 (1979).
3. Milbrandt, J. D., Heintz, N. H., White, W. C., Rothman, S. M., and Hamlin, J. L. Methotrexate-resistant Chinese hamster ovary cells have amplified a 135 kilobase-pair region that includes the dihydrofolate reductase gene. *Proc. Natl. Acad. Sci. USA*, *78*: 6043–6047 (1981).

4. Federspiel, N. A., Beverley, S. M., Schilling, J. W., and Schimke, R. T. Novel DNA rearrangements are associated with dihydrofolate reductase gene amplification. *J. Biol. Chem. 259*:9127–9140 (1984).
5. Ardeshir, F., Guilotto, E., Zieg, J., Brison, O., Liav, W. S. L., and Stark, G. R. Structure of amplified DNA in different Syrian hamster cell lines resistant to *N*-(phosphonacetyl)-L-aspartate. *Mol. Cell. Biol. 3*:2076–2088 (1983).
6. Milbrandt, J.D., Heintz, N. H., White, W. C., Rothman, S. M., and Hamlin, J. L. Methotrexate-resistant Chinese hamster ovary cells have amplified a 135 kb region that includes the dihydrofolate reductase gene. *Proc. Natl. Acad. Sci. USA*, *78*:6043–6047 (1981).
7. Montoya-Zavala, M., and Hamlin, J. L. Similar 150-kilobase DNA sequences are amplified in independently derived methotrexate-resistant Chinese hamster cells. *Mol. Cell. Biol. 5*:619–627 (1985).
8. Looney, J. E., Ma, C., Leu, T.-H., Flintoff, W. F., Troutman, W. B., and Hamlin, J. L. The dihydrofolate reductase amplicons in different methotrexate-resistant Chinese hamster cell lines share at least a 273-kb core sequence, but the amplicons in some cell lines are much larger and are remarkably uniform in structure. *Mol. Cell. Biol. 8*: 5268–5279 (1988).
9. Roninson, I. B. Detection and mapping of homologous, repeated and amplified DNA sequences by DNA renaturation in agarose gels. *Nucleic Acids Res. 11*:5413–5431 (1983).
10. Chang, A. C. Y., Nunberg, J. H., Kaufman, R. J., Erlich, H. A., Schimke, R. T., and Cohen, S. N. Phenotypic expression in *E. coli* of a DNA sequence coding for mouse dihydrofolate reductase. *Nature (London) 275*:617–624 (1978).
11. Milbrandt, J. D., Azizkhan, J. C., Greisen, K. S., and Hamlin, J. L. Organization of a Chinese hamster ovary dihydrofolate reductase gene identified by phenotypic rescue. *Mol. Cell. Biol. 7*:1266–1273 (1983).
12. Heintz, N. H., and Hamlin, J. L. An amplified chromosomal sequence that includes the gene for dihydrofolate reductase initiates replication within specific restriction fragments. *Proc. Natl. Acad. Sci. USA*, *79*:4083–4087 (1982).
13. Heintz, N. H., Milbrandt, J. D., Greisen, K. S., and Hamlin, J. L. Cloning of the initiation region of a mammalian chromosomal replicon. *Nature (London) 302*:439–441 (1983).
14. Looney, J. E., and Hamlin, J. L. Isolation of the amplified dihydrofolate reductase domain from methotrexate-resistant Chinese hamster ovary cells. *Mol. Cell. Biol. 7*: 569–577 (1987).
15. Ford, M., and Fried, M. Large inverted duplications are associated with gene amplification. *Cell*, *45*:425–430 (1986).
16. Ma, C., Looney, J. E., Leu, T.-H., and Hamlin, J. L. Organization and genesis of dihydrofolate reductase amplicons in the genome of a methotrexate-resistant Chinese hamster ovary cell line. *Mol. Cell. Biol. 8*:2316–2327 (1988).
17. Ma, C., Leu, T.-H., and Hamlin, J. L. Multiple origins of replication in the dihydrofolate reductase amplicons of a methotrexate-resistant Chinese hamster cell line. *Mol. Cell. Biol. 10*:1338–1346 (1990).

18. Wahl, G. M., Lewis, K. A., Ruiz, J. C., Rothenberg, B., Zhao, J., and Evans, G. A. Cosmid vectors for rapid genomic walking, restriction mapping, and gene transfer. *Proc. Natl. Acad. Sci. USA*, *84*:2160–2164 (1987).
19. Guilotto, E., Saito, I., and Stark, G. R. Structure of DNA formed in the first step of *CAD* gene amplification. *EMBO J.* *5*:2115–2121 (1986).
20. Trask, B. J., and Hamlin, J. L. Early dihydrofolate reductase gene amplification events in CHO cells usually occur on the same chromosome arm as the original locus. *Genes Dev.* *3*:1913–1925 (1989).

3

Analysis of Gene Amplification by In-Gel DNA Renaturation

Igor B. Roninson *University of Illinois at Chicago, Chicago, Illinois*

I. INTRODUCTION

Frequent occurrence of gene amplification in mammalian cell lines selected for various phenotypes, as well as in tumor cells, provides a unique opportunity for identification and cloning of previously unknown genes responsible for important cellular properties. Several techniques for selective cloning of amplified DNA sequences have been reported in the literature. These techniques are based either on preferential reassociation of amplified sequences [1,2] or on their presence in specific physically separable chromosomal structures (double minutes or unusually large chromosomes with homogeneously staining regions) [3,4]. The in-gel DNA renaturation technique [5] is the only method that allows one to detect the presence of amplified DNA of unknown nature by screening multiple DNA samples. It also provides a particularly useful approach to the isolation of amplified genes, since it allows the investigator to delineate the essential portions within the amplified DNA region (amplicon) prior to cloning. This chapter discusses the principles and applications of the in-gel renaturation methodology. Detailed methodological guidelines for this technique have been presented elsewhere [6,7].

II. THE PRINCIPLE OF THE METHOD

Digestion of mammalian DNA with a restriction enzyme recognizing a 6-bp site gives rise to almost one million restriction fragments, the majority of which contain unique (single copy) DNA sequences. Many of the unique fragments also include short interspersed [repetitive sequence] elements (SINEs), such as the human *Alu*, in addition to single copy sequences. A minority of fragments in the restriction digest are derived entirely from repetitive DNA, including tandemly repeated DNA sequences (e.g., satellite DNA) and long interspersed [repeated] elements (LINEs). After separation of the restriction digest by electrophoresis in an agarose gel and staining with ethidium bromide, digested DNA appears as a smear. The most highly repeated fragments (present at a minimum of 500–1000 copies per haploid genome) form distinctly visible bands within the smear. If the DNA contains an amplicon, each of the amplicon-derived restriction fragments would be present in multiple copies, but the copy number of such fragments is too low (usually between 2 and 100 copies per haploid genome) to be detected by ethidium bromide staining. The abundance of the repeated and amplified fragments relative to single copy fragments can be selectively increased, however, making it possible to detect the corresponding bands even at a moderate degree of amplification. This can be done by denaturing all the DNA fragments in the gel, then allowing them to renature and eliminating the unreannealed fragments by digestion with single strand specific nuclease S1. Since the repeated and amplified fragments are present at a higher local concentration in the gel, they reanneal more efficiently than do single copy fragments. This preferential reannealing forms the principle for detection of amplified DNA by in-gel renaturation [5].

III. DETECTION OF AMPLIFIED DNA

A. The Labeled Tracer Technique

In the original version of the in-gel renaturation technique [5], amplified DNA fragments are detected by labeling a portion of each DNA preparation (tracer DNA) with ^{32}P after restriction enzyme digestion. Tracer DNA fragments are labeled at the termini by replacement synthesis using T4 DNA polymerase. As a result, high specific activity is achieved without significant degradation of tracer DNA fragments. Tracer DNA is then mixed with an excess of unlabeled (driver) DNA digested with the same restriction enzyme, and the mixture is electrophoresed in an agarose gel. After electrophoresis, DNA in the gel is denatured with alkali; then the gel is neutralized and DNA fragments are allowed to renature. Following renaturation, unreannealed DNA is degraded by diffusing single strand specific nuclease S1 into the gel. The entire cycle of denaturation, renaturation, and S1 nuclease digestion is repeated one more time. As a result of this treatment, single copy fragments are almost completely degraded, unless they contain highly repeated SINE elements. These SINE sequences may still reanneal with their counterparts present in other comigrating unique fragments (Fig. 1). The reannealed SINE duplexes, however, are unlikely to contain any associated label unless they happen to be located very close to a restriction site, since the label is confined to a short stretch of 50–100 nucleotides at the ends of the restriction fragments. Thus, only the repeated and amplified fragments that are present at a copy number high enough to efficiently reanneal in the gel can be detected as distinct bands after autoradiography.

When a restriction digest of mammalian DNA without gene amplification is treated by the procedure above, autoradiography reveals a band pattern characteristic for the tandemly repeated or LINE sequences in the studied organism and the restriction enzyme used. The presence of amplified DNA is revealed as an additional set of bands, corresponding to amplicon-derived fragments. Such fragments are best detectable when the background produced by the repeated fragments is at a minimum. We have found that some restriction enzymes provide lower background of repeated fragments (e.g., *Hin*dIII in human DNA or *Bam*HI in mouse or hamster DNA). The limits of sensitivity of this assay are determined primarily by the background and may vary depending on the quality of electrophoresis and other experimental variables [6]. In our hands, the minimum level of amplification necessary for the detection of an amplified fragment by the labeled tracer assay is approximately 25–30 copies for human or hamster DNA, but only 40–50 copies for mouse DNA. Although this sensitivity proved to be sufficient for detection of amplified genes in a number of cases (see below), it was apparent that improved sensitivity would make the in-gel renaturation assay much more useful. We therefore set out to develop more sensitive modifications of this procedure.

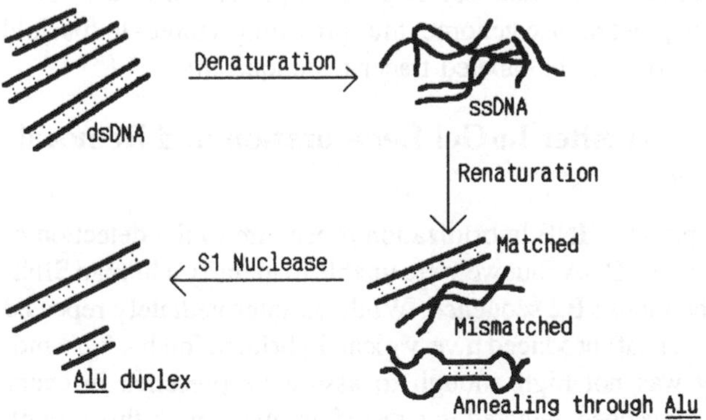

Figure 1 Scheme of in-gel DNA renaturation. (From Ref. 23.)

B. Hybridization with SINE Sequences After In-Gel Renaturation

To overcome the background problem presented by repeated fragments, we decided to substitute labeling of total genomic DNA with a procedure that would detect amplicon-derived fragments preferentially to repeated fragments. We reasoned that a significant proportion of amplicon-derived fragments would include SINE sequences and would therefore hybridize with a SINE-specific probe. In contrast, repeated fragments detectable by in-gel renaturation are derived from other types of repeated sequence, tandem repeats and LINEs, and therefore should not generally hybridize with a SINE-specific probe. Therefore, hybridization of DNA that remains in the gel after in-gel renaturation and S1 nuclease digestion with an appropriate SINE probe should allow us to detect a subset of fragments derived from an amplicon, while giving little if any signal with normal repeated fragments.

A SINE element that can be used as a probe for detection of amplified DNA should be present in the genome at sufficiently high frequency to occur at least once in all amplified units longer than 100 kb. On the other hand, if the frequency of the SINE element is too high, reannealing of the SINE sequences present in comigrating unique restriction fragments would result in an unacceptably high background after Southern hybridization. In the case of mouse DNA, we have been able to find a SINE element that was repeated neither too infrequently nor too often [8]. This was a B2 repeated element, which is ca. 190 bp long and occurs on the average once per 30 kb of DNA. When unlabeled mouse DNA without gene amplification was subjected to two rounds of in-gel renaturation, transferred onto a Southern blot, and hybridized with a B2 repeat probe, only a limited number of bands were visible against a moderate background [8]. In contrast, the presence of amplified DNA was manifested by the appearance of a large number of novel bands. Using DNA preparations with a known copy number of amplified DNA, we have found that B2 hybridization could detect gene amplification at a level as low as 10–15 copies per haploid mouse genome, thus providing a three- to fourfold increase in sensitivity relative to the labeled tracer technique [8].

C. SINE Hybridization After In-Gel Renaturation and Removal of Short Duplexes

We also attempted to apply the SINE hybridization procedure to the detection of amplified sequences in human DNA, but we were unable to identify a human SINE element as suitable as the mouse B2 sequence. While an intermediately repeated human SINE element (0 repeat) produced a very clean hybridization background, its repetition frequency was not high enough to assure its presence in every amplicon (only one band has been detected in a tested amplicon with this probe). In contrast, the highly repeated *Alu* sequence, found on average once every 3–4 kb of human DNA, gave a very high uniform hybridization background, indicating the presence of reannealed *Alu* sequences throughout the lane. We therefore decided to modify the procedure to eliminate from the gel the short *Alu* duplexes remaining after in-gel renaturation and S1 nuclease digestion. In this modification [7], illustrated in Figure 2, an extra gel layer is added on top of the original gel after in-gel renaturation. The resulting gel "sandwich" is placed on its side into an electroblotting apparatus, and DNA is electrophoresed across the "sandwich" at 90° to the original direction of electrophoresis. Electrophoresis is continued until the short (< 1 kb) *Alu* duplexes are completely removed from the gel without concurrent losses of the longer reannealed restriction fragments. After electrophoresis, the remaining fragments are transferred onto a Southern blot, and *Alu*-containing restriction fragments are then detected by hybridization with an *Alu*-specific probe. When applied to human genomic DNA, this assay produced a very low background in the absence of gene amplification. In cell lines carrying a known copy number of amplified DNA, multiple novel bands were readily detectable at an abundance as low as 7–8 copies per haploid genome [7]. The ability to detect such low levels of amplification has been confirmed in studies on gene amplification in human tumors (see Section VI.B).

The level of sensitivity achieved by SINE hybridization after removal of short duplexes is probably close to the absolute limit for any technique that detects amplified DNA sequences on the basis of their preferential reassociation. This technique should be applicable to any genome containing highly repeated SINEs, and it assures that the detected bands are indeed derived from the genome under

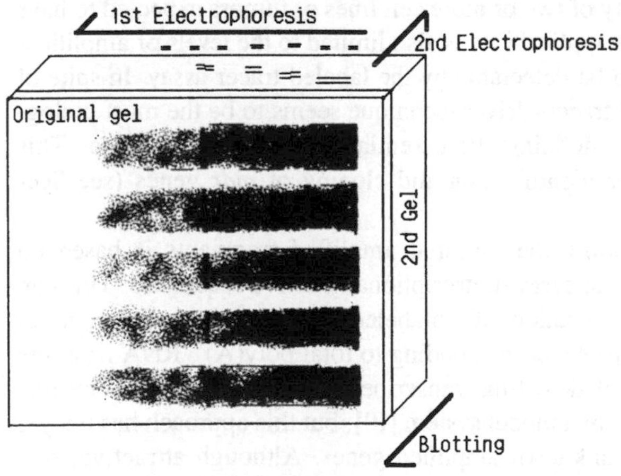

Figure 2 Removal of short duplexes after in-gel renaturation. (From Ref. 23.)

study and do not result, for example, from mycoplasma contamination (a possible artifact of the labeled tracer technique). This procedure therefore provides the optimal screening method for detection of amplified DNA in mammalian cells.

IV. IDENTIFICATION OF ESSENTIAL REGIONS WITHIN THE AMPLICON

Amplicon sizes frequently being on the order of 10^3 kb, it is likely that most of the amplified fragments detectable by in-gel renaturation would represent not the essential gene whose amplification has provided a selective advantage for the tested cell population but, rather, flanking sequences coamplified with this gene. It would be advantageous therefore to identify, prior to cloning, the amplified fragments that are closely linked to the essential gene. It has been shown that both the length and the composition of flanking sequences in cell lines, independently selected for amplification of the same gene, are often highly variable, primarily because of DNA rearrangements that are usually associated with gene amplification [9]. Identification of a subset of amplified fragments that are commonly amplified in different independently derived cell populations therefore allows one to delineate a relatively short region, which is likely to contain the gene of interest. Commonly amplified fragments in different DNA preparations can be identified by the labeled tracer approach, using DNA from one preparation as a tracer and DNA from another preparation as a driver [5]. In-gel renaturation of such tracer–driver mixtures would reveal only fragments that are amplified both in the tracer and in the driver, thus making it possible to define the commonly amplified subset of fragments. The mixed tracer–driver approach, however, has its limitations. First of all, it requires the availability of two or more cell lines or tumors expected to have amplified the same gene. Its applicability is also limited to the levels of amplification that are high enough to be detectable by the labeled tracer assay. In spite of these limitations, the mixed tracer–driver technique seems to be the most biologically justified approach to defining the essential part of the amplicon. This approach has been used for identification and cloning of *mdr* genes (see Section VI.A).

Another approach to finding the essential amplified fragments is based on identification of fragments that carry transcriptionally active sequences. This can be done by Southern hybridization of unlabeled DNA, processed by in-gel renaturation, with a cDNA probe corresponding to total poly(A)$^+$ RNA from the same cells. The feasibility of detecting transcribed amplified fragments by this procedure was demonstrated in a model system [10], but this approach has not yet been used for detection of unknown amplified genes. Although attractive, this assay would not discriminate between the essential gene and other transcribed genes, a number of which are usually present within the flanking sequences.

In addition to the approaches above, a potentially very simple way may be available for locating some essential fragments of the amplicon, taking advantage of a salient feature of the in-gel renaturation methodology that has not been exploited. As discussed below (Section VI.C), amplicon sizes estimated by summation of the sizes of bands observed by in-gel renaturation are usually much smaller than those determined by other methods. This discrepancy can be most readily explained by the existence of an "amplification gradient" within an amplicon, with in-gel renaturation detecting only the central core of the most highly amplified sequences. In all the tested cases, this core has included the essential gene, since cloned gene probes always hybridized with genomic DNA after in-gel renaturation [7,8]. Thus, simple cloning of the most prominent bands detectable by in-gel renaturation may provide clones that are derived from or are reasonably close to the essential gene.

V. CLONING OF AMPLIFIED DNA SEQUENCES

Once a subset of bands representing an essential portion of the amplicon has been identified, one or more of the corresponding fragments can be cloned using in-gel DNA renaturation as a preparative method. The enrichment for the amplified fragment(s) of interest is provided at two steps [11]. The first step of enrichment is achieved by electrophoresis and excision of a gel strip containing the fragment of interest. Further enrichment for the amplified fragment is obtained after denaturation and renaturation of DNA within the gel strip, followed by cloning of the renatured fragments. Digestion with S1 nuclease is not required in this procedure, since only double-stranded restriction fragments with the appropriate sticky ends would ligate to the vector [11]. If desired, S1 nuclease digestion can still be done, provided flush-end fragments are subsequently generated by filling in the ends with Klenow fragment of DNA polymerase I [12]. These steps of enrichment, in addition to the initially elevated copy number of the amplified fragment, result in a high proportion of the recombinant clones containing the desired fragment [11].

After a single fragment has been cloned from the essential amplicon region, the rest of this region can be isolated as a series of overlapping phage or cosmid clones, using the initially cloned fragment as a starting point for "chromosome walking." In the course of chromosome walking, in-gel renaturation can be used as a diagnostic technique to ensure that the isolated genomic clones are not rearranged. This can be done by using total genomic DNA from amplicon-containing cells as a tracer and clone DNA, digested with the same restriction enzyme, as a driver. All the clone-derived fragments, except for those containing or flanking the vector, should protect the genomic tracer DNA from S1 nuclease digestion, unless they are rearranged. DNA rearrangement in the clone is manifested by the loss of some fragments in the autoradiogram after in-gel renaturation [13].

The cloned region of amplified DNA can then be tested for the presence of transcriptionally active sequences, and the identified transcribed fragments can be used as probes for cloning of the corresponding full-length cDNA sequences. Appropriate gene transfer assays can then be designed to test the function of the cloned gene.

VI. APPLICATIONS OF IN-GEL RENATURATION

A. Gene Amplification in Drug-Resistant Cells

The in-gel renaturation technique was initially used to detect gene amplification and to clone amplified genes from multidrug-resistant cells, characterized by cross-resistance to multiple lipophilic cytotoxic drugs, many of which are widely used in cancer chemotherapy [14]. Two multidrug-resistant Chinese hamster cell lines, one selected with adriamycin and the other with colchicine, were analyzed for the presence of amplified DNA by the labeled tracer technique. [11]. Both cell lines were found to contain a set of amplified fragments. Using the mixed tracer–driver assay, some of these fragments were found to be amplified in common between these two cell lines, thus providing the evidence for a common genetic mechanism of resistance in two cell lines that had been independently selected with different drugs. A fragment from the commonly amplified region was then cloned by preparative in-gel renaturation, and the resulting clone was used as a probe for Southern hybridization to show that the degree of amplification of the cloned sequences correlated with the levels of cellular drug resistance [11]. The remainder of the commonly amplified region was then isolated by chromosome walking and shown to contain a gene designated *mdr*1 [13]. The *mdr*1 gene (characterized in more detail in its human and mouse homologues isolated by hybridization with the hamster probe) was subsequently shown to encode P-glycoprotein, a broad-specificity drug efflux pump, responsible for multidrug resistance [14].

Two other groups have also applied the in-gel renaturation methodology to the analysis of multidrug-resistant cell lines [15,16]. In both cases, gene amplification was detected and some of the amplified fragments were cloned. The cloned fragments turned out to belong to the *mdr*1 amplicon.

In-gel renaturation has also been used for the analysis of several other types of drug-resistant cell lines, but no amplified sequences were found [17–19]. These negative results, of course, cannot rule out that the tested cell lines may still have gene amplification at levels below the limits of sensitivity of the assays, especially considering that such limits may differ greatly among the investigators.

Using in-gel renaturation, Mouches et al. [20] have detected gene amplification in a strain of *Culex* mosquito resistant to organophosphate insecticides. The amplified sequences were subsequently found by Southern hybridization to contain a gene for esterase B1 [21].

B. Gene Amplification in Cancer

One of the main applications for which the in-gel renaturation technique was developed is the search for as yet unknown genes that may contribute to tumorigenesis or malignant progression through their amplification in tumor cells. We have used the in-gel renaturation procedure with the removal of short duplexes and SINE hybridization to look for gene amplification in multiple human tumor cell lines and clinical tumor samples [7,22,23]. For this study, we selected tumor types that had not been associated with amplification of known cellular oncogenes. After the detection of amplified DNA by in-gel renaturation, the positive DNA samples were hybridized with a panel of oncogene probes to determine whether the amplified DNA sequences included any of the known oncogenes.

Screening of 16 cell lines derived from various solid tumors and leukemias revealed amplified DNA sequences in only one line, Calu-3 lung adenocarcinoma [7]. After hybridization with oncogene probes, this cell line was found to contain coamplified *neu* and *c-erb* Al genes. However, when one of the amplification-negative cell lines, PC-3 prostatic carcinoma, was selected for in vivo growth in nude mice, amplified DNA sequences became detectable in these cells. The amplified sequences turned out to include the c-*myc* oncogene, which showed no amplification in the parental cell line but was amplified 10- to 12-fold in the in vivo selected cells [7]. The PC-3 cell line carrying the amplified c-*myc* gene was found to have a higher rate of growth and a higher frequency of experimental metastasis in nude mice compared to the PC-3 line without gene amplification, although there was no difference in the in vivo doubling times between these cell lines [24]. These results indicated that c-*myc* amplification provides tumor cells with a selective advantage specific for in vivo growth.

In the initial studies on surgical tumor samples [22], we screened specimens of 29 malignancies of the female genital tract (ovarian, endometrial, and cervical carcinomas), 31 colon carcinomas, 10 non-small-cell lung carcinomas, and one hepatoma, obtained from various sources. No amplified sequences were detectable in any of the samples. Subsequent pathological examination of selected specimens from this set revealed a high proportion of nontumor and necrotic tissue elements, with viable tumor cells accounting for less than 20% of each specimen. It was likely therefore that the negative results obtained in this study could be an artifact resulting from dilution of amplicon-containing DNA from tumor cells with nontumor DNA. We then analyzed a set of specimens derived from eight primary and metastatic ovarian carcinomas, which were carefully excised by a pathologist from the areas of viable tumor. With these preparations, amplified DNA sequences were detected in two out of eight tumors [22]. Hybridization with oncogene probes revealed that both tumors contained an amplified Ki-*ras* gene, which in one case was coamplified with c-*myc*. In one of the tumors, Ki-*ras* was found to be amplified both in the primary tumor and in three different metastatic nodules. The amplified Ki-*ras* genes contained no mutations at codons 12 or 61. The results of our studies pointed out the importance of sample preparation in the analysis of gene amplification in clinical tumor samples and indicated that known oncogenes are likely to be present in a high proportion of amplicons in human tumor cells [22].

A similar study on gastric and esophageal tumors has been reported by Houldsworth et al. [25]. Using the same assay, these investigators have detected gene amplification in 3 of 28 tested tumors. After hybridization with oncogene probes, all three tumors were found to have amplified the *neu* oncogene.

Other investigators have applied in-gel renaturation to the analysis of other tumor types and were able to find three novel oncogenes by this approach. One of these genes was cloned by Vogelstein's group [26] from a glioblastoma cell line. The corresponding amplicon was detected, and several fragments of amplified DNA were cloned in this study by in-gel renaturation (the labeled tracer technique). The bulk of the amplicon was isolated as a series of cosmid clones, using double minute chromosome derived sequences as probes for screening a genomic library. Transcribed sequences within the cosmids were then identified by hybridization with a cellular cDNA probe. These sequences corresponded to a previously unknown gene, designated *gli*, which was shown to be a member of the Kruppel family of zinc finger proteins [27], binding specific DNA sequences in the human genome [28].

Terada's group has analyzed gene amplification in stomach cancer. Using in-gel renaturation, they detected amplified DNA sequences in several tumors [29–31] and cloned a 0.2 kb fragment that was commonly amplified in 3 of 24 surgical specimens and 2 of 13 xenografts of stomach adenocarcinomas [31]. A larger fragment from the human genomic library, hybridizing with the 0.2 kb fragment, was found to contain an exon corresponding to a novel gene, designated K-*sam*, which was expressed not only in stomach cancers but also in small cell lung cancer and germ cell tumors. The product of this gene appears to be a tyrosine kinase homologous to a family of heparin-binding growth factor receptors [32].

Schwab et al. [12] have found by the labeled tracer technique that a cell line derived from genetic melanoma of a platyfish *Xiphonophorus* contained an amplified 7 kb fragment. This fragment was cloned by in-gel renaturation and was found to contain transcribed DNA sequences that have been evolutionarily conserved in all vertebrates. Identification of the product of this gene has not yet been reported.

Samaniego et al. [33] have used the same method to detect gene amplification in a cell line derived from a male germ cell tumor. The amplified sequences failed to hybridize with a panel of 11 commonly amplified oncogene probes, suggesting that they may contain an unknown gene. Interestingly, cytogenetic analysis has suggested that gene amplification in germ cell tumors may be preferentially associated with metastatic extragonadal sites.

Using the labeled tracer assay, Baskin et al. [34] have compared the patterns of reiterated *Eco*RI fragments in various normal and tumor tissues. In addition to detecting tentative cases of gene amplification in a neuroblastoma and a glioma, these investigators reported apparently nonrandom changes in the intensity of some normal repeated fragments between tumor and normal DNA samples from the same individuals. These changes may be indicative of tumor-specific changes in the copy number of the chromosomes or chromosomal regions carrying the corresponding repeated sequences. These results suggest that in-gel DNA renaturation may be used for detection of genetic abnormalities other than gene amplification.

C. Analysis of Amplicon Structure and Replication

In addition to detection and cloning of amplified DNA, the in-gel DNA renaturation technique can be used to analyze the structure of a known amplicon. Several groups have utilized the in-gel renaturation technique to study the amplicon length and to compare amplicon composition in different cell lines. This approach to the analysis of the structure of amplified DNA is very attractive, since it does not require cloning of any amplified sequences, but it is fraught with serious potential pitfalls.

Several factors may affect the estimates of amplicon size obtained by summation of the sizes of bands detectable by in-gel DNA renaturation (the labeled tracer technique). These are the ability to accurately resolve and count comigrating bands, the varying efficiency of in-gel renaturation for fragments of different size and composition, and the frequently observed heterogeneity of amplicon composition within an individual cell line, reflected by different levels of amplification in different parts of the amplicon. The two latter factors are particularly important at moderate (50-fold or lower) levels of amplification, since a decrease in the copy number or renaturation efficiency is likely to render many amplicon-derived fragments undetectable. The extent of the resulting underestimation of the amplicon size is apparent from the comparison of the amplicon sizes estimated initially by in-gel DNA renaturation and subsequently by Southern hybridization of cloned probes with genomic DNA digested with infrequently cutting restriction enzymes and separated by pulsed-field gel (PFG) electrophoresis. Thus, in-gel renaturation analysis of two multidrug-resistant cell lines, each containing approximately 30 copies of the *mdr*1 amplicon, suggested amplicon sizes of 100–150 kb both in a hamster CH^RC5 [11] and a human KB-C4 [15] cell line. In contrast, PFG assays indicated that the sizes of these amplicons are at least 1100 kb in CH^RC5 [35] and 600 kb in KB-C4 [36]. Thus, summation of the sizes of amplified fragments detectable by in-gel DNA renaturation provides a rough size estimate not for the whole amplicon but for the amplicon core, consisting of the most highly amplified sequences.

At very high levels of amplification, in-gel DNA renaturation is unlikely to miss amplified fragments through sensitivity limitations, and resolution of bands in the gel becomes the major constraint. To circumvent this problem, Ma et al. [37] have applied in-gel renaturation to very large restriction fragments separated by PFG. Because of the small number of fragments derived from the *dhfr*-containing amplicons analyzed by these investigators, the estimates of amplicon size obtained were in excellent agreement with those arrived at by cloning an entire amplicon [37].

To compare the amplicon composition in different tumors and tumor-derived cell lines, Kinzler et al. [38] carried out mixed tracer–driver experiments with DNA from neuroblastomas containing N-*myc* amplification and different tumor types with amplification of the c-*myc* gene. The N-*myc* amplicons were found to be highly conserved in different neuroblastomas. This conservation was confirmed by Southern hybridization with a large number of cloned probes [39]. In contrast, c-*myc* amplicons were much more variable in size and in sequence content [38]. A similar analysis on Chinese hamster cell lines containing highly amplified *dhfr* genes has been carried out by Hamlin's group [37,40]. These experiments revealed the presence of a highly conserved amplicon core in several independently derived cell lines.

The same group has also applied in-gel renaturation, using in vivo labeled DNA from synchronized cultures as a tracer, to analyze the movement of the replication fork through the *dhfr* amplicon and to roughly localize the origin of replication within this amplicon [41]. This approach has allowed the investigators to eliminate background labeling from single copy sequences and to quantitate the relative radioactivity in different fragments. Using this method, Ma et al. [42] have detected the presence of an additional replication origin in one of the *dhfr*-amplified cell lines. This origin was cloned by screening a cosmid library with a probe isolated by in-gel renaturation. The usefulness of in-gel DNA renaturation in the analysis of the process of replication suggests that this general methodology may be applicable to a variety of other functional studies.

ACKNOWLEDGMENTS

I express my appreciation to Dr. Manabu Fukumoto, who was responsible for many of the studies described in this chapter and has kindly provided the illustrations. The work in the author's laboratory was supported by grants CA39365 and CA40333 from the National Cancer Institute and a Faculty Research Award from the American Cancer Society.

REFERENCES

1. Brison, O., Ardeshir, F., and Stark, G. R. General method for cloning amplified DNA by differential screening with genomic probes. *Mol. Cell. Biol.* 2:578–587 (1982).

2. Shiloh, Y., Rose, E., Colletti-Feener, C., Korf, B., Kunkel, L. M., and Latt, S. A. Rapid cloning of multiple amplified nucleotide sequences from human neuroblastoma cell lines by phenol emulsion competitive DNA reassociation. *Gene*, *51*:53–59 (1987).
3. George, D. L., and Powers, V. E. Cloning of DNA from double minutes of Y1 mouse adrenocortical tumor cells: Evidence for gene amplification. *Cell*, *24*:117–123 (1981).
4. Kanda, N., Schreck, R., Alt, F., Bruns, G., Baltimore, D., and Latt, S. Isolation of amplified DNA sequences from IMR-32 human neuroblastoma cells: Facilitation by fluorescence-activated flow sorting of metaphase chromosomes. *Proc. Natl. Acad. Sci. USA*, *80*:4069–4073 (1983).
5. Roninson, I. B. Detection and mapping of homologous, repeated and amplified DNA sequences by DNA renaturation in agarose gels. *Nucleic Acids Res.* *11*:5413–5431 (1983).
6. Roninson, I. B. Use of in-gel DNA renaturation for detection and cloning of amplified genes. *Methods Enzymol.* *151*:332–371 (1987).
7. Fukumoto, M., Shevrin, D. H., and Roninson, I. B. Analysis of gene amplification in human tumor cell lines. *Proc. Natl. Acad. Sci. USA*, *85*:6846–6850 (1988).
8. Fukumoto, M., and Roninson, I. B. Detection of amplified sequences in mammalian DNA by in-gel renaturation and SINE hybridization. *Somatic Cell Mol. Genet.* *12*: 611–623 (1986).
9. Federspiel, N. A., Beverly, S. M., Schilling, J. W., and Schimke, R. T. Novel DNA rearrangements are associated with dihydrofolate reductase gene amplification. *J. Biol. Chem.* *259*:9127–9140 (1984).
10. Rose, E. A., and Munro, H. N. Detection of transcribed amplified DNA sequences by a modification of the method of DNA renaturation in an agarose gel. *Nucleic Acids Res.* *15*:8113–8114 (1987).
11. Roninson, I. B., Abelson, H., Housman, D. E., Howell, N., and Varshavsky, A. Amplification of specific DNA sequences correlates with multidrug resistance in Chinese hamster cells. *Nature*, *309*:626–628 (1984).
12. Schwab, M., Oehlmann, R., Bruderlein, S., and Wakamatsu, Y. Amplified DNA in cells of genetic melanoma of *Xiphophorus*. *Oncogene*, *4*:139–144 (1989).
13. Gros, P., Croop, J. M., Roninson, I. B., Varshavsky, A., and Housman, D. E. Isolation and characterization of DNA sequences amplified in multidrug-resistant hamster cells. *Proc. Natl. Acad. Sci. USA*, *83*:337–341 (1986).
14. *Molecular and Cellular Biology of Multidrug Resistance in Tumor Cells* (I. B. Roninson, ed.), Plenum Press, New York, 1991.
15. Fojo, A. T., Whang-Peng, J., Gottesman, M. M., and Pastan, I. Amplification of DNA sequences in human multidrug-resistant KB carcinoma cells. *Proc. Natl. Acad. Sci. USA*, *82*:7661–7665 (1985).
16. Fairchild, C. R., Ivy, S. P., Kao-Shan, C. S., Whang-Peng, J., Rosen, N., Israel, M. A., Melera, P. W., Cowan, K. H., and Goldsmith, M. E. Isolation of amplified and overexpressed DNA sequences from adriamycin-resistant human breast cancer cells. *Cancer Res.* *47*:5141–5148 (1987).
17. Dorr, R. T., Liddil, J. D., Trent, J. M., and Dalton, W. S. Mitomycin C resistant L1210 leukemia cells: Association with pleiotropic drug resistance. *Biochem. Pharmacol.* *36*:3115–3120 (1987).
18. Ziv, Y., Jaspers, N. G., Etkin, S., Danieli, T., Trakhtenbrot, L., Amiel, A., Ravia, Y.,

and Shiloh, Y. Cellular and molecular characteristics of an immortalized ataxia-telangiectasia (group AB) cell line. *Cancer Res. 49*:2495–2501 (1989).

19. Vlahos, N. S., Futscher, B. W., Hora, N. K., Trent, J. M., and Erickson, L. C. Gene amplification affecting *O*(6)-alkylguanine-DNA alkyltransferase activity is not detected in nitrosourea resistant or sensitive human cell lines. *Carcinogenesis, 11*:479–483 (1990).
20. Mouches, C., Fournier, D., Raymond, M., Magnin, M., Berge, J.-B., Pasteur, N., and Georghiou, G. P. Specific amplified DNA sequences associated with organophosphate insecticide resistance in mosquitos of the *Culex-pipiens* complex with a note on similar amplification in the housefly *Musca domestica* L. *C. R. Acad. Sci. Paris Ser. III, 301*:695–700 (1985).
21. Mouches, C., Pasteur, N., Berge, J. B., Hyrien, O., Raymond, M., Robert de Saint-Vincent, B., de Silvestri, M., and Georghiou, G. P. Amplification of an esterase gene is responsible for insecticide resistance in a California *Culex* mosquito. *Science, 233*: 778–780 (1986).
22. Fukumoto, M., Estensen, R. D., Sha, L., Oakley, G. J., Twiggs, L. B., Adcock, L. L., Carson, L. F., and Roninson, I. B. Association of Ki-*ras* with amplified DNA sequences, detected in ovarian carcinomas by a modified in-gel renaturation assay. *Cancer Res. 49*:1693–1697 (1989).
23. Fukumoto, M., Tashiro, H., and Roninson, I. B. Analysis of gene amplification in human tumor cells by in-gel DNA renaturation assay. *Acta Haematol. Japon. 52*: 1463–1470 (1989).
24. Shevrin, D. H., Gorny, K. I., Froelich, C. J., and Roninson, I. B. Growth and metastasis of PC-3 lines with or without c-*myc* amplification. *Proc. Annu. Meet. Am. Assoc. Cancer Res.* A1686 (1989).
25. Houldsworth, J., Cordon-Cardo, C., Ladanyi, M., Kelsen, D. P., and Chaganti, R. S. Gene amplification in gastric and esophageal adenocarcinomas. *Cancer Res. 50*: 6417–6422 (1990).
26. Kinzler, K. W., Bigner, S. H., Bigner, D. D., Trent, J. M., Law, M. L., O'Brien, S. J., Wong, A. J., and Vogelstein, B. Identification of an amplified, highly expressed gene in a human glioma. *Science, 236*:70–73 (1987).
27. Kinzler, K. W., Ruppert, J. M., Bigner, S. H., and Vogelstein, B. The *GLI* gene is a member of the Kruppel family of zinc finger proteins. *Nature, 332*:371–374 (1988).
28. Kinzler, K. W., and Vogelstein, B. The *GLI* gene encodes a nuclear protein which binds specific sequences in the human genome. *Mol. Cell. Biol. 10*:634–642 (1990).
29. Nakatani, H., Tahara, E., Sakamoto, H., Terada, M., and Sugimura, T. Amplified DNA sequences in cancers. *Biochem. Biophys. Res. Commun. 130*:508–513 (1985).
30. Nakatani, H., Tahara, E., Yoshida, T., Sakamoto, H., Suzuki, T., Watanabe, H., Sekiguchi, M., Kaneko, Y., Sakurai, M., and Terada, M. Detection of amplified DNA sequences in gastric cancers by a DNA renaturation method in gel. *Jpn. J. Cancer Res. 77*:849–856 (1986).
31. Nakatani, H., Sakamoto, H., Yoshida, T., Yokota, J., Tahara, E., Sugimura, T., and Terada, M. Isolation of an amplified DNA sequence in stomach cancer. *Jpn. J. Cancer Res. 81*:707–710 (1990).
32. Hattori, Y., Odagiri, H., Nakatani, H., Miyagawa, K., Naito, K., Sakamoto, H.,

Katoh, O., Yoshida, T., Sugimura, T., and Terada, M., K-*sam*, an amplified gene in stomach cancer, is a member of the heparin-binding growth factor receptor genes. *Proc. Natl. Acad. Sci. USA*, *87*:5983–5987 (1990).

33. Samaniego, F., Rodriguez, E., Houldsworth, J., Murty, V. V., Ladanyi, M., Lele, K. P., Chen, Q., Dmitrovsky, E., Geller, N. G., Reuter, V., Jhanwar, S. C., Bosl, G. J., and Chaganti, R. S. K. Cytogenetic and molecular analysis of human male germ cell tumors: Chromosome 12 abnormalities and gene amplification. *Genes, Chromosomes Cancer*, *1*:289–300 (1990).
34. Baskin, F., Grossman, A., Bhagat, S. G., Burns, D., Davis, R. M., Warmoth, L. A., and Rosenberg, R. N. Frequent alterations of specific reiterated DNA sequence abundances in human cancer. *Cancer Genet. Cytogenet.* *28*:163–172 (1987).
35. Van der Bliek, A. M., Van der Velde-Koerts, T., Ling, V., and Borst, P. Overexpression and amplification of five genes in a multidrug-resistant Chinese hamster ovary cell line. *Mol. Cell. Biol.* *6*:1671–1678 (1986).
36. Chin, J. E., Soffir, R., Noonan, K. E., Choi, K., and Roninson, I. B. Structure and expression of the human MDR (P-glycoprotein) gene family. *Mol. Cell. Biol.* *9*:3808–3820 (1989).
37. Ma, C., Looney, J. E., Leu, T. H., and Hamlin, J. L. Organization and genesis of dihydrofolate reductase amplicons in the genome of a methotrexate-resistant Chinese hamster ovary cell line. *Mol. Cell. Biol.* *8*:2316–2327 (1988).
38. Kinzler, K. W., Zehnbauer, B. A., Brodeur, G. M., Seeger, R. C., Trent, J. M., Meltzer, P. S., and Vogelstein, B. Amplification units containing human N-*myc* and c-*myc* genes. *Proc. Natl. Acad. Sci. USA*, *83*:1031–1035 (1986).
39. Zehnbauer, B. A., Small, D., Brodeur, G. M., Seeger, R., and Vogelstein, B. Characterization of N-*myc* amplification units in human neuroblastoma cells. *Mol. Cell. Biol.* *8*:522–530 (1988).
40. Looney, J. E., Ma, C., Leu, T. H., Flintoff, W. F., Troutman, W. B., and Hamlin, J. L. The dihydrofolate reductase amplicons in different methotrexate-resistant Chinese hamster cell lines share at least a 273-kilobase core sequence, but the amplicons in some cell lines are much larger and are remarkably uniform in structure. *Mol. Cell. Biol.* *8*:5268–5279 (1988).
41. Leu, T. H., and Hamlin, J. L. High-resolution mapping of replication fork movement through the amplified dihydrofolate reductase domain in CHO cells by in-gel renaturation analysis. *Mol. Cell. Biol.* *9*:523–531 (1989).
42. Ma, C., Leu, T. H., and Hamlin, J. L. Multiple origins of replication in the dihydrofolate reductase amplicons of a methotrexate-resistant Chinese hamster cell line. *Mol. Cell. Biol.* *10*:1338–1346 (1990).

4

Molecular Cytogenetic Characterization of the Sequential Development of Multidrug Resistance in Two Panels of Colchicine-Resistant Cell Lines

Jeffrey M. Trent, Eckart U. Meese,* and Paul S. Meltzer *The University of Michigan, Ann Arbor, Michigan*

Marilyn L. Slovak *City of Hope National Medical Center, Duarte, California*

Floyd Thompson *University of Arizona, Tucson, Arizona*

Victor Ling *The Ontario Cancer Institute and University of Toronto, Toronto, Ontario, Canada*

I. INTRODUCTION

Two panels of colchicine-resistant (CLC^R) Chinese hamster ovary (CHO) cells have been examined for their coordinated chromosomal and molecular changes. In both panels an increase in the complexity of the karyotype was observed with increasing levels of CLC^R, paralleled with an increase in the level of P-glycoprotein (*PGY*) mRNA and protein. Gene amplification was observed at higher levels of CLC^R associated with recognizable abnormally banding regions (ABRs) or double minute chromosomes (DMs). In one panel by the second step of selection

**Current affiliation*: Department of Human Genetics, University of Saar, Hamburg, Germany.

(sixfold or less relative resistance) a specific translocation was observed which resulted in the duplication of band 1q26 (the chromosomal locus encoding *PGY*). Finally, at the earliest steps of selection in both independent panels, unusual chromosomal structures characteristic of newly amplified *PGY* DNA were observed linked to the long arm chromosome 1, presumably by a process of interstitial duplication. When viewed in light of other recent reports, these results further suggest that in this system amplification may first and frequently arise through recombinational alterations at the locus of the single copy gene.

II. PROPERTIES OF RESISTANT CELL LINES

Two separate CLC^R panels previously derived from the same drug-sensitive parental line (AuxB1) [1] were analyzed cytogenetically in this study. The stepwise clonal selection of CLC^R-resistant cells is presented in Tables 1 and 2. The C5 panel represents a series of three single-step selection mutants (CH^{R}A3, CH^{R}B3, and CH^{R}C5) as well as a subline CH^{R}B30, which was generated from mass culture in cells in high (30 μg/ml) colchicine concentration. The panel members CH^{R}C5 and CH^{R}B30 have been extensively characterized by several groups in addition to the present authors in regard to their molecular biologic and

Table 1 C4 Series CH^R Sublines

Cell line	[drug] μg/ml	Relative resistance	Modal number	Karyotypic changes
AuxB1	0	1	21	Parental (baseline)
CH^{R}2H	0.1	4	21	del (8)(p11)
CH^{R}2HA	0.5 (+EMS)[a]	16	21	del (8)(p11) t(1;8)(q26;q23)[b] del (X)(q23)
CH^{R}C4	5.0	75	21	del (8)(p11) t(1;8)(q26;q23)[b] abnormal (X)[c] DMs (10–20%)
I831[d]	0	3	21/22	del (8)(p11) t(1;8)(q26;q23)[b] abnormal (X)[c] +10 +fragment

[a]Ethylmethyl sulfonate.
[b]With a possible duplication of 1q26.
[c]Augmentation—unknown origin.
[d]Revertant from CH^{R}C4.

Table 2 C5 Series CH^R Sublines

Cell line	[drug] μg/ml	Relative resistance	Modal number	Karyotypic changes
AuxB1	0	1	21	Parental
CH^RA3	0.2 (+EMS)[b]	6	21	[a]
CH^RB3	3.0	20	21	der(7)ABR
CH^RC5	10.0 (+EMS)[b]	170	21	der(7)ABR Z-7ABR
CH^RB30	30.0	>600	21	der(7)ABR t[Z7; der(8)]
I10[c]	0	3		

[a]Slight increase in polyploid cells (AuxB1, 4%; CH^RA3, 12%).
[b]Ethylmethyl sulfonate.
[c]Revertant from C5.

cytogenetic characteristics [1–6]. This chapter provides the initial cytogenetic characterization of the CH^RA3 and CH^RB3 sublines and, additionally, the initial information on a second previously undescribed CLC^R panel which we term the C4 panel. The C4 panel contains two single-step selection mutants (CH^R2H and CH^R2HA) in addition to the mass culture CH^RC4 subline (Fig. 1). As can be observed in Table 1, the relative resistance of the C5 panel range from 6- to more than 600-fold, while the relative resistance of the C4 panel ranges from ~4- to 75-fold resistance.

III. CYTOGENETIC ALTERATIONS ASSOCIATED WITH THE ACQUISITION OF COLCHICINE RESISTANCE

A. General Features of Chromosome Changes

Cytogenetic analysis of both CLC^R panels reveals an increase in the complexity of the karyotype with increasing CLC^R. This included an increase in the range of chromosomes per cell as well as an increase in the percentage of polyploid cells [e.g., AuxB1-modal number 21 (range 16–42), 4% polyploid cells; CH^RB30 modal number 21 (range 20–110), ~50% polyploid cells]. Figure 1 illustrates the chromosomes from the C5 panel at the first through third steps of selection, as well as a revertant (I10) selected by single-step selection from CH^RC5. Shown in Figure 1, an ABR arose at the second step of selection (CH^RB3) associated with a 10-fold amplification of *PGY*, with 2 ABRs observed in the third step (CH^RC5) approximately 50-fold *PGY* amplified subline. Of interest, the revertant from CH^RC5 (I10) lost the ABRs recognizable in the CH^RC5 subline in a single step (Fig. 1). Figure 2 illustrates the karyotype and marker chromosomes of the CH^RC4

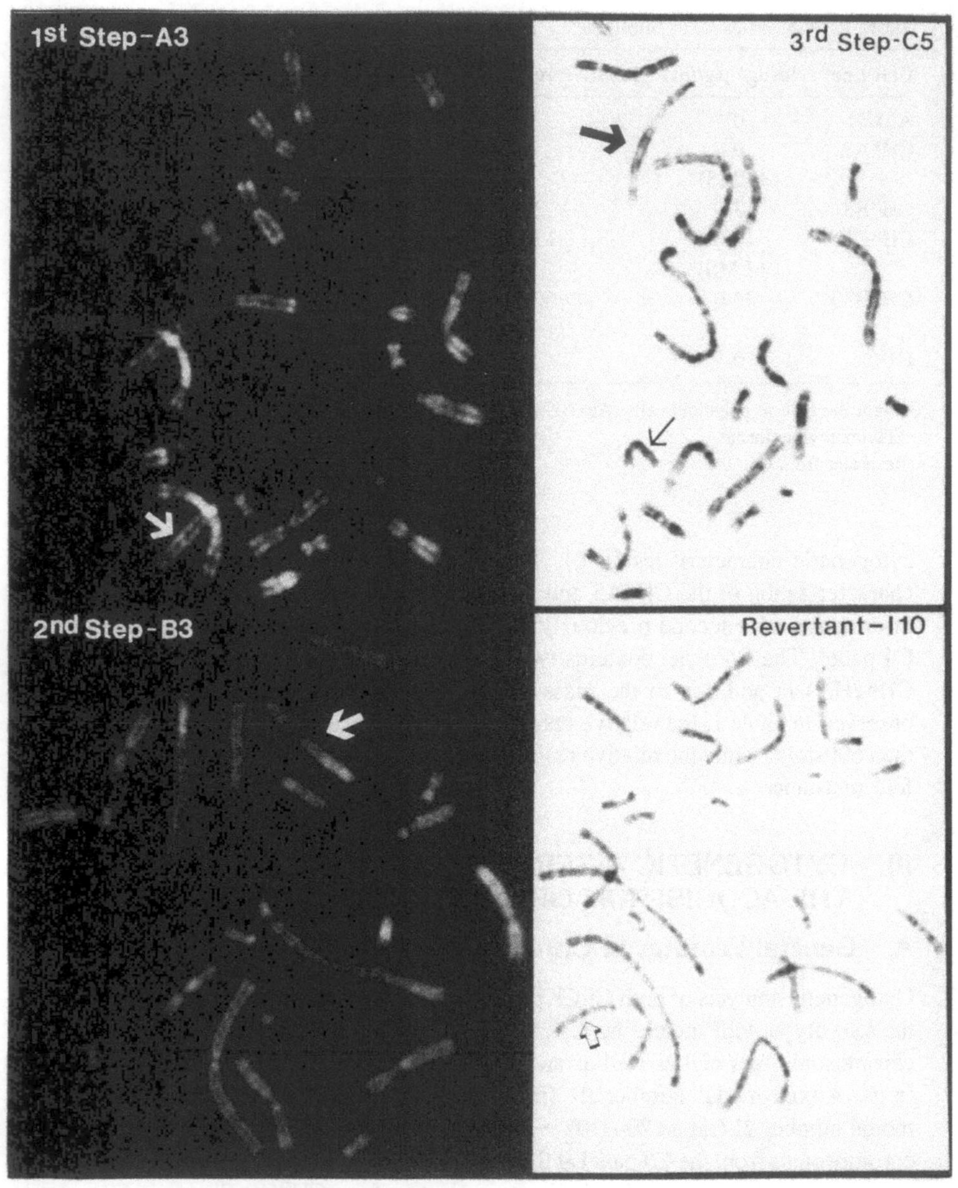
1st Step-A3
3rd Step-C5
2nd Step-B3
Revertant-I10

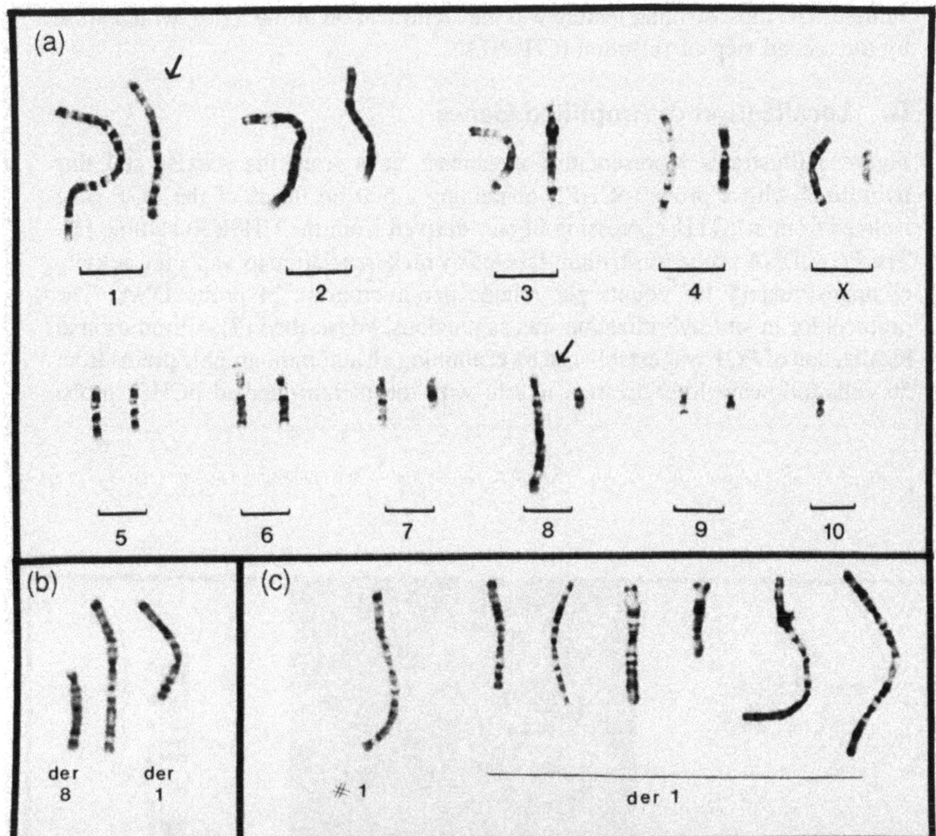

Figure 2 Representative G-banded karyotype of the CH^{R}C4 subline: (a) stem line karyotype of CH^{R}C4, (b) illustration of t(1;8), and (c) examples of derivative chromosome 1's found in the C4 series which have alterations near or at the *PGY* locus (see text).

Figure 1 Q-banded (left) or G-banded (right) metaphases from sublines from the parental AuxB1 and CLC^R resistant sublines of the C5 panel. Upper left: Representative cell from CH^{R}A3; arrow indicates an apparently normal der(7) chromosome. Lower left: Representative cell from CH^{R}B3 demonstrating an extended region at normal banding (ABR) on the distal long arm of the der(7) chromosome (arrow). Upper right: Representative cell from the highly colchicine-resistant subline CH^{R}C5, also demonstrating an extended ABR on the der(7) chromosome (thick arrow) as well as a small ABR on chromosome Z7 (thin arrow). Lower right: Representative cell from the single-step revertant subline I10, demonstrating apparently normal copy of chromosome der(7).

subline. The most striking feature was the identification of the t(1;8), which arose by the second step of selection (CH^{R}2HA).

B. Localization of Amplified Genes

Figure 3 illustrates representative metaphase cells from the AuxB1 cell line hybridized with a probe (pCHP1) containing a 600 bp insert of the *PGY* gene isolated from a λGT11 expression library derived from the CH^{R}B30 subline [3]. The *PGY* cDNA probe was tritium-labeled by nick translation to a specific activity of approximately 10^7 counts per minute per microgram of probe DNA. The protocol for in situ hybridization was as previously described [7]. Chromosomal localization of *PGY* was established by examining all autoradiographic grains from 50 cells following hybridization in situ with the tritium-labeled pCHP1 probe

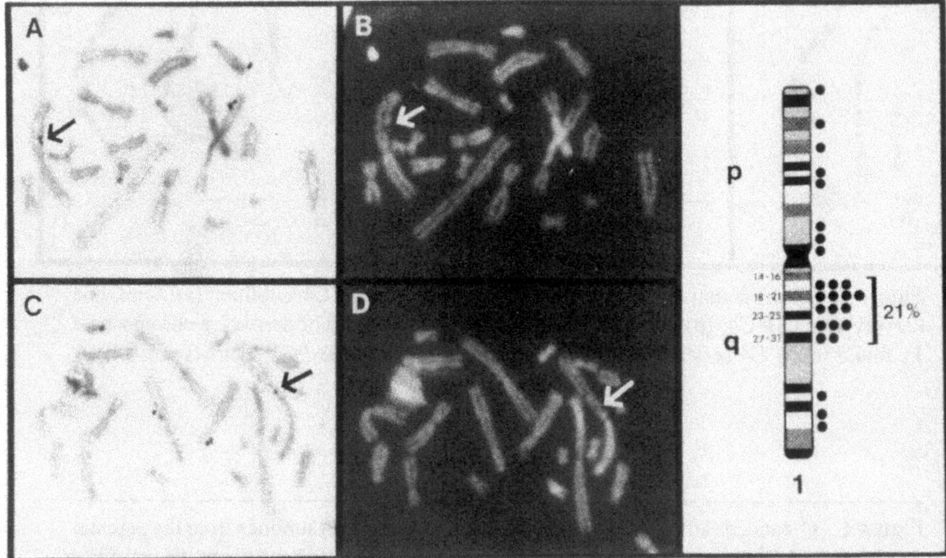

Figure 3 Cytologic evidence for localization of *PGY* sequences of chromosome 1q26 in CHO cells utilizing in situ hybridization. (a) and (c) Representative Wright-stained cells demonstrating autoradiographic grains to 1q26 (arrows), following in situ hybridization with tritium-labeled λCHP1 probe DNA. (b) and (d) The identical cell shown in (a) and (c) stained with quinicrine. (e) Distribution of labeled sites on chromosome 1, demonstrating significant localization of autoradiographic grains flanking chromosome 1q26.

DNA. The total number of chromosomal grains per cell was very low (2.56), with the majority of labeled sites containing single grains. Of the 50 cells examined, 21 (42%) evidenced one or more grains on chromosome 1 with 21% of all chromosomal grains localized to band region 1q18–31 with the majority clustered around 1q26. Examples of grain localization to chromosome 1q are presented in Figure 3. Q-banding prior to in situ hybridization was performed to ensure accurate identification of the chromosomal loci of autoradiographic grains. These results corroborate the results of other investigators who have placed the location of *PGY* at 1q26 [5,8].

Figure 4 documents the analysis of in situ hybridization using the pCHP1 probe DNA to 50 cells each of the resistant sublines CH^RA3, CH^RB3, CH^RB30 from the C5 panel and CH^R2H, CH^R2HA, and CH^RC4 from the C4 panel. Regarding the C5 panel, at the first step of selection no evidence of an ABR was observed, although an increase in the percentage of total grains to chromosome 1 (the region encoding *PGY*) was observed. By the second step of selection a der(7) ABR was observed, which demonstrated a significant increase in autoradiographic grain distribution associated with amplification of *PGY* sequences. Examination of the amplified CH^RB30 cell line also clearly documented an increase in the percentage of grains localized to the ABR. A G-banded example of the der(7) ABR is documented in Figure 4 (insert A).

In contrast to the C5 panel, the clonal sublines of the C4 panels were selected for lower levels of CLC^R, which varied from 4-fold (CH^R2H, 0.1 μg/ml) to 16-fold (CH^R2HA, 0.5 μg/ml) and, finally 75-fold (CH^RCH4, 5.0 μg/ml). As documented in Figure 5, results of Southern blotting demonstrated that a significant level of gene amplification (> 5-fold) of *PGY* was observed only in the CH^RC4 subline (which was the only subline to demonstrate cytologic evidence of gene amplification DMs). By the second step of selection, however, the specific translocation t(1;8)(q26;q23) was observed, which resulted in the duplication of bands around 1q26 (the chromosomal loci demonstrated to encode *PGY*). Photographic documentation of the t(1;8) translocation appears in Figure 4 (inset B) and Figure 2. As further illustrated in Figure 4, by the second step of selection for the C4 panel an increase in autoradiographic signal to chromosome 1 was observed. However, by the highest level of selection (CH^RC4) there was an almost doubling in the percentage of grains to chromosome 1 associated with amplification of *PGY* sequences. Figure 5b illustrates pulsed-field gel analysis electrophoresis (PFGE) performed on C4 panel sublines to determine whether the translocation involving the chromosomal loci of *PGY* sequences had resulted in an altered restriction pattern recognizable by PFGE. As indicated in Figure 3, no evidence for rearrangement of *PGY* sequences was observed by PFGE. Finally, direct evidence of *PGY* protein and mRNA expression was documented in C5 panel sublines by Western blot (Figure 5c) and RNA dot blot (Figure 5d) analysis.

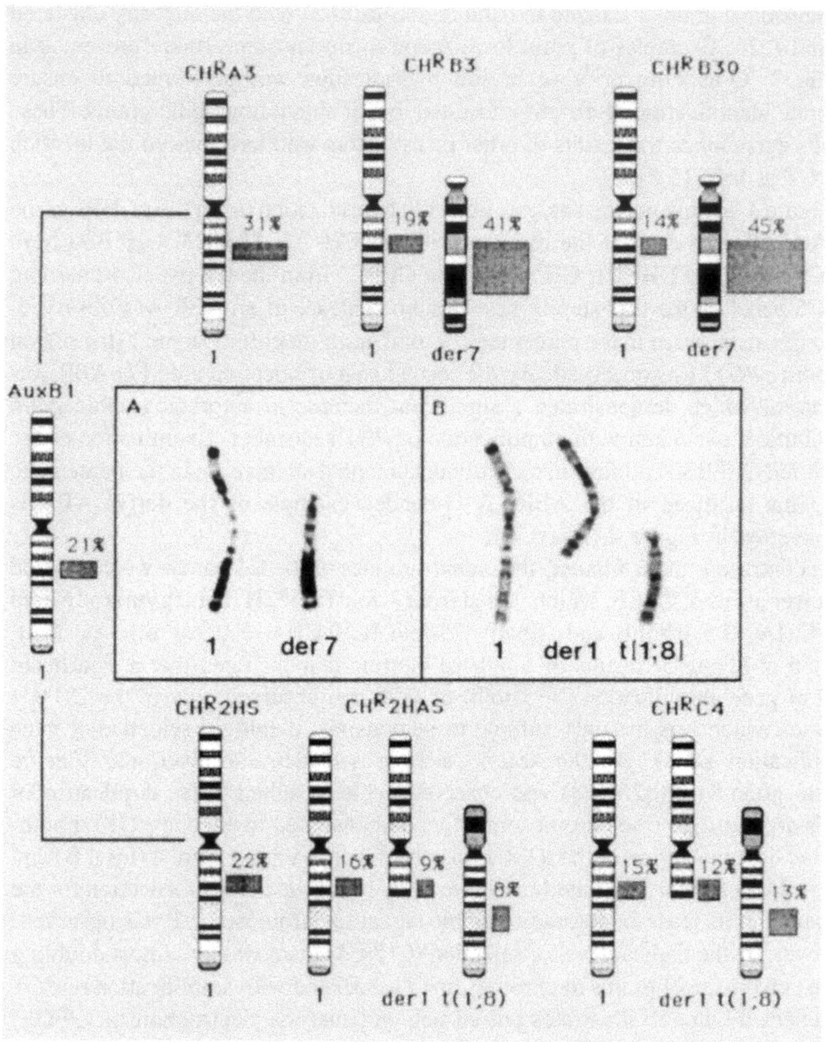
CH^R A3
31%
1
CH^R B3
19%
41%
1
der7
CH^R B30
14%
45%
1
der7
AuxB1
21%
1
A
1
der 7
B
1
der1
t[1;8]
CH^R 2HS
22%
1
CH^R 2HAS
16%
9%
8%
1
der1 t(1;8)
CH^R C4
15%
12%
13%
1
der1 t(1;8)

C. Interstitial Duplication as an Early Event in *PGY* Amplification

Recently, investigators have described and characterized distinctive chromosomal structures formed early in the amplification of two different amplified genes. *CAD* [9] and dihydrofolate reductase *DHFR* [10]. In the *DHFR* study, CHO cells selected for gradually increasing concentrations of methotrexate were shown to contain very large regions of DNA with widely spaced dihydrofolate reductase genes [10]. Of interest, these amplified genes were attached to the end of the same chromosome arm that carried the single copy *DHFR* gene in unselected cells. Very recently work by Stark and colleagues on *CAD*-amplified sublines resistant to *N*-(phosphonacetyl)-L-aspartate (PALA) demonstrated by fluorescence in situ hybridization that newly amplified *CAD* genes were contained in multiple copies of very large DNA regions ($\geqslant$ 10 MB) [9]. As was observed for the *DHFR* gene [10], the amplified-*CAD* DNA was linked to the same chromosome arm (B9) that carries the single copy *CAD* gene in unselected cells. Stark and colleagues have proposed that amplification involves first the transfer of the short arm from one B9 chromosome to another and in subsequent cell cycles this initial duplication expands rapidly through unequal but homologous sister chromatid exchange [9].

As documented in Figure 6, examples of distinctive chromosomal structures that appear cytologically indistinguishable from those by Stark et al [9] have been observed in the earliest steps of selection of both the C5 and C4 panel CLC^R sublines. Of importance, these figures demonstrate that the striking, large regions of DNA link to the long arm of chromosome 1, the region shown to carry *PGY* in unselected cells (Figs. 3, 4; Refs. 5, 8). The finding of these large extended regions of DNA at the site of the single copy gene in early steps in the process of CLC drug selection appears to be entirely analogous to the examples visualized in early steps of *CAD* and *DHFR* amplification [9,10]. Examples of additional derivative

←

Figure 4 Ideograms documenting chromosome 1 abnormalities as well as ABR formation in the C5 (top) and C4 (bottom) CLC^R panels. The AuxB1 parental cell line demonstrated a total of 21% autoradiographic grains to chromosome 1q localizing *PGY* sequences to 1q26 (see text). As illustrated in the C5 panel (top), increasing autoradiographic grain concentration was observed initially at the site of the single copy gene, and subsequently at the site of an ABR on der(7). Inset (A) illustrates the apparently normal G-banded chromosome 1 and the ABR-bearing der(7) chromosome observed in the CH^{R}B3 → CH^{R}B30 members of the C5 panel (see Fig. 1). The C4 panel (bottom) showed no increase in autoradiographic grain distribution at the initial steps of selection (associated with no observed increase in gene copy number). However, by the second step of selection, a specific translocation involving the band encoding-*PGY* was observed with a net increase in the number of autoradiographic grains. Inset B illustrates the t(1;8) and der(1) chromosomes in addition to an apparently normal copy of chromosome 1 retained in the resistant sublines.

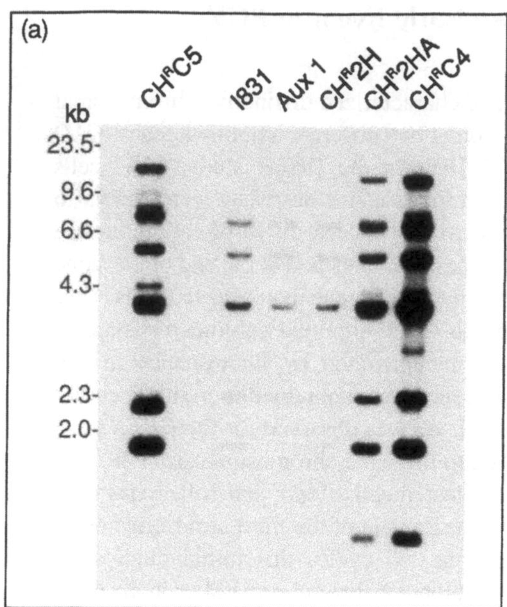
(a)
kb
CH^RC5
I831
Aux 1
CH^R2H
CH^R2HA
CH^RC4
23.5-
9.6-
6.6-
4.3-
2.3-
2.0-

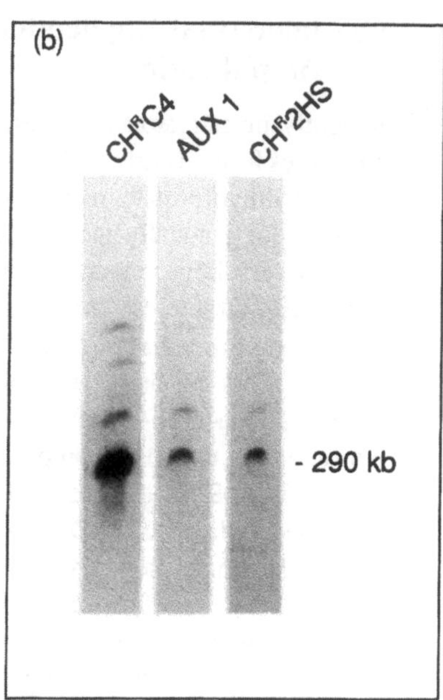
(b)
CH^RC4
AUX 1
CH^R2HS
- 290 kb

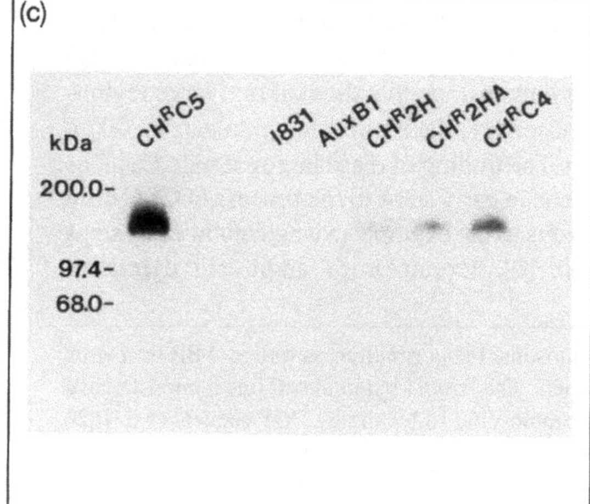
(c)
kDa
CH^RC5
I831
AuxB1
CH^R2H
CH^R2HA
CH^RC4
200.0-
97.4-
68.0-

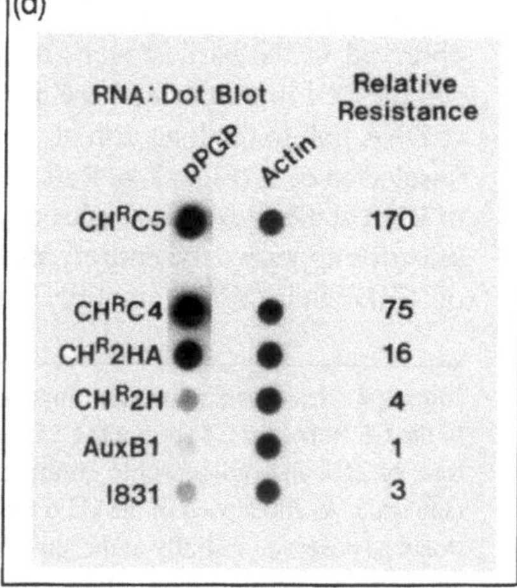
(d)
RNA: Dot Blot
Relative Resistance
pPGP
Actin
CH^RC5 170
CH^RC4 75
CH^R2HA 16
CH^R2H 4
AuxB1 1
I831 3

chromosome 1's from the C4 series showing rearrangements at or near the locus of *PGY* (1q26) are illustrated in Figure 2c.

From our studies, it appears most likely that an interchromosomal recombination event has occurred, leading to a dicentric chromosome that might participate in a bridge–breakage–fusion cycle event. These results (viewed in light of those of *CAD* and *DHFR* amplification) suggest that a recombination-based event resulting in an interstitial duplication followed by expansion through unequal but homologous sister chromatid exchange is a frequent and major mechanism involved in *PGY* gene amplification.

←

Figure 5 Molecular characterization of the C4 series CH^R sublines. (a) Southern blot demonstrating the degree of *PGY* amplification in the C4 panel CH^R2A, CH^R2HA, and CH^RC4. Approximately 40-fold amplified CH^RC5 subline is provided as a control. A single-step revertant selected from the CH^R4 cell line (I831) is also illustrated. (b) PFGE analysis of C4 panel sublines. DNA was digested with *Sfi* I and fractionated by PFGE using a contour-clamped homogeneous electric field. The following parameter was used: 120 V, 55 sec pulse time, 44 hr run time, 1% agarose (BRL) in 0.5 × Tris-borate EDTA. Following alkaline transfer, the blot was hybridized with ^{32}P-labeled pCHP1. As in the standard gel electrophoresis, *PGY* amplification is demonstrated for the subline CH^RC4. However, PFGE analysis does not indicate specific DNA rearrangement in these sublines despite the observation of a translocation near the *PGY* locus on 1q26 (see text). (c) Western blot analysis of the plasma membrane components. Nitrocellulose blots were probed with the C219 monoclonal antibody to P-glycoprotein [3]. CH^RC5 cells markedly overexpress *PGY* and are shown as a positive control. The I831 revertant and parental AuxB1 sublines fail to express significant *PGY*, although the CH^R2H, CH^R2HA, and CH^RC4 lines all overexpress this drug-resistant glycoprotein. (d) P-glycoprotein RNA expression in the CH^RC5 (positive control) and C4 series sublines. Total cellular RNA was extracted from each subline and 10 μg of RNA was applied to nitrocellulose filters. Filters were hybridized with either ^{32}P-labeled cDNA to P-glycoprotein (pCHP1) or ^{32}P-labeled human actin cDNA (an internal control for the amount of RNA present on the filter). Relative resistance of each subline is illustrated in the right-hand panel.

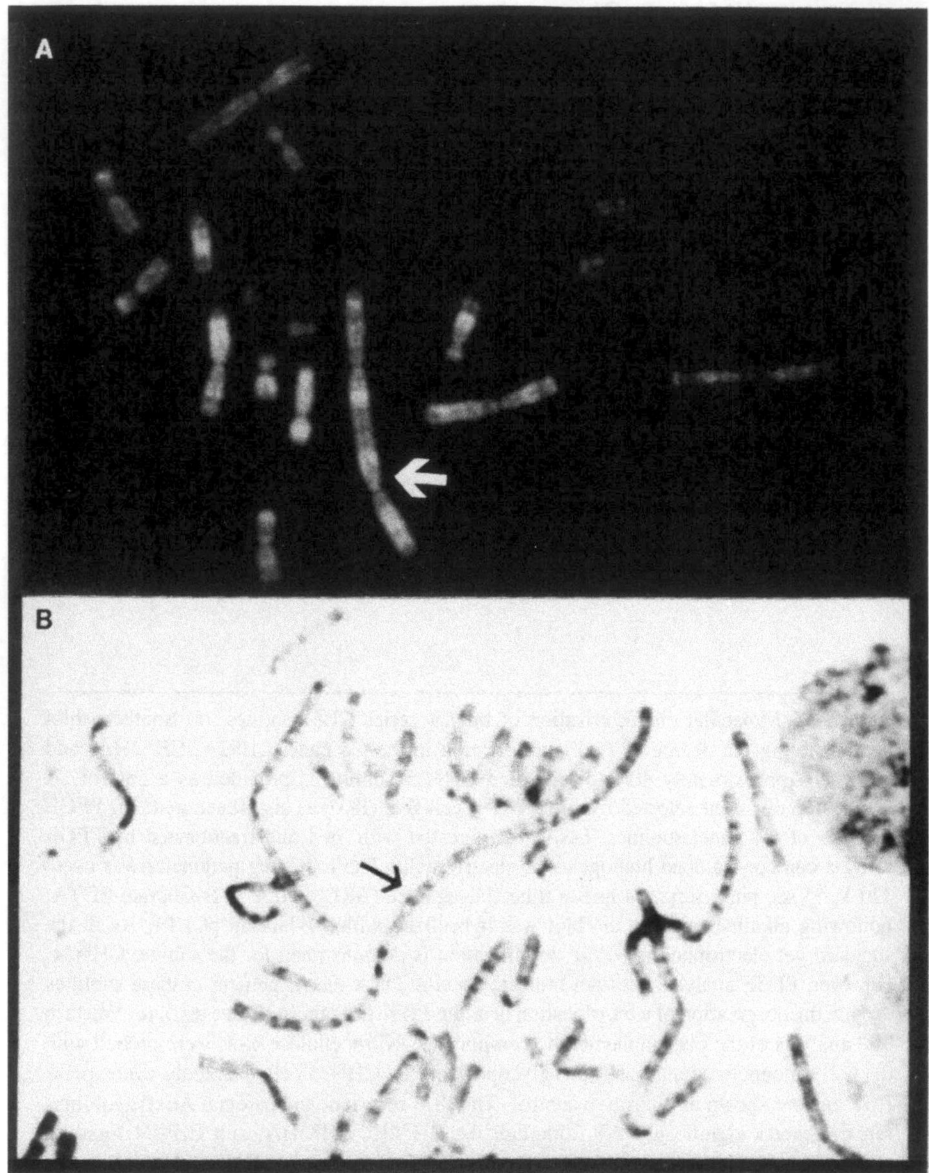

Figure 6 Examples of interchromosomal events leading to dicentric chromosomes with very large regions of DNA attached to the region of chromosome 1 carrying the single copy *PGY* gene (see text).

ACKNOWLEDGMENT

J.M.T. and P.S.M. are supported in part by PHHS grant CA-41183 awarded by the National Cancer Institute, Bethesda, MD.

REFERENCES

1. Ling, V., and Thompson, L. H. Reduced permeability in CHO cells as a mechanism of resistance to colchicine. *J. Cell Physiol.* *83*:103–116 (1974).
2. Riordan, J.R., and Ling, V. Purification of P-glycoprotein from plasma membrane vesicles of Chinese hamster ovary cell mutants with reduced colchicine permeability. *J. Biol. Chem.* *254*:12701–12705 (1979).
3. Riordan, J. R., Deuchars, K., Kartner, N., Alon, N., Trent, J. M., and Ling, V. Amplification of P-glycoprotein genes in multidrug-resistant mammalian cell lines. *Nature*, *316*:817–819 (1985).

3a. Kartner, N., Evernden-Porelle, D., Bradley, G., and Ling, V. Detection of P-glycoprotein in multidrug-resistant cell lines by monoclonal antibodies. *Nature*, *316*:820–823 (1985).

4. Ronison, I. B., Abelson, H. T., Housman, D. E., Howell, N., and Varshavsky, A. Amplification of specific DNA sequences correlates with multidrug resistance in Chinese hamster cells. *Nature*, *309*:626–629 (1984).
5. Jongsma, A. P. M., Spengler, B. A., Van der Blick, A. M., Borst, P., and Biedler, J. L. Chromosomal localization of three genes coamplified in the multidrug-resistant CH^{R}C5 Chinese hamster ovary cell line. *Cancer Res.* *47*:2875–2878 (1987).
6. Borst, P., and Van der Bliek, A. M. Amplification of several different genes in multidrug-resistant Chinese hamster cell lines, in *Molecular and Cellular Biology of Multidrug Resistance in Tumor Cells* (Igor B. Roninson, ed.), Plenum Press, New York, 1991.

6a. Ling, V. Multidrug resistant mutants, in *Molecular Cell Genetics* (M. Gottesman, ed.), Wiley, New York, 1985, pp. 773–787.

7. Trent, J. M., Olson, S., and Lawn, R. M. Chromosomal localization of human leukocyte, fibroblasts, and immune interferon genes by means of in situ hybridization. *Proc. Natl. Acad. Sci. USA,* 79:7809–7813 (1982).
8. Melera, P. W., and Biedler, J. L. Molecular and cytogenetic analysis of multidrug resistance-associated gene amplification in Chinese hamster, mouse sarcoma, and human neuroblastoma cells, in *Molecular and Cellular Biology of Multidrug Resistance in Tumor Cells* (Igor B. Roninson, ed.), Plenum Press, New York, 1991.
9. Smith, K. A., Gorman, P. A., Stark, M. B., Groves, R. P., and Stark G. R. Distinctive chromosomal structures are formed very early in the amplification of *CAD* genes in Syrian hamster cells. *Cell*, *63*:1219–1227 (1990).
10. Trask, B. J., and Hamlin, J. L. Early dihydrofolate reductase gene amplification events on CHO cells usually occur on the same chromosome arm as the original locus. *Genes Dev.* *3*:1913–1925 (1989).

5

Effects of Methotrexate on Recombinant Sequences in Mammalian Cells

Florian M. Wurm *Genentech, Inc., South San Francisco, California*

Maria G. Pallavicini *University of California, San Francisco, California*

I. INTRODUCTION

Recombinant mammalian cells are used increasingly for studies of gene function and expression, and for large-scale production of specific proteins. Chinese hamster ovary (CHO) cells, in particular, are routinely used for protein production in pharmaceutical applications. CHO-DUKX cells, lacking in dihydrofolate reductase gene [1], provided the first mammalian host for a marketed recombinant protein (rtPA, tissue type plasminogen activator), a clot-dissolving human serum protein used in therapy of myocardial infarction [2]. A unique feature associated with CHO-DUKX cells is the possibility of selecting and establishing cell subpopulations and clonal cell lines which, upon treatment with methotrexate (MTX), have amplified transfected expression vector sequences to high copy numbers within their chromosomes [3]. Such cell lines with high copy numbers of recombinant DNA sequences are desirable for research and therapeutic applications, since they often produce the recombinant protein at high rates [2,4–6].

Type of integration and genetic stability of recombinant sequences in transfected CHO cell lines are important issues for the manufacturer, as well as for researchers, requiring stable, characterized model systems for gene expression studies. Although recombinant cells are usually clonal, subtle pressures applied by continuous growth in culture, cell handling, and media manipulations may affect the stability of chromosomally integrated sequences. Hence verification of integration and amplification of recombinant sequences is necessary and may be performed using a variety of DNA-based techniques including Southern DNA blot analysis, cytogenetic analysis of homogeneously staining regions (HSRs), which are often a product of the amplification process [7], and in situ hybridization with radiolabeled probes. Southern DNA blot analysis provides little information about subpopulation heterogeneity and rearrangement of amplified sequences and no information on chromosomal location and number of integration sites. HSRs are detectable as rather large, uniformly staining regions and are recognized as such only when they constitute a major region (i.e., > 10% of the total length) of the chromosome. Although single copy and amplified sequences can be localized using radiolabeled probes, autoradiographic development times are long (weeks to months), and statistical analysis may be required for data interpretation. Furthermore, precise mapping and resolution of integration sites is limited by secondary capture of the emitted isotopic site by an emulsion overlay.

Recently, several groups have reported the detection of unique gene sequences in chromosome mapping using fluorescence in situ hybridization (FISH) techniques with chemically modified nonisotopic probes. These approaches offer high sensitivity with markedly improved speed and spatial resolution [8,9]. We have used FISH to measure integration patterns and distribution and amplification of recombinant sequences in CHO cells in the presence and absence of methotrexate (MTX) [10].

The need for the continued presence of a selective agent (i.e., MTX) in long-term cultures of recombinant CHO cells is controversial. Whereas some studies suggest that MTX is required for stable production of recombinant proteins [5], others indicate that continued culture of a clonal cell line in MTX may not be necessary [11]. We have demonstrated MTX-induced multiple rearrangements of integrated recombinant sequences in CHO cells expressing high levels of c-*myc* protein [6,10]. Although such rearrangements may be associated with increased protein production [10], the genetic instability of cells maintained in MTX may complicate the routine use of recombinant cells as model systems for gene expression and large-scale protein production. In this chapter we describe unique integration patterns in individual clonal recombinant cell lines, heterogeneity in integration patterns in cells cultured in the presence of MTX, and the temporal disappearance of aberrant genotypes in the absence of MTX. We also show stability of recombinant protein production in a nonclonal (heterogeneous) cell population cultured in the absence of MTX for approximately 100 days.

II. RESULTS AND DISCUSSION

A. Unique Integration Patterns of Clonal Recombinant Cell Lines

The loci of integration and the degree of amplification of recombinant sequences in cells are thought to vary between individual clones and thus may be unique descriptors for identification of recombinant cell lines. Integrated recombinant sequences may readily be labeled by FISH with chemically modified DNA probes complementary to the transfected DNA. Biotinylated DNA probes are most commonly used, although other chemical modifications [acetylaminofluorene (AAF), sulfonation, etc.] are also useful, particularly in dual color hybridizations for simultaneous detection of different sequences [12]. Hybridized sequences can be visualized using fluorophore-linked reporter molecules (avidin, anti-AFF antibodies, etc.). We measured the integration pattern of transfected sequences in three individual cloned mammalian cell lines. Each of these clonal cell lines expresses one of three proteins: human recombinant tissue plasminogen activator (rtPA) [2], a truncated version of the human CD4 receptor [13], or the murine c-*myc* protein [6].

The three cell lines were established using CHO-DUKX cells transfected with gene sequences of interest and *DHFR* (dihydrofolate reductase) expression vectors to allow selection of recombinant cells in α-modified medium (lacking glycine, hypoxanthine, and thymidine). Either individual clones or pools of clones were cultivated for extended periods in media containing stepwise-increasing concentrations of MTX. A single individual clone expressing desired levels of recombinant proteins was expanded for long-term growth and recombinant protein production. FISH was performed using biotinylated probes complementary to the recombinant sequences. Prior to hybridization the cells were grown for at least four subcultivations in the absence of MTX. The fluorescence-linked hybridization signals in the rtPA expressing cell line were localized in a submetacentric chromosome, carrying one brightly fluorescent signal in the short arm and a dimmer signal visible as a pair of spots over the center of the chromatids in the long arm (Fig. 1A). In the cell line expressing the truncated CD4 molecule, the signal was found on a small acrocentric chromosome (Fig. 1B). Cells expressing the mouse c-*myc* protein contained amplified recombinant sequences integrated in the long-arm of chromosome 2 (Fig. 1C). These unique integration patterns were found in 95–99% of all metaphases from each cell line and were stable over extended culture periods (60– 160 days) in the absence of MTX.

B. MTX-Induced Heterogeneity of Amplified Sequences

Multiple chromosomal aberrations have been observed in mouse and human cells cultured in the presence of MTX [14]. A similar phenomenon seems to occur in CHO cells [10]. We have shown that continuous application of high concentrations of MTX during cell culture is associated with rearrangement and variable amplification of transfected sequences in a high percentage of CHO cells expressing the mouse c-*myc* gene [10]. Several types of integration patterns were found in these cells: (1) chromosomes with highly extended regions of hybridizing sequences, (2) cells containing transfected DNA integrated into multiple sites on an individual chromosome or (3) into multiple chromosomes, (4) chromosomes joined at amplified regions, (5) circular chromosomes, (6) cells with small derivative chromosomes or fragments, and (7) cells containing a single "normal" integration characteristic of the particular cell line (i.e., Fig. 1A–C). Examples of

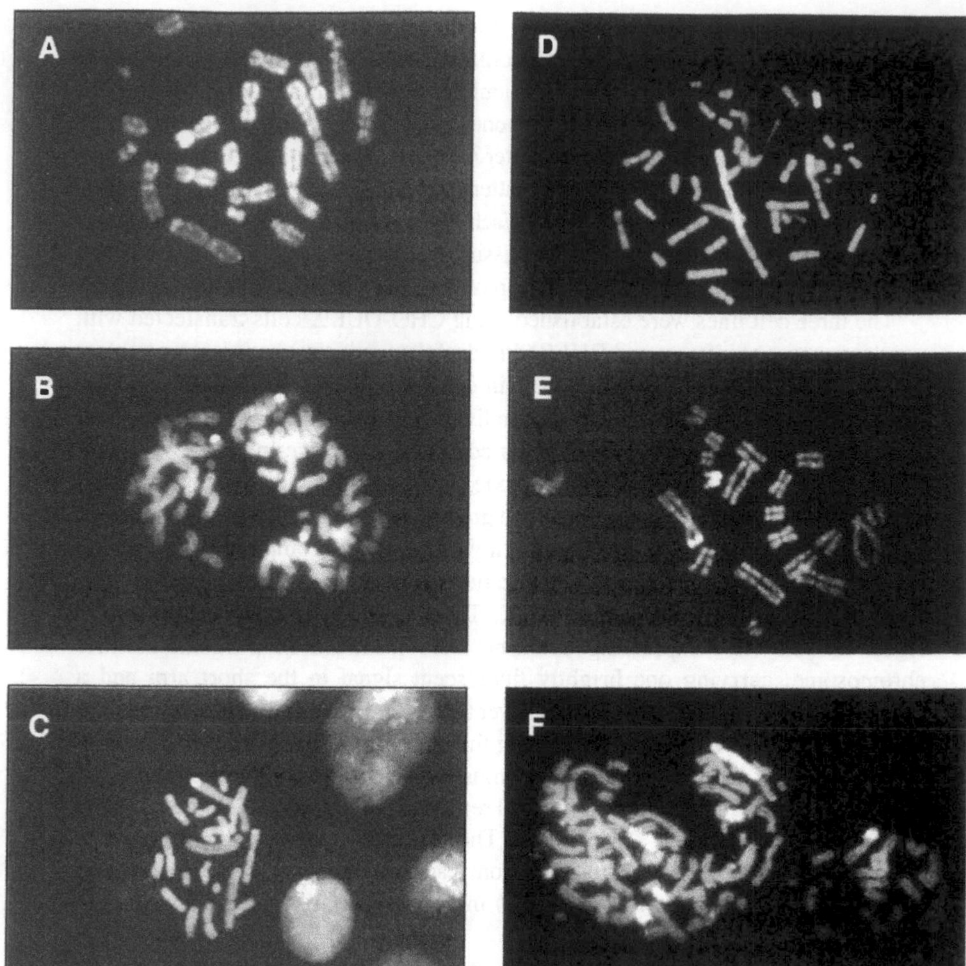

Figure 1 FISH of metaphases of different recombinant CHO cell lines hybridized with DNA probes homologous to recombinant sequences. (A) Cell line producing recombinant tPA. (B) Cell line producing a truncated version of the human CD4 receptor. (C) Metaphase obtained from a cell line expressing the mouse c-*myc* protein. (D)–(F) Hybridization patterns of cells cultured in the presence of MTX: (D) shows a cell line expressing the mouse c-*myc* protein cultured in the presence of MTX, whereas (E) and (F) depict hybridization patterns from cells expressing a chimeric CD4-immunoglobulin protein.

variant hybridization patterns in metaphases obtained from cells producing recombinant c-*myc* cultured in the presence of MTX are shown in Figure 1D. Under this condition, variant hybridization patterns were visible in approximately 40% of c-*myc* CHO metaphases. Similar findings were observed with another recombinant CHO cell line producing a chimeric CD4-immunoglobulin (CD4IgG) protein [15]. When grown in the presence of MTX, CD4IgG-producing cells showed substantial heterogeneity in both chromosomal location and copy number of amplified sequences coding for CD4-immunoglobulin (CD4IgG) (Fig. 1E, F).

The stability of the MTX-induced variant integration patterns in recombinant c-*myc* CHO and CD4IgG-CHO cells was measured following removal of MTX from the culture medium. The temporal decline of MTX-induced variant hybridization patterns in cells cultured without MTX is shown in Figure 2. When c-*myc* expressing CHO cells were grown for 5–10 subcultivations without MTX, the frequency of cells containing multiple and unusual hybridization patterns gradually declined over a 60-day period (Fig. 2a). After three or four subcultivations (approximately 11 days), many of the cells are characterized by a single "normal" integration. The "normal" integration pattern was seen in the majority of cells at 40 days. We verified that this type of population drift occurred with other CHO cell lines, both cloned and uncloned, carrying highly amplified plasmid sequences (Fig. 2b). These studies indicate that many of the structures found in cultures growing in the presence of MTX are genetically unstable.

Two techniques, computer-assisted fluorescence microscopy and immunofluorescence-based flow cytometry, were used to quantitate relative amounts of amplified sequences in genetically unstable chromosomal structures and relative amounts of recombinant protein, respectively, in c-*myc* CHO cells. We demonstrated threefold higher recombinant protein production and threefold higher c-*myc* DNA-linked fluorescence in cells cultured in the presence of MTX [10]. These data suggest a close correlation between relative levels of recombinant sequences and recombinant protein production. Lower recombinant protein productivity of CHO cells cultured in the absence of MTX has been observed by others [5,11] and has contributed to a generalized and oversimplified view that MTX is necessary for maintenance of selectivity and recombinant protein productivity in most recombinant CHO cell lines. Our data clearly show that (1) highly amplified sequences in cells cultured in the presence of MTX are capable of directing syntheses of recombinant proteins, and (2) continued selective pressure is not required for stable production of a variety of recombinant proteins.

C. Recombinant Protein Production by Genetically Heterogeneous Populations in the Absence of MTX

Large-scale production of recombinant proteins for therapeutic applications by transfected mammalian cells requires a stable, reproducible source of cells.

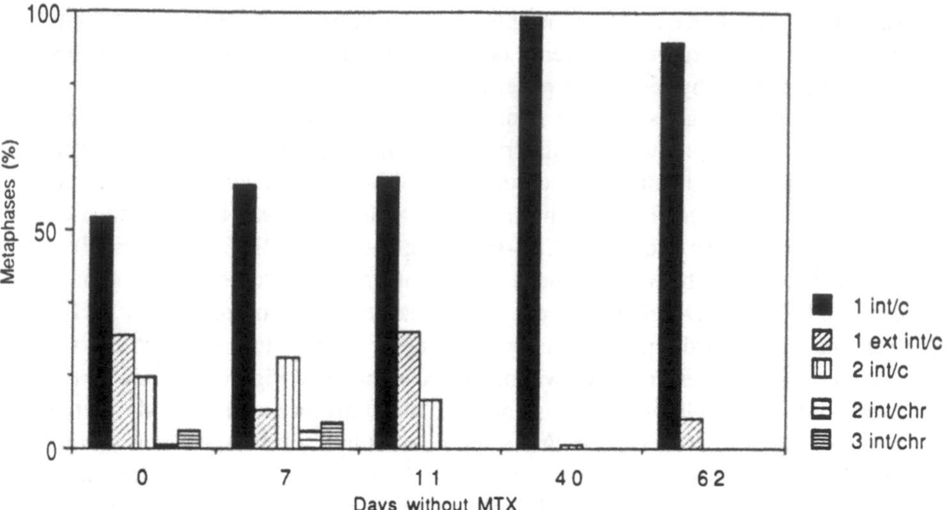

Figure 2a Distribution of integration patterns of amplified recombinant sequences in CHO cells. Temporal disappearance of variant c-*myc* genotypes in CHO cells expressing murine c-*myc*. These cells were adapted to growth in 320 μM MTX (a) after which MTX was removed and the number of metaphases showing 1 and 2 integrations per cell (int/c) as well as multiple integrations per chromosome (int/chr) was measured. Metaphases containing extended integrations (ext int/cell; such as those shown in Fig. 1D) were also enumerated before and after MTX removal.

Several precautions are typically implemented to minimize the opportunity for genetic drift, including establishment of large repositories of the recombinant cells shortly after a producer cell line has been identified (master and working banks) and by imposing time limits (e.g., 100 days) for subcultivation of thawed cells. Cell growth and recombinant protein productivity rates are usually measured at several intervals during culture. The relationship between recombinant protein production, proliferation rate, and genetic stability of integrated recombinant sequences has not been clearly established. Although high copy numbers and extended regions of amplified sequences are often positively correlated with high expression levels in CHO cells, extensive competition of cellular resources for protein synthesis (e.g., of housekeeping proteins) may have adverse consequences on cell proliferation.

To test the hypothesis that high levels of expression are inversely correlated with cell growth rates, we created a cell line with both heterogeneous integration loci and heterogeneous amplification levels of the transfected plasmid sequences. Approximately 300 recombinant CHO colonies, selected in alpha-modified medium after transfection with an expression vector for the chimeric CD4IgG

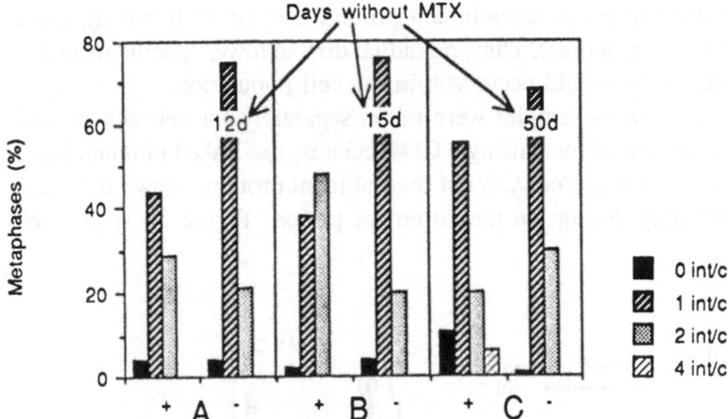

Figure 2b Distribution pattern of integrated sequences in two clones (A and B) expressing CD4IgG, adapted to 500 nm MTX, cultured in the presence and absence of MTX (+/−). The distribution of integrated sequences in a heterogeneous cell line (cell line C) adapted to 250 nm MTX is also shown.

molecule (see above), were pooled and subcultivated in increasing concentrations of MTX over a period of about 3 months. The final cell line was maintained in media containing 250 nm MTX. These cells were then adapted to suspension growth in MTX-containing medium over 10 passages. FISH with probes specific for the transfected CD4IgG plasmid showed that the population was heterogeneous in both number and loci of integration sites and in the degree of amplification of transfected sequences. Each metaphase contained at least one, and often two or more, hybridization signals with no specific integration pattern. The amount of recombinant sequences (estimated by relative probe-linked fluorescence) per cell varied substantially, and it was expected that recombinant protein production per cell would vary accordingly. We used this heterogeneous cell culture to investigate the relationship between protein production and cell growth. Cells were cultivated in a 2-liter vessel equipped for long-term perfusion fermentation. Long-term perfusion fermentation of a heterogeneous population of cells was expected to be the ideal situation for overgrowth of more rapidly proliferating, low- or no-producer subpopulations as has been observed in monoclonal antibody producing hybridomas. This bioreactor configuration allowed the supply of up to 10 volumes of fresh medium per day, while the same volume of spent medium could be removed from the vessel. The cells were grown in this system in the absence of MTX for more than 90 days, and various cultivation parameters (glucose consumption, oxygen uptake rate, lactate production, etc.) were monitored and adjusted daily to maximize growth rates. If the growth rate were inversely correlated with protein production, we would anticipate faster proliferation of cells, producing lower levels of recombinant protein (and possibly having fewer copies of recombinant sequences). Thus, a gradual drift to lower specific recombinant protein productivity should occur within the cell population.

During the culture, daily samples were taken separately for cell density and recombinant protein measurement using a CD4IgG enzyme-linked immunoassay. The specific productivities (pg/cell/day) of recombinant protein, shown in Figure 3a, varied considerably during the measurement period. Figure 3b shows cell

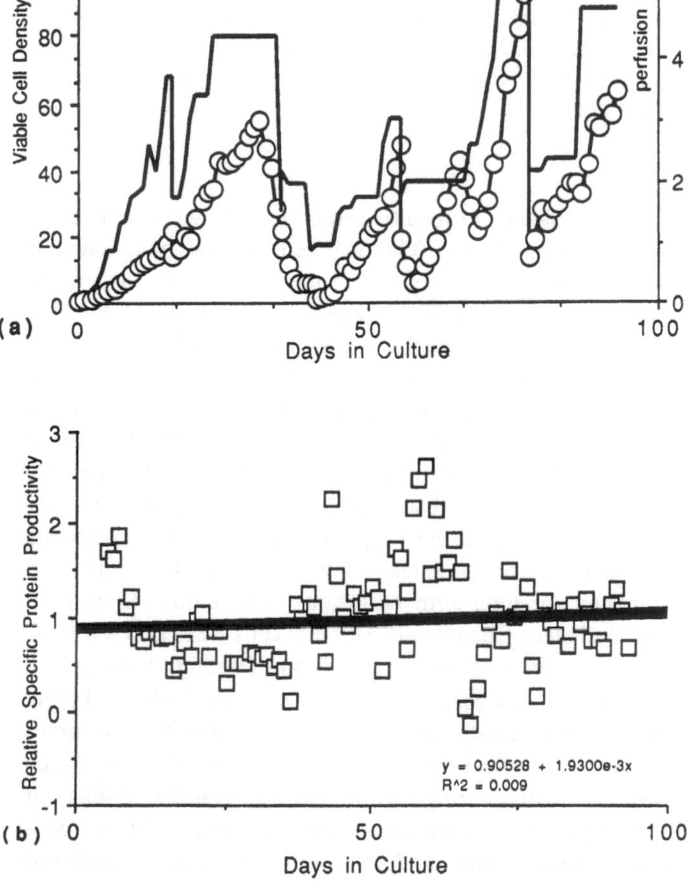

Figure 3 Perfusion cultivation of CD4IgG expressing CHO cells: (a) relative specific productivity, and (b) cell density (viable cells per milliliter) and perfusion rates (vvd = vessel volumes per day).

densities measured during the culture period. Peak densities of 55×10^6 viable cells/ml and 115×10^6 viable cells/ml were seen when perfusion rates were set to about 4.5 and 6 volumes per day, respectively. Changes in environmental conditions resulted in variability of growth rates and cell densities and also influenced specific recombinant protein productivities, which were lowest at times of declining viabilities and cell death. Regression analysis of protein production rates did not indicate a decline, but rather a slight increase of specific recombinant protein productivity throughout the culture.

At the beginning (day 15) and toward the end (day 80) of the culture, the distribution, amplification, and localization of recombinant sequences in this heterogeneous population of cells were measured, using FISH. There was no apparent difference between the distribution patterns of the recombinant sequences in cells measured at these time intervals (data not shown). Thus, these data indicate both cytogenetic and protein production stability in a heterogeneous population of cells cultivated continuously for more than 100 days (> 140 cell generations).

III. SUMMARY

We have presented data about integration patterns and amplification of transfected DNA in CHO cells. FISH was used for routine characterization of transfected sequences. In cells cultured in the presence of MTX, transfected sequences showed variant integration, localization, and amplification patterns. A subset of amplified sequences, typically characterized by multiple or unusual structures, and often having increased length, was unstable and disappeared from cloned and uncloned cell populations within 10–20 days after removal of MTX from the culture medium. The resulting population of cells containing defined, often more modestly amplified sequences, was genetically stable and showed consistent protein production rates over extended culture periods. The patterns of integration of recombinant sequences in the three clonal cell lines that were examined, cultured for more than 10 days without MTX, were sufficiently unique to allow the identification and recognition of each line by its cytogenetic hybridization pattern. Our data also suggest that higher production rates of recombinant proteins are not necessarily correlated with lower population growth rates.

ACKNOWLEDGMENTS

We express our gratitude to Walter Tanaka, Andie Abarquez, Vivek Bajaj, Adriana Johnson, Patricia DeTeresa, and Caridad Rosette for excellent technical assistance and Joe W. Gray and Robert Arathoon for stimulating discussions.

Part of this work was performed under the auspices of the U.S. Department of Energy by the Lawrence Livermore National Laboratory under contract W-7405-ENG-48.

REFERENCES

1. Urlaub, G., and Chasin, L. A. Isolation of Chinese hamster cell mutants deficient in dihydrofolate reductase activity. *Proc. Natl. Acad. Sci. USA*, *77*:4216–4220 (1980).
2. Goedell, D. V., Kohr, W. J., Pennica, D., and Vehar G. A. Human tissue plasminogen activator, pharmaceutical compositions containing it, processes for making it, and DNA and transformed intermediates thereof. United Kingdom Patent 2,119,804 (1983).
3. Wigler, M., Perucho, M., Kurtz, D., Dana, S., Pellicer, A., Axel, R., and Silverstein, S. Transformation of mammalian cells with an amplifiable dominant-acting gene. *Proc. Natl. Acad. Sci. USA*, *77*(6):3567–3570 (1980).
4. Kaufman, R.J., and Sharp, P. Amplification and expression of sequences cotransfected with a modular dihydrofolate reductase complementary DNA gene. *J. Mol. Biol. 159*:601–621 (1982).
5. Weidle, U. H., Buckel, P., and Wienberg, J. Amplified expression constructs for human tissue-type plasminogen activator in Chinese hamster ovary cells: Instability in the absence of selective pressure. *Gene*, *66*:193–203 (1988).
6. Wurm, F. M., Gwinn, K. A., and Kingston, R. E. Inducible overexpression of the mouse c-*myc* protein in mammalian cells. *Proc. Natl. Acad. Sci. USA*, *83*:5414–5418 (1986).
7. Nunberg, J. H., Kaufman, R. J., Schimke, R.T., Urlaub, G., and Chasin, L. A. Amplified dihydrofolate reductase genes are localized to a homogeneously staining region of a single chromosome in a methotrexate-resistant Chinese hamster ovary cell line. *Proc. Natl. Acad. Sci. USA*, *75*:5553–5556 (1978).
8. Pinkel, D., Straume, T., and Gray, W. Cytogenetic analysis using quantitative, high sensitivity fluorescence hybridization. *Proc. Natl. Acad. Sci. USA*, *83*:2934–2938 (1986).
9. Lawrence, J. B., Villnave, C. A., and Singer, R. H. Sensitive, high resolution chromatin and chromosome mapping in situ: Presence and orientation of two closely integrated copies of EBV in a lymphoma line. *Cell*, *52*:51–61 (1988).
10. Pallavicini, M. G., DeTeresa, P. S., Rosette, C., Gray, J. W., and Wurm, F. M. Effects of methotrexate (MTX) on transfected DNA stability in mammalian cells. *Mol. Cell. Biol. 10*:401–404 (1990).
11. Kaufman, R. J., Wasley, L. C., Spiliotes, A. J., Gossels, S. D., Latt, S. A., Larsen, G. R., and Kay, R. M. Coamplification and coexpression of human tissue-type plasminogen activator and murine dihydrofolate reductase sequences in Chinese hamster ovary cells. *Mol. Cell. Biol. 5*:1750–1759 (1985).
12. Nederlof, P. M., van der Flier, S., Wiegart, J., Raap, A. K., Ploem, J. S., and van der Ploeg, M. Multiple fluorescence in situ hybridization. *Cytometry*, *11*:126–131 (1990).
13. Smith, D. H., Byrn, R. A., Marsters, S. A., Gregory, T., Groopman, J. E., and Capon, D. J. Blocking of HIV-1 infectivity by a soluble, secreted form of the CD4 antigen. *Science*, *328*:1704–1707 (1987).
14. Mondello, C., Giorgi, R., and Nuzzo, F. Chromosomal effects of methotrexate on cultured human lymphocytes. *Mutat. Res. 139*:67–70 (1984).
15. Capon, D., Chamow, S. M., Mordenti, J., Marsters, S. A., Gregory, T., Mitsuya, H., Byrn, R. A., Lucas, C., Wurm, F. M., Groopman, J. E., Broder, S., and Smith, D. H. Designing CD4 immunoadhesins for AIDS therapy. *Nature*, *337*:525–531 (1989).

II

Gene Amplification and Drug Resistance

6

Threonyl–tRNA Synthetase Gene Amplification in Borrelidin-Resistant Mammalian Cells

Stuart M. Arfin, Steven C. Gerken,* Kris J. Kontis,† and Matt E. Cruzen *University of California, Irvine, California*

I. INTRODUCTION

Following the initial work of Schimke's and Stark's laboratories, it became apparent that a major class of drug-resistant mutants in mammalian cells were those that overproduced target proteins as a result of amplification of genomic sequences encoding that protein. It appeared reasonable to assume that, given an appropriately specific inhibitor, it should be possible to select for cells that overproduce any protein. Because of our interest in the structure, function, and regulation of aminoacyl–tRNA synthetases and their genes in higher eukaryotes, we screened a number of amino acid analogues and related compounds for their ability to kill Chinese hamster ovary (CHO) cells.

The antibiotic borrelidin, which has been shown to inhibit threonyl–tRNA synthetase (ThrRS) from *Escherichia coli* and yeast, also proved to be highly toxic to CHO cells. This toxicity was antagonized by threonine [1]. Since threonine is an essential amino acid in mammalian cells, mechanisms of resistance based on its increased synthesis as have been reported in bacteria and yeast would not be possible. Thus, apart from mutations leading to decreased transport of the drug or to alterations in the affinity of ThrRS for the drug, resistance would be most

Current affiliations:
*University of Utah, Salt Lake City, Utah.
†University of California, Irvine, California.

likely to arise by overproduction of ThrRS. Therefore, we initiated attempts to amplify the CHO cell ThrRS structural gene by selection for resistance to increasing concentrations of borrelidin.

II. SELECTION OF BORRELIDIN-RESISTANT CHINESE HAMSTER OVARY CELLS

CHO cells were exposed to progressive two- to threefold increments in borrelidin concentration in α-ME medium containing 24 μM threonine (1/20 the normal concentration). Cultures were maintained in each concentration of borrelidin for 3–4 weeks before the next incremental increase. Typically, 90–95% of surviving cells were killed at each new borrelidin concentration. Cells were cloned from populations that survived at various borrelidin concentrations.

CHO cells are very sensitive to borrelidin, and the initial selection for drug-resistant colonies was done at a concentration of 3 ng/ml. Significant increases in ThrRS activity can be seen in cells resistant to 100 ng/ml. Increased ThrRS activity is roughly proportional to borrelidin resistance up to concentrations of 2 μg/ml. Exposure of cells resistant to 2 μg/ml of borrelidin to higher concentrations of the drug results in the death of all cells.

The percentage of the total soluble cellular protein represented by ThrRS in these most highly resistant cell lines (see below) is well below what has been achieved in a variety of other systems. The reasons for the failure to overproduce ThrRS to a greater extent are not immediately clear. It is possible that borrelidin concentrations greater than 2 μg/ml are inhibitory to cellular processes other than the aminoacylation of $tRNA^{Thr}$ or that sequences coamplified with the ThrRS structural gene in some way adversely affect cells.

Alternatively, the markedly elevated ThrRS activity may itself interfere with other cellular processes. One possibility is that high ThrRS activity causes increased misacylation of noncognate tRNA species with threonine, producing poorly functioning or inactive proteins. In *E. coli*, the overproduction of glutaminyl–tRNA synthetase [2] or tyrosyl–tRNA synthetase [3] decreases cell growth and viability. Misacylation of $tRNA^{Tyr}$ with glutamine occurs in *E. coli* cells that overproduce glutaminyl–tRNA synthetase in vivo. If this is a general phenomenon, then the fidelity of protein biosynthesis may be compromised in any cell in which the balance between a specific synthetase and its cognate tRNA isoaccepting family is drastically altered. Since abnormal proteins are generally degraded more rapidly than their normal counterparts, if a significant portion of the proteins in the highly amplified cell lines are structurally altered because of misacylation, the overall rate of protein turnover in these cells may be greater than in wild-type cells. However, pulse chase experiments have failed to detect any such differences. Thus, if misacylation occurs it does not lead to gross alterations in total cellular protein.

A number of aminoacyl–tRNA synthetases are capable of catalyzing the synthesis of p^1,p^4-bis (5′-adenosyl)-tetraphosphate (Ap_4A) in vitro in the presence of the cognate amino acid but in the absence of tRNA. Ap_4A has been detected in many organisms, and it is assumed that its synthesis in vivo is catalyzed by one or more aminoacyl–tRNA synthetases.

Ap_4A has been postulated to be a regulator of cell proliferation, possibly through a direct effect on DNA polymerase, although this hypothesis remains controversial. ThrRS catalyzes Ap_4A formation in vitro, and the rate of synthesis is stimulated sixfold by phosphorylation [4]. ThrRS is phosphorylated in highly resistant cells [5]. This raises the interesting possibility that Ap_4A pools are substantially elevated in the resistant lines, leading to enhanced unscheduled DNA synthesis. This, in turn, could result in rearrangements, duplications, deletions, or other deleterious genetic changes.

Analysis of the cell cycle in wild-type and highly borrelidin-resistant CHO cells by flow cytometry does, in fact, show a twofold increase in the proportion of cells in S phase (18.6% vs. 9.6%) in the resistant line. However, analyses of Ap_4A levels by labeling with [^{32}P]orthophosphate and thin-layer chromatography on polyethyleneimine (PEI)-cellulose sheets do not show any differences among cell lines. Thus, differences in rates of DNA synthesis may exist between wild-type and borrelidin-resistant cells, and these differences may be responsible for the failure to achieve additional amplification. We have no direct evidence, however, to relate enhanced DNA synthesis rates to elevated ThrRS activity.

Cell lines cloned from populations resistant to various concentrations of borrelidin have been cultured for hundreds of generations in the absence of the drug with no resulting loss of resistance or decrease in ThrRS activity. Thus, ThrRS gene amplification appears to be a stable event in CHO cells. Resistant cell lines, however, show no consistent karyotypic alterations, such as double minute chromosomes (DMs) or homogeneously staining regions (HSRs), which are often found in drug-resistant cells. Given the stability of the resistant phenotype in nonselective medium, we would anticipate a chromosomal localization for the amplified genes and therefore, possibly, the presence of a homogeneously staining region. The failure to find an HSR may be a consequence of the relatively small degree of amplification and/or the size of the amplified unit.

III. ThrRS ACTIVITY, PROTEIN AND mRNA LEVELS, AND GENE COPY NUMBER

As shown in Figure 1, the increase in ThrRS enzymatic activity is roughly proportional to the degree of resistance. We have also utilized immunological techniques and Northern and Southern blotting to quantitate specific protein content, mRNA levels, and gene copy number in resistant cells.

When ThrRS protein levels in the resistant lines were measured immuno-

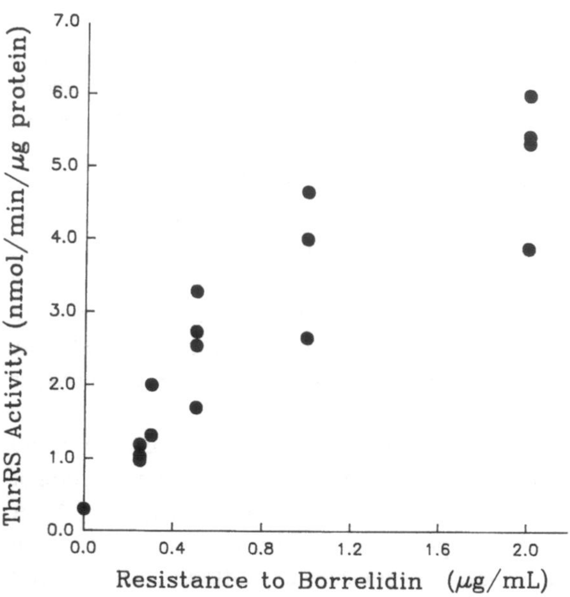

Figure 1 Levels of ThrRS activity in CHO cells resistant to various concentrations of borrelidin. Data are the average activities for independent clones isolated from cell populations resistant to the indicated concentrations of borrelidin.

logically, a surprising result was obtained. Although the most resistant lines had only a 20-fold increase in activity, they had a 60- to 100-fold increase in ThrRS protein levels [6]. The reason for this discrepancy between enzyme activity and enzyme amount remains unclear. The ThrRSs from sensitive and resistant cells do not differ kinetically or in a variety of physical properties. Nevertheless, a structural gene mutation cannot be absolutely excluded. Amplification of such a gene could account for the apparently decreased catalytic efficiency of ThrRS in resistant cell lines.

An alternative possibility is that ThrRS is postranslationally modified to a less active form. If the ability of this catalytically less efficient form of ThrRS to bind borrelidin is unchanged, then the expected phenotype of the cells would be that observed. We were able to demonstrate that ThrRS is, in fact, phosphorylated on one or more serine residues [5]. However, removal of this phosphate in vitro has no effect on activity or sensitivity to inhibition by borrelidin [S. C. Gerken, Ph.D. thesis, University of California, Irvine, 1986]. Similar results have been obtained by Pendergast and Traugh [7] in studies of phosphorylation of purified ThrRS by specific protein kinases.

The levels of ThrRS-specific mRNA and of ThrRS gene sequences in wild-type and borrelidin-resistant cells were determined using a human cDNA probe [8].

Table 1 Relative Levels of ThrRS Protein, mRNA, and Genes in CHO Cell Lines Resistant to Increasing Concentrations of Borrelidin

Cell line	Borrelidin concentration (ng/ml)[a]	ThrRS protein content (fold increase)[b]	ThrRS mRNA content (fold increase)[c]	ThrRS gene copies (fold increase)[d]
GAT^-(wild type)		1	1	1
301/100	150	8	3	4
203/202/250	250	24	10	11
203/202/500	500	42	10	11
203/202/1000	1000	42	16	21
2000A	2000	65	20	34

[a]Cell lines were cloned from populations growing in the indicated concentration of borrelidin.
[b]Determined by quantitative immunoblotting as described in Ref. 6.
[c]Determined by quantitative Northern blot hybridization as described in Ref. 8.
[d]Determined by quantitative Southern blot hybridization as described in Ref. 8.

Some of the results obtained are summarized in Table 1. In general, there is a progressive increase in relative ThrRS protein content, mRNA content, and specific gene sequences in cell lines resistant to increasing concentrations of borrelidin. In addition, high molecular weight DNA isolated from sensitive and resistant cell lines was digested with a variety of restriction endonucleases, subjected to agarose gel electrophoresis, transferred to nylon filters, and hybridized with the human cDNA probe. In all cases the DNA fragments produced by the various nucleases were the same in sensitive and resistant lines, suggesting that amplification is not accompanied by major chromosomal rearrangements [8]. The patterns observed were consistent with the existence of a single ThrRS structural gene having a size of about 20 kb.

IV. REGULATION OF ThrRS EXPRESSION

The expression of many bacterial aminoacyl–tRNA synthetase genes is regulated in response to the extent of charging of their cognate tRNAs [9]. Recent evidence suggests that the levels of some synthetases in mammalian cells may also be increased when the aminoacylation of their cognate tRNAs is restricted [10,11]. We have used the human cDNA probe to quantitate ThrRS-specific mRNA levels in CHO cells subjected to amino acid limitation. Transfer of cells from a complete medium to one lacking any added threonine results in a three- to fivefold increase of ThrRS mRNA in about 24 hr (Fig. 2). When the threonine-free medium also contains 2.5 ng/ml of borrelidin, the increase in ThrRS mRNA levels is more rapid

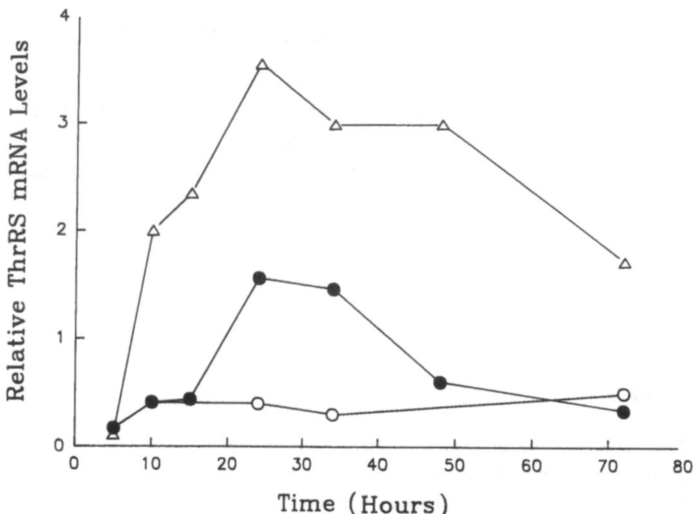

Figure 2 Changes in ThrRS mRNA levels with time in CHO cells grown in the absence of threonine. RNA samples were prepared from cells grown for the indicated times in complete α-ME medium (○), in α-ME medium lacking threonine (●), or in α-ME medium lacking threonine and containing 2.5 ng/ml borrelidin (△). Serial dilutions of the samples were dot-blotted onto Nytran filters and successively hybridized to ^{32}P-labeled probes specific for ThrRS and for actin. Data points represent the relative ThrRS mRNA levels normalized to the actin mRNA levels.

and reaches higher levels (Fig. 2). In both cases ThrRS mRNA levels gradually decrease with time.

The effect of borrelidin suggests that the metabolic "signal" to which the cells are responding is uncharged $tRNA^{Thr}$ rather than decreased levels of threonine per se. The data shown in Table 2 are consistent with this interpretation. ThrRS mRNA levels were determined 24 hr after transfer of CHO cells to medium containing reduced levels of threonine (40 μM; 1/10 the normal concentration) and increasing concentrations of borrelidin. Although the reduced threonine concentration is sufficient to prevent the increase in mRNA levels, in the presence of borrelidin there is a dose-dependent increase in ThrRS mRNA levels.

Similar experiments with CHO cell lines in which the ThrRS structural gene has been amplified show that limitation of $tRNA^{Thr}$ charging in these cell lines also leads to the accumulation of ThrRS mRNA (Fig. 3). Although the mechanism by which expression of the ThrRS gene is regulated is not yet clear, our preliminary evidence suggests that the increased ThrRS mRNA levels result mainly from increased rates of transcription. Thus, it appears that the amplified genes are active and are regulated normally. Furthermore, the data indicate that sufficient tran-

Table 2 Relative Levels of ThrRS mRNA in CHO Cells Grown for 24 hr in Medium Containing Various Concentrations of Borrelidin[a]

Concentration		
Threonine (μM)	Borrelidin (ng/ml)	Relative ThrRS mRNA Content
400	0	1
40	0	1
40	5	2
40	7.5	2
40	10	3.8
40	12.5	5.5
40	15	12.8

[a]CHO cells were grown in monolayer culture in a α-ME medium to about 70% confluency. The medium was then removed, the cells extensively washed with phosphate buffer solution and the medium replaced with α-ME containing the indicated concentrations of threonine and borrelidin. After 24 hr of additional growth, relative ThrRS mRNA content was determined as described in the legend to Figure 2.

scriptional factors are present in CHO cells to support the enhanced transcriptional activity of at least 30 ThrRS genes and must therefore be present in cells at levels well in excess of the amount necessary for the normal transcription of a single ThrRS gene.

V. PRELIMINARY CHARACTERIZATION OF THE ThrRS STRUCTURAL GENE

The structural genes for seven aminoacyl–tRNA synthetases have been mapped to specific human chromosomes. Four of these genes, including that for ThrRS, have been localized to chromosome 5 [12]. The other synthetase genes mapped to this chromosome are those encoding leucyl–, histidyl–, and arginyl–tRNA synthetases. Although the four genes are clearly not linked in the sense that genes which form a bacterial operon are linked, this synteny may indicate an evolutionary or regulatory relatedness. A comparison of the structures of the four genes is, therefore, of considerable interest.

The complete sequence of a human ThrRS cDNA has been determined and the amino acid sequence deduced (M. E. Cruzen and S. M. Arfin, in preparation). Although extensive homology with the ThrRSs from bacteria and yeast was observed, little homology to other aminoacyl–tRNA synthetases was found. In particular, a comparison of the human ThrRS sequence with that of human HisRS [13] showed no evidence of sequence relatedness.

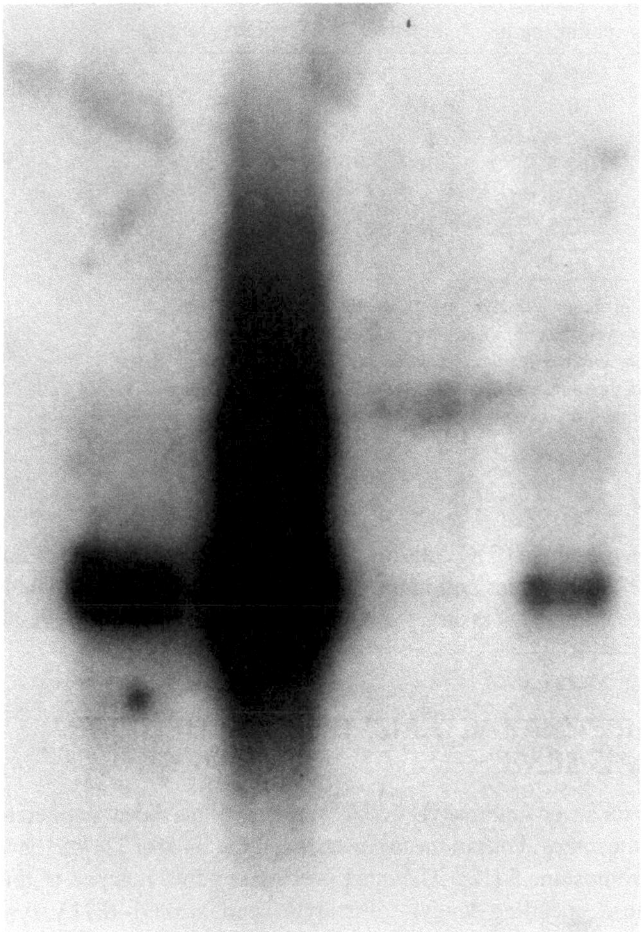

Figure 3 Northern blot of total RNA from wild-type and borrelidin-resistant CHO cells growing in complete medium and in medium lacking threonine. A total of 10 μg of whole-cell RNA from each cell line was denatured with formaldehyde, electrophoresed in 1% agarose, transferred to a Nytran filter, and hybridized with ^{32}P-labeled pTRS1 [8]. Lane 1, 2000A cells grown in complete medium; lane 2, 2000A cells grown in medium lacking threonine for 24 hr; lane 3, GAT^{-} cells grown in complete medium; lane 4, GAT^{-} cells grown in medium lacking threonine for 24 hr.

We have also obtained the sequence of about 1 kb of 5′ flanking sequence from human genomic DNA. This region contains neither a CAAT sequence nor a TATA sequence. The absence of these sequences appears to be characteristic of "housekeeping" genes that are expressed in all cells. TATA and CAAT sequences are also missing from the 5′ flanking sequence of the hamster HisRS gene [14]. Alignments of the 5′ flanking sequences of the human ThrRS and hamster HisRS genes do reveal several common short sequences. The sequence 5′-GGCGGGGTTAG-3′, a potential SP1 binding site, occurs at nucleotide positions -40 to -30 relative to the initiator methionine codon in HisRS and at positions -178 to -168 relative to the initiator methionine codon in ThrRS. The sequences 5′-CTCTGGAAAA-3′ and 5′-GCTCCTCC-3′ also occur in both 5′ flanking regions. Whether any or all of these sequences are important for initiation or regulation of transcription of the two synthetase genes is not yet clear.

VI. DISCUSSION

Borrelidin is a highly specific inhibitor of ThrRS and kills cells by preventing protein synthesis. When cells are exposed to increasing concentrations of the drug in a stepwise manner, amplification of the structural gene for ThrRS occurs. This gene amplification is accompanied by increased steady state ThrRS mRNA and protein levels.

The goal in obtaining cell lines in which the ThrRS genes are amplified was to aid us in preparing the necessary reagents [protein, antibodies, and nucleic acid probes] to study ThrRS gene structure and its expression and protein structure, function, and physiological regulation. All these studies are still very much in progress and some have been summarized above.

It is of interest to note that progress in the characterization of the human and hamster HisRS gene and protein was also aided substantially by the availability of cell lines in which that gene was amplified. These lines were obtained by use of the specific HisRS inhibitor histidinol [15]. Borrelidin and histidinol are both potent inhibitors of the aminoacylation of specific tRNAs but are not themselves attached to tRNA and incorporated into protein. Cell death is therefore due to inhibition of protein synthesis rather than to the formation of altered proteins, and overproduction of the target enzyme could be expected to be a major mechanism by which cells become resistant to such inhibitors. Studies of the substrate specificity of a number of aminoacyl–tRNA synthetases show that the carboxyl moiety of the amino acid substrate can be modified to produce potent competitive inhibitors of these enzymes. Thus, in addition to histidinol, other amino alcohols or amino acid analogues in which the carboxyl group is either removed or replaced by another functional group could prove to be useful in designing schemes for the amplification of specific aminoacyl–tRNA synthetase genes.

REFERENCES

1. Gantt, J. S., Bennett, C. A., and Arfin, S. M. Increased levels of threonyl-tRNA synthetase in a borrelidin-resistant Chinese hamster ovary cell line. *Proc. Natl. Acad. Sci. USA*, *78*:5367–5370 (1981).
2. Swanson, R., Hoben, P., Sumner-Smith, M., Uemura, H., Watson, L., and Soll, D. Accuracy of in vivo aminoacylation requires proper balance of tRNA and aminoacyl–tRNA synthetase. *Science*, *242*:1548–1551 (1988).
3. Bedouelle, H., Guez, V., Vidal-Cros, A., and Hermann, M. Overproduction of tyrosyl–tRNA synthetase is toxic to *Escherichia coli*: A genetic analysis. *J. Bacteriol.* *172*: 3940–3945 (1990).
4. Dang, C. V., and Traugh, J. A. Phosphorylation of threonyl– and seryl–tRNA synthetase by cAMP-dependent protein kinase. *J. Biol. Chem.* *264*:5861–5865 (1989).
5. Gerken, S. C., and Arfin, S. M. Threonyl–tRNA synthetase from Chinese hamster ovary cells is phosphorylated on serine. *J. Biol. Chem.* *259*:1160–1161 (1984).
6. Gerken, S. C., and Arfin, S. M. Chinese hamster ovary cells resistant to borrelidin overproduce threonyl–tRNA synthetase. *J. Biol. Chem.* *259*:9202–9206 (1984).
7. Pendergast, A. M., and Traugh, J. A. Alteration of aminoacyl–tRNA synthetase activities by phosphorylation with casein kinase I. *J. Biol. Chem.* *260*:11769–11774 (1985).
8. Kontis, K. J., and Arfin, S. M. Isolation of a cDNA clone for human threonyl–tRNA synthetase: Amplification of the structural gene in borrelidin-resistant cell lines. *Mol. Cell. Biol.* *9*:1832–1838 (1989).
9. Grunberg-Manago, M. Regulation of the expression of aminoacyl–tRNA synthetases and translation factors, in *Escherichia coli and Salmonella typhimurium, Cellular and Molecular Biology*, Vol. 2 (F. C. Neidhardt, ed.), American Society for Microbiology, Washington, D.C., 1987, pp. 1386–1409.
10. Lazard, M., Mirande, M., and Waller, J.-P. Expression of the aminoacyl–tRNA synthetase complex in cultured Chinese hamster ovary cells. Specific derepression of the methionyl–tRNA synthetase component upon methionine restriction. *J. Biol. Chem.* *262*:3982–3987 (1987).
11. Lazard, M., Mirande, M., and Waller, J.-P. Overexpression of mammalian phenylalanyl–tRNA synthetase upon phenylalanine restriction. *FEBS Lett.* *216*:27–30 (1987).
12. Gerken, S. C., Wasmuth, J. J., and Arfin, S. M. Threonyl–tRNA synthetase gene maps close to leucyl–tRNA synthetase gene on human chromosome 5. *Somatic Cell Mol. Genet.* *12*:519–522 (1986).
13. Tsui, F. W. L., and Siminovitch, L. Isolation, structure and expression of mammalian genes for histidyl–tRNA synthetase. *Nucleic Acids Res.* *15*:3349–3367 (1987).
14. Tsui, F. W. L., and Siminovitch, L. Structural analysis of the 5′ region of the chromosomal gene for hamster histidyl–tRNA synthetase. *Gene*, *61*:349–361 (1987).
15. Tsui, F. W. L., Andrulis, I. L., Murialdo, H., and Siminovitch, L. Amplification of the gene for histidyl–tRNA synthetase in histidinol-resistant Chinese hamster ovary cells. *Mol. Cell. Biol.* *5*:2381–2388 (1985).

7

Amplification of Na,K-ATPase in Ouabain-Resistant HeLa Variants

John F. Ash *University of Utah, Salt Lake City, Utah*

Peter G. Pauw *Gonzaga University, Spokane, Washington*

I. INTRODUCTION: STEP SELECTION FOR RESISTANCE TO OUABAIN

A. Structure and Function of the Na,K-ATPase

The sodium- and potassium-activated adenosinetriphosphatase (Na,K-ATPase) is a ubiquitous and vital enzyme of animal cell plasma membranes. Its structure, function, and regulation have been extensively reviewed (see, e.g., Ref. 1). The enzyme is a heterodimer of membrane spanning catalytic α and glycoprotein β subunits (see Fig. 1). Two α-isoform proteins have been well described, and three to five potential α genes have been identified. Functionally, the enzyme is an electrogenic antiporter, which couples the hydrolysis of ATP to the efflux of three Na^+ and influx of two K^+ ions. The sodium extrusion simultaneously helps relieve the cell of a lethal osmotic water burden and produces an inward electrochemical Na^+ gradient, which energizes transport systems for several metabolites, H^+ and Ca^{2+}. Potassium import and its subsequent diffusion, along with the net export of one positive charge per reaction cycle, produces the resting membrane potential, which is the basis of action potential conduction. Clearly, a perturbation of steady state Na,K-ATPase expression by gene amplification will have interesting consequences for cell physiology. This has been the major focus of our work and this chapter.

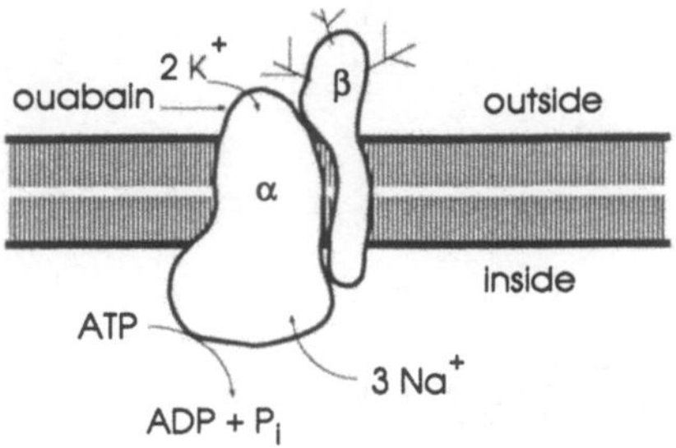

Figure 1 Review of Na,K-ATPase structure and activity. The two subunits are tightly but noncovalently associated and are shown embedded in the plasma membrane bilayer. The α subunit contains the catalytic site for ATP hydrolysis and has a relative molecular mass M_r of approximately 100,000. The glycosylated β subunit has an M_r of approximately 55,000. See text and Ref. 1 for details.

B. Action of Ouabain

The selective agent used in our studies was ouabain, a specific inhibitor of the Na,K-ATPase. Ouabain and related cardiac glycosides bind with high affinity (in HeLa, $K_d = 6 \times 10^{-8}$ M) to an extracellular portion of the α subunit and block a required conformational change in the reaction cycle of the enzyme [1]. Binding of ouabain is stoichiometric (1:1) and is frequently used to measure the number of Na,K-ATPase molecules in the cell membrane. Inhibition of the enzyme by ouabain leads to an increase in $[Na^+]_i$ and, eventually, to cell death. Cardiac glycosides are used clinically to increase arterial blood pressure and for treatment of congestive heart failure. Both effects are apparently mediated by increased cytoplasmic Ca^{2+} concentrations produced indirectly by decreased Na^+/Ca^{2+} antiport activity.

C. Outcome of Step Selection of HeLa with Ouabain

There have been a number of reports of ouabain-resistant mutants that express an Na,K-ATPase with lowered affinity for and a decrease in specific binding of tritium-labeled ouabain obtained by ouabain selection after mutagenesis (e.g., Ref. 2). Mutations affecting ouabain binding are apparently easy to obtain, and small differences in the binding site apparently account for species and isoform differences in ouabain sensitivity [3]. Rodent cells, for example, tolerate ouabain concentrations 10,000-fold higher than those toxic to human cells and rat α subunit isoforms differ 1000-fold in ouabain sensitivity.

However, using step selection procedures and carefully monitoring the specific ouabain binding of resistant colonies, we isolated ouabain-resistant clones of the human carcinoma line HeLa S_3, whose characteristics suggest that resistance is based on gene amplification [4]. Initial selection at 5×10^{-8} ouabain resulted in the isolation of about 10 colonies per 10^6 cells plated. Most of these clones exhibited a doubling in the number of specific ouabain binding sites, consistent with gene amplification, rather than mutation affecting the ouabain binding site. However, most were extremely fragile in culture and grew poorly. A single clone was taken through additional selective steps. Selection at 10^{-7} M ouabain resulted in the isolation of colonies with an additional doubling in ouabain binding site number. These clones attached poorly to culture dishes and required treatment of culture dishes with polyornithine for attachment and growth. Unlike the parental cells, the clones would not grow in suspension. A third-step selection of that clone at 2.25×10^{-7} M ouabain yielded numerous colonies binding 8–10 times as much ouabain as the parental HeLa S_3. Finally, selection for growth on untreated culture dishes yielded a clone we called C^+, which displayed resistance to a 10-fold greater ouabain concentration than HeLa. Scatchard analysis indicated a slightly reduced ouabain affinity ($K_d = 3 \times 10^{-7}$ M vs. 6×10^{-8} M) and approximately 10^7 ouabain binding sites per cell in C^+ (compared to 8×10^5/cell in HeLa). All other ouabain-resistant lines studied in this laboratory are derived from C^+.

II. CHARACTERIZATION OF OUABAIN RESISTANCE IN C^+

A. Stability of Resistance

In the absence of continued selection, ouabain resistance and the increase in specific ouabain binding in C^+ were extremely unstable. Levels of ouabain binding declined significantly after 8 population doublings and reverted to parental HeLa levels by 20 population doublings. The reverted population (C^-) was slightly more sensitive to ouabain than the parental line. Reversion of the ouabain-resistant phenotype was also accompanied by a return to normal HeLa growth characteristics in C^-, including a decrease in population doubling time to 22 hr (vs. 31 hr in C^+) and the ability to grow in suspension culture.

B. Evidence of Increased Expression

In addition to specific ouabain binding, amplified expression of the Na,K-ATPase in C^+ was demonstrated in isolated membranes. Polyacrylamide gel electrophoresis with sodium dodecylsulfate (SDS-PAGE) revealed an approximate 10-fold increase in a band comigrating with the Na,K-ATPase α chain. This peptide could be labeled with [^{32}P]phosphate in the presence of ouabain and Mg^{2+}, confirming its identity as the Na,K-ATPase catalytic subunit [4].

C. Gene Amplification Is the Basis of Increased Na,K-ATPase Expression

The first evidence that the increased Na,K-ATPase expression seen in C^+ is due to gene amplification was the correlation of unstable resistance with the presence of minute chromosomes. C^+ cells contain both double and single minute chromosomes, with an average of 31 minutes. In contrast, minute chromosomes are not present in either HeLa S_3 or C^-.

Subsequently, Levenson and coworkers demonstrated amplification of both α [5] and β [6] chain genes using cDNA probes. Gene copy levels and mRNA levels for the α subunit in C^+ cells grown in 10^{-6} M ouabain were found to be about 100 times HeLa levels, and amplification of the β subunit gene was about 20-fold. In lines derived from C^+ we have observed somewhat lower levels of amplification [7]. The apparent level of amplification of α subunit genes is consistently three- to fivefold higher than that of β subunit genes, suggesting that the two genes are linked in a single amplification unit.

Ouabain resistance in C^+ cells may have required simultaneous amplification of two genes that are not directly linked. In humans three of the Na,K-ATPase genes map to chromosome 1: the α2 and β subunit genes both map to the q arm, but are not known to be in tandem, while the α1 gene maps to the p arm [8]. We have not examined the genesis of the minute chromosomes nor determined the isoform amplified in C^+ cells, and it will be interesting to determine whether the increased K_d for ouabain binding is due to the amplification of a different α isoform. Since all known catalytic and inhibitor binding activities of the enzyme reside in the α subunit, the coordinate amplification of genes for both subunits in the acquisition of ouabain resistance highlights the importance of the β subunit for functional enzyme assembly [9]. A requirement for coordinate amplification would also explain the difficulty we and others have encountered in repeating this selection in other cell lines. The instability of the HeLa genome may have been an important factor in the success of this selection, perhaps by bringing genes for both subunits into proximity that they might be coamplified. It is interesting that an entire chromosome 1 was lost from C^+ cells, while several new marker chromosomes appeared [4].

III. CHARACTERIZATION OF STABLY RESISTANT DERIVATIVES OF C^+

A. Isolation of Stably Amplified Lines

Because of the importance of this enzyme in maintaining cellular homeostasis, we have focused on characterization of the physiological consequences of amplified expression. However, study of such consequences in C^+ was complicated by the requirement for continued selection to maintain the amplified state. Therefore, further characterization of the effects of amplified Na,K-ATPase expression has been performed using stably amplified lines isolated from C^+ during reversion analysis [10].

A population of C^+ was grown in the absence of ouabain for several months.

Cells in the resulting population (C^{-2}) lacked minute chromosomes, but the mean ouabain binding was 2.5-fold higher than that of HeLa. Cloning from C^{-2} revealed two different classes of cells with respect to ouabain binding site number. One class bound ouabain at approximately HeLa levels, while the other class displayed amplified expression of the Na,K-ATPase comparable to C^+. All C^{-2} clones lacked minute chromosomes. Stable resistance was correlated with an abnormal banded region (ABR) on the q (long) arm of one chromosome 17 in two different lines. This ABR was absent in ouabain-sensitive clones of C^{-2} which, like HeLa and C^+, have three normal copies of chromosome 17.

B. Increased Expression of Both α and β Chains in Stably Resistant Lines

Amplified expression of the α chain in stably ouabain-resistant lines was confirmed in Western blots, and increased synthesis of the enzyme was demonstrated by pulse labeling with [^{35}S]methionine [10]. Metabolic labeling with tritiated fucose revealed that expression of a glycoprotein comigrating with the β subunit was eight- to ninefold increased in resistant lines compared to HeLa or a revertant. Preliminary experiments using antibodies produced against synthetic peptides derived from HeLa β subunit sequences also support overexpression of β subunits in stably resistant cells (J. F. Ash, unpublished results).

These characterizations revealed other significant membrane alterations correlated with the isolation by ouabain selection. In particular, expression of several other glycoproteins detected by tritiated fucose labeling is increased in stably resistant lines. Furthermore, increased labeling of two of these glycoproteins was also seen in ouabain-sensitive revertants. Alterations that accompanied the amplification of the Na,K-ATPase but did not revert when amplification was lost probably are secondary changes. Such changes may have permitted cell survival by compensation for some of the initial physiological imbalances caused by the amplified expression of the Na,K-ATPase.

C. Altered Physiological Characteristics of Amplified Lines

In the absence of complicating factors, amplified expression would be expected to lead to an increased sodium electrochemical potential, as a result of the establishment of a lowered steady state $[Na^+]_i$. Such an increase in sodium electrochemical potential would be expected to lead to increased transport rates through sodium-dependent systems. Initial investigations [11] confirmed increased transport rates through the sodium-dependent amino acid transport systems A and ASC in the stably amplified C1 line compared to HeLa. These studies also produced a number of results that complicated interpretation of the consequences of amplification on energization of membrane transport. First, transport through the sodium-independent L system was also increased. Second, under some conditions, transport through the sodium-dependent A and ASC systems was also elevated in the ouabain-sensitive revertant line C4 compared to HeLa. Further experiments suggested that the apparent elevation of system A transport in C4 was due to altered regulation of expression of this transporter, which in HeLa is not significantly expressed unless the cells are starved for amino acids. After starvation in amino acid free medium for 6 hr, system A transport velocities in HeLa and C4 were virtually identical, while transport in C1 was significantly higher. These investigations supported some of the simple predictions we had made about the consequences of amplified Na,K-ATPase expression. But, they also demonstrated that the potential for making meaningful physiological comparisons among parental, amplified,and revertant lines is limited. As pointed out earlier, we have good reason to expect and data to support the selection of secondary changes in cell lines expressing amplification of the Na,K-ATPase genes. Such changes complicate comparisons of amplified lines to either parental or revertant lines.

IV. THE GRADED AMPLIFICATION SERIES STRATEGY: RATIONALE AND APPLICATIONS

A. Heterogeneous Expression of Amplified Genes

Steady state protein expression can be regulated at many levels. After gene amplification, though, feedback mechanisms regulating protein expression may not adequately modulate transcription, translation, or turnover, leading to increased levels of expression of the amplified genes and a new steady state. It is also likely that there will not be a consistent relationship between gene copy number and the level of protein expression. In the absence of continued selection, gradations in expression of amplified genes occur spontaneously. A lack of strict correlation between gene levels and expression in drug-resistant cells has been widely reported [7,12,13]. The variation in expression of amplified genes allows isolation of what we call a graded amplification series (GAS). A GAS is a series of subclonal lines that can serve as a model for studying the consequences of altered expression in a quantitative fashion because their levels of expression of amplified genes have been determined.

B. Exploiting Graded Expression to Model the Potential Consequences of Sodium Pump Regulation

The levels of expression and activity of the Na,K-ATPase must be tightly controlled in most cells to maintain cellular homeostasis. Changes in $[Na^+]_i$ (the cellular parameter most sensitive to the current demand for Na,K-ATPase activity) have been shown to regulate both Na,K-ATPase expression (up-regulation and down-regulation) and Na,K-ATPase activity. Na,K-ATPase expression has also been reported to be under developmental control or to be controlled by hormones including aldosterone, corticosteroids, thyroid hormones, and insulin. In many, if not most, cases such regulation of Na,K-ATPase expression appears to be secondary to alterations in $[Na^+]_i$ (see Ref. 1 for review).

A different kind of regulation of the Na,K-ATPase operating via endogenous factors interacting directly with the enzyme at the external surface of the cell may also be of considerable physiological importance. Studies over the past three decades have suggested that Na,K-ATPase is also regulated by an "endogenous digitalis," a substance that binds to the ouabain binding site and regulates activity in a manner similar to cardiac glycosides (for a review, see Ref. 14). This concept is still controversial, but it is important to understand the potential consequences of this kind of regulation. Regulation of the Na,K-ATPase by such a mechanism could directly modulate $[Na^+]_i$, which would subsequently alter numerous transport and exchange functions that depend on the sodium electrochemical potential.

The GAS strategy provides a novel approach for modeling the steady state consequences of this kind of regulation. Gradations in expression seen in a GAS can model different levels of regulation of Na,K-ATPase activity, producing different levels of $[Na^+]_i$. Comparisons of cellular parameters can then be made among several members of a subclonal series with a precise range of $[Na^+]_i$, rather than only between a parent or revertant and a single mutant.

C. Correlation of Pump Expression with Transport and Steady State Characteristics of a Graded Amplification Series

The GAS strategy has been pursued in our system to study the effect on cell physiology of chronically altering the sodium electrochemical gradient. Central to establishing the validity of this approach was the demonstration that levels of expression of the Na,K-ATPase could be correlated with $[Na^+]_i$. $[Na^+]_i$ was determined by double-label experiments using ^{22}Na and $[^{14}C]$urea (as a measure of cell water volume [15]. As seen in Figure 2a, $[Na^+]_i$ fell as ouabain binding sites per cell increased, while the transmembrane potential roughly becomes more negative (Fig. 2b). These two trends lead to a regular increase in the sodium electrochemical potential, $\Delta\mu_{Na}$, as a function of Na,K-ATPase number (Fig. 2c). The $\Delta\mu_{Na}$ is a key parameter of membrane physiology, providing energy to drive many transport systems, and its variation with Na,K-ATPase level provides the utility of the GAS strategy.

In other studies using six subclones of C1 [7], we found good positive correlations between Na,K-ATPase expression and the uptake rates for tetraphenylphosphonium (TPP^+) and histidine. The former was expected, since TPP^+ distributes across membranes according to the membrane potential. However, histidine transport is not sodium dependent, nor it is driven by the member potential. We proposed that this increased rate of transport reflected exchange

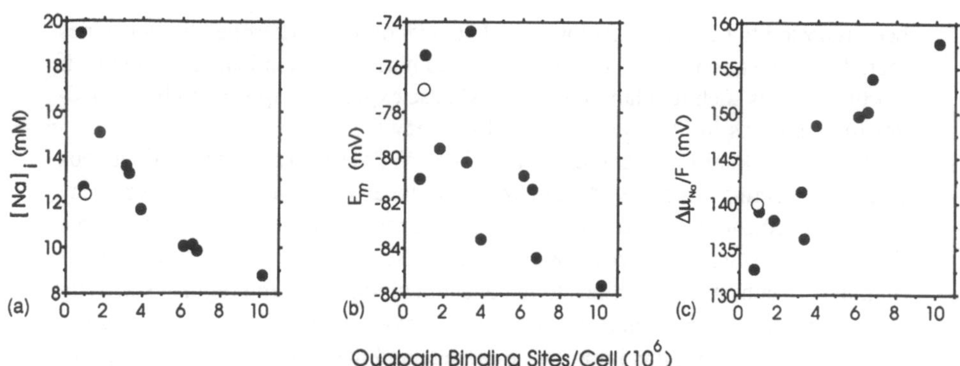

Figure 2 Steady state parameters as a function of Na,K-ATPase levels in 11 cell lines. Specific ouabain binding was used to quantitate enzyme level [4]. The lines include parental HeLa S3 (open circles) and 10 derived lines: seven clonally related members of a graded Na,K-ATPase amplified series, a related line amplified only twofold, and two revertant lines (all shown with solid circles) [15]: (a) internal sodium concentration, (b) transmembrane potential, and (c) sodium electrochemical potential, all varied with ouabain binding sites per cell. Although the membrane potential seen in (b) is not strongly correlated with Na,K-ATPase level, the correlation coefficient for the relationship between sodium electrochemical potential and enzyme level in (c) is 0.84. This result illustrates that a graded amplified series can be used to study steady state physiological relationships. The fact that values vary on both sides of those for parental HeLa cells suggests that secondary changes could accompany Na,K-ATPase amplification. (Figure prepared using data from Table 1 of Ref. 15.)

acceleration, and thus larger intracellular pools of amino acids, which could be exchanged for histidine through the sodium-independent L system.

We examined amino acid pool sizes to test the prediction that they would be larger in lines with increased expression of Na,K-ATPase [15]. Acid-soluble pools of 14 amino acids were quantitated in cell lysates by conventional amino acid analysis. These pools varied over nearly a twofold range as a linear function of $\Delta\mu_{Na}$ in the cells examined. Pool sizes were large for all amino acids in amplified cells, except for the cationic amino acids lysine and arginine.

Finally, we examined what may be one metabolic consequence of the increased amino acid pools associated with amplified expression of the sodium pump, a decrease in the secretion of lactate in amplified lines [15]. Lactate production was negatively correlated with amino acid pool size, especially considering just the Glu + Gln pools (correlation coefficient, 0.95). The oxidation of glucose in HeLa ends with the production of lactate, with no glucose carbon entering the Krebs cycle. Mitochondrial oxidative phosphorylation, which normally produces about 50% of the ATP in HeLa, is driven entirely by oxidation of amino acids, principally Glu and Gln, both of which increased with $\Delta\mu_{Na}$ (Fig. 3b). The decreased lactate production by the amplified cells could represent the inhibition of glycolysis due to increased mitochondrial ATP synthesis driven by larger pools of Glu and Gln: a kind of Pasteur effect produced by gene amplification.

D. Conclusions

The construction of series of amplified subclones differing in enzyme expression allows the simultaneous identification and titration of related cellular activities (Fig. 4). The quantitative relationships we have observed emphasize the coupling between the Na,K-ATPase and transport functions and should eventually help uncover the detailed nature of this coupling. Observations of systematic steady state alterations in cellular parameters as a function of sodium pump expression demonstrate the value of this approach to modeling sodium pump regulation and encourage the development of similar graded amplification series in other cell lines.

The study of gene amplification has focused primarily on the molecular biology of amplification and the role this process plays in drug resistance and cell transformation. Our results demonstrate that gene amplification can also be employed as a tool to manipulate cell physiology.

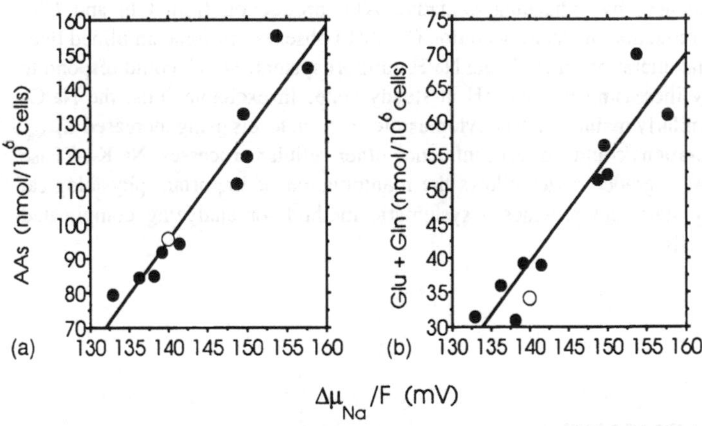

Figure 3 Steady state amino acid levels as a function of sodium electrochemical potential. (a) Acid-soluble levels of 14 amino acids (AAs) were determined for the same lines shown in Figure 2 (correlation coefficient, 0.97). (b) Pools of glutamine (Gln) and glutamate (Glu), which account for nearly 40% of the total, were seen to vary as a function of $\Delta\mu_{Na}$ (correlation coefficient, 0.88). Oxidation of these two amino acids in amplified lines could explain their reduced level of lactate secretion (see text). (Figure prepared using data from Table 1 of Ref. 15 and unpublished data.)

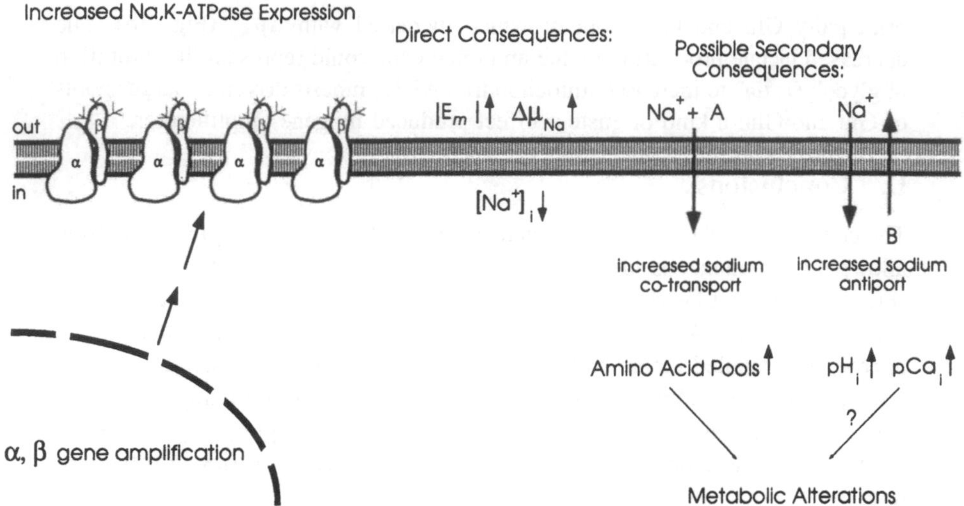

Figure 4 Gene amplification can alter cell physiology. Amplified levels of Na,K-ATPase α and β genes, mRNA, and proteins have been measured in ouabain-resistant cell lines. The increased enzyme expression directly affects membrane properties, decreasing internal sodium levels and increasing the magnitudes of the membrane potential and sodium electrochemical gradient. These alterations, in turn, can modulate other transport and steady state processes. We have demonstrated increases in sodium-dependent amino acid transport and in amino acid pool sizes. These increased pools could then alter metabolic pathways—for instance, by enhancing oxidative ATP production from Gln and Glu, accounting for the reduction in lactate secretion (> 40%) observed in some amplified lines [15]. Other potential alterations include the Na/H antiport system, which could respond to increased $\Delta\mu_{Na}$ by increasing internal pH at steady state. In excitable cells, the Na/Ca antiporter could similarly maintain lower cytoplasmic calcium levels using increased $\Delta\mu_{Na}$. Either of these alterations could, in turn, influence other cellular processes. Na,K-ATPase amplification across a graded series allows the manipulation of important physiological variables at steady state and provides a systematic method for analyzing complicated systems in intact cells.

ACKNOWLEDGMENTS

The authors thank their many colleagues and friends at the University of Utah for their continued help and support and Dr. Robert Levenson and his colleagues at Yale University for their collaboration and plasmids. This work was supported by National Science Foundation grants DCB-8904388 (J. F. A.) and DCB-8901988 (P. G. P.).

REFERENCES

1. *The Na+,K+-Pump Part A: Molecular Aspects*, Part B: *Cellular Aspects* (J. C. Skow, J. G. Nørby, A. B. Maunsbach, and M. Esmann, eds.), *Progress in Clinical and Biological Research*, Vols. 268A and 268B, A. R. Liss, New York, 1988.
2. Soderberg, K., Rossi, B., Lazdunski, M., and Louvard, D. Characterization of ouabain-resistant mutants of a canine kidney cell line, MDCK. *J. Biol. Chem. 258*: 12300–12307 (1983).
3. Canfield, V., Emanuel, J. R., Spickofsky, N., Levenson, R., and Margolskee, R. F. Ouabain-resistant mutants of the rat Na,K-ATPase α2 isoform identified by using an episomal expression vector. *Mol. Cell. Biol. 10*:1367–1372 (1990).
4. Ash, J. F., Fineman, R. M., Kalka, T., Morgan, M., and Wire, B. Amplification of sodium- and potassium-activated adenosinetriphosphatase in HeLa cells by ouabain step selection. *J. Cell Biol. 99*:971–983 (1984).
5. Emanuel, J. R., Garetz, S., Schneider, J., Ash, J. F., Benz, E. J., Jr., and Levenson, R. Amplification of DNA sequences coding for the Na,K-ATPase α-subunit in ouabain-resistant C^+ cells. *Mol. Cell. Biol. 6*:2476–2481 (1986).
6. Mercer, R. W., Schneider, J. W., Savitz, A., Emanual, J., Benz, E. J., Jr., and Levenson, R. Rat brain Na,K-ATPase β-chain gene: Primary structure, tissue-specific expression, and amplification in ouabain-resistant HeLa C^+ cells. *Mol. Cell. Biol. 6*:3884–3890 (1986).
7. Pauw, P. G., and Ash, J. F. Graded amplification of the Na,K-ATPase across a sub-clonal series: Effects on membrane physiology. *J. Cell Physiol. 130*:199–206 (1987).
8. Burns, G. A. P., and Sherman, S. L. Report of the committee on the genetic constitution of chromosome 1. *Human Gene Mapping 10 (1989): Tenth International Workshop on Human Gene Mapping. Cytogenet. Cell Genet. 51*:67–90 (1989).
9. Takeyasu, K., Tamkun, M. M., Renaud, K. J., and Fambrough, D. M. Ouabain-sensitive ($Na^+ + K^+$)-ATPase activity expressed in mouse L cells by transfection with DNA encoding the α-subunit of an avian sodium pump. *J. Biol. Chem. 263*:4347–4354 (1988).
10. Pauw, P. G., Johnson, M. D., Moore, P., Morgan, M., Fineman, R. M., Kalka, T., and Ash, J. F. Stable gene amplification and overexpression of sodium- and potassium-activated ATPase in HeLa cells. *Mol. Cell. Biol. 6*:1164–1171 (1986).
11. Johnson, M. D., Ash, J. F., and Pauw, P. C. Alterations in amino acid transport in Na,K-ATPase amplified HeLa cells. *J. Biol. Chem. 261*:40–43 (1986).
12. Wolman, S. R., Craven, M. L., Grill, S. R., Domin, B. A., and Cheng, Y.-C. Quantitative correlation of homogeneously stained regions on chromosome 10 with dihydrofolate reductase enzyme in human cells. *Proc. Natl. Acad. Sci. USA, 80*:807–809 (1983).
13. Scotto, K. W., Biedler, J. W., and Melera, P. W. Amplification and expression of genes associated with multidrug resistance in mammalian cells. *Science, 232*:751–755 (1986).
14. DeWardener, H. E., and Clarkson, E. M. Concept of natriuretic hormone. *Physiol. Rev. 65*:658–759 (1985).
15. Pauw, P. G., Sheck, R. N., and Ash, J. F. Steady-state physiological variations across a graded series of Na,K-ATPase-amplified cells. *Mol. Cell. Biol. 9*:116–123 (1989).

8

Asparagine Synthetase Gene Amplification in Albizziin-Resistant Hamster and Human Cell Lines

Michael T. Barrett and Irene L. Andrulis *Mount Sinai Hospital and University of Toronto, Toronto, Ontario, Canada*

I. INTRODUCTION

The asparagine synthetase (AS) system has many advantages for somatic cell genetic studies, since the same parental background can be used to select mutants that have various levels of AS activity, ranging from complete absence [1] to 300-fold overproduction [2] of AS. Cells of most types exhibit AS activity, making asparagine a nonessential amino acid. Although *AS* is a housekeeping gene, the activity of wild-type AS is regulatable (for review, see Ref. 3). The enzyme is expressed at a basal level in the presence of asparagine, and the specific activity is increased two- to threefold by removal of asparagine from the medium or decreased charging of tRNA.

Until recently *AS* was believed to act solely as a housekeeping gene for the biosynthesis of asparagine. However, in addition to its function in amino acid biosynthesis, AS may play a role in another aspect of cellular physiology. Basilico and colleagues transfected human DNA into a BHK line (ts11) with a temperature-sensitive mutation in cell cycle control and isolated a gene capable of complementing the defect [4]. From DNA sequence comparisons it is clear that this gene is *AS* [5], suggesting that AS expression is required for cell cycle progression. However it remains to be proved whether the temperature-sensitive mutation in the ts11 cell line is in *AS* itself or in another gene whose function may be rescued by expression of the transfected *AS* gene.

II. ASPARAGINE SYNTHETASE OVERPRODUCING CHINESE HAMSTER OVARY CELL LINES

To isolate mutants of Chinese hamster ovary (CHO) and human cell lines that have alterations in the expression of AS activity, amino acid analogues of the substrates of AS were used [2,6–9]. Mutants resistant to either β-aspartyl hydroxamate (β-AHA), an aspartic acid analogue, or albizziin (Alb), a glutamine analogue, exhibit elevated levels of AS activity (Table 1). However, the level of drug resistance and the amount of AS overproduction that can be achieved differ with the two drugs. Alb^R mutants selected in a single step are up to 40-fold more resistant to Alb than the parental line and express levels of AS activity up to 17-fold greater than that of wild-type cells [2,8]. Using multiple rounds of selection, lines that are highly resistant to Alb can be obtained which exhibit up to 300-fold increases in AS activity [2]. On the other hand, even the most highly resistant multistep β-AHA^r lines express increases in AS activity only 20-fold over that of the parental line [6].

Overexpression of AS in the β-AHA^r and Alb^R mutant lines appears to be due to different mechanisms. In single-step Alb^R cell lines, the *AS* gene was found to be amplified from three- to ninefold over the parental line [10], while multistep lines had up to 82-fold increases in gene copy number. Alb^R mutants share many characteristics with drug-resistant lines isolated in other amplification systems, including selection protocol, codominance in hybrids, greatly increased copy number in highly resistant cells, and karyotypic changes. In contrast, using β-AHA, which affects the same target, we have not been able to isolate *AS* amplification mutants. β-AHA resistance appears to be due to increases in *AS* mRNA [10] and protein [7] without amplification of the DNA [10].

Table 1 Asparagine Synthetase Overexpression and Amplification in CHO and Human Mutant Lines

Cell line	Selective drug	AS activity (fold increase)	Relative DNA copy number increase
Wild-type CHO	None	1	1
Single-step Alb^R	Albizziin	11–17	3–9
Multistep Alb^R	Albizziin	≤300	≤82
β-AHA	β-AHA	6–11	1
Parental human	None	1	1
Multistep HT4	Albizziin	17	≤10
L-$Asparaginase^r$	L-Asparaginase	~10	1

III. ASPARAGINE SYNTHETASE OVEREXPRESSING HUMAN CELL LINES

Certain tumor cells are asparagine auxotrophs, since they exhibit little or no AS activity and rely on the surrounding medium as a source of exogenous asparagine. These tumor cells can be selectively killed by treatment with the chemotherapeutic drug L-asparaginase, which hydrolyzes circulating asparagine. Childhood acute lymphoblastic leukemia (ALL) represents one of the most sensitive of such human malignancies, and asparaginase is now routinely used for treatment of children with this disease. The mechanism responsible for the lack of activity of AS in the sensitive leukemic cells and overproduction in the asparaginase-resistant variants is of clinical importance for the treatment of ALL patients both initially and during relapse. To study the development of drug resistance, human cell lines resistant to L-asparaginase or albizziin were isolated by multistep selection of HT1080 fibrosarcoma and MIA PaCa-2 pancreatic carcinoma cells [9]. Mutants are cross-resistant to both drugs, but more resistant to the drug used for selection. L-Asparaginase resistant and Alb-resistant cell lines express elevated levels of AS activity and protein, up to 17-fold over that of the parental cells [9] (Table 1). As observed for CHO cells, enzyme overproduction is due to gene amplification in the human Alb^R cells. However, L-asparaginase resistant cells do not contain amplified copies of *AS* [9].

In human Alb-resistant cells, the level of amplification observed was low relative to that obtained in CHO cells (Table 1), even though selection in progressively increasing concentrations of albizziin was imposed for more than a year. It is possible that the differences in maximal degree of amplification observed in human and hamster cells may reflect differences in the chromosomal environment of the *AS* gene or in the biochemical machinery involved in the process of amplification.

IV. ALBIZZIIN AND GENE AMPLIFICATION

Gene amplification has been shown to occur in the absence of the selective agent (for review, see Ref. 11), suggesting that amplification can occur spontaneously in cells adapted to culture and that selection identifies these events. Our results on AS amplification in CHO and human cells suggest that this may not always be the case. Cell lines selected in β-AHA or L-asparaginase overproduce AS, but this is not due to amplification of the gene, even though these cell lines are cross-resistant to Alb. Only cells directly selected in Alb contain amplified copies of the *AS* gene. If gene amplification is a process that occurs continuously and independently of the selective agent, one would expect to observe amplification in the gene in cells selected in L-asparaginase or β-AHA. One explanation might be that the effect of Alb is related to its ability to inhibit DNA synthesis. One major difference between Alb and β-AHA and L-asparaginase is that Alb as a glutamine analogue inhibits two enzymes in purine biosynthesis, PRPP and FGAR amidotransferases, (5-phosphoribosyl-1-pyrophosphate and *N*-formylglycinamide ribonucleotide) thus affecting DNA synthesis; whereas β-AHA and L-asparaginase do not.

V. ASPARAGINE SYNTHETASE OVEREXPRESSION IN EMBRYONIC STEM CELLS

Although gene amplification does occur as a normal event during differentiation in *Xenopus* and development in *Drosophila*, this process is probably rare in "normal" human cells [12,13]. It is possible that gene amplification requires some form of genomic instability, consistent with the ability to select amplification mutants in established cell lines and the common observation of amplified genes in tumors. The availability of the embryonic stem (ES) cell system offers a unique method of examining the question of amplification in "normal" cells. These cells can be grown in culture and subjected to selection procedures. The important features of the system are that the genetically altered ES cells will remain pluripotent, they retain a normal karyotype if grown properly, and they can be introduced into blastocysts to produce chimeric mice. We are using the ES cell system to examine whether Alb-resistant lines, with *AS* gene amplification, can be derived from ES cells. We can then ask whether the Alb-resistant ES cells maintain a normal karyotype and are capable of reconstituting a mouse, or whether we have selected for cells that have undergone events such that the cells are no longer "normal" and cannot function to produce a chimeric mouse. It is possible that only ES cells that have undergone gross chromosomal alterations will have amplified the *AS* gene. We have worked out selection conditions for Alb-resistant ES cells and have obtained a number of independent clones of Alb-resistant ES cells. Although some of the mutants overexpress *AS* mRNA, we have not yet observed any ES amplification mutants (Wei and Andrulis, unpublished observations).

VI. DNA METHYLATION OF THE ASPARAGINE SYNTHETASE GENE

Studies on expression of AS in the CHO mutants indicated that Alb^R mutants had increases in expression of AS that exceeded the increases in DNA copy number, and β-AHA^R lines had elevated levels of mRNA without amplification [10]. We examined the DNA methylation status of the *AS* gene in the mutants, since methylation of cytosines affects gene expression in some systems, and studies by Harris and colleagues [15,16] had suggested the involvement of DNA methylation in the regulation of AS. The latter found that rat and hamster cell lines, which are auxotrophic for asparagine and express no AS activity, could be reverted to prototrophy (*AS*$^+$) at a high frequency by treatment with 5-azacytidine [14,15], which inhibits DNA methylation. We examined the importance of DNA methylation at the molecular level by determining the DNA methylation pattern of the gene from wild-type cells, from asparagine-requiring N3 cells, and from the drug-resistant mutants. We also selected a number of *AS*+ spontaneous revertants of this line to produce an isogenic series of cell lines going from WT (*AS*+) → N3 (*AS*−) → N3 revertants (*AS*+), which differ only in the expression of the *AS* gene.

In CHO cells the gene for *AS* was found to contain a cluster of potential sites for CpG methylation in a 1 kb region surrounding the first exon. Fourteen of the sites that can be assayed for methylation by *Msp* I/*Hpa* II digestions were found in this region, with an additional 9 *Msp* I sites spread throughout the remainder of the gene. The level of expression of AS activity showed a direct correlation with the extent of methylation of a subset of *Msp* I sites found in the 5′ region of the gene [10] (Fig. 1). The rest of the gene was completely methylated in most cell lines. Wild-type cells, which express a basal level of AS activity, were partially demethylated in the 5′ region. Asparagine-requiring N3 cells, which lacked detectable *AS* mRNA, were methylated throughout the entire gene, whereas the spontaneous revertants of N3 exhibited extensive hypomethylation of the *AS* gene. Alb^R cell lines were extensively demethylated in the 5′ region. In addition, overexpression of AS in β-AHA^R lines was also correlated with DNA hypomethylation [10]. The data indicate that although the Alb^R lines were isolated in a single step of selection, two events may be involved in the overproduction of AS in these lines. We have shown that increased expression was due to 5′ regional demethylation as well as amplification of the *AS* gene in each of these mutants. In contrast, β-AHA resistance was associated with demethylation of the *AS* gene without amplification. The results indicate that DNA methylation affects AS expression and suggest that an increase in transcription of the gene may be a necessary, but not sufficient, initial event for gene amplification.

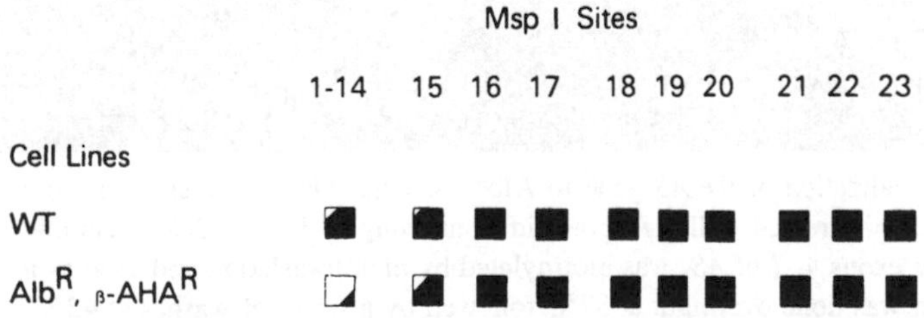

Figure 1 Summary of the DNA methylation status of the 23 *Msp* I sites in the *AS* gene from the wild-type (WT) and AS over producing Alb^R and β-AHA^R lines.

VII. IN SITU HYBRIDIZATION

To address the question of mechanism of amplification, it is important to determine the location of the amplified sequences, since some predictions concerning models can be derived from this information. In some drug resistance systems the amplified genes are on the same chromosome as the resident gene; however, in many others there is movement associated with amplification which can involve deletion and extrachromosomal events (for review, see Ref. 16). Cytogenetic analysis of the CHO Alb^R mutants showed that each line had specific karyotypic

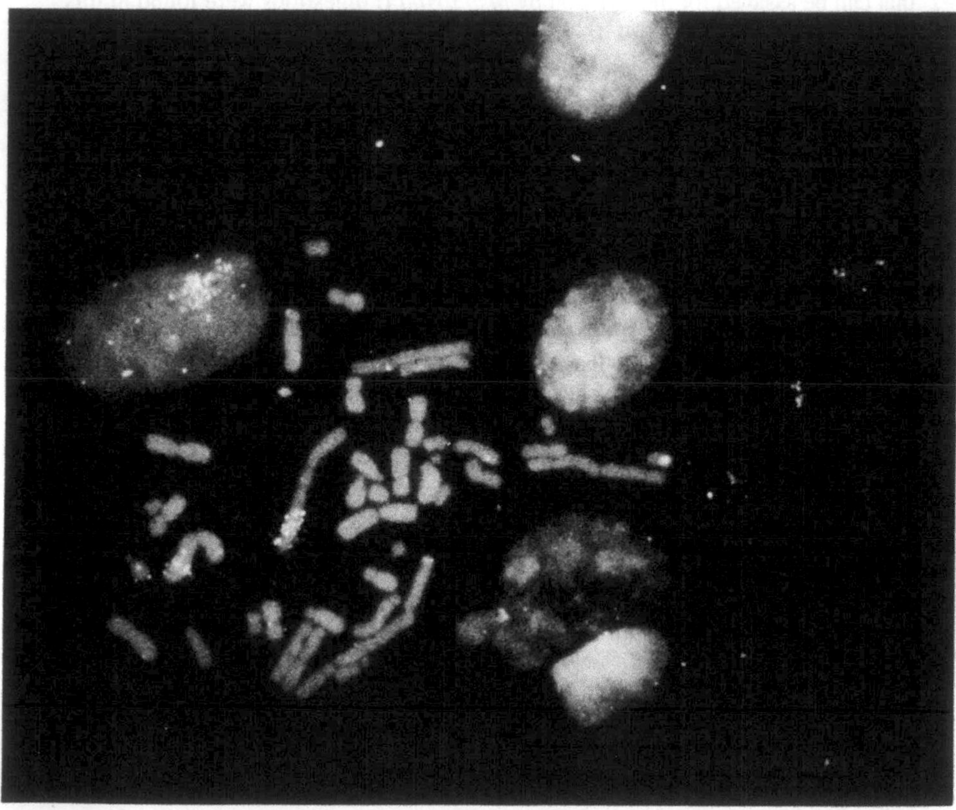

Figure 2 In situ localization of the *AS* gene in Alb^R 24 cells. Metaphase spreads were prepared from Colcemid-treated cells. A plasmid containing 6 kb of CHO genomic sequences, including exons 4–7 of *AS*, was biotinylated by nick translation and used as a probe. Hybridization was done overnight at 37°C followed by a series of washes at 42°C. Signal was detected and amplified using avidin conjugated to fluorescein and biotin-conjugated goat antiavidin. Slides were stained in 4′6-Diamidino-2-phenyl-indole dihydrochloride (DAPI) followed by propidium iodide and viewed under fluorescence.

alterations, including breaks, rearrangements, and other abnormal staining regions as well as homogeneously staining regions (HSRs) [2]. In each mutant these abnormalities affect the long arm of chromosome 1 [2], although in some of the highly resistant lines other chromosomes are also altered. It is possible that the native CHO gene is located on chromosome 1 and that the amplification event is associated with a chromosomal break in this region. To determine whether the hamster *AS* amplification units have remained at the site of the native gene or have moved to other locations, we have been performing in situ hybridizations to map the *AS* genes. In the multistep Alb^R 24 line, which has an HSR on chromosome 1, this region contains the amplified sequences (Fig. 2). The most highly resistant cell line, Alb^R 52, contains an HSR on chromosome 7 which hybridizes with *AS*-specific probes. This analysis is being extended to the single-step Alb^R mutants and to wild-type CHO.

VIII. ASPARAGINE SYNTHETASE AMPLIFICATION UNITS

To address the question of the size and complexity of the amplification units in the single-step lines and how they compare with the amplicons in multistep lines, we have used cross-field gel (XFG) electrophoresis for the long-range analysis. A large (1.5 Mb) *Not* I fragment is detected in wild-type cells and in all the single-step lines, indicating that most of the amplicons in these cell lines extend beyond this size (Fig. 3). One single-step line (Alb^R 27) is the single exception, since it also contains a second fragment of 800 kb. The multistep lines showed a limited heterogeneity in this region. Alb^R 24 and Alb^R 43 have patterns similar to the single-step lines while in Alb^R 52, *AS* maps predominantly to an 800 kb *Not* I piece, suggesting that most of its amplicons are smaller than the other cell lines.

A feature of gene amplification is the occurrence of novel restriction fragments created at amplicon–amplicon or amplicon–wild-type boundaries. Using the same XFG analysis, we have identified a unique novel joint (NJ) in Alb^R 27 on a 50 kb *Nru* I fragment (Fig. 4). We are presently cloning this region to determine the structure and the origin of this unique fragment.

To extend the long-range map in these cell lines, we have taken advantage of a cluster of rare cutting enzyme sites located in the CpG island around exon 1 [10]. Enzymes that cut within this site allow simultaneous mapping in 5′ and 3′ directions using probes that span or flank the enzyme sites. A 1.2 Mb *Sac* II fragment seen in wild-type and the single-step lines extends 3′ from *AS* (Fig. 5). Preliminary studies indicate that there is a 1.5 Mb 5′ piece in both the wild-type and single-step mutants. The multistep lines show limited heterogeneity: for example, the Alb^R 52 DNA contains an 800 kb *Sac* II fragment, which hybridizes with both 5′- and 3′-specific probes, indicating that a predominant amplicon is contained on this fragment.

These results indicate that there are very large amplicons containing the *AS*

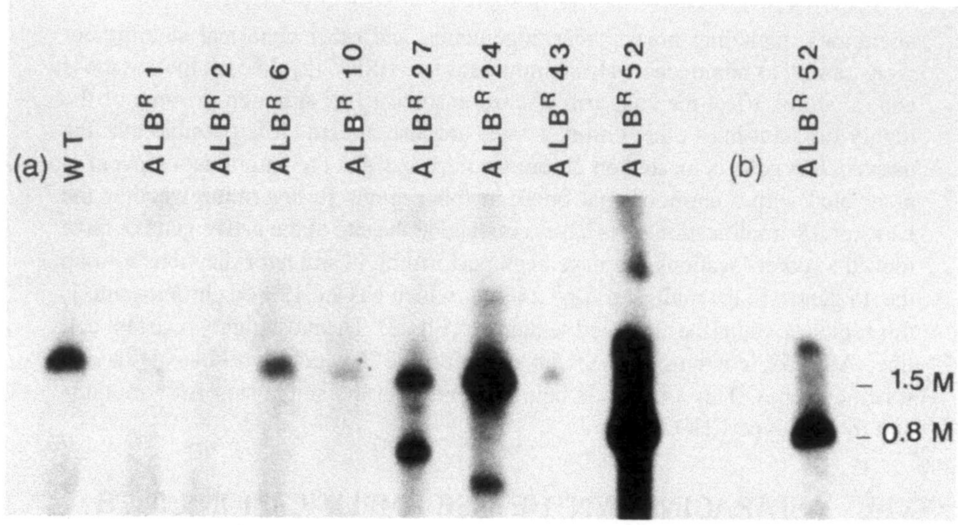

Figure 3 Southern blot analysis of *Not* I restricted DNA from wild-type (WT) and albizziin, resistant (Alb^R) cell lines probed with a genomic *AS* probe. Agarose plugs containing DNA (approximately 3 μg) were digested with 30 units of *Not* I, then run 168 hr at 42 volts 15°C in 0.5 TBE (Tris-Borate-EDTA) with constant 30 min switch times followed by 48 hr at 42 volts 15°C with constant 5 min switch times. Exposures were overnight (A) and for 8 hr (B). Size markers were chromosomes from yeast strains *S. cerevisiae* AB1380 and *H. wingii*. Differences in signals may be due to quality of DNA plugs.

gene in the single-step CHO cell lines. Since we have not detected the end of the amplification units, we cannot yet estimate the sizes. Nonetheless, our preliminary analysis suggested that the CHO *AS* amplification units are much more homogeneous than those observed in other systems. We detect patterns of limited heterogeneity, but this is in the highly selected multistep mutants. As more amplification systems are examined, we should be able to ascertain whether these features are unique to the *AS* gene, the nature of the selection, or the parental CHO cell line from which the mutants were isolated.

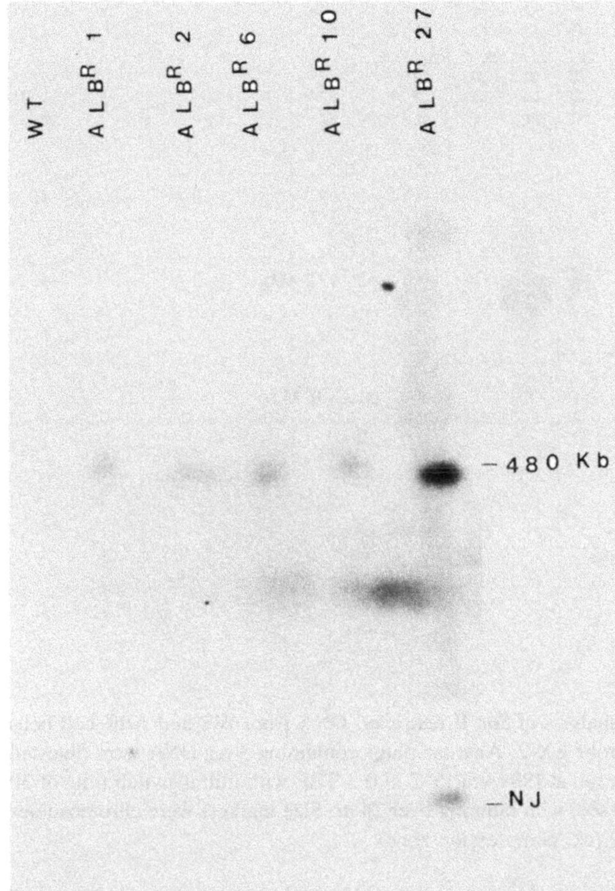

Figure 4 Southern blot analysis of *Nru* I restricted DNA from WT and Alb^R cell lines probed with EX-2. Agarose plugs containing 3 μg DNA were digested with 30 units of *Nru* I, then run at 190 volts 15°C in 0.5 TBE with initial switch time of 30 sec, final switch time of 80 sec, with ramping over 24 hr. Size markers were chromosomes from *S. cerevisiae* AB1380 and λ concatemers (NJ, novel joint).

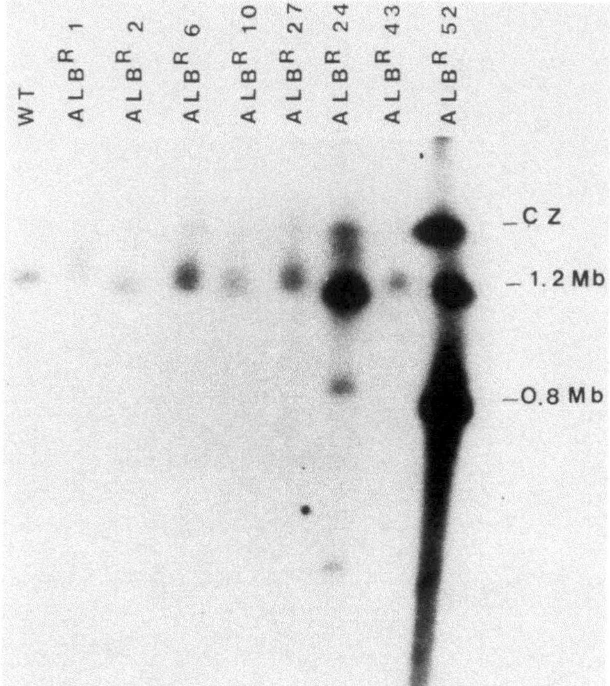

Figure 5 Southern blot analysis of *Sac* II restricted DNA from WT and Alb^R cell liens probed with *AS* genomic probe EX-2. Agarose plugs containing 3 μg DNA were digested with 30 units of *Sac* II then run at 190v/95i 15°C in 0.5 TBE with initial switch time of 30 sec, final switch time of 80 sec, with ramping over 24 hr. Size markers were chromosomes from *S. cerevisiae* AB1380 (cz, compression zone).

ACKNOWLEDGMENTS

This work was supported by the Medical Research Council and the National Cancer Institute of Canada.

REFERENCES

1. Waye, M. M. Y., and Stanners, C. P. Isolation and characterization of CHO cell mutants with altered asparagine synthetase. *Somatic Cell Genet.* *5*:625–639 (1979).
2. Andrulis, I. L., Duff, C., Evans-Blackler, S., Worton, R., and Siminovitch, L. Chromosomal alterations associated with overproduction of asparagine synthetase in albizziin-resistant Chinese hamster ovary cells. *Mol. Cell. Biol.* *3*:391–398 (1983).
3. Andrulis, I. L. Regulation and amplification of asparagine synthetase, in *Molecular Cell Genetics* (M. M. Gottesman, Ed.), Wiley, New York, 1985, pp. 489–518.

4. Greco, A., Gong, S. S., Ittman, M., and Basilico, C. Organization and expression of the cell cycle gene, ts11, that encodes asparagine synthetase. *Mol. Cell. Biol.* *9*:2350–2359 (1989).
5. Andrulis, I. L., Chen, J., and Ray, P. N. Isolation of human cDNAs from asparagine synthetase and expression in Jensen rat sarcoma cells. Mol. Cell Biol. 7:2435–2443 (1987).
6. Andrulis, I. L., and Siminovitch, L. Isolation and characterization of Chinese hamster ovary cell mutants resistant to the amino acid analog β-aspartyl hydroxamate. *Somatic Cell. Genet.* *8*:533–545 (1982).
7. Gantt, J. S., Chiang, C.-S., Hatfield, G. W., and Arfin, S. M. Chinese hamster ovary cells resistant to β-aspartyl hydroxamate contain increased levels of asparagine synthetase. *J. Biol. Chem.* *255*:4808–4813 (1980).
8. Andrulis, I. L., Evans-Blackler, S., and Siminovitch, L. Characterization of single-step albizziin-resistant Chinese hamster ovary cell lines with elevated levels of asparagine synthetase activity. *J. Biol. Chem.* *260*:7523–7527 (1985).
9. Andrulis, I. L., Argonza, R., and Cairney, A. E. L. Molecular and genetic characterization of human cell lines resistant to L-asparaginase and albizziin. *Somatic Cell Mol. Genet.* *16*:59–65 (1990).
10. Andrulis, I. L., and Barrett, M. T. DNA methylation patterns associated with asparagine synthetase expression in asparagine-overproducing and -auxotrophic cells. *Mol. Cell. Biol.* *9*:2922–2927 (1989).
11. Stark, G. R., and Wahl, G. M. Gene amplification. *Annu. Rev. Biochem.* *53*:447–491 (1984).
12. Wright, J. A., Smith, H. S., Watt, F. M., Hancock, M. C., Hudson, D. L. and Stark, G. R. DNA amplification is rare in normal human cells. *Proc. Natl. Acad. Sci. USA*, *87*:1791–1795 (1990).
13. Tlsty, T. D. Normal diploid human and rodent cells lack a detectable frequency of gene amplification. *Proc. Natl. Acad. Sci. USA*, *87*:3132–3136 (1990).
14. Sugiyama, R. H., Arfin, S. M., and Harris, M. Properties of asparagine synthetase in asparagine-independent variants of Jensen rat sarcoma cells induced by 5-azacytidine. *Mol. Cell. Biol.* *3*:1937–1942 (1983).
15. Harris, M. Induction and reversion of asparagine auxotrophs in CHO-K1 and V79 cells. *Somatic Cell. Mol. Genet.* *12*:459–466 (1986).
16. Stark, G. R., Debatisse, M., Giulotto, E., and Wahl, G. M. Recent progress in understanding mechanisms of mammalian DNA amplification. *Cell*, *57*:901–908 (1989).

9

Overproduction of Damage Recognition Proteins in Human Cells Resistant to Cisplatin and Ultraviolet Radiation

Chuck C.-K. Chao *Chang Gung Medical College, Taoyuan, Taiwan, Republic of China*

I. INTRODUCTION

cis-Diamminedichloroplatinum(II), or cisplatin, is widely used for the treatment of a variety of tumors. In rare cases, however, tumor cells become resistant to it, especially following long-term exposure to the drug, causing a major limitation to cancer treatment. These clinically selected resistant cells can be mimicked in cell cultures by the stepwise exposure of cells to increasing concentrations of cisplatin. Indirect evidence suggests that cisplatin resistance (CPR) can be due to enhanced cellular ability in eliminating cisplatin-DNA adducts, including intra- and interstrand crosslinks (reviewed in Ref. 1). This material is supported by the observation that reactivation of genes on transfected plasmid DNA is enhanced in CPR cells [2–4]. The repair of cisplatin-modified DNA (CMD) involves a variety of enzymes and accessory proteins that presumably must have access to the damaged DNA for effective repair to occur. It is reasonable to think that one or more of these factors can be identified through DNA binding activity. In fact, it has been documented that nuclear extracts isolated from mammalian and yeast mutants in DNA repair fail to interact with damaged DNA [5,6]. Thus, for clinical applications, the DNA damage mediated sensitivity and resistance of cancer cells to cisplatin may potentially be monitored using a DNA binding assay.

To test this possibility, we applied the gel mobility shift assay [7] and identified CMD as well as UV-modified DNA (UVMD) recognition proteins from the cells. The results indicate that there is overproduction of damage recognition proteins

(DRPs) and enhancement of DNA repair (reflected from plasmid reactivation) in CPR cells. Our findings also suggest these DRPs as potential indicators of sensitivity or resistance of human cells to agents capable of damaging DNA. Typical results are summarized below.

II. ENHANCEMENT OF PLASMID REACTIVATION IN RESISTANT CELLS

pRSV*cat* plasmid carrying a known degree of cisplatin or UV damage was introduced into cells by electroporation as described elsewhere [3]. During the period of optimal expression of the transfected plasmids, cell extracts were prepared for *CAT* assays. The *CAT* activity reflects the level of DNA repair of the cells because its expression depends on the intactness of the plasmid. *CAT* activity decreases with increasing cisplatin damage (Fig. 1a, lanes 2–5; refer to the figure legend for details). The calculated *CAT* activity (relative to undamaged pRSV*cat* plasmid) is shown in Figure 1b. There is a threefold enhancement of *CAT* activity (cf. 50% inhibition) in resistant cells compared to parental cells. A similar experimental strategy was also carried out for UV damage. As shown in Figure 2, a greater *CAT* activity expressed from UV-damaged pRSV*cat* was detected in CPR cells. Quantitation of the relative *CAT* activity is presented in Figure 2b. A twofold increase of *CAT* activity was observed in CPR cells. The data indicate that host reactivation of cisplatin- or UV-damaged plasmid was enhanced in CPR cells, suggesting an increase in DNA repair capability in the resistant cells. The advantages of this strategy are (1) the repair machinery of the tested cells remains intact, and (2) the degree of DNA damage inflicted in the plasmid in vitro is known. However, this method does not fully reflect the damage and repair environment inside the cells, because the "naked" plasmid DNA does not truly represent the chromosomal DNA, which in an actual case is complexed with cellular proteins.

III. IDENTIFICATION OF DAMAGE RECOGNITION PROTEINS

The recognition of damaged DNA is a key step in DNA repair of the cells. DNA–protein binding is potentially useful for the identification of DRPs. The dG.dC-rich DNA fragment f103 (103 bp) and the dA/dT-rich f130 (130 bp) were treated with cisplatin and UV, respectively, and used as a probe for gel mobility shift assay [7]. A CMD DRP was identified in the HeLa nuclear extracts (indicated with arrowhead in Fig. 3). The specificity of this DRP was demonstrated by competition assay (Fig. 3a). For example, significant competition was seen by the addition of 10-fold competitor (i.e., 3 ng cisplatin-modified f103) (cf. lanes 3 and 5). Figure 3b shows the dose–response behavior of the damage binding. The CMD binding increases with increasing damage, as indicated on top. A UVMD DRP was also

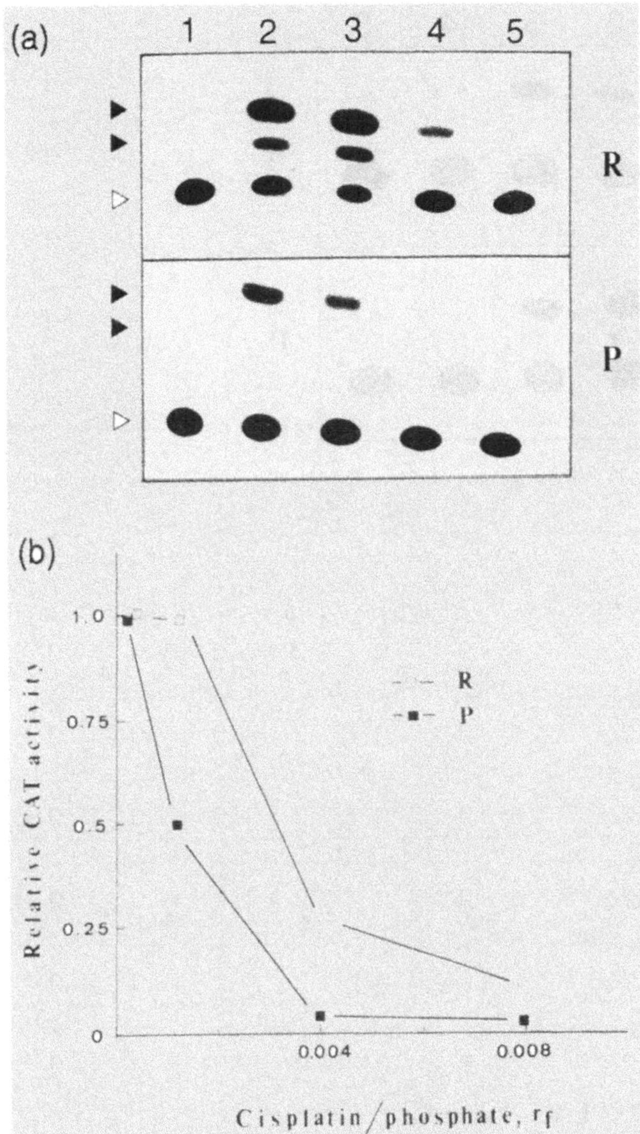

Figure 1 Host cell reactivation of cisplatin-treated pRSV*cat* plasmid in HeLa-CPR (R) and parental (P) cells. (a) *CAT* activity shown by silica thin-layer chromatography: lane 1, mock transfection; lanes 2–5, plasmid with cisplatin treatment r_f = 0, 0.0008, 0.004, and 0.008, respectively. Open arrowhead, chloramphenicol; solid arrowhead, modified chloramphenicol. (b) Relative *CAT* activity as a function of r_f.

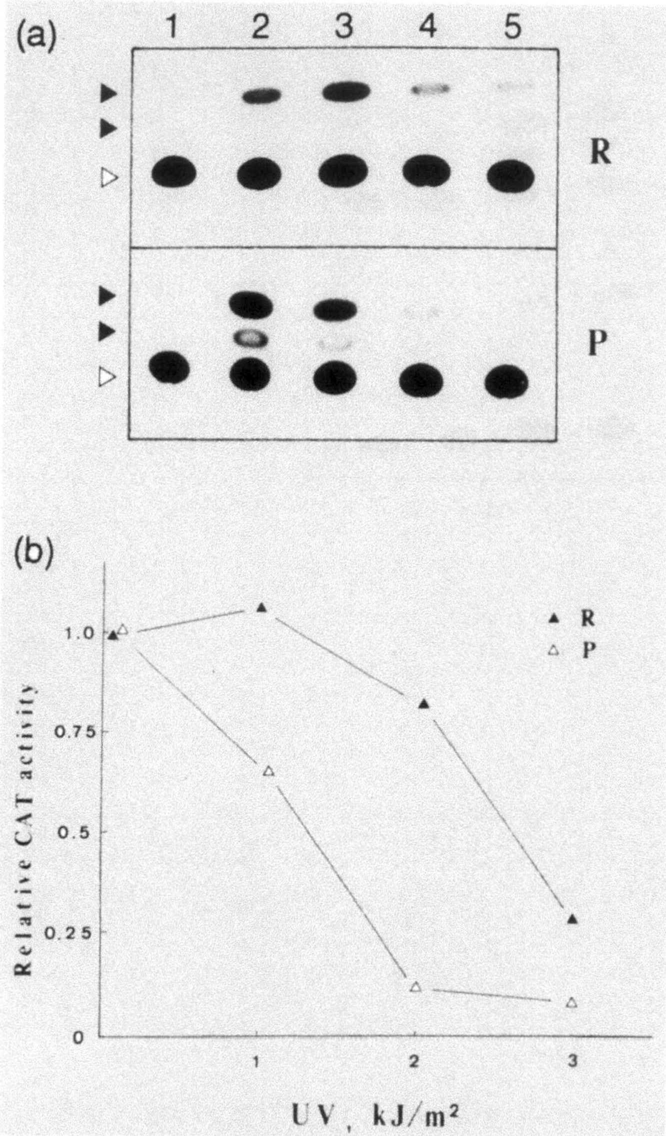

Figure 2 Host cell reactivation of UV-irradiated pRSV*cat* plasmid in HeLa-CPR (R) and parental (P) cells. (a) *CAT* activity shown by silica thin-layer chromatography: lane 1, mock transfection; lanes 2–5, plasmid with UV irradiation 0, 1, 2, and 3 kJ/m^2, respectively. Open arrowhead, chloramphenicol; solid arrowhead, modified chloramphenicol. (b) Relative *CAT* activity as a function of UV fluence.

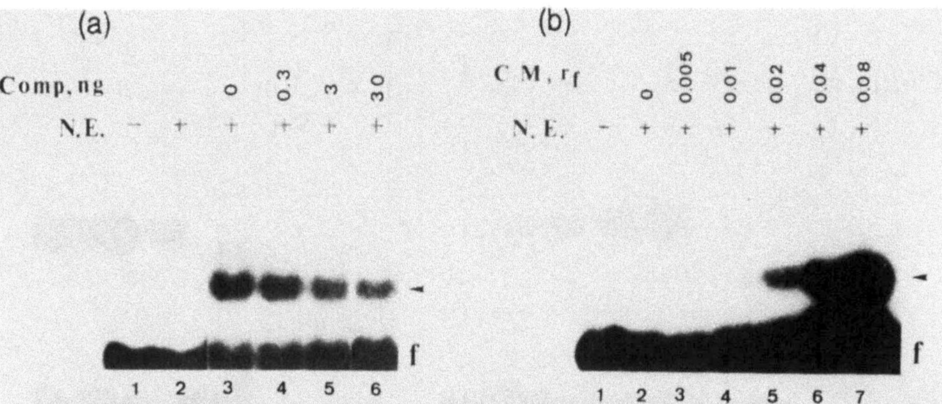

Figure 3 Specificity of CMD DRP. (a) Competition of CMD binding activity with indicated amount of cold CMD. Typical reaction includes 2 μg of HeLa nuclear extracts (N.E.), 2 μg of poly(dI.dC), and 0.3 ng of cisplatin-treated f103 probe ($r_f = 0.08$). Control lanes are without N.E. (lane 1) and untreated f103 probe (lane 2). (b) Dose–response of CMD binding activity. N.E. was incubated with f103 probe carrying the indicated amount of cisplatin modification. Control lane is untreated f103 probe (lane 1); f, unbound free DNA probe; arrowhead, protein-bound DNA probe.

identified. Competition and dose–response behavior of UVMD DRP are shown in Figure 4. For example, 10-fold competitor (i.e., 3 ng UV-modified f130) significantly competes off the UVMD binding (Fig. 4a: cf. lanes 3 and 5). UVMD binding increases with increasing UV damage (Fig. 4b). The results indicate that the two identified DRPs are specific for CMD and UVMD.

IV. OVERPRODUCTION OF DAMAGE RECOGNITION PROTEINS IN RESISTANT CELLS

It will be of great interest to see whether the level of expression of DRPs is associated with cellular sensitivity to DNA damage. Cells featuring different extents of DNA repair capability and sensitivity to cisplatin and UV were assayed for their nuclear DRPs. As shown in Figure 5a, CMD DRP varies slightly between cells. On the other hand, there is a three- to fourfold increase in the level of UVMD DRP in CPR cells compared to the parental cells (Fig. 5b: cf. lanes 3 and 4). A revertant cell line of CPR cells (lane 5) showed a DRP level similar to parental cells (lane 3). Repair-competent VA13 cells (a WI-38 derivative) and repair-defective cells from XP groups E and A were also included in this study. UVMD DRPs were greatly reduced in XP cells compared to VA13 or HeLa cells. The data suggest UVMD DRP as a potential indicator of the sensitivity or resistance of cells to DNA

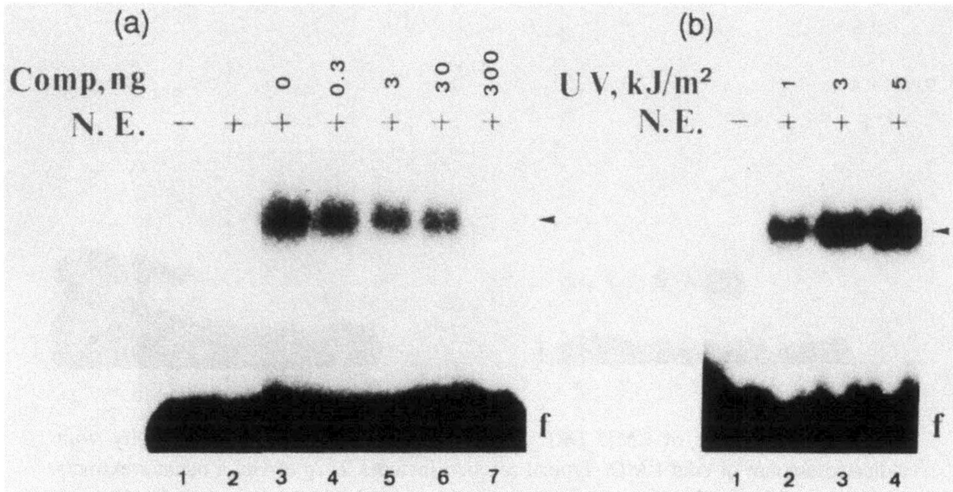

Figure 4 Specificity of UVMD DRP. (a) Competition of UVMD binding activity with indicated amount of cold UVMD. Typical reaction includes 2 μg of HeLa nuclear extracts (N.E.), 2 μg of poly(dI.dC), and 0.3 ng of 2 kJ/m^2 UV-irradiated f130 probe. Control lanes are without N.E. (lane 1) and unirradiated f130 probe (lane 2). (b) Dose–response behavior of UVMD binding activity. N.E. was incubated with UV-modified f130 probe carrying the indicated fluence of UV. Control lane is unirradiated f130 probe (lane 1). Symbols are the same as in Figure 3.

damage. However, we cannot explain the lack of association of the CMD DRPs with cellular sensitivity to DNA damage, and further study is needed (see Section V). The results also suggest that the mechanism of cisplatin-induced DNA damage and repair is more complicated than that for UV.

V. DISCUSSION

We have shown an enhanced plasmid reactivation in human cells resistant to cisplatin and UV. XP cells also display reduced host cell reactivation of plasmid DNA damaged by either cisplatin or UV. Like several other investigators (e.g., see Ref. 8), we have shown that UVMD DRP is associated with cellular resistance to UV damage. One may then speculate about DRPs as potential indicators of the sensitivity or resistance of cells to DNA-damaging agents. However, evidence showing the association of CMD DRP with CMD repair and sensitivity of cells to DNA damage is contradictory. This is not surprising because it is known that XP cells show differential sensitivity to different DNA-damaging agents (reviewed in Ref. 9). The response of XP cells to these DNA-damaging agents depends on the type of damage to DNA and the extent to which repair of the damage requires

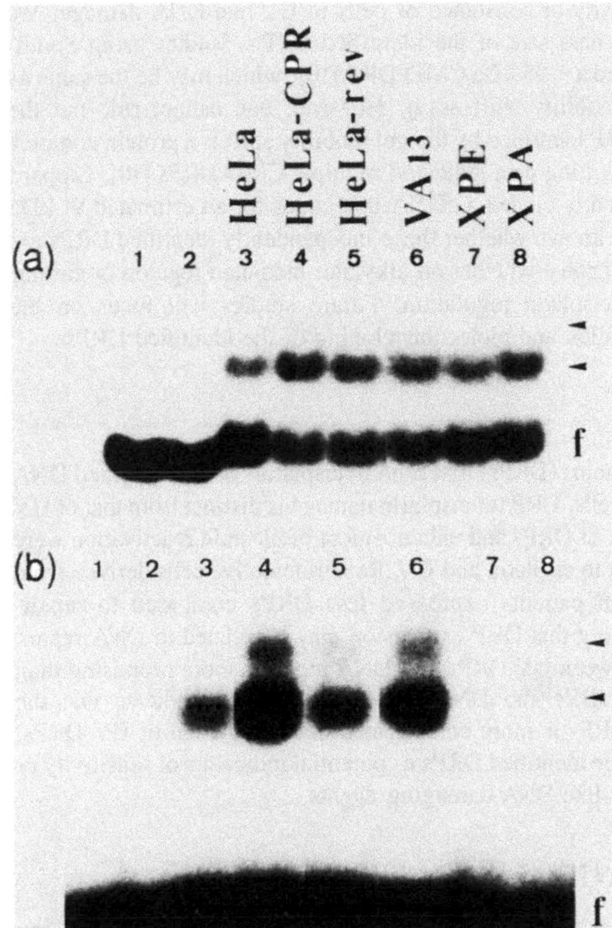

Figure 5 Overexpression of DRPs in resistant cells. (a) CMD binding activity. (b) UVMD binding activity. Typical binding reaction includes 2 μg of nuclear extracts of parental (HeLa), resistant (CPR), revertant (HeLa-rev), VA13, XP group E (XPE), or group A (XPA) cells. Symbols are the same as in Figure 3.

enzymes that are deficient in XP cells. Different DNA-damaging agents inflict DNA lesions of various kinds, each of which might be repaired by a different type of excision repair. Since specific knowledge of the DNA damage recognition and DNA repair pathways involved in cisplatin resistance is still limited, we can only suggest that the excision repair of damage induced by UV involves the UVMD DRP(s) that are deficient in XP cells, whereas excision repair of cisplatin-induced lesions bypasses these deficiencies. Therefore, we propose that DRPs may serve as indicators of the sensitivity or resistance of cells to UV-like DNA damage. We do not know the exact mass size of the identified DRPs. Studies using Southwestern blotting indicated a ~95 kDa CMD DRP [10], which may be the same as that identified by gel mobility shift assay. However, one cannot rule out the possibility that CMD DRP identified by the gel mobility shift is a protein complex because Southwestern blotting data indicated multiple CMD DRPs [10]. Lippard and colleagues have recently cloned a cDNA that codes for an estimated 91 kDa CMD DRP [11]. It is not known whether these independently identified DRPs are inducible. It is also not known whether an alkylator-mediated regulon or similar system is involved in cisplatin regulation. Future studies will focus on the biochemical characterization and molecular cloning of the identified DRPs.

VI. SUMMARY

Damage-recognition proteins (DRPs) that bind to cisplatin- or UV-modified DNA were identified in HeLa cells, DRP for cisplatin damage is distinct from that of UV damage. Overproduction of DRPs and enhancement of plasmid reactivation were detected in cells resistant to cisplatin and UV. Repair-defective cells derived from xeroderma pigmentosum patients expressed less DRPs compared to repair-competent cells, suggesting that DRP expression may be related to DNA repair. However, correlation between UV DRPs and DNA repair is more promising than that between cisplatin DRPs and DNA repair. The results indicate that the regulation of cisplatin DRPs is more complicated than regulation of UV DRPs. These data also suggest the identified DRPs as potential indicators of sensitivity or resistance of cells to UV-like DNA-damaging agents.

ACKNOWLEDGMENTS

The author thanks Drs. P. C. Huang (Johns Hopkins University), Victor Ling (Ontario Cancer Institute), and Mary Helen L. Young for helpful discussions. He is also grateful to Drs. Delon Wu and Chau-Hsiung Chang for encouragement and S.-L. Huang and P.-W. Cheng for technical assistance. This work was supported in part by grants from Chang Gung Medical College (CMRP256 and NMRP031) and the National Science Council, Republic of China (NSC79–0412–B182–50).

REFERENCES

1. Hospers, G. A. P., Mulder, N. H., and De Vries, E. G. E. Mechanisms of cellular resistance to cisplatin. *Med. Oncol. Tumor Pharmacother.* *5*:145–151 (1988).
2. Sheibani, N., Jennerwein, M. M., and Eastman, A. DNA repair in cells sensitive and resistant to *cis*-diamminedichloroplatinum(II): Host cell reactivation of damaged plasmid DNA. *Biochemistry*, *28*:3120–3124 (1989).

3. Chao, C. C.-K., Lee, Y.-L., and Lin-Chao, S. Penotypic reversion of cisplatin resistance accompanies reduced host cell reactivation of damaged plasmid. *Biochem. Biophys. Res. Commun. 170*:851–859 (1990).
4. Chao, C. C.-K., Lee, Y.-L., Cheng, P.-W., and Lin-Chao, S. Enhanced host cell reactivation of damaged plasmid DNA in HeLa cells resistant to *cis*-diammine-dichloroplatinum(II). *Cancer Res. 51*:601–605 (1991).
5. Chu, G., and Chang, E. Xeroderma pigmentosum group E cells lack a nuclear factor that binds to damaged DNA. *Science, 242*:564–567 (1988).
6. Patterson, M., and Chu, G. Evidence that xeroderma pigmentosum cells from complementation group E are deficient in a homolog of yeast photolyase. *Mol. Cell. Biol. 9*:5105–5112 (1989).
7. Chao, C. C.-K., Huang, S.-L., Huang, H., and Lin-Chao, S. Cross-resistance to UV radiation of a cisplatin-resistant human cell line: Overexpression of cellular factors that recognize UV-modified DNA. *Mol. Cell. Biol. 11*:2075–2080 (1991).
8. Chu, G., and Chang, E. Cisplatin-resistant cells express increased levels of a factor that recognizes damaged DNA. *Proc. Natl. Acad. Sci. USA, 87*:3324–3327 (1990).
9. Friedberg, E. C. *DNA Repair*. W. H. Freeman, New York, 1985.
10. Chao, C. C.-K., Huang, S.-L., Lee, L.-Y., and Lin-Chao, S. Identification of inducible damage-recognition proteins that are overexpressed in HeLa cells resistant to *cis*-diamminedichloroplatinum(II). *Biochem. J. 277*:875–878 (1991).
11. Toney, J. H., Dohanue, B. A., Kellett, P. J., Bruhn, S. L., Essigmann, J. M., and Lippard, S. J. Isolation of cDNAs encoding a human protein that binds selectively to DNA modified by the anticancer drug *cis*-diamminedichloroplatinum(II). *Proc. Natl. Acad. Sci. USA, 86*:8323–8332 (1989).

10

Ornithine Decarboxylase Gene Amplification

Philip Coffino *University of California, San Francisco, California*

I. INTRODUCTION

The gene for ornithine decarboxylase (ODC) is one among many that have been amplified by purposeful selection in cultured animal cells. Given the diversity of genes for which this has been accomplished, it is likely that the prerequisites for success are not gene specific, but rather that a concurrence of the following elements is required: (1) the gene encodes something, usually an enzyme, without which the cell does poorly, but an excess of which is not too toxic, (2) there is available an inhibitor that specifically reduces activity of the gene product, and (3) amplification of the desired gene happens more frequently or is globally less disturbing than alternative perturbations capable of raising activity. Not all these prerequisites need be fulfilled to perfection. In the case of ODC, the inhibitor is pharmacologically near perfect but overproduction is not without toxicity, and the cell seemingly has ready recourse to means other than amplification to achieve overproduction. Nonetheless, as will be described, amplification of both endogenous and transfected ODC genes is readily attained.

Polyamines are both ubiquitous in living organisms and essential for life [1]. Ornithine decarboxylase is the first and commonly rate-limiting step in the pathway committed to polyamine synthesis. Its product, putrescine, is the precursor of the polyamines spermidine and spermine. Genetic and pharmacological evidence demonstrates that cells cannot live without polyamines. Furthermore, cells that are stimulated to divide greatly increase their level of ODC activity within hours of mitogenic stimulation. Consequently, pharmacologists interested in chemotherapy of diseases in which cells proliferate inappropriately have long perceived ODC as a potential target and have devoted considerable resources to the design of inhibitors of the enzyme. These endeavors have resulted in drugs with still unestablished therapeutic efficacy but remarkable pharmacological potency and specificity [2]. Among such drugs is α-difluoromethylornithine (DFMO), an enzyme-activated inhibitor that, upon decarboxylation, produces a product that covalently binds and inactivates ODC. Efficiency of this process is high; about one in three decarboxylations of DFMO results in enzymatic suicide.

II. INHIBITOR SPECIFICITY

When the use of DFMO to amplify the ODC gene was contemplated, it had already been well established that the drug profoundly inhibits proliferation of cultured cells when used at a several micromolar concentration and that putrescine, the product of ODC, could fully prevent or reverse the growth inhibition [2]. However, for purposes of effective selection of high level overproduction and gene amplification it is important to be able to use an agent at very high levels, while still maintaining specificity of toxicity. Using S49 mouse lymphoma cells [3], we found that several micromolar putrescine could fully reverse the growth inhibitory affect of even 40 mm DFMO, the highest concentration we tested, as long as compensatory adjustments were made in salt to keep the culture medium isotonic (L. McConlogue and P. Coffino, unpublished data). This provided some assurance that the inhibitor retains sufficient specificity even at very high concentrations to allow "head room" for selection of high level overproducers.

III. MOUSE LYMPHOMA CELLS THAT OVERPRODUCE ODC

S49 cells that overproduce ODC were selected by growing them in medium containing a sufficient concentration of the drug to slow growth [4,5]. The concentration of DFMO was incremented when the doubling time of the population returned to the normal value, about 16 hr. A resultant population of cells that grew well in 100 μM DFMO was cloned in soft agar containing the same concentration of drug. Clonal isolates were propagated in medium free of DFMO and screened for ODC activity [5]. Several clones with the highest activities were further incrementally selected in concentrations of drug up to 50 mm. A clonal isolate was obtained from the resistant population by growth in soft agar containing that concentration of DFMO and characterized with respect to ODC synthesis. Polyacrylamide gel electrophoresis with sodium dodecyl sulfate (SDS PAGE) analysis of [^{35}S]methionine metabolically labeled cell lysates indicated that about 22% of newly synthesized cytoplasmic protein was ODC in the resistant cells [4]. Figure 1 shows the results of such an analysis of wild-type S49 cells and of cells

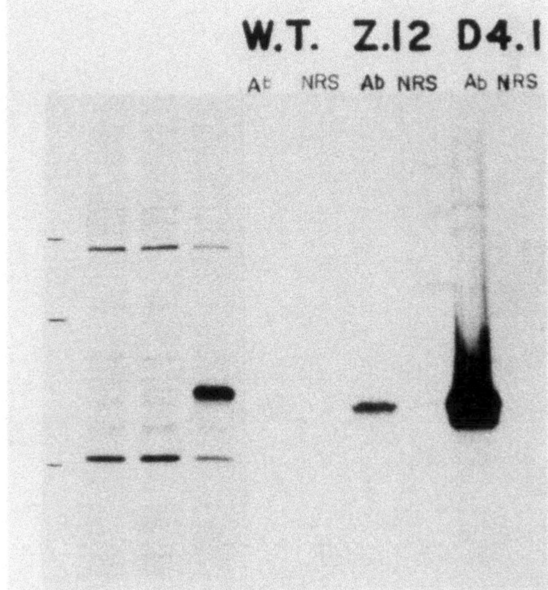

Figure 1 ODC synthesis in wild-type and overproducing S49 cells. Wild-type, Z.12, and D4.1 cells (lanes 1, 2, and 3, respectively) were pulse-labeled with [^{35}S]methionine and cytoplasmic extracts prepared and analyzed by SDS-PAGE. Extracts were also immunoprecipitated with normal rabbit serum or with serum specific for mouse ODC and similarly analyzed. (Reproduced from Ref. 4.)

resistant to 100 μM DFMO (Z.12) cells) and to 50 mm DFMO (D4.1 cells). Because the rate of ODC synthesis in wild-type cells is low, it is difficult to determine its absolute amount by these methods. The synthesis of ODC in the D4.1 overproducers is, however, at least 500-fold greater than in the wild-type precursor cells.

Northern and Southern blot analysis of wild-type and D4.1 S49 cells using a mouse ODC cDNA probe [6] showed a vast increase in the amount of ODC mRNA and an increased copy number of an ODC gene in the D4.1 cells (Fig. 2). I shall return below to the question of the relative increase in ODC protein synthesis, mRNA, and gene copy number.

IV. OTHER CELLS THAT OVERPRODUCE ODC

Effective selection for overproduction of ODC in cultured cells has been attained by several laboratories [7–15]. Not only mouse cells, but human [12,14] and hamster [7,8,10,11,15] cells as well have been made to overproduce ODC. In

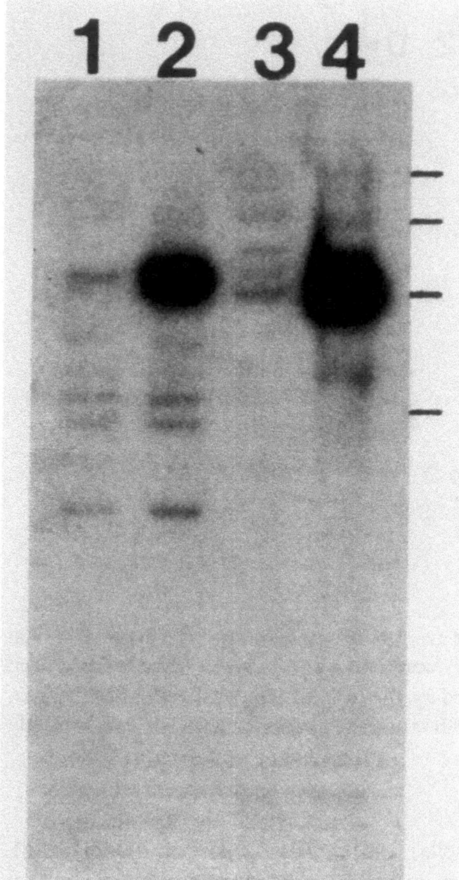

Figure 2 Southern blot analysis of ODC gene in wild-type and D4.1 overproducing cells. DNA from wild-type (lanes 1 and 3) and from D4.1 (lanes 2 and 4) cells were digested with *Eco*R1 (lanes 1 and 2) or with *Bam*H1 (lanes 3 and 4) and sequences hybridizing to a mouse ODC cDNA probe visualized. The DNA amplified in the D4.1 cells is the active mouse ODC gene. Some or all of the background bands are ODC pseudogenes. (Reproduced from Ref. 6.)

addition to the use of a drug that directly inhibits ODC, two other strategies have accomplished overproduction. One of these made use of a cell line deficient in arginase to limit pools of intracellular ornithine, the substrate of ODC, thereby causing starvation for polyamines [10]. In still another case, selection for amplification was applied to ribonucleotide reductase (in hamster cells) and the linked ODC gene was fortuitously coamplified [11]. It appears reasonable to conclude, therefore, that, although pharmacological inhibition of ODC is the most direct and general of available selections, other means that limit polyamines can, if available, also result in selection for ODC overproduction and that increased copy number of the ODC gene can be achieved as a genetically linked unselected marker.

V. ODC OVERPRODUCTION NOT DEPENDENT ON GENE AMPLIFICATION

Because in D4.1 cells selected for resistance to 50 mm DFMO, ODC appeared to be overproduced to a higher degree than could be accounted for by the change in gene copy number, this relationship was examined in S49 cells resistant to lesser levels of drug [16]. Among four such populations, ODC activity was increased between 13- and 81-fold and ODC mRNA levels between 2- and 16-fold, but the number of copies of the active ODC gene (there are additional pseudogenes in the mouse) was increased no more than 2-fold. These and related results indicate that resistance to DFMO can occur by increases in mRNA steady state levels that are independent of gene copy number and also by changes in the efficiency with which that mRNA is translated into active enzyme. Under the selection conditions described, cells first utilized mechanisms other than ODC gene amplification to enhance ODC production. Later, as conditions of selection were made more stringent, cells utilized amplification of the ODC gene to attain still higher levels of enzyme synthesis.

VI. CELLULAR TOXICITY OF ODC OVERPRODUCTION

S49 cells selected for resistance to DFMO grow better in the presence of the drug than in its absence (L. McConlogue and P. Coffino, unpublished data). The basis for this difference is not known, but plausible guesses can be made. It is likely that overproducing cells are dependent on DFMO for optimum growth because high level uninhibited expression of ODC activity is toxic. In seeming contradiction to this view, addition of putrescine to culture medium, even to very high levels, is benign, suggesting that the product of ODC cannot itself cause cellular toxicity. However, polyamine transport into cells can be regulated, and it is likely that cells can limit internal polyamine pools by regulating influx. When the source of putrescine is a high level of ODC within cells, that means of control is not available. If putrescine can become toxic, it is likely that it does so after conversion to spermadine and spermine. These last two polyamines participate in a metabolic pathway that results in their acetylation and subsequent oxidation. There are indications that the products of such metabolic oxidations, which include hydrogen peroxide and aldehydes, can cause cellular toxicity [17]. An alternate view of the mechanism of toxicity produced by ODC overproduction is that ornithine, the substrate of ODC, becomes diverted from participation in other important metabolic pathways. This is unlikely, because providing an excess of ornithine to ODC-overproducing cells does not relieve their dependence on DFMO for optimum growth.

VII. AMPLIFICATION OF TRANSFECTED ODC GENES

Experience in using the ODC gene as an amplifiable element in transfection experiments has been limited but encouraging [15,18,19]. Mouse ODC cDNA sequences have been transfected into recipient-cultured animal cells and amplified, not only in ODC$^-$ Chinese hamster ovary (CHO) cells but also in recipient cells that are wild type with respect to ODC activity [18]. The ODC gene, therefore, offers promise as a dominant, selectable marker. Among the recipients in which a transfected ODC gene has been successfully amplified to copy numbers in excess of 1000 are Chinese hamster ovary, African green monkey CV1, and mouse NIH 3T3 cells [18]. In some of these experiments a dihydrofolate reductase (DHFR) gene was used as an unselected marker and, when linked to ODC in an expression vector, was found to be coamplified after selection with DFMO. For the selected (ODC) and unselected (DHFR) genes, mRNA levels were approximately commensurate with gene copy number. The ODC gene, coupled with a technically facile selection using DFMO, appears, therefore, to show promise of general utility as a means for amplification of transfected genes to copy numbers in excess of 1000.

VIII. SUMMARY

Polyamine starvation causes cellular growth arrest and toxicity. Depletion of polyamines can be induced by potent and very specific pharmacological inhibitors of ornithine decarboxylase; cells capable of increased ODC synthesis can survive such treatment. The specificity of the ODC inhibitor DFMO is so great that concentrations of inhibitor vastly in excess of those required to prevent growth of normal cells can readily be used to select cells that overproduce ODC more than 1000-fold. Cells have means of overproduction other than gene amplification available, but very high level expression is associated with amplification. Excess synthesis of polyamines may itself become toxic and require maintenance of overproducing cells in medium containing inhibitor. Expression vectors containing the ODC gene have been transfected and selected for amplification, suggesting that the gene may be exploited as a dominant amplifiable element for cotransfection and high level expression of other desired genes.

REFERENCES

1. Tabor, C. W., and Tabor, H., Polyamines. *Annu. Rev. Biochem. 53*:749–790 (1984).
2. Bey, P., Danzin, C., and Jung, M. Inhibition of basic amino acid decarboxylases involved in polyamine biosynthesis, in *Inhibition of Polyamine Metabolism* (P. P. McCann, A. E. Pegg, and A. Sjoerdsma, eds.), Academic Press, San Diego, 1987, pp. 1–31.
3. van Daalen Wetters, T., and Coffino, P. Cultured S49 mouse T lymphoma cells, in

Molecular Genetics of Mammalian Cells: Methods in Enzymology, Vol. 151 (M. M. Gottesman, ed.), Academic Press, San Diego, 1987, pp. 9–19.
4. McConlogue, L., and Coffino, P. A mouse lymphoma cell mutant whose major protein product is ornithine decarboxylase. *J. Biol. Chem. 258*:12083–12086 (1983).
5. McConlogue, L., and Coffino, P. Ornithine decarboxylase in difluoromethyl-ornithine-resistant mouse lymphoma cells. *J. Biol. Chem. 258*:8384–8388 (1983).
6. McConlogue, L., Gupta, M., Wu, L., and Coffino, P. Molecular cloning and expression of the mouse ornithine decarboxylase gene. *Proc. Natl. Acad. Sci. USA, 81*:540–544 (1984).
7. Choi, J., and Scheffler, I. E. A mutant of Chinese hamster ovary cells resistant to α-methyl- and α-difluoromethylornithine. *Somatic Cell Genet. 7*:219–233 (1981).
8. Choi, J., and Scheffler, I. E. Chinese hamster ovary cells resistant to α-difluoromethylornithine are overproducers of ornithine decarboxylase. *J. Biol. Chem. 258*: 12601–12608 (1983).
9. Kahana, C., and Nathans, D. Isolation of cloned cDNA encoding mammalian ornithine decarboxylase. *Proc. Natl. Acad. Sci. USA, 81*:3645–3649 (1984).
10. Pohjanpelto, P., Hölttä, E., Jänne, O. A., Knuutila, S., and Alitalo, K. Amplification of ornithine decarboxylase gene in response to polyamine deprivation in Chinese hamster ovary cells. *J. Biol. Chem. 260*:8532–8537 (1985).
11. Srinivasan, P. R., Tonin, P. N., Wensing, E. J., and Lewis, W. H. The gene for ornithine decarboxylase is co-amplified in hydroxyurea-resistant hamster cells. *J. Biol. Chem. 262*:12871–12878 (1987).
12. Leinonen, P., Alhonen-Hongisto, L., Laine, R., Jänne, O. A., and Jänne, J. Human myeloma cells acquire resistance to difluoromethylornithine by amplification of ornithine decarboxylase gene. *Biochem. J. 242*:199–203 (1987).
13. Pegg, A. E., Madhubala, R., Kameji, T., and Bergeron, R.J. Control of ornithine decarboxylase activity in α-difluoromethylornithine-resistant L1210 cells by polyamines and synthetic analogues. *J. Biol. Chem. 263*:11008–11014 (1988).
14. Hirvonen, A., Eloranta, T., Hyvönen, T., Alhonen, L., and Jänne, J. Characterization of difluoromethylornithine-resistant mouse and human tumour cell lines. *Biochem. J. 258*:709–713 (1989).
15. Hsieh, J. T., Denning, M. F., Heidel, S. M., and Verma, A. K. Expression of human chromosome 2 ornithine decarboxylase gene in ornithine decarboxylase-deficient Chinese hamster ovary cells. *Cancer Res. 50*:2239–2244 (1990).
16. McConlogue, L., Dana, S. L., and Coffino, P. Multiple mechanisms are responsible for altered expression of ornithine decarboxylase in overproducing variant cells. *Mol. Cell. Biol. 6*:2865–2871 (1986).
17. Coffino, P., and Poznanski, A. Killer polyamines? *J. Cell Biolchem. 45*:54–58 (1991).
18. Chang, T. R., and McConlogue, L. Amplification and expression of heterologous ornithine decarboxylase in Chinese hamster cells. *Mol. Cell. Biol. 8*:764–769 (1988).
19. Moshier, J. A., Gilbert, J. D., Skunca, M., Dosescu, J., Almodovar, K. M., and Luk, G. D. Isolation and expression of a human ornithine decarboxylase gene. *J. Biol. Chem. 265*:4884–4892 (1990).

11

Amplification of the IMP Dehydrogenase Gene in Mammalian Cells

Frank R. Collart and Eliezer Huberman *Argonne National Laboratory, Argonne, Illinois*

I. BIOLOGICAL ROLE OF IMP DEHYDROGENASE

Inosine-5′-monophosphate dehydrogenase (EC 1.1.1.205, IMPDH) catalyzes the formation of xanthylic acid from inosinic acid (IMP) in the purine de novo synthetic pathway. IMPDH is a branch point enzyme in the synthesis of adenine and guanine nucleotides and is the rate-limiting enzyme in the de novo synthesis of guanine nucleotides [1]. This enzyme has an essential role in providing necessary precursors for DNA and RNA biosynthesis, which can be verified by the abrupt cessation of DNA synthesis observed when cells are treated with IMPDH inhibitors and the circumvention of this effect by the addition of exogenous guanosine [2].

In general, normal tissues that exhibit increased cell proliferation also exhibit increased IMPDH activity [3]. A similar relationship between increased proliferation and elevated enzyme activity has also been found in rat hepatomas having varied growth rates [1]. However, these hepatomas manifested IMPDH activities that were disproportionately higher than those of normal tissue [1], suggesting that increased IMPDH activity may be associated not only with growth control, but also with malignant transformation or tumor progression. Another facet of IMPDH activity is demonstrated by the ability of IMPDH inhibitors to induce terminal cell differentiation in a variety of human tumor cell types (Table 1) [4–11]. This induction is a consequence of the inhibitor-evoked reduction in guanine nucleotide levels, because exogenous guanosine prevents the acquisition of the differentiated

Table 1 Differentiation of Human Tumor Cells Caused by Inhibitors of IMP Dehydrogenase

Cell line	Markers	Ref.
Human myeloid leukemia	Nitroblue tetrazolium reduction; surface antigen; nonspecific esterase	4, 5
Human erythroid leukemia	Hemoglobin synthesis	6, 7
Human T-lymphoid leukemia	Surface antigens	8
Human breast epithelial	Lipid and casein formation	9, 10
Human melanoma	Tyrosinase; CF21 antigen; melanin synthesis	11

phenotype in treated cells. These studies indicate that the level of IMPDH activity and of guanine nucleotides affects cell growth and differentiation, and they suggest that treatment with IMPDH inhibitors may provide a useful means of overcoming the differentiation block that exists in tumor cells. In this context, it is interesting to note that tiazofurin, an IMPDH inhibitor, is currently used in clinical trials for the treatment of myeloid leukemia [12].

II. SELECTION OF MYCOPHENOLIC ACID RESISTANT CELLS

To assess the role of IMPDH in the regulation of cell growth and differentiation, we isolated the IMPDH cDNA and prepared polyclonal and monoclonal antibodies directed against the enzyme. Because most cells have low levels of IMPDH, we selected variants from Chinese hamster V79 cells that had higher IMPDH levels. This strategy was facilitated by mycophenolic acid (MPA), a specific IMPDH inhibitor. MPA has a K_i in the nanomolar range and does not inhibit other cellular enzymes at submillimolar concentrations [13]. Variants were obtained by treating V79 cells with the mutagen *N*-methyl-*N*′-nitro-*N*-nitrosoguanidine and selecting colonies of cells resistant to 1 μg/ml of MPA [14]. The resistance level of one

Table 2 Characteristics of the V79 Cell Line and Its MPA-Resistant Variants

Cell type	Resistance to MPA (μg/ml)	Enzyme activity (mU/mg of total protein)
V79	0.1	0.7
VM1	5.0	4.0
VM2	10.0	4.7
VM3	25.0	6.3
VM4	50.0	7.4

Source: From Ref. 15.

variant was further increased by a stepwise selection procedure in the presence of increasing concentrations of MPA [15]. After several steps, four variants, VM1 through VM4, were obtained that were resistant to 5, 10, 25, and 50 μg/ml of MPA, respectively; the parental V79 cells were resistant to only 0.1 μg/ml of MPA (Table 2). The increased resistance of the variants to MPA was associated with increased IMPDH activity in the cell homogenates. VM1 cells exhibited about a 6-fold increase in IMPDH activity over the parental cells, and the VM2, VM3, and VM4 cells exhibited about 7-, 9-, and 11-fold increases, respectively.

III. ASSOCIATION OF MYCOPHENOLIC ACID RESISTANCE WITH IMP DEHYDROGENASE GENE AMPLIFICATION

To elucidate the basis for the increased enzyme activity in the MPA-resistant cells, we isolated and characterized the enzyme from the VM2 variant [16]. In denaturing gel electrophoresis, the purified protein migrated as a single species with an apparent molecular mass of 56 kDa. Two proteins of identical molecular mass were detected by two-dimensional gel electrophoresis and Coomassie Blue staining of the gel (Fig. 1). However, immunoblot analysis of whole-cell homogenate

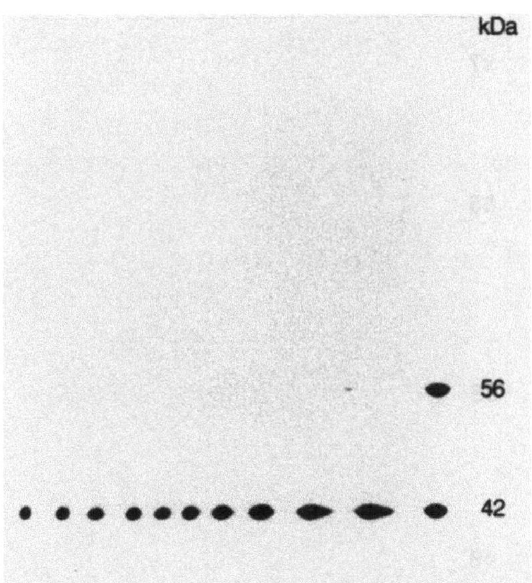

Figure 1 Two-dimensional gel electrophoresis pattern of cellular proteins from the mycophenolic acid resistant VM2 cells. Total cellular proteins were visualized by staining with Coomassie Brilliant Blue R-250. The pattern is oriented with the acid side to the left and the basic side to the right. (From Ref. 16.)

preparations using monoclonal antibodies prepared from the purified protein indicated the presence of at least three forms of IMPDH. These minor forms, constituting less than 10% of the total amount of IMPDH, are presumably either charge-modified forms of the enzyme or other minor species of IMPDH [17]. The K_m values of the VM2 enzyme for IMP and NAD were similar to those reported for partially purified IMPDH from V79 cells [14]. Moreover, the VM2 enzyme retained a high sensitivity to MPA, having a K_i in the nanomolar range. The

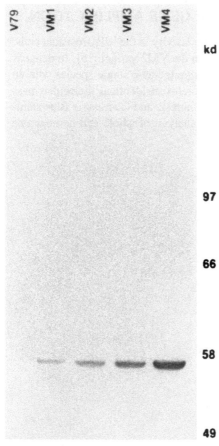

Figure 2 Immunoblot analysis of proteins of the V79 cell line and the MPA-resistant cell variants. Cellular proteins (25 μg) were electrophoresed through a 7.5% polyacrylamide gel and transferred to nitrocellulose. Blots were incubated with the polyclonal anti-IMPDH antibody, and antibody binding was detected with a peroxidase-based detection system. (From Ref. 15.)

similarity of the kinetic parameters for the VM2 and parental V79 enzymes indicates that the gene coding for the enzyme in the VM2 cells has not been altered in a way that changes the affinity of the enzyme for the substrate or its sensitivity to MPA. This suggests that the basis for the increased resistance in the VM2 cells is an alteration in the regulation of IMPDH.

The purified IMPDH from VM2 cells was used to prepare an IMPDH-specific polyclonal antibody. This antibody, in conjunction with immunoblot analysis, was used to determine the amount of enzyme in the parental V79 cells and in the MPA-resistant variants. The antiserum reacted, in both the V79 and variant cells, with a protein of 56 kDa corresponding to IMPDH (Fig. 2). Furthermore, the amount of enzyme in the MPA-resistant cell variants was positively correlated with the degree of resistance and was more than one order of magnitude higher than the amount of enzyme detected in the parental V79 cells. The amount of enzyme in the variant cells was 14- to 27-fold higher than the amount detected in V79 cells.

To analyze the relationship between the degree of resistance and IMPDH mRNA levels, we used a mouse IMPDH probe to characterize the steady state levels of IMPDH mRNA. The mouse cDNA probe of 750 base pairs was isolated from a mouse expression library using the IMPDH polyclonal antiserum [16]. Hybridization signals corresponding to a 2.2 kb message were detected in both total cellular RNA and poly(A) RNA samples from V79 and MPA-resistant variant cells (Fig. 3). The steady state level of this message in the variant cells correlated with their MPA resistance.

The IMPDH cDNA probe was also used to analyze for alterations or changes in the copy number of the IMPDH gene. Southern blot analysis of genomic DNA isolated from the parental V79 and the MPA-resistant cell variants revealed that the gene for IMPDH is amplified in the drug-resistant cells (Fig. 4). Furthermore, the degree of gene amplification in the cell variants was positively correlated to the level of resistance to MPA and closely paralleled the amounts of cellular enzyme. The gene copy level in the most resistant cell variant, VM4, was approximately 20-fold higher than the level in the parental V79 cells. Southern blot analysis of genomic DNA digested with other restriction enzymes showed a similar variation in the intensity of the hybridization signals for V79 and variant cell DNA. For each restriction enzyme, identical restriction fragment patterns were obtained for each of the five cell lines, suggesting that no major internal rearrangements of the IMPDH gene had occurred.

Thus, amplification of the IMPDH gene appears to be closely involved in establishing resistance to MPA in the V79 cell variants, with each selection step at increased MPA concentration resulting in a correspondingly increased level of IMPDH gene amplification. Similar results have been obtained by other investigators who used MPA to select for resistant cells. Some of these variants have altered regulation of IMPDH, as indicated by increased expression of the enzyme [18–20]. In two cases, a cDNA probe for IMPDH was used to establish gene

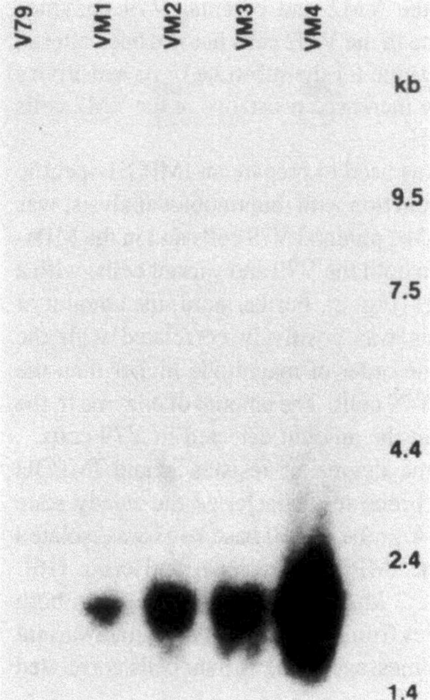

Figure 3 Northern blot analysis of RNA from V79 and the MPA-resistant cell lines. Total RNA (12 μg) was electrophoresed in denaturing formaldehyde–agarose gels and blotted onto nylon membrane. The filters were hybridized with the mouse IMPDH cDNA probe, washed, and exposed to autoradiographic film with an intensifying screen. (From Ref. 15.)

amplification as the basis for the resistance to MPA [15,18]. Gene amplification, however, is not the only mechanism of resistance to MPA. In some cases MPA resistance is attributable to altered kinetic properties of the enzyme, suggesting the variants have an alteration in the structural gene coding for IMPDH [14,21]. Additionally, a MPA-resistant variant derived from the parasite *Tritrichomonas foetus* can salvage xanthine more efficiently from a mixture of purines and thus bypass the MPA-mediated block of IMPDH [22]. The availability of the different variant cells should be useful in studying the effects of IMPDH activity on cell growth.

IV. CONSERVATION OF THE IMPDH GENE

The Chinese hamster cell variants also provided the means to prepare antibody and cDNA probes and to subsequently isolate the IMPDH cDNA from human [16,17],

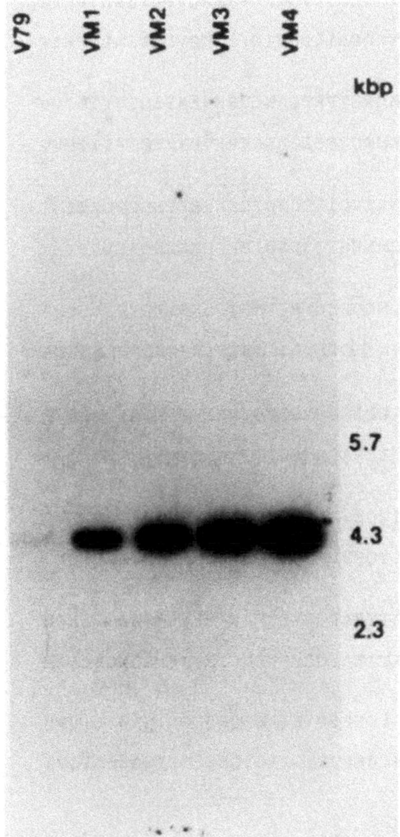

Figure 4 Southern blot analysis of DNA from V79 and the MPA-resistant cell lines. Genomic DNA was digested with the *Hin*dIII restriction enzyme, electrophoresed through a 0.8% agarose gel, and blotted onto a nylon membrane. The filters were hybridized with a fragment of the mouse IMPDH cDNA. (From Ref. 15.)

rodent [16], and protozoan [18] sources. All the eukaryotic clones thus far characterized specify a protein of 514 amino acids with an approximate molecular mass of 56 kDa. Comparison of the deduced amino acid sequences of the human and Chinese hamster proteins indicates a high level of sequence conservation (Fig. 5). Of the 514 amino acids, only eight differences are noted between the human and Chinese hamster proteins; five of these changes are conservative in terms of the chemical nature of the amino acid [16]. The mouse and Chinese hamster cDNAs differ in only five amino acid residues [23]. The leishmanial IMPDH gene, isolated from a MPA-resistant strain of *Leishmania donovani* [18],

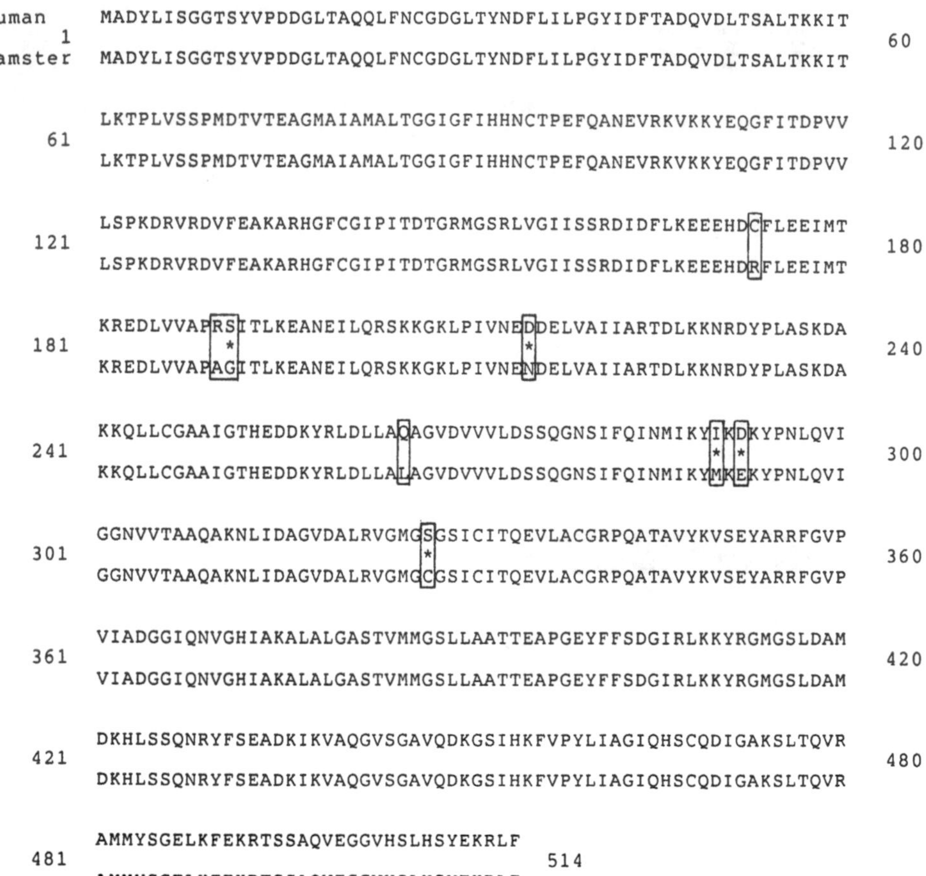

Figure 5 Comparison of the human and Chinese hamster deduced protein sequences. The nonmatching amino acids are boxed. The unmatched amino acids that have similar chemical properties are denoted with asterisks. (From Ref. 16.)

specifies a protein having a 53% identity with the mammalian (either human, hamster, or mouse) enzymes. Considering conservative substitutions between similar amino acids, the leishmanial and mammalian IMPDH proteins match at more than 70% of the residues.

This conservation in amino acid sequence is reflected in the characteristics of IMPDH from a variety of sources. The molecular mass, kinetic parameters, sensitivity to inhibition by MPA, and hydrophobic nature are similar in all the eukaryotic IMPDH enzymes thus far characterized. The conservation in amino acid sequence further indicates an essential role for IMPDH in cell growth and differentiation. The availability of the antibodies and cDNA probes should contribute to an increased understanding of the role of this enzyme in cell growth and differentiation and gene in amplification.

ACKNOWLEDGMENT

This work was supported by the U.S. Department of Energy, Office of Health and Environmental Research, under contract W-31-109-ENG-38.

REFERENCES

1. Weber, G. Biochemical strategy of cancer cells and the design of chemotherapy. *Cancer Res. 43*:3466 (1983).
2. Cohen, M. B., and Sadee, W. Contribution of the depletions of guanine and adenine nucleotides to the toxicity of purine starvation in mouse T-lymphoma cells. *Cancer Res. 43*:1589–1591 (1983).
3. Cooney, D. A., Wilson, Y., and McGee, E. A straight-forward radiometric technique for measuring IMP dehydrogenase. *Anal. Biochem. 130*:339–345 (1983).
4. Knight, R. D., Mangum, J., Lucas, D. L., Cooney, D. A., Khan, E. C., and Wright, D. G. Inosine monophosphate dehydrogenase and myeloid cell maturation. *Blood, 69*: 634–639 (1987).
5. Collart, F. R., and Huberman, E. Expression of IMP dehydrogenase in differentiating HL-60 cells. *Blood, 75*:570–576 (1990).
6. Yu, J., Lemas, V., Page, T., Connor, J. D., and Yu, A. L. Induction of erythroid differentiation in K562 cells by inhibitors of inosine monophosphate dehydrogenase. *Cancer Res. 49*:5555–5560 (1989).
7. Olah, E., Natsumeda, Y., Ikegami, T., Kotek, Z., Horanyi, M., Szelenyi, J., Paulik, E., Kremmer, T., Holan, S. R., Sugar, J., and Weber, G. Induction of erythroid differentiation and modulation of gene expression by tiazofurin in K–562 leukemia cells. *Proc. Natl. Acad. Sci. USA, 85*:6533–6537 (1988).
8. Kiguchi, K., Collart, F. R., Henning-Chubb, C., and Huberman, E. Cell differentiation and altered IMP dehydrogenase expression induced in human T-lymphoblastoid leukemia cells by mycophenolic acid and tiazofurin. *Exp. Cell Res. 187*:47–53 (1990).
9. Sidi, Y., Panet, C., Wasserman, L., Cyjon, A., Novogrodsky, A., and Nordenberg, J. Growth inhibition and induction of phenotypic alterations in MCF-7 breast cancer cells by an IMP dehydrogenase inhibitor. *Br. J. Cancer, 58*:61–63 (1988).
10. Bacus, S., Kiguchi, K., Chin, D., King, C. R., and Huberman, E. Differentiation of cultured human breast cancer cells (AU-565 and MCF-7) associated with loss of cell surface HER-2/*neu* antigen. *Mol. Carcinog. 3*:350–362 (1990).
11. Kiguchi, K., Collart, F. R., Henning-Chubb, C., and Huberman, E. Induction of cell differentiation in melanoma cells by inhibitors of IMP dehydrogenase: Altered patterns of IMP dehydrogenase expression and activity. *Cell Growth Differ. 1*:259–270 (1990).
12. Tricot, G. J., Jayaram, H. N., Lapis, E., Natsumeda, Y., Nichols, C. R., Kneebone, P., Heerema, N., Weber, G., and Hoffman, R. Biochemically directed therapy of

leukemia with tiazofurin, a selective blocker of inosine 5′-monophosphate dehydrogenase activity. *Cancer Res.* *49*:3696–3701 (1989).

13. Lee, H.-J., Pawlak, K., Nguyen, B. T., Rogins, R. K., and Sadee, W. Biochemical differences among four inosinate dehydrogenase inhibitors, mycophenolic acid, ribovirin, tiazofurin, and selenazofurin, studied in mouse lymphoma cell culture. *Cancer Res.* *45*:5512 (1985).
14. Huberman, E., McKeown, C. K., and Friedman, J. Mutagen-induced resistance to mycophenolic acid in hamster cells can be associated with increased inosine 5′-phosphate dehydrogenase activity. *Proc. Natl. Acad. Sci. USA*, *78*:3151–3154 (1981).
15. Collart, F. R., and Huberman, E. Amplification of the IMP dehydrogenase gene in Chinese hamster cells resistant to mycophenolic acid. *Mol. Cell. Biol.* *7*:3328–3331 (1987).
16. Collart, F. R., and Huberman, E. Cloning and sequence analysis of the human and Chinese hamster inosine 5′-monophosphate dehydrogenase. *J. Biol. Chem.* *263*: 15769–15772 (1988).
17. Natsumeda, Y., Ohno, S., Kawasaki, H., Konno, Y., Weber, G., and Suzuki, K. Two distinct cDNAs for human IMP dehydrogenase. *J. Biol. Chem.* *265*(9):5292–5295 (1990).
18. Wilson, K., Collart, F. R., Huberman, E., Stringer, J. R., and Ullman, B. Amplification and molecular cloning of the IMP dehydrogenase gene of *Leishmania donovani*. *J. Biol. Chem.* *266*:1665–1671 (1991).
19. Hodges, S. D., Fung, E., McKay, D. J., Renaux, B. S., and Snyder, F. F. Increased activity, amount, and altered kinetic properties of IMP dehydrogenase from mycophenolic acid-resistant neuroblastoma cells. *J. Biol. Chem.* *264*:18137–18144 (1989).
20. Cohen, M. B. Selection and characterization of mycophenolic acid-resistant leukemia cells. *Somatic Cell. Mol. Genet.* *13*:627–633 (1987).
21. Ullman, B. Characterization of mutant murine lymphoma cells with altered inosinate dehydrogenase activities. *J. Biol. Chem.* *258*:523–528 (1983).
22. Headstrom, L., Cheung, K. S., and Wang, C. C. A novel mechanism of mycophenolic acid resistance in the protozoan parasite *Tritrichomonas foetus*. *Biochem. Pharmacol.* *39*:151–160 (1990).
23. Tiedeman, A. A., and Smith, J. M. Isolation and sequence of a cDNA encoding mouse IMP dehydrogenase. *Gene*, *97*:289–293 (1991).

12

Amplification and Analysis of the Mouse Dihydrofolate Reductase Gene

Gray F. Crouse *Emory University, Atlanta, Georgia*

I. DIHYDROFOLATE REDUCTASE GENE AMPLIFICATION

Dihydrofolate reductase (Dhfr) is a ubiquitous enzyme that is essential for the de novo synthesis of purines, thymidylate, and glycine. There are specific inhibitors of Dhfr such as methotrexate, which are toxic to cells by virtue of Dhfr inhibition and have proved to be extremely valuable as chemotherapeutic agents for the treatment of various cancers. One of the problems associated with methotrexate therapy is that many patients eventually develop resistance to the drug. Beginning in the early 1960s, in order to study the phenomenon of drug resistance, various researchers were able to develop lines of cultured cells that were highly resistant to inhibitors of dihydrofolate reductase. It was only in 1978 that many of these lines were shown to be methotrexate resistant as a result of amplification of the *Dhfr* gene [1]. Although it is now becoming clear that amplification of the *Dhfr* gene is only one possible mechanism for patient resistance to methotrexate, this seminal work sparked substantial interest in the phenomenon of gene amplification, and in subsequent years many other genes discussed in this book were shown to be capable of amplification under certain selective conditions.

One of the remarkable aspects of *Dhfr* gene amplification is the high copy numbers that can be obtained in cell lines: copy numbers as high as 1000 have been observed both of the wild-type gene [2] and from introduced *Dhfr* minigenes [3]. During development of methotrexate-resistant cell lines, several properties of the process were noted: *Dhfr* amplification occurred only during selection with gradually increasing doses of methotrexate; there was usually a linear relationship between the level of methotrexate resistance and the *Dhfr* gene copy number; and some cell lines developed stable methotrexate resistance, whereas other cell lines rapidly lost methotrexate resistance in the absence of continued selection. There appeared to be some species difference in the stability of the amplified genes: hamster cells usually gave rise to stable lines, while mouse cell lines seemed to be predominantly unstable, although stable amplified cell lines could be obtained. The difference in stability of the amplified *Dhfr* gene was determined to be due to chromosomal location: stable *Dhfr* genes were chromosomally located, while amplified *Dhfr* genes that were readily lost were found to be on extrachromosomal fragments known as double minute chromosomes [4].

There are two important points to note which are often overlooked or misunderstood:

1. Although amplification of *Dhfr*, as well as many other genes, has been observed in cells of many different types, gene amplification by selection has not been observed in completely normal mammalian cells [5,6].
2. There is nothing about the structure of the gene that directs its amplification.

This second point is most clearly shown by the findings that amplification of either *Dhfr* minigenes [3] or cDNA constructs [7] transfected into $Dhfr^-$ cell lines can be selected by treatment with methotrexate.

II. *Dhfr* GENE STRUCTURE

The existence of cell lines containing multiple copies of the *Dhfr* gene has made many aspects of the cloning and analysis of the gene much easier. For example, mouse *Dhfr* cDNA made from amplified cell lines was abundant enough to permit screening cDNA clones by complementation of *Escherichia coli*, as well as by standard colony screening with enriched cDNA probes [8]. Purified cDNA clones were then used to screen mouse lambda genomic libraries so that a complete map of the gene and its flanking regions could be assembled [9]. Although there are only 558 nt of coding sequence for the mouse dihydrofolate reductase protein, these nucleotides were found to be present in six exons spread over 31 kb of DNA. The structure of the mouse *Dhfr* gene is shown in Figure 1. At the time the work was done, the *Dhfr* gene was one of the largest genes known, and even though there are many larger ones known now, the ratio of noncoding to coding sequence is still exceptionally large. The significance, if any, of this finding is not known, but the human and hamster *Dhfr* genes have approximately the same size and structure [10–12].

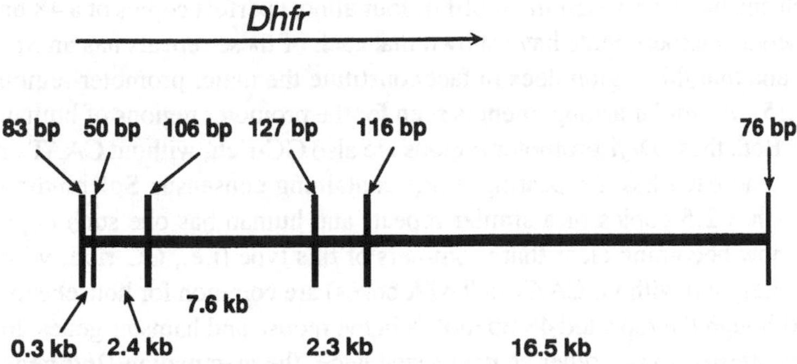

Figure 1 Mouse *Dhfr* gene organization. Sizes of the coding region of each exon appear above the gene, and sizes of each intron below.

III. *Dhfr* GENE REARRANGEMENTS DURING AMPLIFICATION

The mouse *Dhfr* gene was originally cloned from the S180 M500 cell line, which was known to carry the *Dhfr* genes on double minute chromosomes [9]. More than one type of gene organization in this cell line was found during the cloning and mapping of the *Dhfr* gene. Two different types of variant *Dhfr* genes in addition to the wild-type gene were discovered, neither of which was present in other, independently amplified, cell lines [9]. One of the variant structures involved DNA 5′ of the structural gene, but the other variant involved deletion of the first three exons of the gene as well as part of the third intron, presumably rendering these variant genes inactive. Both these variants were amplified to levels of 10–30% of that of the amplified wild-type gene in this cell line [9]. Subsequent work uncovered even more variant patterns in the *Dhfr* gene region of this cell line [13].

These results indicate that gene amplification can be a relatively "sloppy" process, accompanied by numerous gene rearrangements. Some of these rearrangements can, at least under certain circumstances, inactivate the gene being selected. The fact that inactive genes are also amplified must mean that double minute chromosomes can carry more than one copy of the *Dhfr* gene, so that selection for the inactive copy is maintained by linkage to an active gene. The existence of differing frequencies of the variants implies that not all double minute chromosomes in a cell have the same DNA content. Also the extent of amplification and arrangement of the amplified gene can vary markedly from one cell line to another.

IV. THE *Dhfr* PROMOTER REGION

When the 5′ flanking region of *Dhfr* was sequenced, the location of the promoter was not obvious, since no CAAT or TATA box, which was expected to be a part of an RNA polymerase II transcribed gene, was observed [9]. Instead, the region was very GC-rich and had a very striking motif of four almost perfect copies of a 48 bp repeat. Subsequent experiments have shown that each of these repeats has an Spl binding site and that this region does in fact constitute the major promoter region for *Dhfr* [14,15]. A similar arrangement is seen for the promoter regions of human and hamster. Both these *Dhfr* promoter regions are also GC-rich, without CAAT or TATA boxes, and each has a repeating motif containing consensus Spl binding sites; hamster has 2.5 copies of a similar repeat, and human has one such copy [11,16]. It is now becoming clear that promoters of this type (i.e., GC-rich, with Spl binding sites, and without CAAT or TATA boxes) are common for housekeeping genes. Although the repeated 48 bp motifs in the mouse and hamster genes do not have a counterpart in any other characterized gene, the mammalian *Dhfr* gene can be considered to have a typical housekeeping gene promoter.

The most surprising aspect of the mouse *Dhfr* promoter region was the discovery that it was acting bidirectionally [17]. The promoter regions of the human and hamster *Dhfr* genes also produce divergent transcripts [18,19]. Analysis of the mouse *Dhfr* promoter region showed that the 5′ ends of the divergent transcript were heterogeneous, with the longest beginning in the first copy of the 48 bp repeat; the divergent transcripts extended for at least 15 kb and were transcribed by RNA polymerase II [17]. Later work has shown that the promoter region is even more complex, for there are two promoters for *Dhfr*: the major promoter, and a minor promoter lying 300–400 bp upstream of the major promoter. Both these promoters are bidirectional, with the minor promoter contained in the first intron of the divergent transcripts produced from the major promoter [14]. This organization is illustrated in Figure 2.

Although each of the promoter regions is bidirectional, it is not clear whether the same elements are responsible for transcription in both directions. A number of mutations were made in the major promoter in an attempt to answer this question [15]. In general, different sequences appear to be involved in promotion of transcription in each direction, since mutations that affected transcription tended to exert most of their effect on transcription in one direction. However, several mutations affected both sets of transcripts. In addition, when the number of 48 bp repeats was reduced from 4 to 1, there was still bidirectional transcription, although at a greatly reduced level. This result implies that even one 48 bp repeat contains the potential for bidirectional transcription. There are also a number of different transcription factors that bind to the major promoter region, including Spl, AP-2, AP-3, CTF/NF-1, and probably E2F and/or HIP1 [15,20]. The large number of factors binding to this region offers the possibility of various types of transcriptional regulation.

It is interesting that although overproduction of Dhfr can lead to methotrexate resistance, in no case has resistance been found to be due to a change in either transcription rate or mRNA stability. This observation could be attributed to several factors other than a failure to find existing mutations. The promoter region,

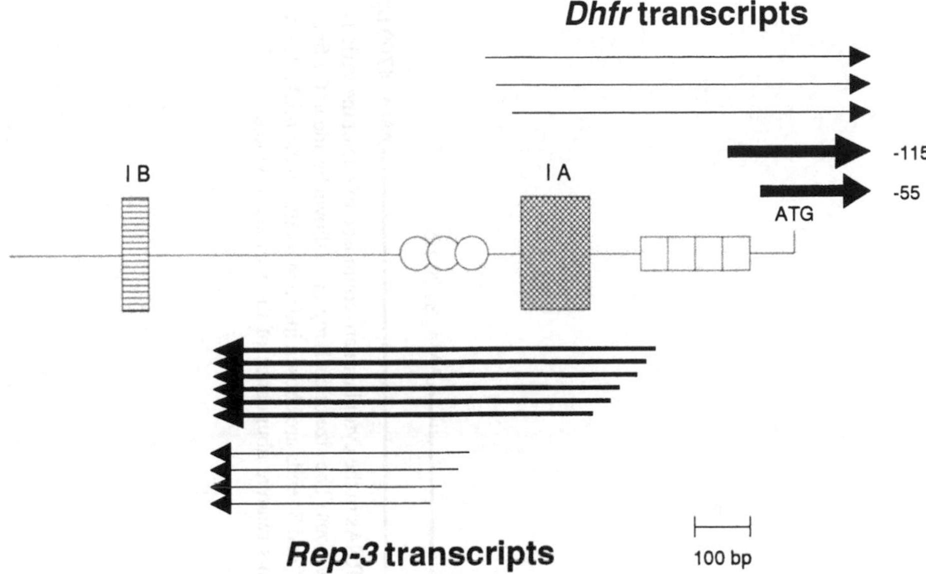

Figure 2 Mouse *Dhfr/Rep-3* promoter region: small squares represent the four 48 bp repeats in the major promoter region, and the circles represent consensus Spl binding sites in the minor promoter region. The mouse *Dhfr* coding region begins at the ATG and the coding regions of exons IA and IB of *Rep-3* are shown. The thickness of the lines representing transcripts is an approximation of their relative abundance.

as a result of its structure, might be fairly resistant to change in transcription rate by simple mutation. Even though a mutation might lead to low levels of methotrexate resistance, even moderate levels of methotrexate resistance would require an additional mechanism, such as gene amplification. This line of reasoning would suggest that some amplified lines could contain mutated genes, which were then later amplified; the only such case that has been reported involved a mutation in the coding region to give a methotrexate-resistant enzyme [21]. Given the lack of evidence for any mutational increase in transcription of the *Dhfr* gene, the most likely possibility is that amplification is a more frequent event than a mutation that increases the levels of *Dhfr* mRNA.

V. TRANSCRIPTION OF THE *Dhfr* GENE

All the mammalian *Dhfr* genes that have been analyzed have been found to have a very complex transcriptional pattern. The mouse, hamster, and human genes all have multiple 5′ and 3′ ends. The variation in the 5′ end is due to variable transcription initiation and the multiple 3′ ends are due to alternative polyadenyla-

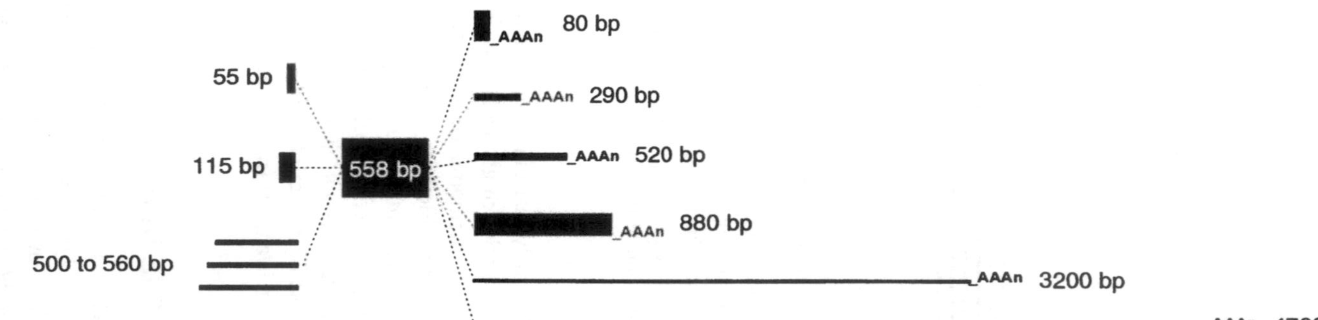

Figure 3 Composition of multiple *Dhfr* mRNA products. The mouse *Dhfr* mRNAs in the cytoplasm are composed of one of the various 5′ ends, the 558 bp of coding sequence, and one of the various 3′ untranslated regions. The size of each region is shown beside it; the final mRNA has an additional poly(A) tail of approximately 200 bp. It appears that 5′ and 3′ ends are chosen independently, so the final mRNA can be any combination of these products; the thickness of the lines represents relative abundance of the various products.

tion sites; there is no evidence that any of the transcripts have different coding sequences. It also appears that the 5′ and 3′ ends are chosen independently, so that the total number of transcripts made is a product of the number of ends [22]. The array of transcripts produced in mouse is illustrated in Figure 3.

In spite of all the work that has been done on various mammalian *Dhfr* genes, it is still not clear why there should be so many different *Dhfr* transcripts. More than 90% of the mouse *Dhfr* transcripts initiate at either -55 or -115 bp from the ATG, and there appears to be some tissue specificity in utilization of these two major *Dhfr* 5′ ends [14]. A hairpin loop with a calculated ΔG of -13.4 kcal has been found in the region just upstream of -55 [23]. This loop would therefore be present in the -115 but not the -55 transcript, and so might affect translation. Although there are at least six different polyadenylation sites [22], the majority of transcripts utilize either the first or fourth poly(A) site to produce mRNAs of 950 and 1800 nt. The shorter of these mRNAs has only 80 nt of 3′ untranslated sequences, whereas the longer mRNA has 880 bp. There are regions of AU-rich sequences in this longer 3′ region, and it is possible that there could be a regulatory significance to the longer sequence.

In the past, there has been some confusion in the distinction between mRNA polyadenylation and transcription termination. The widely spaced *Dhfr* polyadenylation sites made it possible to determine whether there were multiple transcription termination sites related to the multiple polyadenylation sites, or whether there was one site of transcription termination. Experiments were performed to measure transcription throughout the *Dhfr* gene, with the results shown in Figure 4. These experiments quite clearly showed that the level of transcription is essentially constant through the *Dhfr* gene and that termination occurs within a relatively short region about 1 kb 3′ of the sixth polyadenylation site [24].

VI. REGULATION OF THE *Dhfr* GENE

Dhfr is regulated by a number of different agents, including amino acid starvation, serum stimulation, change of cellular growth states, and induction by polyoma virus [25–30]. In addition, Dhfr has for many years been considered to be a cell-cycle-regulated enzyme, although the basis for that belief rested mainly on experiments that released cells from quiescence. A recent report suggests that Dhfr is in fact not regulated during the cell cycle [31]. This work illustrates the substantial differences between cell cycle regulation and processes that occur upon release of cells from quiescence, a difference that has not always been appreciated. There has been considerable controversy about the factors responsible for growth regulation of Dhfr. A variety of experiments examining the regulation of the endogenous mouse *Dhfr* gene [26,27] as well as modular *Dhfr* minigenes [32] suggested that posttranscriptional mechanisms were responsible for at least most of the observed changes in *Dhfr* mRNA levels. Although there have been various

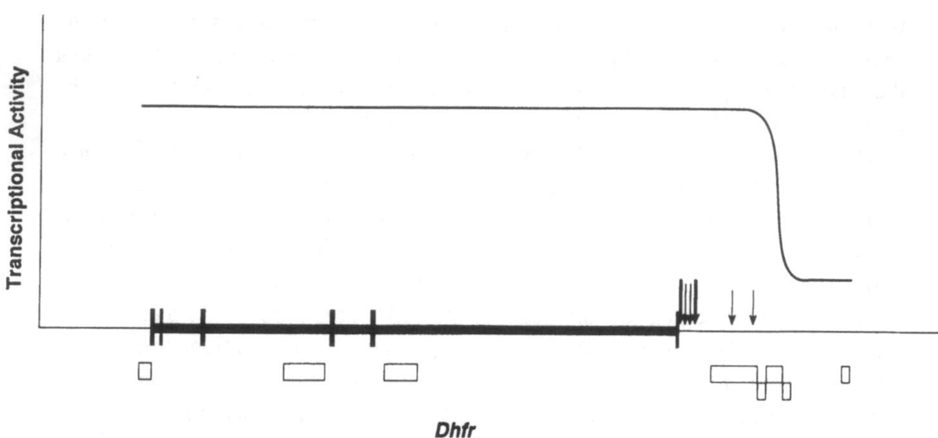

Figure 4 Transcriptional activity of the *Dhfr* gene. A relative measure of transcriptional activity is displayed as a function of position along the mouse *Dhfr* gene. Below the gene (with introns shown as a thick line and exons denoted by vertical bars) are open boxes, which represent the locations of probes used to measure transcriptional activity. Polyadenylation sites are noted by arrows, whose thickness and size indicate relative utilization.

reports that transcriptional events were primarily responsible for growth regulation, it now appears that serum stimulation of *Dhfr* (and probably growth regulation as well) is due mainly to posttranscriptional events [31].

It is still not certain what sequences might mediate the regulation of *Dhfr*. The implications from the work with *Dhfr* minigenes was that sequences in the 3′ untranslated region of the mRNA were important [32]; this idea is attractive, for it helps explain why there are some major species of *Dhfr* mRNA in all observed mammalian species that have very long 3′ untranslated regions. Additional evidence for this suggestion is still lacking, and various other cis-acting sequences may be involved in regulation as well.

As mentioned previously, there are multiple 5′ and 3′ ends of *Dhfr* mRNA, and much less attention has been paid to regulation involving differential production of these different mRNAs. There appears to be tissue specificity in the utilization of the -55 and -115 start sites of *Dhfr* mRNA [14], and so it is possible that these sites are used differentially in different growth stages as well, although there is as yet no published evidence for this suggestion. Another report suggests that 5-fluorouracil treatment can change the ratio of the usage of the two most commonly used polyadenylation sites [33]. Therefore there are at least some indications that the multiple 5′ and 3′ ends of *Dhfr* mRNA may be differentially regulated, presumably because of functional differences in at least some of the various transcripts.

VII. THE DIVERGENTLY TRANSCRIBED GENE, *Rep-3*

The identity of the gene being transcribed divergently from the *Dhfr* promoter was of considerable interest, since it was so close to the *Dhfr* gene and since the minor *Dhfr* promoter was found to lie within this other gene. cDNA clones of this gene were obtained and sequenced [14]. Comparison of the predicted amino acid sequence with sequences in GenBank revealed substantial similarities with two bacterial proteins, MutS and HexA [14]. Both MutS and HexA are known to be involved in DNA mismatch repair, and so the best current hypothesis is that the mouse gene is also involved in DNA mismatch repair [14]. The name of the mouse gene, originally *Rep-1*, was changed to *Rep-3* when it was discovered that *Rep-1* and *Rep-2* had already been used to describe other mouse genes [15].

Transcription of the *Rep-3* gene is quite complex. Like *Dhfr*, the *Rep-3* gene has a major and a minor promoter, but unlike *Dhfr*, each promoter is associated with its own exon I, which is spliced to the same exon II [14]. Each of the alternate exon I sequences contains a coding sequence, so that the alternate exons presumably yield proteins with different sequences. In addition, there is evidence for other alternative splicing in the *Rep-3* transcript, giving the potential for several alternative proteins to be produced [14].

The *Rep-3* gene is very large. The approximately 4 kb of mRNA sequence is contained in over 130 kb of DNA and at least 22 exons (Kang Liu and Gray F. Crouse, manuscript in preparation). The map of this locus is shown in Figure 5. As far as is known, amplification of the *Rep-3* gene confers no selective advantage on cells growing in the presence of methotrexate; therefore analysis of the coamplification of *Rep-3* with *Dhfr* should give some indication of the amount of DNA faithfully amplified. Four independent mouse cells with amplified *Dhfr* genes were examined for the presence of the major 4 kb *Rep-3* transcript. Three of the four lines contained detectable levels of this transcript, and in two of these lines the 4 kb transcript was the major species [14]. This result implies that in two of the four cell lines, most of the amplified *Dhfr* genes were in units that extended at least 130 kb 5′ of the *Dhfr* gene. The third cell line contained only a minor amount of the 4 kb transcript, with most of the *Rep-3* transcripts being much smaller. This result suggests that most of the amplified *Dhfr* genes were in units that had rearrangements within 130 kb 5′ of the gene. The fourth cell line showed no amplified levels of *Rep-3*, suggesting either that the gene was not coamplified or that there were rearrangements in the 5′ region that led to low levels of expression of any form of *Rep-3*.

VIII. *Dhfr* AS AN AMPLIFIABLE MARKER

Dhfr has many advantages as an amplifiable marker: there are inhibitors that are very specific to be used for selection; one can either use cell lines that lack Dhfr

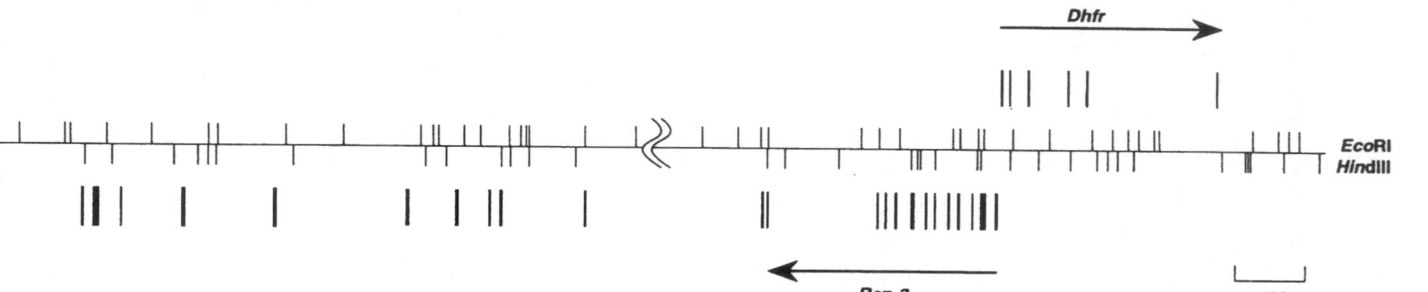

Figure 5 Map of the *Dhfr*/*Rep-3* locus. A map of *Eco*RI and *Hin*dIII restriction sites is shown, with the location of *Dhfr* exons above the line and *Rep-3* exons below. The wavy lines indicate the one region of the *Rep-3* gene, estimated to be 15 kb, that has not yet been cloned.

or use mutant *Dhfr* genes as a dominant marker; if one selects with low levels of the inhibitor, *Dhfr* amplification is the major cause of inhibitor resistance; and one can achieve very high levels of gene amplification. If a gene of interest is linked to the *Dhfr* marker, it will generally be possible to amplify it to high levels. The potential problems associated with this approach have been indicated in the foregoing discussion of amplification in the *Dhfr* locus. The standard method of obtaining coamplified products is to transfect the construct containing the gene of interest and the *Dhfr* marker into a cell line and select for transfectants, which will have integrated randomly into the genome. Subsequent amplification of the integrant will result in amplification of DNA flanking the integration site, and products of genes in this region can also be overexpressed, as *Rep-3* has been in certain cell lines, and rearranged products can also be expressed. Certain integration sites can prove to be refractory to amplification because of flanking sequences. These are considerations that would presumably apply to the use of any amplifiable marker.

ACKNOWLEDGMENTS

Research support from the National Institutes of Health and the American Cancer Society is gratefully acknowledged. I regret that it has been impossible to cite all the important work on the *Dhfr* gene and its amplification. A more complete review on the mammalian *Dhfr* locus has recently been published [34] and should be consulted for further information.

REFERENCES

1. Alt, F. W., Kellems, R. E., Bertino, J. R., and Schimke, R. T. Selective multiplication of dihydrofolate reductase genes in methotrexate-resistant variants of cultured murine cells. *J. Biol. Chem. 253*:1357–1370 (1978).
2. Milbrandt, J. D., Heintz, N. H., White, W. C., Rothman, S. M., and Hamlin, J. L. Methotrexate-resistant Chinese hamster ovary cells have amplified a 135-kilobase-pair region that includes the dihydrofolate reductase gene. *Proc. Natl. Acad. Sci. USA, 78*:6043–6047 (1981).
3. Crouse, G. F., McEwan, R. N., and Pearson, M. L. Expression and amplification of engineered mouse dihydrofolate reductase minigenes. *Mol. Cell. Biol. 3*:257–266 (1983).
4. Schimke, R. T., Brown, P. C., Kaufman, R. J., McGrogan, M., and Slate, D. L. Chromosomal and extrachromosomal localization of amplified dihydrofolate reductase genes in cultured mammalian cells. *Cold Spring Harbor Symp. Quant. Biol. XLV*:785–797 (1981).
5. Wright, J. A., Smith, H. S., Watt, F. M., Hancock, M. C., Hudson, D. L., and Stark, G. R. DNA amplification is rare in normal human cells. *Proc. Natl. Acad. Sci. USA, 87*:1791–1795 (1990).
6. Tlsty, T. D. Normal diploid human and rodent cells lack a detectable frequency of gene amplification. *Proc. Natl. Acad. Sci. USA, 87*:3132–3136 (1990).
7. Ringold, G., Dieckmann, B., and Lee, F. Co-expression and amplification of

dihydrofolate reductase cDNA and the *Escherichia coli* XGPRT gene in Chinese hamster ovary cells. *J. Mol. Appl. Genet. 1*:165–175 (1981).
8. Chang, A. C.Y., Nunberg, J. H., Kaufman, R. J., Erlich, H. A., Schimke, R. T., and Cohen, S. N. Phenotypic expression in *E. coli* of a DNA sequence coding for mouse dihydrofolate reductase, *Nature*, *275*:617–624 (1978).
9. Crouse, G. F., Simonsen, C. C., McEwan, R. N., and Schimke, R. T. Structure of amplified normal and variant dihydrofolate reductase genes in mouse sarcoma S180 cells. *J. Biol. Chem. 257*:7887–7897 (1982).
10. Carothers, A. M., Urlaub, G., Ellis, N., and Chasin, L. A. Structure of the dihydrofolate reductase gene in Chinese hamster ovary cells. *Nucleic Acids Res. 11*: 1997–2012 (1983).
11. Chen, M., Shimada, T., Moulton, A. D., Cline, A., Humphries, R. K., Maizel, J., and Nienhuis, A. W. The functional human dihydrofolate reductase gene. *J. Biol. Chem. 259*:3933–3943 (1984).
12. Milbrandt, J. D., Azizkhan, J. C., Greisen, K. S., and Hamlin, J. L. Organization of a Chinese hamster ovary dihydrofolate reductase gene identified by phenotypic rescue. *Mol. Cell. Biol. 3*:1266–1273 (1983).
13. Federspiel, N. A., Beverley, S. M., Schilling, J. W., and Schimke, R. T. Novel DNA rearrangements are associated with dihydrofolate reductase gene amplification. *J. Biol. Chem. 259*:9127–9140 (1984).
14. Linton, J. P., Yen, J.-Y. J., Selby, E., Chen, Z., Chinsky, J. M., Liu, K., Kellems, R. E., and Crouse, G. F. Dual bidirectional promoters at the mouse *dhfr* locus: Cloning and characterization of two mRNA classes of the divergently transcribed *Rep-1* gene. *Mol. Cell. Biol. 9*:3058–3072 (1989).
15. Smith, M. L., Mitchell, P. J., and Crouse, G. F. Analysis of the mouse *Dhfr/Rep-3* major promoter region by using linker-scanning and internal deletion mutations and DNase I footprinting. *Mol. Cell. Biol. 10*:6003–6012 (1990).
16. Ciudad, C. J., Urlaub, G., and Chasin, L. A. Deletion analysis of the Chinese hamster dihydrofolate reductase gene promoter. *J. Biol. Chem. 263*:16274–16282 (1988).
17. Crouse, G. F., Leys, E. J., McEwan, R. N., Frayne, E. G., and Kellems, R. E. Analysis of the mouse *dhfr* promoter region: Existence of a divergently transcribed gene. *Mol. Cell. Biol. 5*:1847–1858 (1985).
18. Mitchell, P. J., Carothers, A. M., Han, J. H., Harding, J. D., Kas, E., Venolia, L., and Chasin, L. A. Multiple transcription start sites, DNase I-hypersensitive sites, and opposite-strand exon in the 5′ region of the CHO *dhfr* gene. *Mol. Cell. Biol. 6*:425–440 (1986).
19. Fujii, H., and Shimada, T. Isolation and characterization of cDNA clones derived from the divergently transcribed gene in the region upstream from the human dihydrofolate reductase gene. *J. Biol. Chem. 264*:10057–10064 (1989).
20. Means, A. L., and Farnham, P. J. Transcription initiation from the dihydrofolate reductase promoter is positioned by HIP1 binding at the initiation site. *Mol. Cell. Biol. 10*:653–661 (1990).
21. Haber, D. A., and Schimke, R. T. Unstable amplification of an altered dihydrofolate reductase gene associated with double-minute chromosomes. *Cell*, *26*:355–362 (1981).

22. Yen, J.-Y. J. and Kellems, R. E. Independent 5′- and 3′-end determination of multiple dihydrofolate reductase transcripts. *Mol. Cell. Biol.* *7*:3732–3739 (1987).
23. Sazer, S., and Schimke, R. T. A re-examination of the 5′ termini of mouse dihydrofolate reductase RNA. *J. Biol. Chem.* *261*:4685–4690 (1986).
24. Frayne, E. G., Leys, E. J., Crouse, G. F., Hook, A. G., and Kellems, R. E. Transcription of the mouse dihydrofolate reductase gene proceeds unabated through seven poly(A) sites and terminates near a region of repeated DNA. *Mol. Cell. Biol.* *4*: 2921–2924 (1984).
25. Kellems, R. E., Alt, F. W., and Schimke, R. T. Regulation of folate reductase synthesis in sensitive and methotrexate-resistant sarcoma 180 cells, *J. Biol. Chem.* *251*:6987–6993 (1976).
26. Leys, E. J., Crouse, G. F., and Kellems, R. E. Dihydrofolate reductase gene expression in cultured mouse cells is regulated by transcript stabilization in the nucleus. *J. Cell Biol.* *99*:180–187 (1984).
27. Leys, E. J., and Kellems, R. E. Control of dihydrofolate reductase messenger ribonucleic acid production. *Mol. Cell. Biol.* *1*:961–971 (1981).
28. Kellems, R. E., Morhenn, V. B., Pfendt, E. A., Alt, F. W., and Schimke, R. T. Polyoma virus and cyclic AMP-mediated control of dihydrofolate reductase mRNA abundance in methotrexate-resistant mouse fibroblasts. *J. Biol. Chem.* *254*:309–318 (1979).
29. Santiago, C., Collins, M., and Johnson, L. F. In vitro and in vivo analysis of the control of dihydrofolate reductase gene transcription in serum-stimulated mouse fibroblasts. *J. Cell Physiol.* *118*:79–86 (1984).
30. Collins, M. L., Wu, J. S., Santiago, C. L., Hendrickson, S. L., and Johnson, L. F. Delayed processing of dihydrofolate reductase heterogeneous nuclear RNA in amino acid-starved mouse fibroblasts. *Mol. Cell. Biol.* *3*:1792–1802 (1983).
31. Feder, J. N., Assaraf, Y. G., Seamer, L. C., and Schimke, R. T. The pattern of dihydrofolate reductase expression through the cell cycle in rodent and human cultured cells. *J. Biol. Chem.* *264*:20583–20590 (1989).
32. Kaufman, R. J., and Sharp, P. A. Growth-dependent expression of dihydrofolate reductase mRNA frmm modular cDNA genes. *Mol. Cell. Biol.* *3*:1598–1608 (1983).
33. Armstrong, R. D., Lewis, M., Stern, S. G., and Cadman, E. C. Acute effect of 5-fluorouracil on cytoplasmic and nuclear dihydrofolate reductase messenger RNA metabolism. *J. Biol. Chem.* *261*:7366–7371 (1986).
34. Hamlin, J. L., and Ma, C. The mammalian dihydrofolate reductase locus. *Biochim. Biophys. Acta*, *1087*:107–125 (1990).

13

Amplification of the Adenylate-Deaminase 2 and Linked Genes in Chinese Hamster Lung Fibroblasts

Michelle Debatisse, Franck Toledo, Bruno Robert De Saint Vincent, and Gérard Buttin *Institut Pasteur, Paris, France*

I. INTRODUCTION

Adenylate deaminase (AMPD) is a ubiquitous enzyme, with several isoforms, in mammalian tissues. Its physiological importance is indicated by the identification of inherited deficiency states in man. Recently the existence of two genes for AMPD has been disclosed in rat and man: *AMPD1*, expressed in the skeletal muscle of the adult rat, and *AMPD2*, predominantly expressed in other adult tissues and in embryonic tissues, including muscle [1]. During investigations on purine metabolism in the CCL39 line of Chinese hamster fibroblasts and its deoxycytidine-kinase deficient GMA32 derivative, we isolated, from GMA32, mutants overproducing *AMPD2* through gene amplification. In this survey, we summarize the methods exploited for the isolation and characterization of these mutants. We also present information obtained on the organization of the amplified units, on the structure of their junctions, and on the dynamics of the amplification process in this system.

II. THE POLYGENIC AMPLIFIED DOMAIN AND ITS GENETIC MAP

A. Selective Medium

Coformycin is a tight binding inhibitor of adenosine deaminase (ADA); at higher concentrations it also inhibits AMPD [2]. The two enzymes control the channeling of purines from the adenylic to the guanylic compartment. When inosine monophosphate (IMP) biosynthesis is blocked by azaserine, cells can be rescued by exogenous adenylic purines such as deoxyadenosine or adenine (Fig. 1). Deoxyadenosine (dAdo) can be converted to hypoxanthine (Hx) through the sequential operation of ADA and NP, then Hx is phosphoribosylated to IMP by HGPRT. Thus, in medium containing azaserine and deoxyadenosine, inhibition of ADA by coformycin leads to cell death due to starvation of IMP and guanylic derivatives.

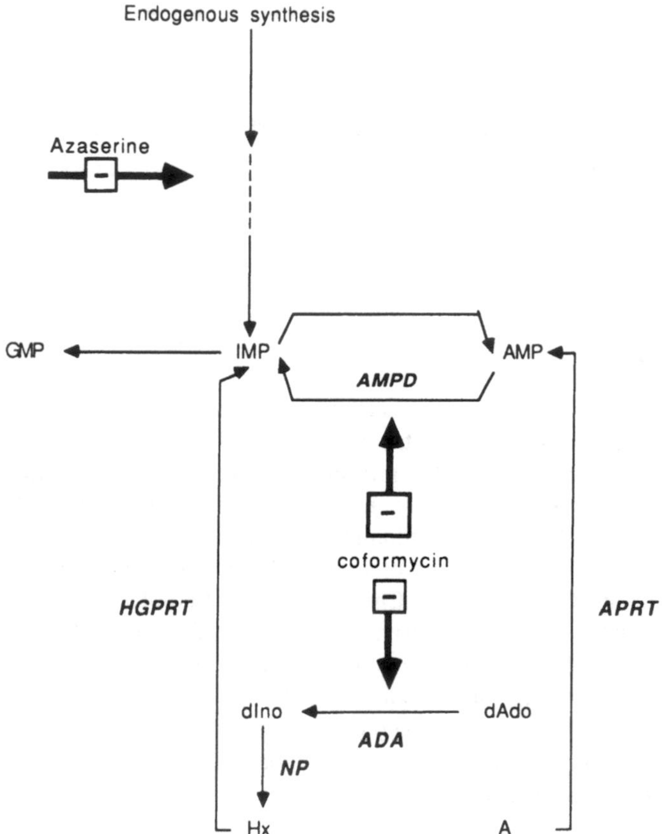

Figure 1 IMP starvation arising from the simultaneous shutoff of the "de novo" biosynthesis pathway and of the two adenylic purine deamination pathways: boxed minus signs indicate inhibitor action; A, adenine; dAdo, deoxyadenosine; dIno, deoxyinosine; Hx, hypoxanthine; AMP, adenosine monophosphate; IMP, inosine monophosphate; GMP, guanosine monophosphate; AMPD, adenylate deaminase; APRT, adenine phosphoribosyl transferase; ADA, adenosine deaminase; HGPRT, hypoxanthine–guanine phosphoribosyl transferase; NP, nucleoside-phosphorylase.

Adenine is also converted to IMP after phosphoribosylation by APRT and deamination by AMPD. Using this purine source, AMPD-dependent cell growth can also be prevented by coformycin addition. Resistant mutants have been selected in each of these two complex media and were shown to overexpress either ADA [3] or AMPD [4]. The AMPD selective medium has been successfully used by others with human and rat cells [5]. AMPD-overproducing mutants were further extensively studied.

B. Stepwise Isolation of Variants with Increasing Coformycin Resistance and AMPD Activity

The plating efficiency of GMA32 in the presence of adenine as a purine source is reduced to about 10^{-5} if 0.5 μg/ml of coformycin is added (Fig. 1). Three resistant clones, designated 4, 5, and 6, were recovered and further analyzed. All three exhibited a 7- to 10-fold increase in AMPD-specific activity and were resistant only to the coformycin concentration that was used in their selection. During prolonged growth in regular medium, most cells of these clones returned to wild-type drug sensitivity [4]. Two clones capable of growth in medium containing a 10 times higher drug concentration (5 μg/ml), were isolated from each of the first-step mutants. In these six second-step lines, AMPD-specific activity was 30- to 40-fold higher than in the wild-type line. Like their first-step parents, these second-step mutants remained sensitive to coformycin concentrations above that chosen for their isolation. In a third step, we derived from each of them one third-step mutant resistant to 25 μg/ml of coformycin; a 100- to 150-fold increase in AMPD activity was reached in these mutants compared to the wild-type line. We thus derived three independent families of mutants (Fig. 2), with well-known lineage [6].

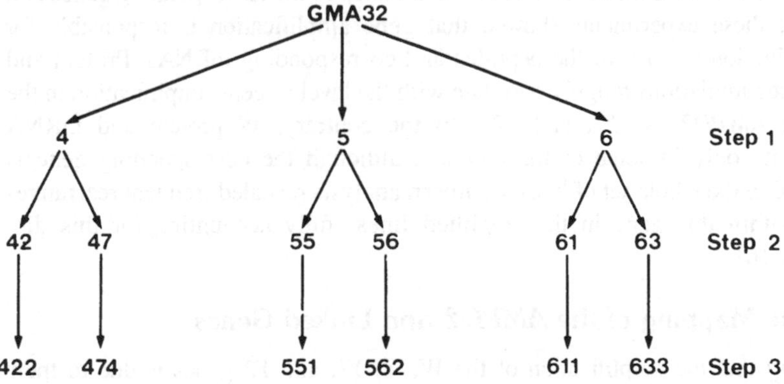

Figure 2 Derivation of cell lines (nomenclature simplified from that used in earlier publications).

C. Characterization of Third-Step Mutants: Several Genes Are Overexpressed and Amplified

Gel electrophoresis analysis of extracts made from third-step mutants displayed a remarkable pattern: at least five peptides can accumulate in mutant lines [6]. One of them, abundant in every mutant, was identified as AMPD. AMPD was easily purified from a resistant line (line 611) because the enzyme remains bound to phosphocellulose under conditions at which most other proteins are eluted: a 150-fold purification was achieved in a single step. This purified AMPD was used as an electrophoretic marker and as an immunogen to raise antibodies in rabbits [4]. Four other proteins designated X, Y1, Y2, and W are each abundant only in a subset of the "third-step" mutants; for example, Y1 accumulates at various degrees in all lines but 611 and 633, while X, Y2, and W accumulate in lines 611, 633, and 422, poorly in 551 and 562, but not at all in 474 [6]. A similar distribution of these four peptides was found among the "in vitro" translation products of mRNAs made from the different mutants. Clones corresponding to X, Y1, Y2, and W were isolated from cDNA libraries constructed from 611 and 474 mRNAs [7]; these two lines were chosen because they overproduce RNAs encoding complementary sets of coaccumulable proteins. The nucleotide sequences of *Y1* and *X* cDNA clones were determined and allowed the identification of these genes: the coding sequence of *Y1* is essentially identical to mammalian glutathione *S*-transferase (GST) of the μ family [8]; the *X* nucleotide sequence is significantly homologous to a bacterial and to a yeast seryl–tRNA synthetase gene (manuscript in preparation). The identification of these two genes was further supported by the observation that the accumulation of *X* and *Y1* products was accompanied by an increased activity of seryl–tRNA synthetase and GST, respectively.

X, *Y1*, *Y2*, and *W* cDNAs and a rat *AMPD2* cDNA (kindly provided by Dr. Sabina) were used to titrate the relative abundance of the corresponding genes and mRNAs: these experiments showed that gene amplification is responsible for overproduction of each of the peptides and corresponding mRNAs. Protein and mRNA accumulations roughly correlate with the level of gene amplification in the cases of *AMPD2*, *X*, *Y1*, and *Y2*. On the contrary, W protein and mRNA accumulate only in some of the mutants, although the corresponding gene is amplified in the whole set of lines. Southern analysis revealed frequent rearrangements within this gene in the amplified lines, fully accounting for this discrepancy [6].

D. The Mapping of the *AMPD2* and Linked Genes

Assuming that the amplification of the *W*, *X*, *Y1*, and *Y2* genes is due to their physical linkage to the selected gene, the *W* gene should be very close to the *AMPD2* gene because at least part of it is coamplified in every "third-step"

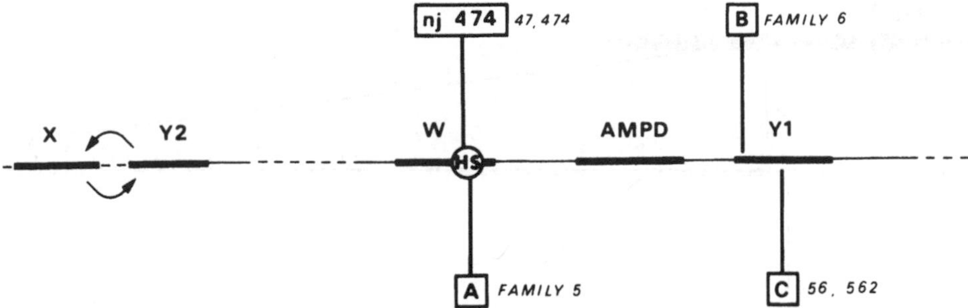

Figure 3 Map of the *AMPD2* region in the wild-type line, indicating location of the four rearrangements analyzed: HS, recombination hot spot; A, B, C, NJ474, rearrangements; on the right are indicated the cell lines in which they exist.

mutant. The segregation pattern of the other three genes in the different "third-step" lines allowed us to draw a genetic map of the amplifiable locus (Fig. 3) on which *AMPD2* lies between *W* and *Y1*, and *X* and *Y2* beyond *W* [6]. To confirm this map, two cosmid libraries were constructed from the DNAs of lines 56 and 61 [9]. We chose these two mutants because in line 61, most amplified units include the *W* gene, whereas in line 56 a highly amplified novel joint (A) interrupts this gene. Thus, the A joint can be used to orientate the *W* gene with regard to the *AMPD2* gene, during chromosome walking experiments. In line 56, the *Y1* gene is highly amplified, suggesting that the wild-type sequence arrangement between genes *AMPD2* and *Y1* is conserved. On the contrary, the *Y1* gene is amplified very weakly in line 61, suggesting that another joint (B) must be located between the *AMPD2* and *Y1* genes in this mutant. A first screening with *W* or *Y1* cDNA allowed us to clone the previously identified A joint and the putative B joint from the libraries constructed from 56 and 61 DNAs, respectively. In a one-step walk from *W* toward *AMPD2* we selected cosmids, some of which hybridize to *Y1* cDNA. Therefore, the *W* and *Y1* genes are physically linked and lie about 50 kb apart in the wild-type DNA. Moreover, the *AMPD2* gene maps, as expected, between them (Fig. 3).

III. STRUCTURAL AND DYNAMIC ASPECTS OF AMPLIFICATION

A. Analysis of Four Amplified Rearrangements

1. *Structure of the Junctions*

Figure 4b shows the restriction map of the B rearrangement, aligned along the wild-type map. It can be seen that the rearranged sequence contains an inverted duplication (arrows) organized around a 8 kb nonpalindromic center. This non-duplicated sequence was subcloned and used to probe overlapping cosmids

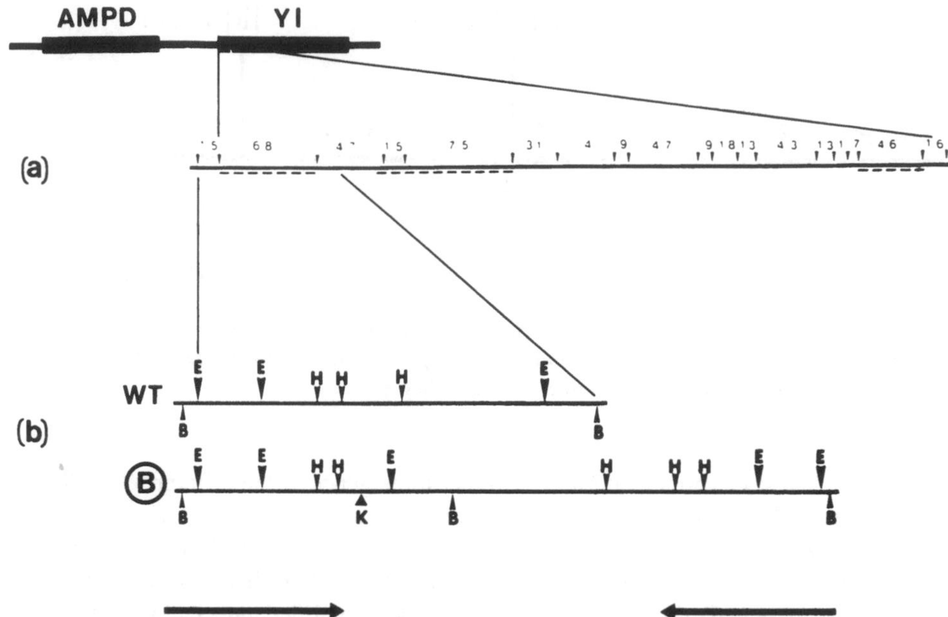

Figure 4 Localization and structure of the B rearrangement. (a) *Eco*RI map of the wild-type *Y1* region contributing to the B rearrangement. (b) Compared restriction map of wild-type and rearranged sequences: E, *Eco*RI; H, *Hin*dIII; B, *Bam*HI; K, *Kpn* I.

covering the cloned *AMPD2* domain; it hybridized to four *Eco*R1 fragments underlined in Figure 4a. Thus, the nonpalindromic center of the duplication is a patchwork of sequences spanning some 40 kb of the wild-type genome. A precise analysis of the B rearrangement indicates that at least four events of recombination or deletion must be postulated to explain its formation (to be published). The A rearrangement is also complex and results from at least two recombination or deletion events. The first one links a sequence of the *W* gene to a sequence that is not included in the 150 kb of cloned DNA but is part of the amplifiable domain, since it is slightly amplified in some independent mutants; the corresponding novel joint has been called NJ551–562 [10]. The second recombination event links this coamplifiable sequence to a sequence that is amplified only in family 5 mutants and has been mapped, by "in situ" hybridization, some 5 Mb away from the *AMPD2* gene on chromosome 1 (to be published).

We analyzed two other novel joints, designated as C and NJ474 [11]. Like the B rearrangement, they link sequences in an inverted orientation with a nonpalindromic central sequence of 7.5 and 0.862 kb long in C and NJ474, respectively. In both cases, a single illegitimate recombination event can account for the formation of the junction (Fig. 3, Table 1).

Table 1 Properties of the Four Analyzed Rearrangements

Rearrangement	Inverted	Minimal number of recombination events	First observed: Step of selection	First observed: Copy number
A	No	2	First	Single
B	Yes	4	First	Single
C	Yes	1	Second	Multiple
NJ474	Yes	1	Second	Multiple

Thus, three out of the four rearrangements studied so far link DNA stretches in inverted orientation. In these three cases, they link two amplified units, since no other amplified rearrangement has been detected between each of these joints and the *AMPD2* gene. These examples add to the growing evidence that the amplified DNA is frequently arranged as inverted duplications in mammalian cells [12].

2. *Mode of Formation*

Because the lineage of the sequentially amplified mutants is known, we could reconstruct the history of each of the four cloned joints (Fig. 3). The A and B rearrangements are present in every member of families 5 and 6, respectively. They were first detected in the corresponding first-step mutants at single copy number. The C joint was observed in lines 56 and 562, where it is repeated some 10 times; it is not present in the parent 5. The NJ474 joint was first observed in a second-step mutant (mutant 47) directly at a high copy number; it is present on every copy accumulated during the second step of amplification. It is then reamplified in line 474, the third-step mutant derived from 47 [11].

Thus, two main categories of junctions were observed in this panel of mutants. Complex rearrangements (A and B) were identified in first-step mutants at single copy number. Simple junctions were formed and amplified during the second step of selection (Table 1).

B. Preferentially Reamplified Structures

The first-step mutant 5 was subcloned; each subclone tested contained about 15 amplification units, one of which was labeled by the A rearrangement as in the original clone. One second-step mutant was derived from each of five subclones: in these independent second-step derivatives, only the unit identified by the A rearrangement was involved in the reamplification events [9]. Because the random probability of one particular unit being amplified is about 1 in 15, amplification of the unit identified by the A joint in five of five independent mutants studied is a result expected with a probability of about 10^{-6}. Thus, the choice of the reamplified unit is not random in this case. Preferential reamplification of the structure identified by the B rearrangement was also obvious, although it has been studied less extensively.

C. Identification of a Hot Spot for Novel Joints Associated with Amplification

Using cosmid probes covering some 120 kb of wild-type coamplifiable sequence out of the *W* gene, we detected only four novel joints—including B and C—in the DNA of the six highly amplified cell lines. On the contrary, 12 different junctions—clustered within 2.6 kb of a *W* intron—including A and NJ474, have been detected in the same six lines, and none in 8.1 kb in the immediate surrounding. The frequency of novel joints in this short region of the *W* gene is therefore much higher than in the rest of the gene and is about 150-fold higher than in the studied sequence outside this gene. The *W* "hot spot" has been sequenced: it is a mosaic of *Alu*-like repetitive sequences and pseudopalindromes, it also contains long A + T-rich segments. Each of these structures is known to favor genomic instability and recombination [10].

There are similarities between the crossover sites of NJ551-562 (in the A rearrangement) and NJ474. The single NJ551-562 break point located in the hot spot lies at the center of a 70 bp palindrome, which could form a significantly stable stem–loop structure. Both NJ474 break points are located in the hot spot; they also lie at the top of potential stem–loop structures. Moreover, one of the NJ474 break points, like that of NJ551-562, is associated with a putative topoisomerase I cleavage site. Thus, a related mechanism could have been implicated in these two independent recombination events.

D. Cytological Analysis

The panel of the established amplified lines (Fig. 2) was analyzed using in situ hybridization of tritiated probes on banded chromosomes (unpublished results). In first- and second-step mutants, clusters of amplified copies were most frequently located close to the end of the long arm of a chromosome 1; they were frequently found on several chromosomes in third-step mutants. The single copy genes that were not detected in these experiments can now be observed using fluorescent probes. We recently localized the single copies of the *AMPD2*, μ glutathione transferase (*Y1*), and *W* genes on the long arm of chromosomes 1 of GMA32. Thus, in first-step and second-step lines, amplified copies are usually integrated in one of the homologues that bears the unamplified gene in the parental line. When mutants selected to resist to the concentration of coformycin used for the isolation of the first-step lines (0.5 μg/ml) were examined as soon as possible after an amplification event according to the method described by Smith et al. [13], a regular ladder of a few large (50 Mb) units was frequently observed on a chromosome 1 while the single copy of the marker gene remained visible on the other chromosome 1. Within the same clone, at the early stages, cells with these regular ladders coexist with cells exhibiting a variety of more condensed structures, which finally prevail in clones allowed to expand further. Thus, the condensed structure specific of each so-called first-step line is already the product of complex events following the generation of the first amplified copies.

IV. CONCLUDING REMARKS AND PERSPECTIVES

As previously observed for a variety of other systems, the chromosomal sequences amplified in cells overproducing AMPD2 are far larger than the selected gene. Their size in mutants accumulating 100–150 copies of the *AMPD2* gene—designated above as third-step mutants—is not known but exceeds 100 kb, and these units include several expressed genes. Mapping DNA sequences of that size is a common difficult problem and requires the demanding technology of chromosome walking or jumping. It is therefore of particular interest that, as exemplified in this system, the presence of an amplifiable gene can be exploited to order a cluster of genes by studying the segregation pattern of the sequences flanking the selected gene in independent amplified mutants.

Our initial attempts to reconstruct the mechanisms of gene amplification were carried out on established cell lines obtained through defined steps of increasing resistance to the selective agent. A molecular analysis of this material disclosed two features of the amplification process, which have been also observed in other model systems: (1) the relative homogeneity of the amplified units accumulated within the cells of each highly amplified mutant, contrasting with the marked differences observed from clone to clone, and (2) the frequent organization of this material as clusters of units linked "head to head."

When care was taken to examine amplification units in cells selected under conditions favoring the observation of recent amplification events, a significant proportion of the cells were characterized by the distribution of the *AMPD2* and flanking genes within regular tandems of very large (~ 50 Mb) units located on one copy of chromosomes 1 which bear the original *AMPD2* genes. As discussed by Smith et al. [13], who observed such structures in the *CAD* amplification model, and by Trask and Hamlin [14], they are likely to be the products of unequal sister chromatid exchanges linking the first amplified units in a "head-to-tail" configuration. A rapid evolution during clonal expansion can be explained if the large ladders are highly unstable and if, among the possible rearrangements occurring in the structures, some permit better cell growth under the actual environmental conditions. The molecular mechanisms leading to shortening and to "head-to-head" arrangement of amplified units is presently unknown. Work is in progress to determine whether intrachromosomal rearrangements and deletions take place, and whether transient extracellular structures are generated and reintegrated as is the case in some other systems [15]. Whatever the possible complexity of this critical stage, the present analysis of this model system suggests that (1) a segregative mechanism, probably unequal sister chromatid exchange, leads to different outcomes because unstable structures are generated in the primary event, and (2) rearrangements, created by chance during this molecular "tinkering" step, may allow different mechanisms—including replicative ones—to operate. Such switches occur in few cells of the population and are not per se amenable to direct study, unless the cells have a clear selective advantage and overgrow the population. Then subclones are recovered in which a particular rearrangement and a particular mechanism are cloned. In two of our families of mutants, such preferentially reamplified units are identified: the molecular basis of their selective advantage is presently under study.

ACKNOWLEDGMENTS

We thank M. Mahérou for careful typing of the manuscript. This work was supported in part by the Université Pierre et Marie Curie, the Ligue Nationale Française contre le Cancer, the Fondation pour la Recherche Médicale Française, and the Ministère de la Recherche et de la Technologie (F.T.).

REFERENCES

1. Morisaki, T., Sabina, R. L., and Holmes, E. W. Adenylate deaminase: A multigene family in humans and rats. *J. Biol. Chem.* *265*:11482–11486 (1990).
2. Debatisse, M., and Buttin, G. The control of cell proliferation by preformed purines: A genetic study. II. Pleiotropic manifestations and mechanism of a control exerted by adenylic purines on PRPP synthesis. *Somatic Cell. Genet.* *3*:513–527 (1977).
3. Fernandez-Mejia, C., Debatisse, M., and Buttin, G. Adenosine-resistant Chinese hamster fibroblast variants with hyperactive adenosine-deaminase: An analysis of the protection against exogenous adenosine afforded by increased activity of the deamination pathway. *J. Cell Physiol.* *120*:321–328 (1984).
4. Debatisse, M., Berry, M., and Buttin, G. Stepwise isolation and properties of unstable Chinese hamster cell variants that overproduce adenylate deaminase. *Mol. Cell. Biol.* 2:1346–1353 (1982).
5. Sabina, R. L., Marquetant, R., Desai, N. M., Kaletha, K., and Holmes, E. W. Cloning and sequence of rat myoadenylate deaminase cDNA: Evidence for tissue-specific and developmental regulation. *J. Biol. Chem.* *262*:12397–12400 (1987).
6. Debatisse, M., Hyrien, O., Petit-Koskas, E., Robert de Saint Vincent, B., and Buttin, G. Segregation and rearrangement of coamplified genes in different lineages of mutant cells that overproduce adenylate desaminase. *Mol. Cell. Biol.* *6*:1776–1781 (1986).
7. Debatisse, M., Robert de Saint Vincent, B., and Buttin, G. Expression of several amplified genes in an adenylate-deaminase overproducing variant of Chinese hamster fibroblasts. *EMBO J.* *3*:3123–3124 (1984).

8. Robert de Saint Vincent, B., Hyrien, O., Debatisse, M., and Buttin, G. Coamplification of μ class glutathione *S*-transferase genes and an adenylate desaminase gene in coformycin-resistant Chinese hamster fibroblasts. *Eur. J. Biochem. 193*:19–24 (1990).
9. Debatisse, M., Saito, I., Buttin, G., and Stark, G. R. Preferential amplification of rearranged sequences near amplified adenylate deaminase genes. *Mol. Cell. Biol. 8*:17–24 (1988).
10. Hyrien, O., Debatisse, M., Buttin, G., and Robert de Saint Vincent, B. A hot spot for novel amplification joints in a mosaic of *Alu*-like repeats and palindromic A + T-rich DNA. *EMBO J. 6*:2401–2408 (1987).
11. Hyrien, O., Debatisse, M., Buttin, G., and Robert de Saint Vincent B. The multicopy appearance of a large inverted duplication and the sequence at the inversion joint suggest a new model for gene amplification. *EMBO J. 7*:407–417 (1988).
12. Stark, G. R., Debatisse, M., Giulotto, E., and Wahl, G. M. Recent progress in understanding mechanisms of mammalian DNA amplification. *Cell*, *57*:901–908 (1989).
13. Smith, K. A., Gorman, P. A., Stark, M. B., Groves, R. P., and Stark, G. R. Distinctive chromosomal structures are formed very early in the amplification of *CAD* genes in Syrian hamster cells. *Cell*, *63*:1219–1227 (1990).
14. Trask, B. J., and Hamlin, J. L. Early dihydrofolate reductase gene amplification events in CHO cells usually occur on the same chromosome arm as the original locus. *Genes Dev. 3*:1913–1925 (1989).
15. Ruiz, J. C., and Wahl, G. M. Chromosomal destabilization during gene amplification. *Mol. Cell. Biol. 10*:3056–3066 (1990).

14

Organization and Amplification of Metallothionein Genes in Eukaryotic Cells

C. Edgar Hildebrand, Deborah L. Grady, and Raymond L. Stallings *Los Alamos National Laboratory, Los Alamos, New Mexico*

I. INTRODUCTION

Metallothioneins (MTs) are low molecular weight, cysteine-rich metal-binding proteins (reviewed in Ref. 1). Mammalian MTs contain 60–61 amino acids, of which 20 are cysteine, and have two evolutionary conserved metal-binding domains [1]. These metalloproteins are involved in cellular detoxification of heavy metals (Cd^{2+}, Hg^{2+}) and play a putative role in regulation of essential trace metals (Zu^{2+}, Cu^{2+}) [1–3]. MTs in lower eukaryotes differ markedly from mammalian MTs [yeast (53 amino acids), and neurospora (25 amino acids)], although the neurospora MT is precisely homologous to the amino-terminal 25 amino acids of mammalian MTs [1–3]. In mammalian organisms, human *MT* genes are present as multigene families (see Refs. 1–5 and this review). Individual *MT* genes are under inducible regulation by metal ions and/or glucocorticoids at the transcriptional level [1,5].

The role of MTs in cellular protection against heavy metal toxicity has been exploited to select metal-resistant cell lines. In general, cell lines selected for metal resistance overexpress MTs and have amplified *MT* genes [1,6]. Amplification of eukaryotic *MT* genes has been described in cellular systems ranging from yeasts to human (summarized in Refs. 1, 2, and 6 and this review).

II. CELLULAR METAL RESISTANCE AND MT OVERPRODUCTION

Overexpression of *MT*s has been achieved in yeast [7], mouse cells [8], hamster cells [6,9], and in human–rodent cell hybrids (Crawford, unpublished observations) by growing cells in culture in the presence of elevated concentrations of Cu^{2+} (for yeast) [7] or the toxic heavy metal cadmium (for mammalian cells [6,8,9].

In the yeast *Saccharomyces cerevisiae*, copper-MT encoded by the *CUP1* gene [7] is overproduced in copper-resistant variants selected by stepwise increases of copper concentrations in growth medium.

Early studies of metal resistance and *MT* expression in both mouse [8] and hamster [6,9] mammalian cells revealed an unexpected sensitivity to Cd toxicity and an absence of detectable MT production [6,8,10]. However, continued culture in Cd^{2+}-containing medium yielded variant Cd-resistant, metallothionein-producing cell lines [8,10]. Further studies of this first step in selection of Cd-resistant rodent cell lines showed that, in Cd-sensitive *MT*-nonexpressing cell lines, the *MT* genes are hypermethylated [11,12]. In contrast, *MT* genes were hypomethylated in *MT*-expressing, Cd-resistant variants selected from the parental Cd-sensitive cells [11,12]. Treatment of Cd-sensitive cell lines with the hypomethylating agent 5-azacytidine dramatically increased the frequency of appearance of Cd-resistant variants [11,12]. In Chinese hamster ovary (CHO) cells, the first level of selection for Cd resistance produced variants inducible for *MT* but retaining the same (single copy) number of *MT* gene copies as the Cd-sensitive parent [12]. Stepwise increases in Cd concentration in the selection procedure yielded variants with elevated levels of Cd resistance correlated with increased MT production, *MT* mRNA expression, and *MT* gene copy number [5,6,10].

III. *MT* GENE ORGANIZATION AND AMPLIFICATION

A. Yeast

Amplification of the copper thionein *CUP1* gene in copper-resistant yeast can be explained by a two-step process: (1) an initial duplication of the *CUP1* gene followed by (2) homologous recombination and unequal crossing over, sister chromatid exchange, or gene conversion. This action results in amplification, or contraction, of the *CUP1* region [7,13,14]. However, DNA sequence analysis did not reveal regions of sequence homology, which were predicted if homologous recombination events were involved in the amplification process [1,13,14]. Results of genetic analyses of an amplified *CUP1* gene through the mating of copper-resistant and copper-sensitive yeast strains suggest that gene conversion is the primary mechanism for meiotic *MT* gene amplification in yeast [1].

B. Mammalian Cells

Mouse variant cell lines derived from a Cd-sensitive Friend erythroleukemia cell line selected by stepwise increases in Cd concentration in growth medium showed approximately 14-fold increased expression of *MT* mRNA (*MT* 1) and a 6-fold increase in *MT* I gene copy number [8]. Coamplification of the mouse *MT* I and *MT* II genes was not demonstrated directly. However, the close linkage of *MT* I and *MT* II coupled with the observation of an amplified 55 kb fragment containing the *MT* I gene suggests that the *MT* genes are coamplified [8,15].

In CHO cells, *MT* I and *MT* II genes are arranged in tandem in the same transcriptional orientation separated by approximately 6 kb of DNA (Fig. 1) [4]. Molecular analyses of *MT* overexpressing Cd-resistant cell lines (Fig. 2) revealed coordinate amplification (up to 10–20-fold amplification) of both *MT* I and II genes [2]. Cytogenetic analyses of the Cd-resistant cell line, Cd^{r}200T1 demonstrated a consistent chromosomal rearrangement involving an interchromosomal

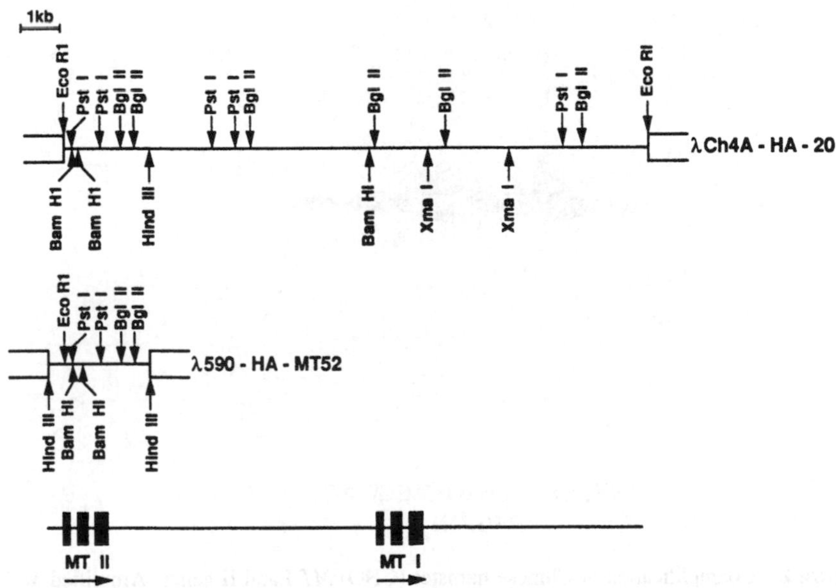

Figure 1 Chinese hamster *MT* locus. λCh 4A-HA-20: restriction enzyme map of a clone containing the Chinese hamster *MT* I gene from a Charon 4A *Eco*RI total digest library prepared from a Chinese hamster cell line amplified for the *MT* genes. The clone consists of 13 kb of hamster DNA and contains the entire *MT* I gene and a portion of the *MT* II gene. λ 590-HA-MT52: restriction enzyme map for the *MT* II gene region present on the cloned 2.2 kb *Hin*dIII fragment. The map of the *MT* locus (shown at the bottom) indicates the coding regions of the exons (dark boxes). Thin dark lines represent intronic regions and thick dark lines the intergenic spaces. Arrows indicate the transcriptional orientation.

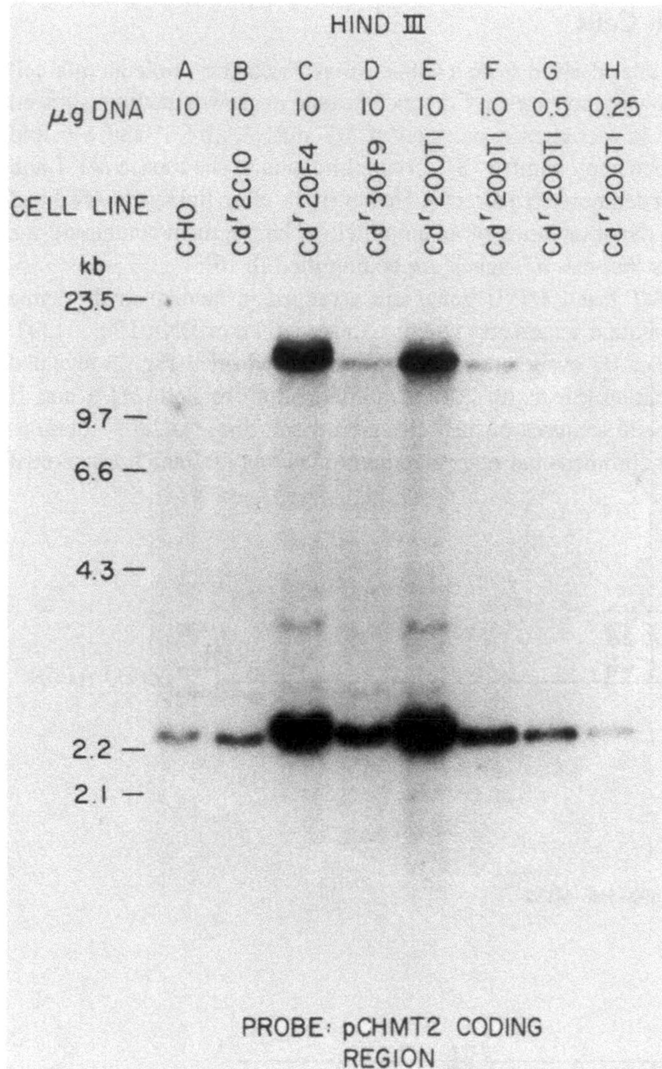

Figure 2 Coamplification of Chinese hamster (CHO) *MT* I and II genes. Amplified *MT* genes were detected by Southern filter hybridizations of *Hin*dIII-digested DNA from Cd-sensitive CHO cells and from cell lines having increasing levels of Cd resistance (Cd^r 2C10, Cd^r 20F4, Cd^r 30F9, and Cd^r 200T1, lanes B–E), electrophoresed in an agarose gel, and probed with a CHO *MT* II cDNA probe (pCHMT2) [6]. By varying the loads of Cd^r 200T1 DNA (lanes E–H), it is possible to compare the copy number of *MT* sequences in Cd^r 200T1 with those in CHO and the other variants.

translocation event near the amplified *MT* gene cluster identified by in situ hybridization (Fig. 3). These studies localized the amplified *MT* cluster near a break point involving the short arm of Chinese hamster chromosome 3 [2]. To determine whether this chromosomal rearrangement was fortuitously associated with the aneuploid CHO karyotype, a new euploid Chinese hamster cell line (Dede) was subjected to Cd selection, and Cd-resistant variants were isolated [16]. Molecular analysis of Cd-resistant variants of the Dede line demonstrated coordinate amplification of both *MT* I and II genes [16]. Cytogenetic analysis of the same variants revealed consistent karyotypic rearrangements involving the short arm of chromosome 3 (Fig. 3) [16]. Independently, a Chinese hamster × mouse hybrid panel was used to map the *MT* gene cluster in Chinese hamsters to chromosome 3 [16]. Collectively, these studies in Chinese hamster cell lines strongly implicate chromosomal rearrangements involving chromosome 3p in the region of the *MT* gene cluster in *MT* gene amplification [2,16]. In the case of the mouse *MT* gene amplification, Cd-resistant mouse cell lines were found to have chromosomal aneuploidies characterized by the presence of additional chromosomes [1]. It is noteworthy that the *MT* locus maps to mouse chromosome 8 in an extended region that is syntenic with Chinese hamster chromosome 3 and human chromosome 16.

Attempts to study human *MT* gene amplification by Cd selection of human *MT*-inducible diploid human fibroblast cultures were unsuccessful. However, fusion of Cd-sensitive (*MT*-noninducible) CHO cells with an *MT*-inducible diploid human fibroblast followed by stepwise selection in increasing concentrations of $CdCl_2$ in the culture medium yielded a population of Cd-resistant hybrid cells (Crawford, unpublished results). Several individual Cd-resistant clones were isolated and characterized for expression of hamster and human MTs. These Cd-resistant hybrids expressed only human MTs and no hamster MT as determined by nondenaturing polyacrylamide gel electrophoresis of cytoplasmic extracts (see Ref. 6, and Hildebrand, unpublished results). Molecular analysis of human *MT* gene sequences via quantitative DNA spot blotting and Southern gel transfer revealed approximately four- to fivefold amplification of all expected fragments of the functional human *MT* gene cluster (Ref. 3, and unpublished results). Thus, these analyses demonstrated coamplification of all functional human *MT* genes localized in human chromosome 16 (and spanning approximately 60 kb of DNA) [3], while the Chinese hamster *MT* genes remained unexpressed and at single copy level.

Cytogenetic analyses using G-banding techniques on Cd-resistant hybrid cell lines having amplified human *MT* genes demonstrated rearrangement of human chromosome 16 (Stallings, unpublished observations). The combined results from studies of *MT* gene amplification and Chinese hamster, mouse, and hamster × human hybrid cells strongly suggest a mechanistic relationship between amplification of *MT* genes and chromosomal rearrangements. A relationship between gene amplification and chromosomal rearrangements involving the amplified locus has been reported for other genes (e.g., *CAD* and *DHFR* loci) (see Biedler and

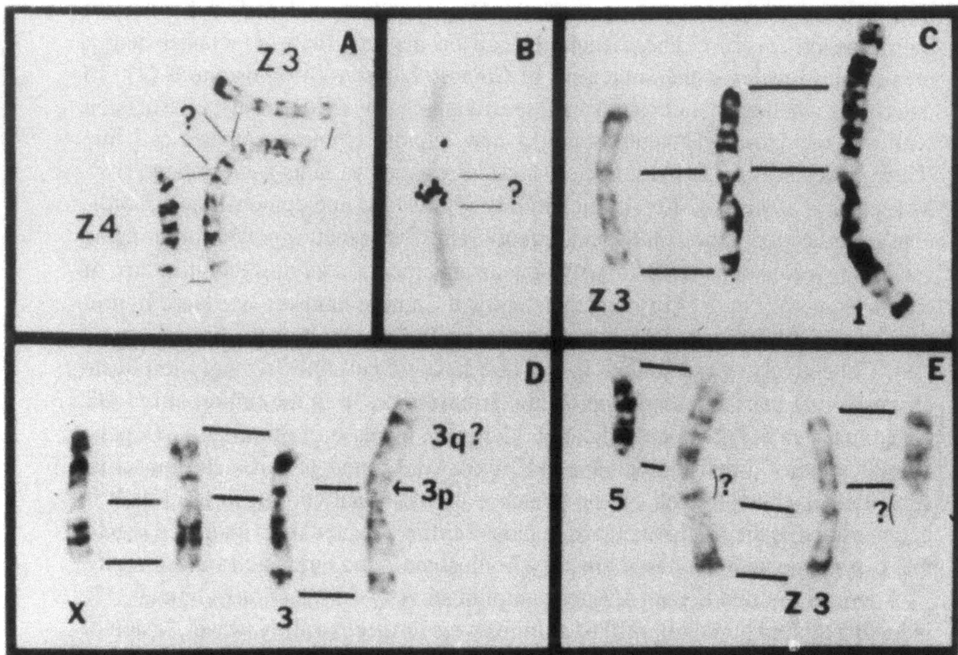

Figure 3 G-Banded metaphase chromosomes from CHO and Dede cell lines resistant to Cd^{2+}. (A) In the Cd^r 200T1 variant, a large submetacentric chromosome was derived by a translocation between wild-type CHO chromosomes Z3 and Z4. Both chromosomes Z3 and Z4 originated from normal Chinese hamster chromosome 3. (B) In situ hybridization analysis of Cd^r 200T1 with an ^{125}I-labeled *MT* II coding sequence probe revealed a single site of hybridization located at the translocation junction between the Z3 and Z4 chromosomes. (C) In the variant Cd^r 75T8, a large portion of chromosome Z3p was translocated to chromosome Z7 (not shown) and replaced by a segment from chromosome 1 (shown upside-down to facilitate comparison). The break point on Z3p is very similar to that observed in Cd^r 200T1. (D) The Cd-resistant cell line, Dede 5-2, derived from the euploid Chinese hamster line Dede, contains one normal X chromosome (upside-down), a 3p:Xp translocation chromosome, one normal chromosome 3, and a chromosome consisting of 3q, a small segment of 3p (p1→ 17), and a large segment of unidentified origin (the more distal segment is similar to 3q in banding pattern). (E) A translocation between chromosome 5 (upside-down to facilitate comparison) and Z3p has occurred in the variant Cd^r 20F4. The translocation chromosome has a short arm composed of chromosome 5q and a long arm composed primarily of Z3p. The question mark denotes a G-band of uncertain origin in the region derived from Z3p. The last chromosome on the right was derived from Z3q. The short arm of this chromosome (marked by ?) could have originated from 5p (by reciprocal translocation) or from the proximal portion of Z3p. (See Ref. 16 for more detail.)

Spengler, Chapter 25, this volume, and Wahl et al., Chapter 40, this volume). Amplification of loci near translocation break points also may play a key role in karyotypic rearrangements observed in cancer (reviewed in this volume: see Section Four).

IV. SUMMARY

New technologies in cytogenetics and molecular biology (e.g., quantitative fluorescence in situ hybridization [17,18], pulsed-field gel electrophoresis [19,20], two-dimensional (neutral/neutral and neutral/alkaline) gel electrophoresis [21], and polymerase chain reaction [22], to note a few) provide the tools to describe in molecular detail the dynamics of the amplification process and the structure(s) of the amplified units as well as associated changes in chromosomal organization (e.g., rearrangements, translocations) coincident with amplification of *MT* genes. Furthermore, the emergence of detailed long-range physical maps of the human genome [23] as well as the genomes of other eukaryotic species (mouse, *Drosophila*, nematode, yeast) will provide valuable resources for more detailed molecular and cytogenetic studies of extended regions encompassing not only the *MT* gene locus but also the many other amplifiable mammalian loci.

ACKNOWLEDGMENTS

The authors thank Darlene Lee for assistance in the preparation of this manuscript. The authors also are indebted to the numerous collaborators, whose work is cited herein, who have contributed to the studies of cadmium resistance and *MT* gene amplification in mammalian cell lines over the course of the past 15 years. This work was performed under the auspices of the U.S. Department of Energy, Office of Health and Environmental Research, and the Los Alamos National Laboratory.

REFERENCES

1. Hamer, D. H. Metallothionein, *Annu. Rev. Biochem.* *55*:913–951 (1986).
2. Grady, D. L., Moyzis, R. K., and Hildebrand, C. E. Molecular and cellular mechanisms of cadmium resistance in cultured cells, in *Metallothionein II* (J. H. R. Kagi and Y. Kogima, eds.), *Experientia* Suppl. Vol. 52, Birkhauser Verlag, Basel, 1987, pp. 447–456.
3. Grady, D. L., Robinson, D. L., and Hildebrand, C. E. Genomic sequence of the Chinese hamster *MT* I gene. *Nucleic Acids Res.* *18*:7149 (1990).
4. West, A., Stallings, R., Hildebrand, C. E., Chiu, R., Karin, M., and Richards, R. I. Human metallothionein genes: Structure of the functional locus at 16q13. *Genomics*, *8*:513–518 (1990).
5. Enger, M. D., Rall, L. B., and Hildebrand, C. E. Thionein gene expression in Cd^{2+}-resistant variants of the CHO cell correlation of thionein synthesis rates with

translatable mRNA levels during induction, deinduction and superinduction. *Nucleic Acids Res.* *7*:271–288 (1979).

6. Crawford, B. D., Enger, M. D., Griffith, B. B., Griffith, J. K., Hanners, J. L., Longmire, J. L., Munk, A. C., Stallings, R. L., Tesmer, J. G., Walters, R. A., and Hildebrand, C. E. Coordinate amplification of metallothionein I and II genes in cadmium–resistant Chinese hamster cells: Implications for mechanisms regulating metallothionein gene expression. *Mol. Cell Biol.* *5*:320–329 (1985).
7. Karin, M. E., Najarian, R., Haslinger, A., Valenzuela, P., Welch, J., and Fogel, S. Primary structure and transcription of an amplified genetic locus: The *CUP1* locus of yeast. *Proc. Natl. Acad. Sci. USA*, *81*:337–341 (1984).
8. Beach, L. R., and Palmiter, R. D. Amplification of the metallothionein I gene in cadmium-resistant mouse cells. *Proc. Natl. Acad. Sci. USA*, *78*:2110–2114 (1981).
9. Gick, G. G., and McCarty, K. S. Amplification of the metallothionein in cadmium and zinc resistant Chinese hamster ovary cells. *J. Biol. Chem.* *257*:9049–9053 (1982).
10. Enger, M. D., Hildebrand, C. E., Seagrave, J., and Tobey, R. A. Cellular resistance to cadmium, in *Handbook of Experimental Pharmacology: Cadmium*, Vol. 80 (E. C. Foulkes, ed.), Springer-Verlag, New York, 1986.
11. Compere, S. J., and Palmiter, R. D. DNA methylation controls the inducibility of mouse metallothionein I in lymphoid cells. *Cell*, *25*:233–240 (1981).
12. Stallings, R. L., Crawford, B. D., Tobey, R. A., Tesmer, J., and Hildebrand, C. E. 5-Azacytidine-induced conversion to cadmium resistance correlates with early S phase replication of inactive metallothionein genes in synchronized CHO cells. *Somatic Cell. Mol. Genet.* *12*:423–432 (1986).
13. Fogel, S., and Welch, J. S. Tandem gene amplification mediates copper resistance in yeast. *Proc. Natl. Acad. Sci. USA*, *79*:5342–5346 (1982).
14. Butt, T. R., Sternberg, E. J., Gorman, J. A., Clark, P., Hamer, D., Rosenberg, M., and Crooke, S. T. Copper metallothionein of yeast, structure of the gene, and regulation of expression. *Proc. Natl. Acad. Sci. USA*, , *81*:3332–3336 (1984).
15. Searle, P. F., Davison, B. L., Stuart, G. W., Wilke, T. M., Norstedt, G., and Palmiter, R. D. Regulation, linkage, and sequence of mouse metallothionein I and II genes. *Mol. Cell Biol.* *4*:1221–1230 (1984).
16. Stallings, R. L., Munk, A. C., Longmire, J. L., Hildebrand, C. E., and Crawford, B. D. Assignment of genes encoding metallothionein I and II to Chinese hamster chromosome 3: Evidence for the role of chromosome rearrangement in gene amplification. *Mol. Cell. Biol.* *4*:2932–2936 (1984).
17. Pinkel, D., Straume, T., and Gray, J. W. Cytogenetic analysis using quantitative, high-sensitivity, fluorescence hybridization. *Proc. Natl. Acad. Sci. USA*, *83*:2934–2938 (1986).
18. Lichter, P., Tang, C. C., Call, K., Hermanson, G., Evans, G. A., Housman, D., and Ward, D. C. High-resolution mapping of human chromosome 11 by in situ hybridization with cosmid clones. *Science*, *247*:64–69 (1990).
19. Schwartz, D. C., and Cantor, C. R. Separation of yeast chromosome-sized DNAs by pulsed-field gel electrophoresis. *Cell*, *37*:67–75 (1984).
20. Cantor, C. R., Smith, C. L., and Mathew, M. K. Pulsed-field gel electrophoresis of very large DNA molecules. *Annu. Rev. Biophys. Chem.* *17*:287–304 (1988).

21. Vaughn, J. P., Dijkwel, P. A., and Hamlin, J. L. Replication initiates in a broad zone in the amplified dihydrofolate reductase domain. *Cell*, *61*:1075–1087 (1990).
22. Ehrlich, H. A., Gelfand, D., and Sninsky, J. J. Recent advances in the polymerase chain reaction. *Science*, *252*:1643–1651 (1991).
23. Stallings, R. L., Torney, D. C., Hildebrand, C. E., Longmire, J. L., Deaven, L. L., Jett, J. H., Doggett, N. A., and Moyzis, R. K. Physical mapping of human chromosomes by repetitive sequence fingerprinting. *Proc. Natl. Acad. Sci. USA*, *87*: 6218–6222 (1990).

15

Thymidylate Synthetase Gene Amplification in Drug-Resistant Mammalian Cells

Lee F. Johnson *The Ohio State University, Columbus, Ohio*

I. BIOCHEMICAL PROPERTIES OF THYMIDYLATE SYNTHETASE

Thymidylate synthetase (EC 2.1.1.45, TS) catalyzes the reductive methylation of deoxyuridylate to form thymidylate in the *de novo* biosynthetic pathway (Fig. 1). The cofactor N^5,N^{10}-methylene-tetrahydrofolate (CH_2FH_4) is the methyl donor and is oxidized to dihydrofolate (FH_2) during the course of the reaction [1]. The active enzyme is a homodimer with a subunit molecular weight of approximately 35 kDa in mammalian cells. TS enzyme level is much higher in rapidly proliferating cells than in quiescent cells [2], in keeping with the role of the enzyme in DNA replication.

TS is an essential enzyme in proliferating mammalian cells that are not supplied with exogenous thymidine. For this reason, TS is an important target enzyme in cancer chemotherapy and has been the subject of intensive enzymatic and structural analyses during the past two decades [1]. These studies revealed that a key reaction intermediate is a covalent ternary complex between deoxyuridylate (dUMP), cofactor, and a cysteine at the active site of the enzyme (Fig. 2). The recent determination of the three-dimensional structure of bacterial TS [3] will greatly facilitate future biochemical analyses of the enzyme and its interaction with substrates and inhibitors.

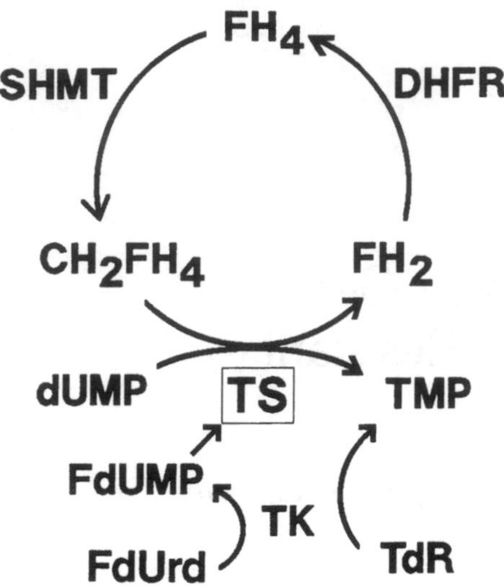

Figure 1 Biosynthesis of thymidylic acid. Key intermediates and enzymes are indicated. Enzyme abbreviations: DHFR, dihydrofolate reductase; SHMT, serine hydroxymethyl transferase; TK, thymidine kinase; TS, thymidylate synthetase.

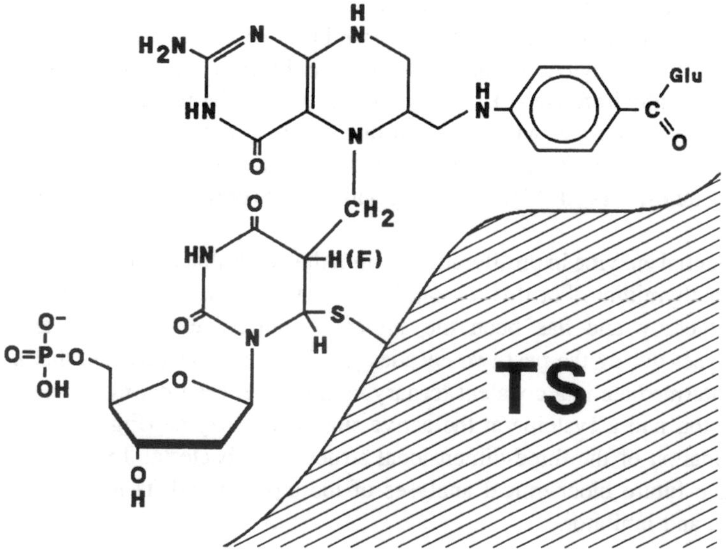

Figure 2 Structure of the covalent ternary complex between TS, dUMP (or FdUMP), and CH_2FH_4.

II. INHIBITORS OF TS ACTIVITY

A variety of TS inhibitors have been developed and shown to be useful as anticancer drugs. The earliest include the nucleotide analogues 5-fluorouracil (FUra) and 5-fluorodeoxyuridine (FdUrd). The mechanism by which these analogues inhibit TS activity has been elucidated [1]. The analogues are first converted to the nucleotide 5-fluorodeoxyuridylate (FdUMP). The metabolic activation of FdUrd (Fig. 1) proceeds via the salvage enzyme thymidine kinase (TK), whereas the activation of FUra involves several metabolic steps. FdUMP is identical to the normal substrate, dUMP, except for the presence of fluorine at the 5 position of the pyrimidine ring. FdUMP is recognized by TS and proceeds through the normal course of the reaction up to the formation of the covalent ternary complex (Fig. 2). However, the fluorine cannot be displaced from the pyrimidine ring and the reaction is blocked at this stage. FdUrd is an attractive chemotherapeutic drug, since it is a highly specific inhibitor that is toxic to proliferating cells but not quiescent cells.

TS inhibitors that are analogues of the CH_2FH_4 cofactor have also been developed. For example, N^{10}-propargyl-5,8-dideazofolic acid (CB3717) is a quinazoline-based cofactor analogue that is a specific, tight-binding inhibitor of TS enzyme activity both in vitro and in vivo [4]. The compound has been shown to have significant antitumor activity. An advantage of this compound (relative to FUra or FdUrd) is that it does not require metabolic activation, thus eliminating a possible mechanism of resistance.

III. SELECTION OF FdUrd-RESISTANT CELL LINES

The therapeutic effectiveness of the TS inhibitors (as well as many other anticancer agents) is limited by the emergence of drug-resistant clones within the tumor cell population. For this reason, the mechanisms by which cells develop resistance to chemotherapeutic drugs in general, and TS inhibitors in particular, is of practical as well as theoretical interest. Cells can develop resistance to the toxic effects of FdUrd by several different mechanisms. For example, cells that lack TK activity are unable to convert FdUrd to FdUMP and are resistant to high concentrations of FdUrd. Other mechanisms of resistance include alterations in the primary structure of TS, which lead to decreased affinity for FdUMP or increased rates of catabolism of the nucleoside analogues. In addition, several laboratories have shown that resistance to fluoropyrimidines can be achieved through overproduction of TS [5–8].

To further understand the mechanisms by which mammalian cells develop resistance to chemotherapeutic drugs through enzyme overproduction (as well as to facilitate the analysis of *TS* gene expression), we set out to isolate a mouse cell line that was resistant to high concentrations of FdUrd. Our strategy was to select for cells that were able to proliferate in gradually increasing concentrations of FdUrd, which was similar to the strategy used in the isolation of cells that overproduced dihydrofolate reductase (DHFR). To reduce the toxicity from metabolic products of FdUrd (such as 5-fluorouridine), which may affect cell viability by mechanisms unrelated to TS, uridine and cytidine were included in the medium. To reduce the possibility that resistance would be achieved through loss of TK activity, the cells were periodically selected for the presence of TK activity by growing them in medium containing hypoxanthine, aminopterin, and thymidine (HAT medium).

A line of mouse 3T6 fibroblasts that was resistant to 0.3 μM FdUrd was developed in the following manner [7]. Cells were seeded in medium supplemented with 0.06 μM FdUrd, 1 mM uridine, and 1 mM cytidine, which was lethal to most of the cells in the population. After the surviving cells had adapted to the drug, which took several weeks, they were selected in HAT medium to eliminate cells deficient in TK activity, then further selected in medium containing gradually increasing concentrations of FdUrd as well as uridine and cytidine. Several months of continuous selection were required for the cells to achieve resistance to 0.3 μM FdUrd. After a final selection in HAT medium, clones of cells resistant to 0.3 μM FdUrd were isolated and further characterized. We found that TS was overproduced up to 20-fold in some of the clones. Unfortunately, in spite of considerable effort, we were unable to isolate 3T6 cells that could grow at higher concentrations of FdUrd.

We applied a similar FdUrd selection strategy to a line of 3T6 derivatives (M50L3), which we had previously selected for resistance to high concentrations of methotrexate by virtue of *DHFR* gene amplification. We reasoned that amplification of the *TS* gene might occur more readily n a cell line that had been selected for its ability to amplify another gene. In other words, the ability to carry out gene amplification might be a selectable trait. In line with this hypothesis (although certainly not verifying it), we found that the M50L3 cells adapted much more rapidly and to much higher concentrations of FdUrd than the original 3T6 cell line. For example, it took 5 months of selection for the 3T6 cells to adapt to 0.3 μM FdUrd, whereas the M50L3 cells adapted to 2 μM FdUrd in only 2 months. In addition, TS was usually overproduced to a greater extent in the M50L3-derived cells. In the clone designated LU3-7, TS was overproduced approximately 50-fold as compared with the parental M50L3 (or 3T6) cell line [7]. Most of our subsequent analyses were performed with this clone.

The specific activity of TK was significantly lower in all of the FdUrd-resistant mouse fibroblasts than in the parental cells. The mechanism responsible for this decrease is not known. The TK deficiency may partially contribute to the resistance phenotype.

TS normally represents only a tiny fraction of total cellular protein, as expected for a housekeeping enzyme. However, TS was one of the most abundant proteins in

LU3-7 cells. TS from the overproducing cells had the same molecular weight as the wild-type enzyme 3T6 cells, as determined by one-dimensional sodium dodecyl sulfate polyacrylamide gel electrophoresis. In addition, TS isolated from the overproducing cell line was inactivated at the same concentration of FdUMP as wild-type TS. Therefore, at least by these criteria, the overproduced TS from the LU3-7 cells is indistinguishable from wild-type mouse TS [7].

IV. MOLECULAR CLONING OF MOUSE *TS* cDNA

Detailed molecular analysis of the *TS* gene and its expression depended on the cloning of cDNA corresponding to *TS* mRNA. This was accomplished by constructing a cDNA library corresponding to mRNA from LU3-7 cells and screening replicas of the library with two different labeled cDNA probes. The first probe was complementary to LU3-7 mRNA, and the second was complementary to 3T6 mRNA. Clones that contained *TS* cDNA gave a strong signal with the first probe but not the second. This screen, though tedious, eventually led to the isolation of a cDNA clone (pMTS-3) that contained the entire coding region of mouse *TS* mRNA [9].

V. AMPLIFICATION OF THE *TS* GENE IN FdUrd-RESISTANT MOUSE CELLS

To determine whether the *TS* gene was amplified in the TS-overproducing cells, DNA was isolated from wild-type 3T6 cells as well as the FdUrd-resistant derivatives and analyzed by the Southern blot procedure using ^{32}P-labeled *TS* cDNA as the probe. Figure 3 shows that the restriction fragments detected by the probe were much more intense in the LU3-7 cells than in the parental cell line. Dilution analyses revealed that the *TS* gene was amplified approximately 50-fold in the LU3-7 cells. All the amplified fragments in LU3-7 cells were also detected in the parental cell line, which indicates that major gene rearrangements had not occurred in the vicinity of the gene during the amplification process [10]. Since there was a parallel increase in the number of copies of the gene and the amount of TS enzyme and mRNA in the LU3-7 cells, it is likely that most (if not all) of the amplified genes are transcriptionally active.

It is also noteworthy that the expression of the amplified *TS* genes is regulated in the same manner in LU3-7 cells as in 3T6 cells [11]. Therefore the LU3-7 cell line represents a convenient model system for analyzing the regulation of *TS* gene expression. We found that when quiescent LU3-7 cells were stimulated to proliferate, the amount and rate of labeling of *TS* mRNA increased at least 10-fold. However, the rate of *TS* gene transcription changed by less than threefold. Therefore it appears that *TS* gene expression is controlled primarily at the posttranscriptional level in growth-stimulated mouse fibroblasts [12].

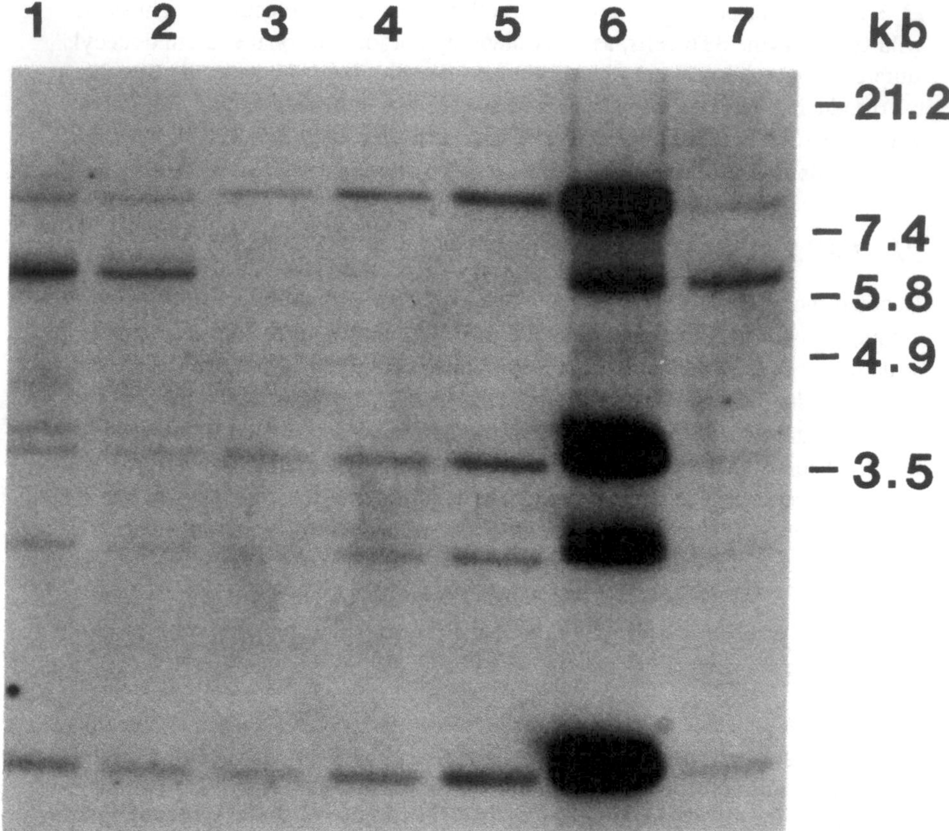

Figure 3 *TS* gene amplification in FdUrd-resistant LU3-7 cells. DNA isolated from LU3-7 and 3T6 cells was digested with *Eco*RI, fractionated by electrophoresis on a 1% agarose gel, transferred to nitrocellulose, and probed with cloned ^{32}P-labeled *TS* cDNA. Lane 1 contained 5μg of 3T6 DNA. Lanes 2 and 7 contained 4 μg of M50L3 DNA. The remaining lanes contained LU3-7 DNA: lane 3, 0.05 μg; lane 4, 0.1 μg; lane 5, 0.2 μg; lane 6, 5 μg. (From Ref. 10.)

When LU3-7 cells were grown in medium that lacked FdUrd, the extent of TS overproduction decreased with a half-life of approximately 3 weeks [7]. To determine whether this was due to a decrease in the number of copies of the *TS* gene, DNA was isolated from cultures grown for different times in the absence of FdUrd and subjected to Southern blot analysis. As shown in Figure 4, *TS* gene copy number decreased with the same kinetics as that of the enzyme [10]. The

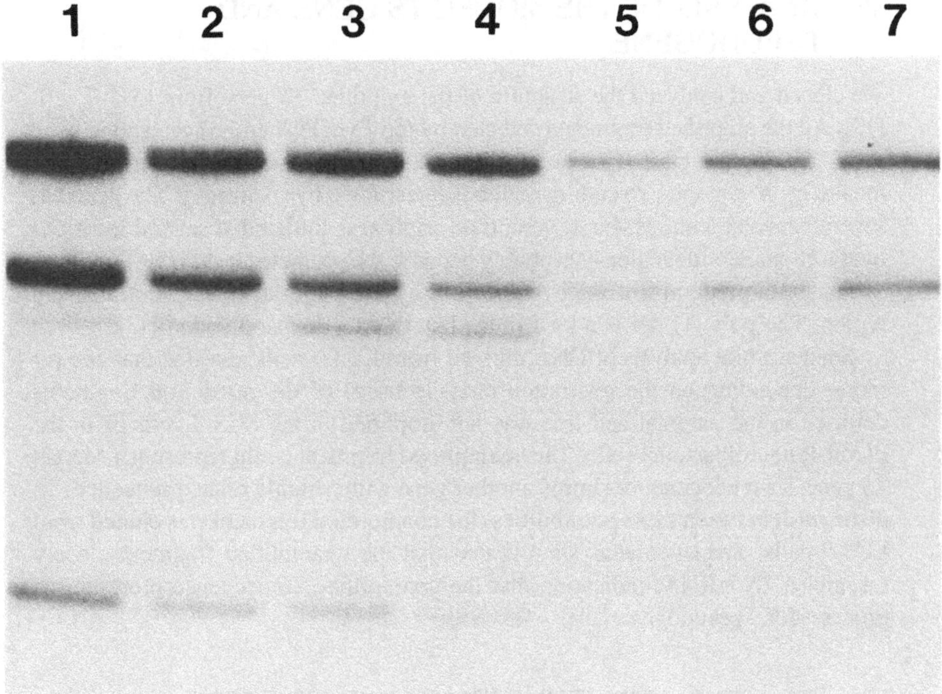

Figure 4 Loss of amplified *TS* genes following growth in the absence of selective pressure. LU3-7 cells were maintained in the absence of FdUrd for 0 months (lane 1), 1 month (lane 2), 2 months (lane 3), and 3 months (lane 4); 5 μg of DNA was isolated from the cells and digested with *Hin*dIII. Lane 5, 6, and 7 represent 0.1, 0.25, and 0.5 μg (respectively) of *Hin*dIII-digested DNA isolated from LU3-7 cells maintained in the presence of FdUrd. DNA was analyzed as described in Figure 3. (From Ref. 10.)

instability of the amplified *TS* genes suggests that they are maintained as extrachromosomal elements (double minute chromosomes).

We also determined that the *TS* gene was amplified in a C-46 mouse neuroblastoma cell line that was selected for resistance to 0.4 μM FdUrd [10]. This cell line, which was isolated by Baskin and coworkers [5], overproduced TS approximately 12-fold. Southern blot analyses revealed that the *TS* gene was amplified about 30-fold in the resistant cell line. The pattern of amplified fragments was indistinguishable from that observed with DNA from LU3-7 cells when the DNA was digested with *Hin*dIII. However, differences were observed with *Eco*RI-digested DNA. The differences appear to be the result of restriction site polymorphisms [10].

VI. ANALYSIS OF THE MOUSE *TS* GENE AND PSEUDOGENE

We cloned and analyzed the structure of the amplified *TS* gene from LU3-7 cells [13]. All the amplified fragments detected by the *TS* cDNA probe correspond to an exon-containing region of the gene. The gene extends 12 kb from the first exon to the last exon, and the 1 kb coding region is interrupted by six introns. The gene has several unusual features. First, gene transcription is initiated at several locations over a 60-nucleotide region—probably because the promoter lacks a TATAA box. Second, the predominant mRNA encoded by the gene lacks a 3′ untranslated region. The poly(A) tail is added immediately after the stop codon [13].

Southern blot analysis of DNA derived from LU3-7 cells revealed that one (or more, depending on the restriction enzyme used) of the restriction fragments detected in the parental cell line was not amplified in the LU3-7 cells or in the FUdR-R neuroblastoma cells. The unamplified fragment could represent a second *TS* gene, a pseudogene, or simply another gene with a highly related sequence. To distinguish between these possibilities, the unamplified fragment was cloned from LU3-7 cells. Sequence analysis revealed that the unamplified fragment closely resembled *TS* mRNA, indicating that the unamplified *TS* sequence represents a processed *TS* pseudogene [14].

VII. *TS* GENE AMPLIFICATION IN MOUSE CELLS RESISTANT TO FOLATE ANALOGUES

Jackman and coworkers [4] found that when the mouse lymphoblastoid cell line L1210 was selected for survival in gradually increasing concentrations of the cofactor analogue CB3717, highly resistant cells could be selected. One such cell line, L1210:C15, was resistant to 500 μM CD3717 and was found to overproduce TS enzyme level approximately 45-fold. An unexpected observation was that the CB3717-resistant cells did not display cross-resistance to FdUrd. This appears to be the result of depletion of the reduced folate pool in cells exposed to FdUrd.

DNA and RNA blot analyses revealed that there was a corresponding increase in the amount of *TS* mRNA and the number of copies of the *TS* gene in the drug-resistant cells [15]. There was no indication that structural alterations had occurred during the amplification process. The degree of *TS* gene amplification decreased by a factor of only 2 when the drug-resistant cells were grown for one year in the absence of selective pressure. This is in marked contrast to the 3-week half-life observed in the LU3-7 fibroblasts [10] and suggests that in the L1210:C15 cells, the amplified *TS* genes are stably integrated into chromosomes rather than maintained as extrachromosomal elements.

VIII. AMPLIFICATION OF THE *TS* GENE IN HUMAN CELLS

TS gene amplification has also been observed in a human cell line that was selected for resistance to high concentrations of FdUrd. Berger and coworkers [8] used a novel FdUrd selection procedure to isolate a HEp-2 cell line that overproduced TS approximately 100-fold. Their selective medium contained folinic acid and deoxyinosine in addition to FdUrd. Folinic acid renders cells more sensitive to FdUrd by shifting the equilibrium between free and FdUMP-bound TS. Deoxyinosine was added to reduce the catabolism of FdUrd via thymidine phosphorylase. A cell line was isolated (designated HEp-2/500) that was resistant to 500 nM FdUrd and that overproduced TS approximately 80-fold. In contrast to the situation with the mouse 3T6 fibroblasts, there was no reduction in TK activity in the FdUrd-resistant cell population. Therefore it was not necessary to select for TK(+) cells during the course of the drug selection.

DNA and RNA blot analysis (using the mouse *TS* cDNA as a probe) revealed that the amount of *TS* mRNA and the number of copies of the *TS*gene increased approximately 100-fold in the resistant cell line. The pattern of restriction fragments detected with the cDNA probe was the same in the amplified and parental cell lines, indicating that no rearrangements had occurred during the amplification process. One restriction fragment that was detected in the parental cell line was not amplified in the overproducing cell line. This fragment, which was subsequently shown to have a chromosomal location different from that of the functional human *TS* gene [16], probably corresponds to a pseudogene. As in the mouse LU3-7 cell line, the level of overproduction and degree of gene amplification diminished after prolonged growth in the absence of selective pressure (half-life, 3 weeks), suggesting that the amplified *TS* genes are present on extra-chromosomal elements.

IX. *TS* GENE AMPLIFICATION IN DRUG-RESISTANT TUMOR CELLS

Fluoropyrimidine antimetabolites have been widely used in cancer chemotherapy. Unfortunately, a clinical response is observed in only a low percentage of patients. Furthermore, the responses are generally transient and are followed by the emergence of drug-resistant derivatives. Clark and coworkers [17] have shown that the *TS* gene is amplified approximately fourfold in a colonic tumor that developed resistance following prolonged 5-FUra/leukovorin combination chemotherapy. Curiously, the TS enzyme level was approximately the same as in other tumors in which the *TS* gene was not amplified. The reason for the discrepancy is not known. These results suggest that amplification of the *TS* gene is one of several mechanisms (though probably not the predominant mechanism) by which tumor cells develop resistance to fluoropyrimidine antimetabolites. This fact should certainly be considered when designing therapeutic strategies that involve TS inhibitors.

ACKNOWLEDGMENTS

Studies in the author's laboratory were supported by grants from the National Institute for General Medical Sciences (GM29356), the National Science Foundation (PCM-8312017), and the National Cancer Institute (CA16058).

REFERENCES

1. Danenberg, P. V. Thymidylate synthetase—A target enzyme in cancer chemotherapy. *Biochim. Biophys. Acta*, *473*:73–92 (1977).
2. Navalgund, L. G., Rossana, C., Muench, A. J., and Johnson, L. F. Cell cycle regulation of thymidylate synthetase gene expression in cultured mouse fibroblasts. *J. Biol. Chem.* *255*:7386–7390 (1980).
3. Hardy, L. W., Finner-Moore, J. S., Montfort, W. R., Jones, M. O., Santi, D. V., and Stroud, R. M. Atomic structure of thymidylate synthase: Target for rational drug design. *Science*, *235*:448–455 (1987).
4. Jackman, A. L., Alison, D. L., Calvert, A. H., and Harrap, K. R. Increased thymidylate synthase in L1210 cells possessing acquired resistance to *N*-10-propargyl-5,8-dideazafolic acid (CB3717): Development, characterization, and cross-resistance studies. *Cancer Res.* *46*:2810–2815 (1986).
5. Baskin, F., Carlin, S. C., Kraus, P., Friedkin, M., and Rosenberg, R. N. Experimental chemotherapy of neuroblastoma. II. Increased thymidylate synthetase activity in a 5-fluorodeoxyuridine-resistant variant of mouse neuroblastoma. *Mol. Pharmacol.* *11*:105–117 (1975).
6. Priest, D. G., Ledford, B. E., and Doig, M. T. Increased thymidylate synthetase in 5-fluorodeoxyuridine resistant cultured hepatoma cells. *Biochem. Pharmacol.* *29*:1549–1553 (1980).
7. Rossana, C., Rao, L. G., and Johnson, L. F. Thymidylate synthetase overproduction in 5-fluorodeoxyuridine-resistant mouse fibroblasts. *Mol. Cell. Biol.* *2*:1118–1125 (1982).
8. Berger, S. H., Jenh, C.-H., Johnson, L. F., and Berger, F. G. Thymidylate synthase overproduction and gene amplification in fluorodeoxyuridine-resistant human cells. *Mol. Pharmacol.*, *28*:461–467 (1985).
9. Geyer, P. K., and Johnson, L. F. Molecular cloning of DNA sequences complementary to mouse thymidylate synthase messenger RNA. *J. Biol. Chem.*, *259*:7206–7211 (1984).
10. Jenh, C.-H., Geyer, P. K., Baskin, F., and Johnson, L. F. Thymidylate synthase gene amplification in fluorodeoxyuridine-resistant mouse cell lines. *Mol. Pharmacol.* *28*:80–85 (1985).
11. Jenh, C.-H., Rao, L. G., and Johnson, L. F. Regulation of thymidylate synthase enzyme synthesis in 5-fluorodeoxyuridine-resistant mouse fibroblasts during the transition from the resting to growing state. *J. Cell. Physiol.* *122*:149–154 (1985).

12. Jenh, C.-H., Geyer, P. K., and Johnson, L. F. Control of thymidylate synthase mRNA content and gene transcription in an overproducing mouse cell line. *Mol. Cell. Biol. 5*: 2527–2532 (1985).
13. Deng, T., Li, D., Jenh, C.-H., and Johnson, L. F. Structure of the gene for mouse thymidylate synthase: Locations of introns and multiple transcriptional start sites. *J. Biol. Chem. 261*:16000–16005 (1986).
14. Li, D., and Johnson, L. F. A mouse thymidylate synthase pseudogene derived from an aberrantly processed RNA molecule. *Gene, 82*:363–370 (1989).
15. Imam, A. M. A., Crossley, P. H., Jackman, A. L., and Little, P. F. R. Analysis of thymidylate synthase gene amplification and of mRNA levels in the cell cycle. *J. Biol. Chem. 262*:7368–7373 (1987).
16. Nussbaum, R. L., Walmsley, R. M., Lesko, J. G., Airhart, S. D., and Ledbetter, D. H. Thymidylate synthase-deficient Chinese hamster cells: A selection system for human chromosome 18 and experimental system for the study of thymidylate synthase regulation and fragile X expression. *Am. J. Hum. Genet. 37*:1192–1205 (1985).
17. Clark, J. L., Berger, S. H., Mittelman, A., and Berger, F. G. Thymidylate synthase gene amplification in a colon tumor resistant to fluoropyrimidine chemotherapy. *Cancer Treatment Rep. 71*:261–265 (1987).

16

Adenosine Deaminase: A Dominant Amplifiable Genetic Marker

Rodney E. Kellems, Rueyling Lin, and A. Craig Chinault *Baylor College of Medicine, Houston, Texas*

I. THE BIOLOGY OF ADENOSINE DEAMINASE

Adenosine deaminase (ADA) is an enzyme of purine metabolism that is widely distributed among mammalian cells and tissues [1,2]. In mice the highest levels are found at the maternal–fetal interface throughout gestation and in the gastrointestinal tract of adult animals (Table 1). During prenatal development ADA levels undergo a pronounced developmentally regulated increase within the implantation chamber between days 7 and 9 of gestation (see Ref. 2 and references cited therein). The enzyme occurs predominantly in the maternal cells of the antimesometrial decidua (a maternal tissue that surrounds the midgestation embryo), where it is one of the most abundant soluble proteins. Pharmacological inhibition of ADA activity on days 7 or 8 of gestation results in nearly 100% resorption or malformation, suggesting that maternal ADA activity is critical to normal embryogenesis during this period. With continued gestation the synthesis of ADA at the maternal–fetal interface shifts to the developing placenta, where it is primarily associated with extraembryonic spongiotrophoblasts. In postnatal murine tissues the highest levels of ADA protein are observed within the epithelium of the tongue, esophagus, forestomach, and intestine [1]. In these tissues ADA levels are low at birth but increase during the morphogenesis of the gastrointestinal epithelium. By 5 weeks ADA is the major gene product and accounts for over 20% of the soluble protein.

The physiological importance of ADA presumably stems from the physiologi-

Table 1 Tissue Distribution of ADA Activity in Mice

Tissue	ADA activity (nmol/min/mg)[a]
Gastrointestinal tract	
Tongue	5200 ± 1100 (4)
Esophagus	2200 ± 600 (4)
Forestomach (nonglandular)	4200 ± 950 (4)
Glandular stomach	150 ± 47 (4)
Duodenum	4200 ± 2000 (4)
Proximal jejunum	1900 ± 660 (4)
Ileum	360 ± 190 (4)
Large intestine (cecum)	50 ± 20 (4)
Genitourinary tract	
Embryo/decidual unit, day 9	1100 ± 90 (8)
Placenta, day 13	590 ± 50 (6)
Emryo, day 13	23 ± 5 (5)
Uterus	24 ± 16 (4)
Vagina	31 ± 18 (8)
Bladder	37 ± 4 (2)
Skin	
Ear pinnae	26 ± 17 (5)
Foot pad	19 ± 13 (4)
Back	7 ± 1 (4)
Miscellaneous	
Thymus	130 ± 6 (4)
Liver	63 ± 7 (3)
Kidney	63 ± 1 (2)
Spleen	44 ± 6 (4)
Lung	7 ± 1 (3)
Muscle	
Cardiac	4 ± 1 (2)
Skeletal	4 ± 1 (2)

[a]Values are the mean ± the standard deviation; sample size is shown in parentheses.
Source: Adapted from Ref. 1, with permission.

cal impact of its substrates (reviewed in Ref. 3). One substrate, adenosine, affects a variety of physiological processes by interaction with cell surface adenosine receptors, which in turn control intracellular cyclic AMP concentrations. Specific effects include neurological, cardiopulmonary, and a variety of metabolic actions. Adenosine receptor agonists are effective inhibitors of growth and organogenesis in mouse embryos [4]. The other ADA substrate, deoxyadenosine, exerts marked lymphocytotoxic effects. Elevated levels of deoxyadenosine are believed to be responsible for the severe combined immunodeficiency disease that is associated with ADA deficiency in humans [3].

Although ADA is found in virtually all mammalian cells and tissues and is essential for proper functioning of a mammalian organism, the enzyme is not required for the growth of cells in culture under standard cell culture conditions. However, as explained below, it is possible to devise cell culture conditions in which ADA is required for growth.

II. SELECTIVE OVERPRODUCTION OF ADA IN CULTURED MOUSE CELLS

To address a number of questions concerning the physiology and molecular biology of ADA, it is necessary to have adequate quantities of pure enzyme and monospecific antibody. To facilitate the purification of ADA, Yeung et al. [5] isolated murine cell lines that produce very large quantities of the enzyme. These cells were isolated as a result of a two-part selection scheme. The first part involved the identification of cell culture conditions in which ADA expression is required for growth. The important components of the selection scheme are alanosine (50 μM), adenosine (1.1 mM), and uridine (1.0 mM). Alanosine is an aspartic acid analogue (Fig. 1) that is metabolized by succinylaminoimidazolecarboxamide synthetase to alanosylaminoimidazolecarboxamide, an inhibitor of adenylosuccinate synthetase. As a result, de novo AMP synthesis is blocked; see Figure 2 for relevant metabolic pathways. Adenosine is included in the selection medium to circumvent this block and provide a salvage route for AMP synthesis via the adenosine kinase pathway (Fig. 2, reaction 3). However the high concentration of adenosine used is cytotoxic and is believed to interfere with several metabolic pathways [3]. For example, synthesis of uridylate (UMP) is blocked at the level of orotic acid, ribonucleotide reductase is inhibited as a result of the accumulation of dATP, and methylation reactions mediated by *S*-adenosylmethionine (SAM) are inhibited as a result of the accumulation of *S*-adenosylhomocysteine (Fig. 2). Uridine is added to provide a salvage route for UMP synthesis via the uridine kinase reaction. Increased ADA levels are required to reduce the remaining metabolic burden of adenosine. Thus, this selection scheme selects for the presence of three enzymes: adenosine kinase to provide a salvage route for AMP synthesis, uridine kinase to provide a salvage route for UMP synthesis, and ADA to reduce the metabolic burden of adenosine. The requirement for ADA activity when cells are grown under these conditions is illustrated by the fact that cells are killed by the addition of 10 nM deoxycoformycin, a potent and specific inhibitor of ADA (Fig. 1). Under normal conditions the presence of deoxycoformycin is not fatal to cells.

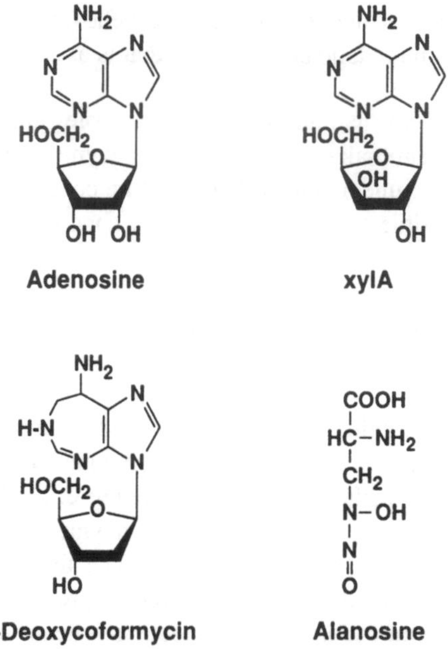

Adenosine xylA

2'-Deoxycoformycin Alanosine

Figure 1 Chemical structures of compounds used in ADA selection protocols. Adenosine is a natural substrate for ADA. Xylofuranosyl-adenine (xyl-A), a cytotoxic analogue of adenosine, is a substrate for ADA and adenosine kinase; 2′-deoxycoformycin is a transition state analogue inhibitor of ADA (K_i = 2.5 × 10^{-12} M). Alanosine is an aspartic acid analogue. See Figure 2 for relevant metabolic pathways.

The second part of the selection scheme involves the use of deoxycoformycin in conjunction with alanosine, adenosine, and uridine. Selection for resistance to increasing concentrations of deoxycoformycin under these conditions results in the isolation of cells with corresponding increases in ADA levels. The relationship between drug resistance and ADA levels is linearly related over a range of several thousandfold (Fig. 3), indicating that increased deoxycoformycin resistance is fully accounted for by a corresponding increase in the intracellular levels of ADA. This selection strategy resulted in the isolation of cells with an 11,400-fold increase in ADA activity [6], in which the enzyme accounts for approximately 75% of the soluble cellular protein (Fig. 4). The enzyme has been purified from these cells in high yield and large quantity using a simple two-step purification procedure. The resulting enzyme is indistinguishable from that derived from the parental cells as judged by a number of biochemical criteria [6]. Monospecific anti-mouse ADA antibody was raised in sheep against the purified enzyme. The antibody has been used successfully in a variety of immunocytochemical studies to determine the cellular localization of ADA in mice (see Refs. 1,2, and references cited therein).

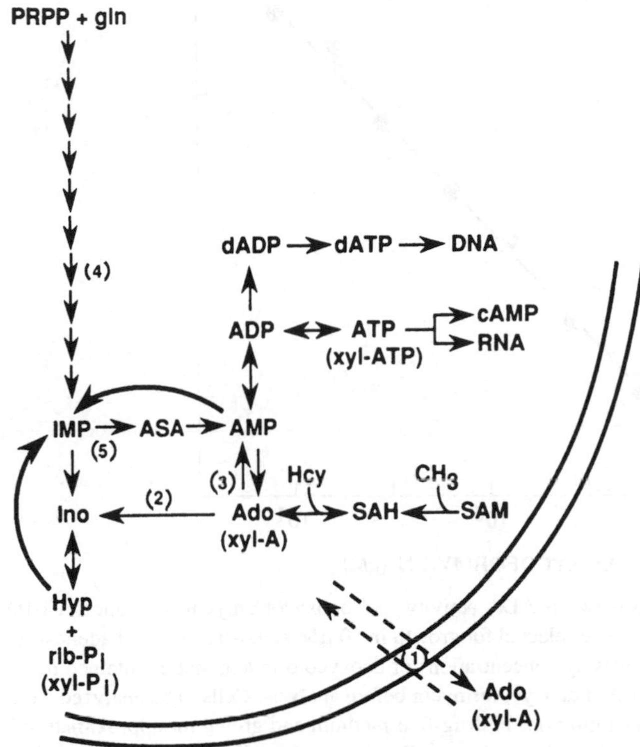

Figure 2 Metabolism of adenosine, xylofuranosyl-adenine, and alanosine. Extracellular adenosine (Ado) and xylofuranosyl-adenine (xyl-A) enter cells via a high affinity nucleoside transport system (1). Once inside, these nucleosides may be deaminated by ADA (2) or phosphorylated by adenosine kinase (3). Ado is also in equilibrium with *S*-adenosylhomocysteine (SAH), which is derived from *S*-adenosylmethionine (SAM) in transmethylation reactions. Alanosine is metabolized by succinylaminoimidazolecarboxamide synthetase (4), to alanosylaminoimidazolecarboxamide, an inhibitor of adenylosuccinate synthetase (5).

III. AMPLIFICATION AND MOLECULAR CLONING OF MURINE *ADA* GENE SEQUENCES

A prerequisite to the analysis of *ADA* gene structure and regulation is the isolation of full- length functional *ADA* cDNAs. Such clones not only provide probes for the isolation and characterization of the *ADA* gene, but also provide the starting material for constructing functional *ADA* minigenes for use in gene transfer and amplification (see below). Using mRNA from deoxycoformycin-resistant cells Yeung et al. [7] were able to construct a cDNA library vastly enriched for *ADA*

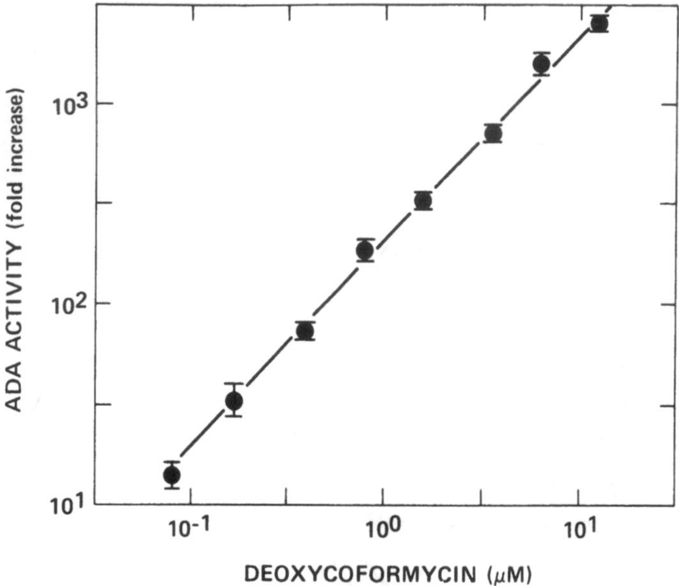

Figure 3 Relationship between ADA activity and deoxycoformycin resistance. Cl-1D cells (murine fibroblasts) were selected for growth in 50 μM alanosine, 1.1 mM adenosine, 1.0 mM uridine, and increasing concentrations of deoxycoformycin and maintained for at least 2 months at each level of deoxycoformycin before analysis. Cells to be analyzed were passaged one time (1:8 cell dilution) in drug-free medium and grown for approximately 1 week before harvesting cells and preparing cells extracts for ADA assay. The drug-free medium was replaced three times during the week in an effort to remove deoxycoformycin. The values shown represent the mean and standard deviation of at least three determinations. (Reprinted with permission from Ref. 5.)

cDNA sequences. Subtractive hybridization strategies were used to obtain probes suitable for library screening and led to the identification of *ADA* cDNA clones [7]. Such clones were used to show that the increased levels of ADA that accompany deoxycoformycin resistance result from a corresponding amplification of *ADA* genes. Full-length *ADA* cDNAs were subsequently identified by functional complementation of ADA-deficient *Escherichia coli* [8]. Sequence analysis of one such clone revealed an open reading frame of 1056 nucleotides, encoding a deduced protein of 352 amino acids with a molecular weight of approximately 41,000. An *ADA* minigene containing the entire open reading frame was constructed and shown by transient expression in monkey kidney cells to encode authentic mouse *ADA* [8].

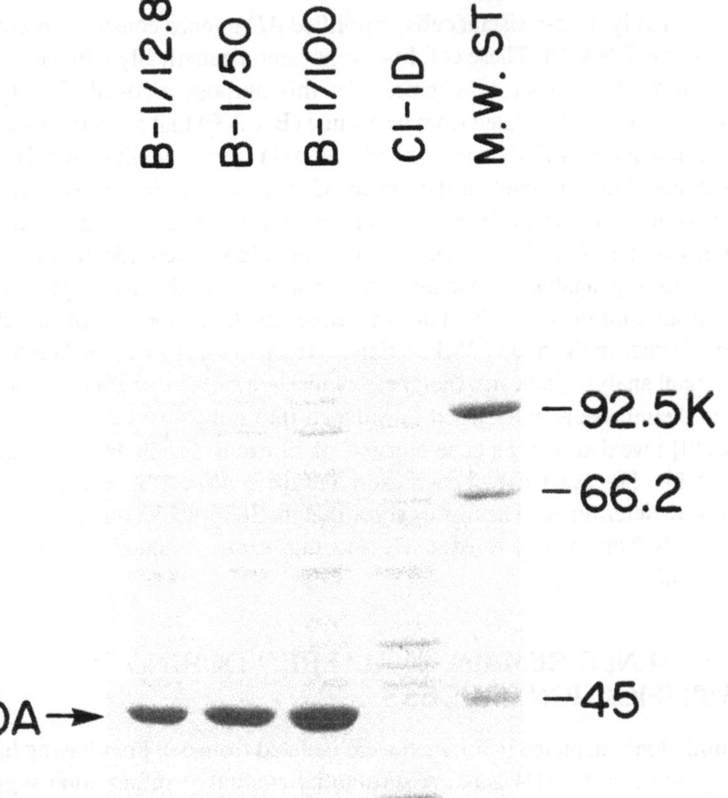

Figure 4 Massive overproduction of ADA in highly drug-resistant cells. Sodium dodecyl sulfate polyacrylamide gel electrophoretic analysis of the soluble protein from Cl-1D cells and populations of Cl-1D cells (termed B-1) selected for resistance to 50 μM alanosine, 1.1 mM adenosine, 1.0 mM uridine, and high concentrations of deoxycoformycin. Numbers following B-1 refer to micromolar levels of 2′-deoxycoformycin resistance. Protein was visualized by staining with Coomassie Brilliant Blue. (Reprinted with permission from Ref. 6.)

IV. MOLECULAR CLONING OF INTACT *ADA* GENES FROM AMPLIFIED CELL LINES

In the most highly drug-resistant cells, amplified *ADA* genes constitute more than 5% of genomic DNA [9]. These cell lines represent a genetically enriched source from which to clone intact *ADA* genes. For this purpose a cosmid library was constructed using DNA isolated from a cell line (B-1/200) that overproduces ADA by approximately 11,400-fold as a result of *ADA* gene amplification [10]. To identify clones likely to contain the entire *ADA* gene, the library was screened using synthetic oligonucleotide probes complementary to regions in the extreme 5′ and 3′ ends of the *ADA* cDNA sequence. Cosmid clones were identified that, by Southern blotting analysis, appeared to contain the entire *ADA* gene and a substantial amount of 5′ and 3′ flanking sequence. Overlapping cosmid clones extending further in the 5′ and 3′ directions were also identified, see below (Fig. 6). Functional analysis indicated that these clones are capable of encoding murine adenosine deaminase activity when introduced into human cell lines. Structural analysis [10] revealed that the gene consists of 12 exons distributed over approximately 25 kb. The exact size of each exon and the sequence of each exon/intron junction were determined. The results show that the 1056-nucleotide open reading frame extends from exon 1 to exon 11, and that exon 12 contains untranslated sequences only.

V. *ADA* GENES REMAIN UNALTERED DURING THE AMPLIFICATION PROCESS

The cosmid clones depicted in Figure 5 were isolated from cell lines having highly amplified copies of the *ADA* gene. A substantial amount of information suggests that the *ADA* genes remained structurally and functionally unaltered during the process of amplification. Southern blotting analysis using *ADA* cDNA as a hybridization probe revealed no differences in the restriction digestion patterns between the *ADA* genes present in the genetically enriched DNA from *ADA*-amplified cell lines and in DNA from nonamplified cell lines or normal mouse tissue. For a more detailed comparison, three overlapping *ADA* structural gene clones were isolated from a normal mouse spleen library. The largest clone, L7.1, is a 15 kb segment of genomic DNA that includes the 5′ portion and flanking region of the *ADA* gene. Comparison of the restriction map of this lambda clone with that of the cosmid clones that share the same region revealed no differences (Fig. 5). The sequence of approximately 300 nucleotides immediately 5′ of the ATG translation start codon was determined for L7.1 and is identical to that determined for the comparable region from the cosmid clones. The results shown in Figure 5 indicate that the electrophoretic mobility of *ADA* encoded by the cosmid clones is indistinguishable from that of *ADA* produced by a mouse thymoma cell line, EL-4, providing evidence that C2.2 and C3.5 encode normal

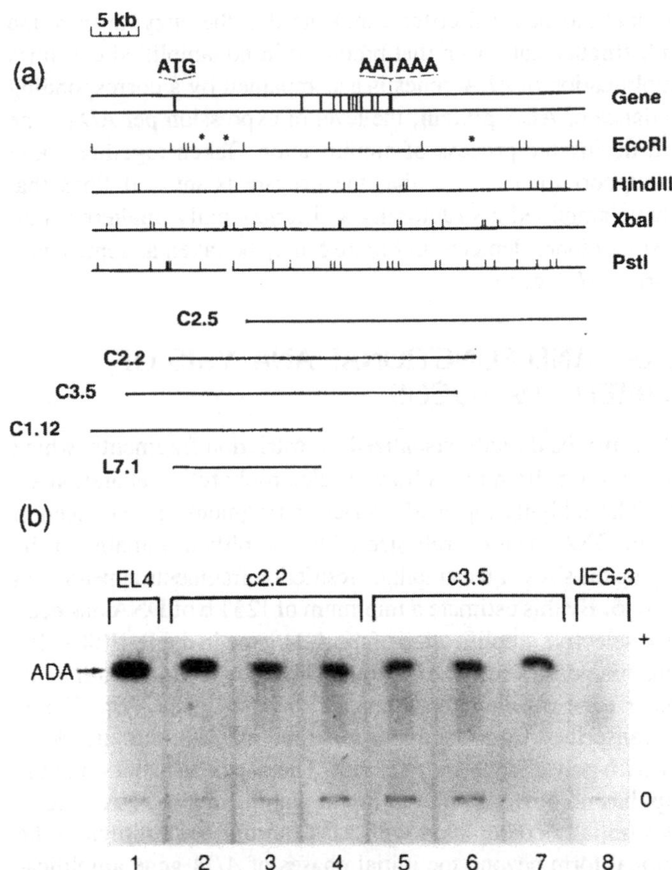

Figure 5 Structural and functional analysis of the murine *ADA* gene. (a) The structure of the murine *ADA* gene. The positions and relative sizes of the 12 exons of the *ADA* gene are indicated by the solid vertical bars. Restriction maps are provided for four enzymes. The position of one lambda clone and four cosmid clones are shown. The small gap in the *Pst* I map is a region that was not analyzed for *Pst* I sites. *Eco*RI fragments that contain highly repetitive DNA sequence elements are marked by asterisks. (b) Functional analysis of cosmid clones c2.2 and c3.5. As a control for transfection efficiency, 25 μg of each cosmid, along with 3.6 μg pRSVβ-*gal*, was introduced into human JEG-3 cells. After 65 hr, cells were harvested and extracts were prepared and adjusted to contain equal concentrations of β-galactosidase activity. The adjusted cell extracts were applied to agarose gels for electrophoretic fractionation and histochemical detection of ADA activity. Lane 1 contains extract from the mouse T-cell line EL4: this serves as a standard to indicate the mobility of authentic mouse ADA. Lanes 2–4 contain extracts from JEG-3 cells transfected with three different preparations of cosmid c2.2. Lanes 5–7 contain extracts from JEG-3 cells transfected with three different preparations of cosmid c3.5. Lane 8 contains cell extract from nontransfected JEG-3 cells. The low level of ADA activity present in these cells is undetectable by histochemical analysis. (Reprinted with permission from Ref. 10.)

murine *ADA*. Additional biochemical criteria indicate that the enzyme from the amplified cells is indistinguishable from that produced in nonamplified cell lines [6]. Because the amplification of *ADA* genes is accompanied by a corresponding increase in the abundance of ADA protein, the level of expression per *ADA* gene remained unchanged during the process of amplification. Taken together, these results lend strong support to our view that the drug-resistant cell lines that overproduce ADA have amplified a structurally and functionally unaltered *ADA* gene and that the cosmid clones depicted in Figure 5 may be taken as representative of a normal murine *ADA* gene.

VI. STRUCTURAL AND FUNCTIONAL ANALYSIS OF THE AMPLIFIED *ADA* LOCUS

Highly amplified DNA may be directly visualized as restriction fragments, which stain intensely with ethidium bromide following electrophoretic separation on agarose gels (Fig. 6). These highly repeated restriction fragments are not seen in the parental C1-1D cell DNA. The overall size of the amplified domain can be estimated by summing up the sizes of the amplified restriction fragments appearing in the gel shown in Figure 6. By this estimate a minimum of 125 kb of DNA has been coamplified during the selective amplification of the *ADA* gene in the B-1/100 cells.

While mapping the boundaries of the *ADA* transcription unit, Al-Ubaidi et al. discovered that another gene overlaps the 3′ end of the *ADA* gene [10]. The 3′ overlapping gene is transcribed from the opposite strand as *ADA* and encodes a relatively abundant 1.3 kb polyadenylated transcript. The tissue distribution of the overlapping transcript has no obvious relationship to that of *ADA* mRNA. The 3′ overlapping gene was initially coamplified with *ADA* but did not continue to be coamplified in functional form beyond the initial phases of *ADA* gene amplification. A lack of faithful coamplification of closely linked genes is a common phenomenon. For example, a divergent transcription unit that exists immediately upstream of the dihydrofolate reductase gene is coamplified in some, but not all, methotrexate-resistant cell lines [11]. Lack of faithful coamplification of linked genes is a problem often encountered in efforts to cotransfer and coamplify nonselectable genes of academic or commercial value using amplifiable genetic markers such as ADA or dihydrofolate reductase [12]. As a result of DNA recombination, rearrangement, and/or other genetic alterations that accompany gene amplification, linked genes with no selective advantage are commonly lost during the amplification process. In many cases the linked genes may even provide a selective disadvantage and for that reason be selected against.

VII. *ADA* MINIGENES FUNCTION AS DOMINANT SELECTABLE MARKERS

The selection procedures described above have the ability to isolate from a mixed population of cells those having the highest levels of ADA activity. This feature

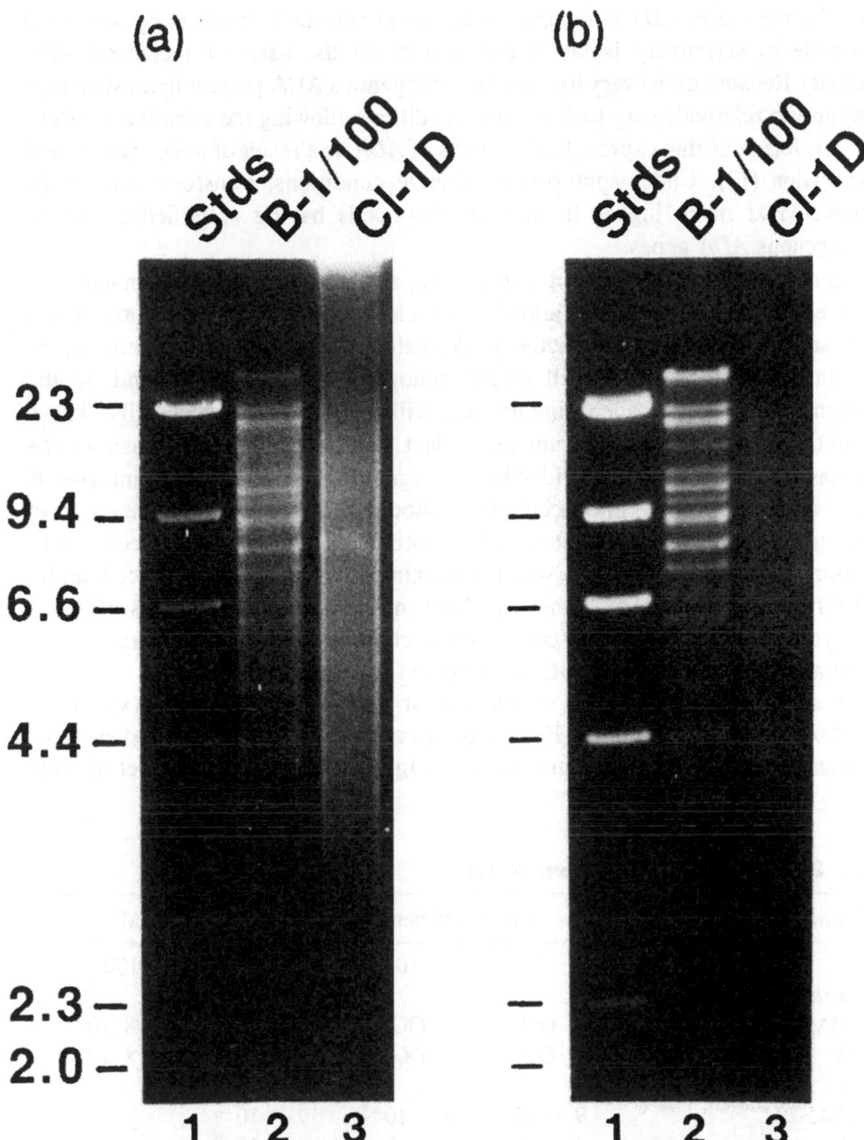

Figure 6 Direct visualization of highly amplified DNA at the *ADA* locus. (a) Agarose gel electrophoretic pattern of genomic DNA from a parental mouse line (C1-1D) and a cell line with high level amplification of the *ADA* locus (B-1/100) following digestion with *Eco*RI. (b) Same gel after being subjected to an in-gel renaturation technique, which selectively removes nonamplified sequences (see Chapter 3, this volume).

enables the use of *ADA* minigenes as dominant selectable markers by making it possible to selectively isolate transformants on the basis of increased ADA activity. Because of the very low level of endogenous ADA present in most cells, it has proven relatively easy to determine conditions allowing the selective recovery of transformants that express higher levels of *ADA* as a result of gene transfer and expression [12]. Under appropriate selective conditions, transformants can be recovered at much higher frequencies than cells having amplified copies of endogenous *ADA* genes.

In conjunction with alanosine, adenosine, and uridine, deoxycoformycin may be used to establish conditions allowing the selective recovery of transformants on the basis of increased ADA activity. Typically, a concentration of deoxycoformycin is identified which will inhibit endogenous ADA activity and, in the presence of alanosine/adenosine/uridine, kill all nontransformed cells. These conditions provide the opportunity to select transformants on the basis of the increased *ADA* expression provided by an exogenously supplied *ADA* minigene. A range of deoxycoformycin concentrations should be tested to determine the level best suited for the cell line of interest [12]. Optimal conditions allow no colonies to appear in mock transfections when approximately 10^7 cells are placed under selective conditions (e.g., Table 2, alanosine/adenosine/uridine plus 100 nM deoxycoformycin). Under such conditions all colonies appearing in the transfected population contain functional *ADA* minigenes.

A selection scheme based on the use of xylofuranosyl-adenine (xyl-A), a cytotoxic adenosine analogue (Fig. 1), can frequently provide a useful alternative to alanosine, adenosine, and uridine. As originally shown by Yeung et al. [13]

Table 2 *ADA*: A Dominant Selectable Marker[a]

Selection	Plating efficiency in deoxycoformycin (nM)			
	3	10	30	100
Alanosine/adenosine/uridine				
p9ADA5-29	OG	OG	3×10^{-4}	2×10^{-4}
Mock	OG	OG	1×10^{-5}	$<2 \times 10^{-7}$
xyl-A				
p9ADA5-29	9×10^{-5}	3×10^{-5}	$7 - 10^{-6}$	
Mock	4×10^{-5}	6×10^{-7}	$<2 \times 10^{-7}$	

[a]Dihydrofolate reductase deficient Chinese hamster ovary cells (DUKX-B11) were transfected with p9ADA5-29, using calcium phosphate, and subcultured 48 hr later into selection medium (50 μM alanosine, 1.1 mM adenosine, 1.0 mM uridine or 4 μM xyl-A) in the presence of various concentrations of deoxycoformycin. Colonies were stained and counted 12 days after subculturing into selective medium. The frequency of colony formation, expressed as the number of colonies relative to the number of cells plated into selective medium (5×10^6), is shown. (Mock) refers to similarly treated cells that received no DNA; OG, overgrowth of cells.
Source: Adapted from Ref. 14 with permission.

cells will not grow in the presence of micromolar concentrations of xyl-A if ADA activity is inhibited. These observations indicated that it should be possible to devise selective conditions based on the use of xyl-A and deoxycoformycin to isolate transformants having increased *ADA* expression. Kellems et al. [12] and Kaufman et al. [14,15] have shown that it is possible to use 4 μM xyl-A in combination with deoxycoformycin to selectively isolate transformants based on increased ADA activity encoded by *ADA* minigenes. It is possible to determine a deoxycoformycin concentration that completely eliminates the appearance of background colonies in mock transfections, yet allows the recovery of transformants with elevated ADA (Table 2, 4 μM xyl-A, 30 nM deoxycoformycin). Transformants recovered under these conditions show elevated ADA activity, approximately 10–20-fold, due to the expression of the introduced *ADA* minigenes. Although the xyl-A/deoxycoformycin selection scheme just described is a good strategy for the isolation of transformants with elevated ADA, we recommend against the long-term use of xyl-A selection because the likelihood of isolating mutants deficient in adenosine kinase becomes high.

VIII. COAMPLIFICATION USING *ADA* MINIGENES

ADA minigenes, along with any nonselectable gene that has been cotransferred, may be selectively coamplified by growth of transformants in alanosine, adenosine, uridine, and increasing concentrations of deoxycoformycin [12,14,15]. The level of coamplification may exceed several hundredfold. The preferential selection of cells that have amplified input *ADA* minigenes, in contrast to endogenous *ADA* genes, presumably results from the greater selective advantage provided by the former. *ADA* expression from the minigene locus is typically much greater than that from the endogenous *ADA* gene for one or more of the following reasons: (1) *ADA* minigenes are usually constructed using a promoter that is stronger than the natural *ADA* promoter, (2) the input *ADA* minigene may have integrated into a locus favorable for expression and/or amplification, and (3) it may have integrated in multiple tandemly arrayed copies. For these reasons the *ADA* minigene locus provides for much more ADA production, and therefore greater selective advantage, than do the endogenous *ADA* loci.

The selection protocols described here have a number of highly desirable features that facilitate the use of *ADA* as a dominant amplifiable genetic marker. First, the selection scheme is quite versatile in that, in the presence of xyl-A or alanosine/adenosine/uridine, it is possible to determine a deoxycoformycin concentration that allows the isolation of cells with increased ADA. This feature, coupled with the relatively low levels of ADA present in most of the commonly used cells, makes it possible to selectively isolate transformants on the basis of increased ADA activity. Second, extremely high levels of ADA are not toxic to cells, as indicated by the existence of cells with highly amplified *ADA* genes in which ADA enzyme accounts for 75% of the soluble protein. Thus, the level to which a nonselectable gene may be coamplified is not likely to be limited by toxicity resulting from increased ADA. Third, the selection scheme is highly specific in yielding only cells with the desired increase in ADA. This feature reduces the burden associated with the presence of cells with unwanted phenotypes, such as transport mutants or drug-resistant ADA mutants. Fourth, ADA selection protocols are compatible with other amplification schemes, such as methotrexate selection for amplified dihydrofolate reductase genes. This allows for the simultaneous or alternating use of ADA and dihydrofolate reductase selection protocols to accomplish a number of objectives. In many cases it will be desirable to have the ability to consecutively and independently cotransfer and coamplify different genes in the same mammalian cell. Fifth, a sensitive and reliable histochemical assay is available to detect ADA activity following electrophoretic separation on agarose gels (see Fig. 5b). This assay readily distinguishes human, mouse, hamster, and bacterial ADA activity in cell extracts, making it possible to use combinations of *ADA* minigenes and transformed cells, allowing ADA encoded by gene transfer vectors to be easily distinguished from endogenous ADA.

ACKNOWLEDGMENTS

Financial support was provided by the National Institutes of Health (AI25255) and the Robert A. Welch Foundation (Q-893). Alanosine was provided by the Drug Synthesis and Chemistry Branch, and 2′-deoxycoformycin, by the Natural Products Branch, both of the Developmental Therapeutics Program, Division of Cancer Treatment, National Cancer Institute. We are also grateful to Parke-Davis Pharmaceutical Research Division of Warner-Lambert Company for providing Pentostatin (2′-deoxycoformycin).

REFERENCES

1. Chinsky, J. M., Ramamurthy, V., Fanslow, W. C., Ingolia, D. E., Blackburn, M. R., Shaffer, K. T., Higley, H. R., Trentin, J. J., Rudolph, F. B., Knudsen, T. B., and Kellems, R. E. Developmental expression of adenosine deaminase in the upper alimentary tract of mice. *Differentiation*, *42*:172–183 (1990).
2. Knudsen, T. B., Blackburn, M. R., Chinsky, J. M., Airhart, M. J., and Kellems, R. E. Ontogeny of adenosine deaminase in the mouse decidua and placenta: Immunolocalization and embryo transfer studies. *Biol. Reprod.* *44*:171–184 (1991).
3. Kredich, N. M., and Hershfield, M. S. Immunodeficiency diseases caused by adenosine deaminase deficiency and purine nucleoside phosphorylase deficiency, in *The Metabolic Basis of Inherited Disease* (C. R. Scriver, A. L. Beaudet, W. S. Sly, and D. Valle, eds.), McGraw-Hill, New York, 1989, pp. 1045–1075.
4. Knudsen, T. B., and Elmer, W. A. Evidence for negative control of growth by adenosine in the mammalian embryo: Induction of $Hm^{x/+}$ mutant limb outgrowth by adenosine deaminase. *Differentiation*, *33*:270–279 (1988).

5. Yeung, C.-Y., Ingolia, D. E., Bobonis, C., Dunbar, B. S., Riser, M. E., Siciliano, M. J., and Kellems, R. E. Selective overproduction of adenosine deaminase in cultured mouse cells. *J. Biol. Chem. 258*:8338–8345 (1983).
6. Ingolia, D. E., Yeung, C.-Y.., Orengo, I. F., Harrison, M. L., Frayne, E. G., Rudolph, F. B., and Kellems, R. E. Purification and characterization of adenosine deaminase from a genetically enriched mouse cell line. *J. Biol. Chem. 260*:13261–13267 (1985).
7. Yeung, C.-Y.., Frayne, E. G., Al-Ubaidi, M. R., Hook, A. G., Ingolia, D. E., Wright, D. A., and Kellems, R. E. Amplification and molecular cloning of murine adenosine deaminase gene sequences. *J. Biol. Chem. 258*:15179–15185 (1983).
8. Yeung, C.-Y., Ingolia, D. E., Roth, D. B., Shoemaker, C., Al-Ubaidi, M. R., Yen, J.-Y., Ching, C., Bobonis, C., Kaufman, R. J., and Kellems, R. E. Identification of functional murine adenosine deaminase cDNA clones by complementation in Escherichia coli. *J. Biol. Chem. 260*:10299–10307 (1985).
9. Ingolia, D. E., Al-Ubaidi, M. R., Yeung, C.-Y., Bigo, H., Wright, D. A., and Kellems, R. E. Molecular cloning of the murine adenosine deaminase gene from a genetically enriched source: Identification and characterization of the promoter region. *Mol. Cell. Biol. 6*:4458–4466 (1986).
10. Al-Ubaidi, M. R., Ramamurthy, V., Maa, M.-C., Ingolia, D. E., Chinsky, J. M., Martin, B. D., and Kellems, R. E. Structural and functional analysis of the murine adenosine deaminase gene. *Genomics, 7*:476–485 (1990).
11. Linton, J. P., Yen, J.-Y. J., Selby, E., Chen, Z., Chinsky, J. M., Liu, K., Kellems, R. E., and Crouse, G. F. Dual bidirectional promoters at the mouse *dhfr* locus: Cloning and characterization of two mRNA classes of the divergently transcribed *Rep-1* gene. *Mol. Cell. Biol. 9*:3058–3072 (1989).
12. Kellems, R. E., Johnstone, E. M., Ward, K. E., and Little, S. P. Adenosine deaminase: A dominant amplifiable genetic marker for use in mammalian cells, in *Genetics and Molecular Biology of Industrial Microorganisms* (C. L. Hershberger, S. W. Queener, and G. Hegeman, eds.), American Society for Microbiology, Washington, D.C., 1989, pp. 215–225.
13. Yeung, C.-Y., Riser, M. E., Kellems, R. E., and Siciliano, M. J. Increased expression of one of two adenosine deaminase alleles in a human choriocarcinoma cell line following selection with adenine nucleosides. *J. Biol. Chem. 258*:8330–8337 (1983).
14. Kaufman, R. J., Murtha, P. Ingolia, D. E., Yeung, C.-Y., and Kellems, R. E. Selection and amplification of heterologous genes encoding adenosine deaminase in mammalian cells. *Proc. Natl. Acad. Sci. USA, 83*: 3136–3140 (1986).
15. Israel, D. I., and Kaufman, R. J. Retroviral-mediated transfer and amplification of a functional human factor VIII gene. *Blood, 75*:1074–1080 (1990).

17

Amplification of the *HPRT* Gene

David W. Melton *Edinburgh University, Edinburgh, Scotland*

I. INTRODUCTION

The enzyme hypoxanthine–guanine phosphoribosyltransferase (HPRT) catalyzes the phosphoribosylation of the purine bases hypoxanthine and guanine to form inosine and guanosine monophosphate. This is the first step in the purine salvage pathway in mammalian cells. The enzyme is expressed constitutively in all tissues and cell types. Two features have led to the *HPRT* locus being widely used in somatic cell genetics. First, it is possible to select both for and against *HPRT* expression in cultured cells. Second, the gene is X chromosome linked, so there is only one copy in male cells and a single functional copy in female cells, due to X chromosome inactivation. The *HPRT* gene is, in addition, of medical interest, since *HPRT* deficiency results in the severe neurological disorder Lesch–Nyhan syndrome (see Ref. 1 for a review of the *HPRT* gene and its expression).

There have been several reports of *HPRT* gene amplification in cultured mammalian cells [2–5]. However, given the great deal of work carried out on the *HPRT* locus, reports of gene amplification have been infrequent. This is because selection for *HPRT* gene amplification is provided only in a restricted set of circumstances, which are considered in the next section. Even where these conditions apply, the extent of *HPRT* gene amplification reported has been low compared to levels obtained, for example, with the dihydrofolate reductase gene. For these reasons the *HPRT* locus has not proved suitable for studies on the mechanism of gene amplification itself. However, *HPRT* gene amplification did make possible the original isolation of cloned *HPRT* sequences [2].

II. SELECTION FOR GENE AMPLIFICATION

A. *HPRT* Selection Systems

Cells with HPRT enzyme activity are selected in HAT medium [6], which contains aminopterin (A) to block de novo purine synthesis. In the presence of aminopterin, cells are dependent on functional salvage pathways for survival. Hypoxanthine (H) provides the substrate for the purine salvage pathway and thymidine (T) the substrate for pyrimidine salvage, which is required because aminopterin also blocks de novo pyrimidine synthesis. For cells to survive in HAT medium, 5–10% of normal HPRT enzyme activity is sufficient [7]. For this reason and because it has not been possible to make selection more stringent, to require progressively higher levels of HPRT activity for survival, HAT selection on cells containing a wild-type HPRT protein does not provide selection for gene amplification, in the way that selection for increasing resistance to methotrexate does for the dihydrofolate reductase gene.

Cells lacking HPRT enzyme activity are selected in the presence of base analogues, such as 6-thioguanine, which are phosphoribosylated by HPRT with fatal consequences for the cell. Revertants with HPRT enzyme activity, arising in populations of HPRT-deficient mutants, may be selected in HAT medium. All reported instances of *HPRT* gene amplification have occurred in revertants of *HPRT*-deficient mutants with a similar mutant phenotype.

B. Phenotype of *HPRT*-Deficient Mutants Revertible by Gene Amplification

HPRT gene amplification was first reported in revertants of the HPRT-deficient mouse neuroblastoma cell line NB^- [2]. Extracts from NB^- cells grown at 37°C contained no measurable enzyme activity and no immunologically detectable HPRT protein (Fig. 1a). However, HPRT protein was readily detectable following in vitro translation of mRNA extracted from NB^- cells (Fig. 1b). Northern hybridization analysis revealed equivalent levels of *HPRT* mRNA in NB^- cells and their wild-type progenitor, the NB^+ line (Fig. 2). Thus, NB^- cells produce an altered HPRT protein with an extreme temperature-sensitive phenotype in vivo. The properties of this altered protein are considered in more detail in Section III.A. The other *HPRT*-deficient mutants that have given rise to revertants with gene amplification are all derivatives of the Chinese hamster lung cell line, V79. The mutant RJK10 has a similar extreme temperature-sensitive phenotype to NB^- [4], while the two other mutants have a less extreme phenotype. The mutant RJK526 grows normally at 33°C in HAT medium and cell extracts contain 10% of

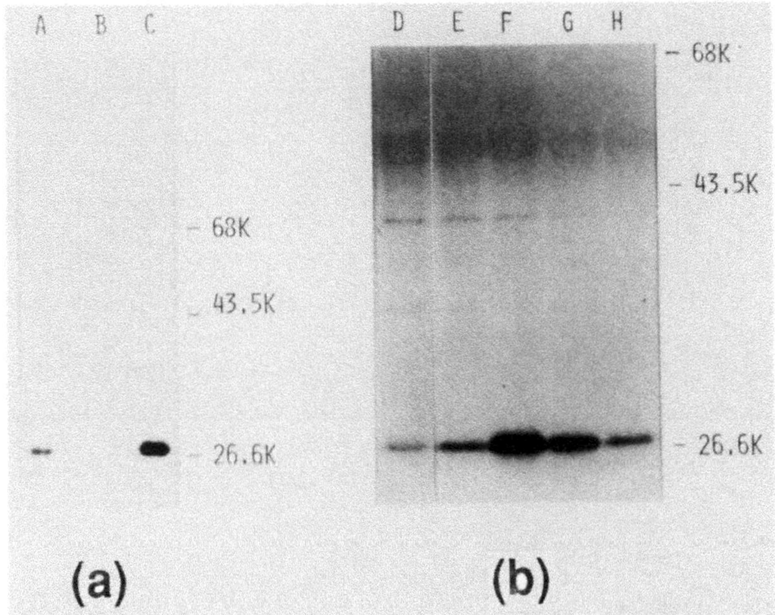

Figure 1 (a) HPRT protein levels in mouse neuroblastoma cells. Cells were labeled in vivo with [^{35}S]methionine and HPRT protein was immunopurified from cell extracts and electrophoresed on a denaturing polyacrylamide gel. Lanes: A, NB$^+$; B, NB$^-$; and C, NBR4. All extracts contained equivalent amounts of material precipitable in trichloroacetic acid. The mobility of molecular weight marker proteins run on the same gel is shown. (b) In vitro translation of mRNA from mouse neuroblastoma cells: mRNA purified from neuroblastoma cells was translated in a rabbit reticulocyte lysate system and the products were immunopurified and electrophoresed. Lanes: D, NB$^-$ (product from translation of 3.7 μg mRNA); E, NB$^+$ (4.6 μg); and F–H, NBR4 (0.06, 0.03, and 0.015 μg mRNA respectively). (From Ref. 9.)

wild-type HPRT enzyme activity [3]. There is no growth in HAT medium at 39°C and no detectable enzyme activity. The half-life at 37°C of the enzyme from RJK526 cells is 10-fold less than for the wild-type enzyme. Similarly, the mutant RJK531 will grow at 32°C, but not at 39°C in HAT medium [5].

Reversion of such *HPRT* mutants could occur by two distinct mechanisms. The first would involve an additional mutational event, either to repair the original lesion or at a second site, to generate a protein that can function sufficiently well to allow growth in HAT medium. Alternatively, cells could acquire HAT resistance by producing elevated levels of the mutant protein. This could be achieved by *HPRT* gene amplification; mutants with the most extreme phenotype would be

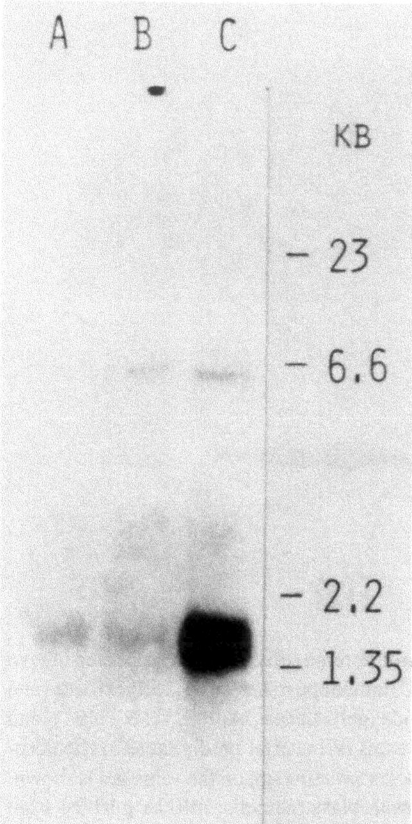

Figure 2 *HPRT* mRNA levels in mouse neuroblastoma cells: mRNA (4 μg) was glyoxylated, electrophoresed, transferred to nitrocellulose, and probed with a radiolabeled *HPRT* cDNA clone. Lanes: A, NB^-; B, NB^+; and C, NBR4. The mobility of molecular weight standards is shown. (Modified from Ref. 2.)

expected to require a greater degree of gene amplification to acquire HAT resistance than mutants with a less severely affected HPRT protein.

III. EXAMPLES OF *HPRT* GENE AMPLIFICATION

A. NBR4, a Revertant of NB^-

Revertants of the NB^- line arose at an overall frequency of 2×10^{-6} and were of two distinct types. One group of revertants contained normal levels of HPRT protein with the same heat stability, electrophoretic mobility, and kinetic properties as the protein from wild-type cells. These revertants appear to have arisen by correction of the original mutation in NB^- cells. The second group all contained an enzyme with altered electrophoretic mobility and extreme heat lability, the half-life of the enzyme at 80°C being 40-fold lower than wild type. Within this group the revertant with the highest enzyme activity was NBR4. The NBR4 enzyme had half the affinity for hypoxanthine of the wild type, and a 20-fold lower affinity for 5′-phosphoribosyl-1-pyrophospate. *HPRT* cDNA sequences isolated from NBR4 cells revealed a missense mutation at amino acid residue 200, changing the codon for Asp to Asn [8]. It is believed that all revertants of this type and the NB^- line itself contain this same mutant protein. Extracts from NBR4 cells grown at 37°C contained 5–10 times more immunologically detectable HPRT protein than wild-type cells, while no cross-reacting material was evident in NB^- extracts [7] (see Fig. 1a). In vitro translation of mRNA extracted from NBR4 cells produced 25- to 50-fold more HPRT protein than wild type [9] (see Fig. 1b). The lower level of HPRT protein detectable in NBR4 cells in vivo is presumably due to the extreme lability of the protein.

Northern analysis indicated that NBR4 cells contained approximately 20 times more *HPRT* mRNA than wild-type cells (Fig. 2). The Southern analysis shown in Figure 3 proves that this is due to *HPRT* gene amplification. The organization of the mouse *HPRT* gene is shown in Figure 4. There are nine exons, extending over 33 kb of the mouse X chromosome. Restriction fragments containing particular exon elements, detected with an *HPRT* cDNA probe, are indicated in Figure 3. All restriction fragments containing *HPRT* exon elements present in wild-type cells are amplified approximately 20-fold in NBR4 cells. Other revertants with the same mutant protein, but lower HPRT activity than NBR4, showed 5–10-fold amplification. The only *HPRT* sequence not amplified in NBR4 cells constitutes a processed *HPRT* pseudogene [10]. There is no evidence that any of the amplified gene copies in NBR4 cells have been rearranged to generate novel restriction fragments detectable with *HPRT* cDNA probes. The size of the amplified unit has not been precisely determined. It extends at least 12 kb 5′ and 13 kb 3′ of the gene, making a minimum size of 48 kb, but does not extend to adjacent markers located some 2 cM away on the X chromosome.

All neuroblastoma revertants with amplified *HPRT* genes contained a large marker chromosome, which was not present in NB^+ and NB^- cells (Fig. 5). The mouse X chromosome in both NB^+ and NB^- cells occurs as a centric fusion with mouse chromosome 12 (Fig. 5E). In the revertants the X-containing arm of the fusion chromosome has undergone a triplication, so that there are now three copies of the mouse X, with the middle copy inverted. The amplified genes have been localized to the third copy of the mouse X by in situ hybridization [1]. The amplification has not affected the normal appearance of the mouse X, and the

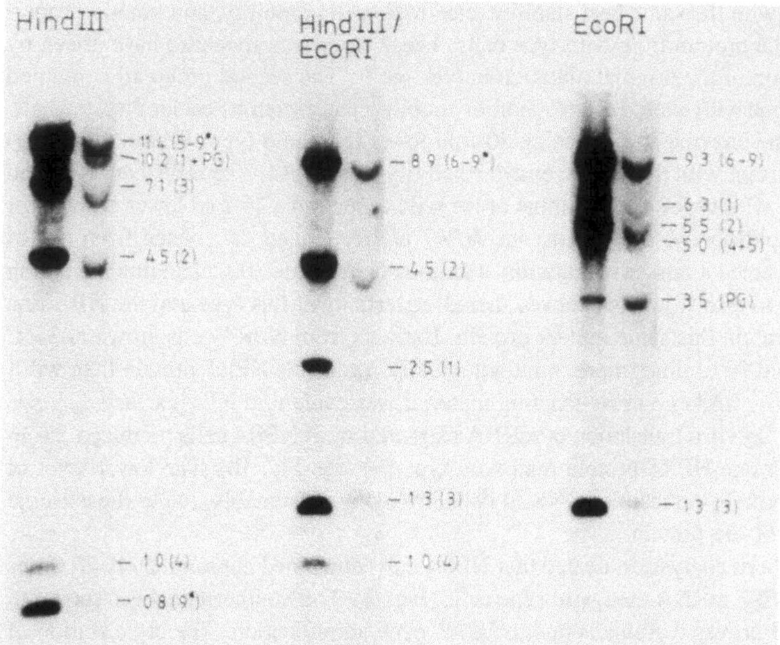

Figure 3 Amplification of the *HPRT* gene in NBR4 cells. DNA from NBR4 (left) and NB$^+$ (right) cells was digested with *Hin*dIII, *Hin*dIII/*Eco*RI, and *Eco*RI and subjected to Southern analysis with an *HPRT* cDNA probe. The size (kb) of each hybridizing fragment is given, along with the exons present in each fragment (in parentheses). Asterisks indicate that exon 9 sequences occur on two discrete *Hin*dIII fragments, owing to the presence of a *Hin*dIII site within this exon (see Fig. 4). The 10.2 kb *Hin*dIII fragment is a doublet, one fragment containing exon 1 and the second an *HPRT* pseudogene (*PG*). (From Ref. 1, by permission of the Oxford University Press.)

location of the amplified sequences (Fig. 5G) is consistent with the *HPRT* gene having undergone amplification at its normal location. This situation differs from several other examples of gene amplification, where amplified genes occur at homogeneously staining chromosomal regions or on acentric chromosome fragments, but it should be remembered that the degree of *HPRT* gene amplification is much lower than in these other systems.

HPRT gene amplification in the neuroblastoma revertants is chromosomal, but when HAT selection was removed, 6-thioguanine-resistant derivatives appeared rapidly in the population. These derivatives retained the marker chromosome but had lost the amplified genes.

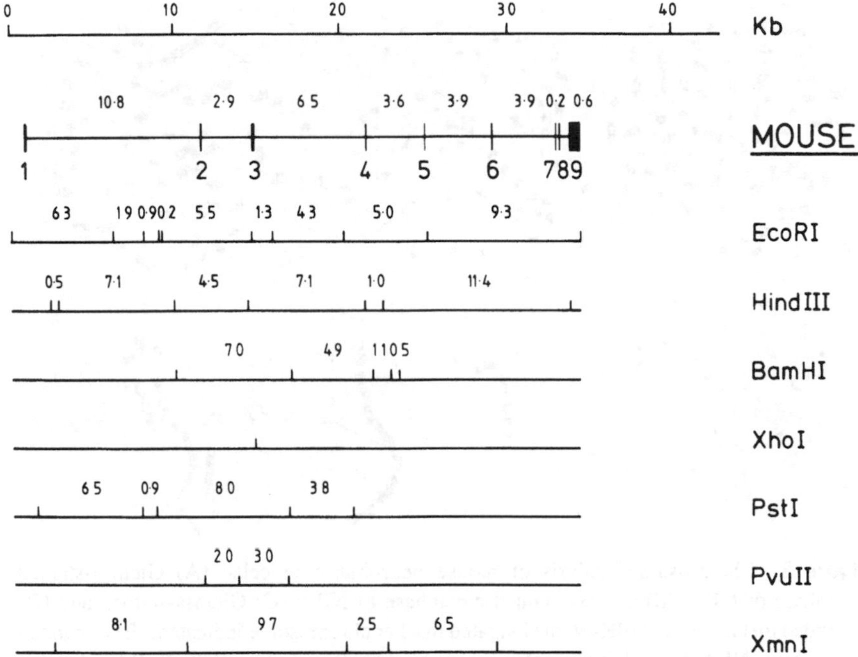

Figure 4 Organization of the mouse *HPRT* gene. *HPRT* exons are depicted as numbered vertical bars, with the size (kb) of each intervening sequence indicated. The positions of restriction sites mapped within the gene are indicated, and the size of each fragment is given. (From Ref. 1, by permission of the Oxford University Press.)

B. Revertants Derived from Chinese Hamster Lung Cell Mutants

The situation with reversion of the mutant RJK10 is very similar to that described for NB⁻. Revertants also fell into two groups, with one group containing a normal HPRT protein and the second group overproducing a mutant protein as a consequence of 20-fold gene amplification. Revertants of mutants RJK526 and RJK531 occurred at a higher frequency (1×10^{-4} for RJK531) and contained the same temperature-sensitive protein as the mutants themselves. Revertants of RJK526 showed up to fivefold amplification, with no evidence of chromosomal alteration, and twofold amplification occurred in the RJK531 revertants. In both cases the amplification was unstable and 6-thioguanine-resistant derivatives arose at high frequency.

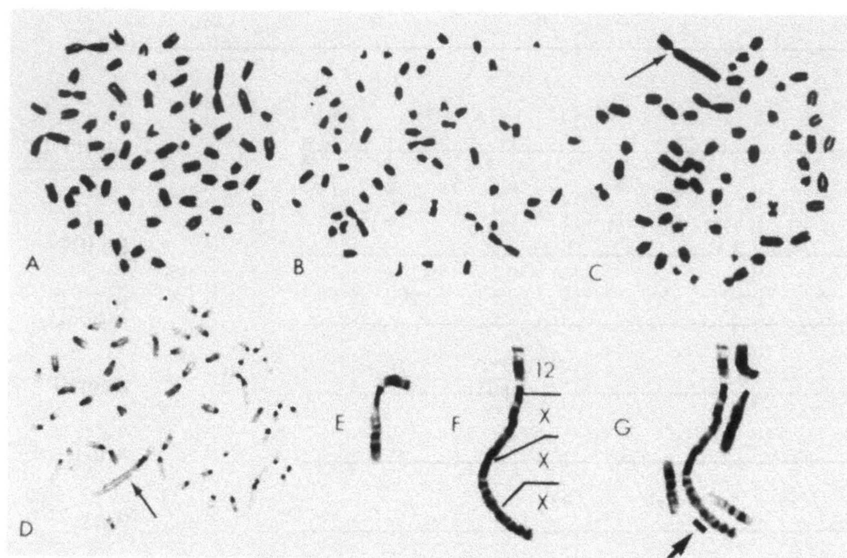

Figure 5 Chromosome analysis of mouse neuroblastoma cells: (A) Giemsa-stained metaphase of NB$^+$; (B) Giemsa-stained metaphase of NB$^-$; (C) Giemsa-stained and (D) C-banded metaphase of NBR4 with elongated marker chromosome indicated; (E) G-banded marker from NB$^+$, comprising a centric fusion of chromosome 12 and the X; (F) G-banded elongated marker from NBR4, with the p arm corresponding to chromosome 12, while the q arm consists of three copies of the X, and the middle segment is inverted; and (G) reconstruction of the NBR4 marker shown in (F); the X:12 translocation from NB$^+$ (E) has been cut at the centromere and three copies of the q arm are aligned to demonstrate homology with the NBR4 marker; the location of the amplified *HPRT* genes, mapped by in situ hybridization is indicated by the bar. (Modified from Ref. 9.)

REFERENCES

1. Melton, D. W. *HPRT* gene organization and expression, in *Oxford Surveys on Eukaryotic Genes*, Vol. 4 (N. Maclean, ed.), Oxford University Press, Oxford, 1987, pp. 34–76.
2. Brennand, J., Chinault, A. C., Konecki, D. S., Melton, D. W., and Caskey, C. T. Cloned cDNA sequences of the hypoxanthine phosphoribosyltransferase gene from a mouse neuroblastoma cell line found to have amplified genomic sequences. *Proc. Natl. Acad. Sci. USA*, *79*:1950–1954 (1982).
3. Fuscoe, J. C., Fenwick, R. G., Jr., Ledbetter, D. H., and Caskey, C. T. Deletion and amplification of the *HGPRT* locus in Chinese hamster cells. *Mol. Cell Biol. 3*:1086–1096 (1983).
4. Fenwick, R. G., Jr., Fuscoe, J.C ., and Caskey, C. T. Amplification versus mutation as a mechanism for reversion of an *HGPRT* mutation. *Somatic Cell Mol. Genet. 10*:71–84 (1984).

5. Zownir, O., Fuscoe, J. C., Fenwick, R. G., and Morrow, J. Gene amplification as a mechanism of reversion at the *HPRT* locus in V79 Chinese hamster cells. *J. Cell. Physiol. 119*:341–348 (1984).
6. Littlefield, J. W. Selection of hybrids from matings of fibroblasts in vitro and their presumed recombinants. *Science, 145*:709 (1964).
7. Melton, D. W. Cell fusion-induced mouse neuroblastoma HPRT revertants with variant enzyme and elevated HPRT protein levels. *Somatic Cell Genet. 7*:331–344 (1981).
8. Melton, D. W., Konecki, D. S., Brennand, J., Chinault, A. C., and Caskey, C. T. Structure, expression and mutation of the hypoxanthine phosphoribosyltransferase gene. *Proc. Natl. Acad. Sci. USA, 81*:2147–2151 (1984).
9. Melton, D. W., Konecki, D. S., Ledbetter, D. H., Hejtmancik, J. F., and Caskey, C. T. In vitro translation of hypoxanthine phosphoribosyltransferase mRNA: Characterization of a mouse neuroblastoma cell line that has elevated levels of hypoxanthine phosphoribosyltransferase protein. *Proc. Natl. Acad. Sci. USA, 78*:6977–6980 (1981).
10. Isamat, M., Maclcod, K. F., King, A., McEwan, C., and Melton, D. W. Characterization, evolutionary relationships and chromosome location of a processed mouse *HPRT* pseudogene. *Somatic Cell Mol. Genet. 14*:359–369 (1988).

18

Amplification of the Gene for *N*-Acetylglucosaminyl Phosphate Transferase, an Enzyme Catalyzing an Early Step in Asn-Linked Glycosylation, in Chinese Hamster Ovary Cells Resistant to Tunicamycin

Jane R. Scocca and Sharon S. Krag *The Johns Hopkins University School of Hygiene and Public Health, Baltimore, Maryland*

I. INTRODUCTION

In the synthesis of asparaginyl glycoproteins by eukaryotic cells, the initial transfer of carbohydrate to asparagine residues of newly synthesized polypeptide chains occurs in the endoplasmic reticulum. The first carbohydrate transferred is an oligosaccharide previously assembled on a polyisoprenoid lipid carrier, dolichyl phosphate, by the stepwise transfer of 14 simple sugar residues (*N*-acetylglucosamine, GlcNAc; mannose, Man; and glucose, Glc) from activated donors to the growing saccharide chain [1] (Fig. 1). The biosynthetic enzymes involved in the synthesis of the oligosaccharide are present in small amounts in the endoplasmic reticular membrane. The purification of these enzymes from normal cells has been tedious, and in most cases isolation of homogeneous enzyme has thus far been impossible.

II. TARGET ENZYME AND DRUG

In 1971 a specific inhibitor of asparagine-linked glycosylation was first identified [2]. This drug, tunicamycin (TM), blocked glycosylation of Asn-linked glycopro-

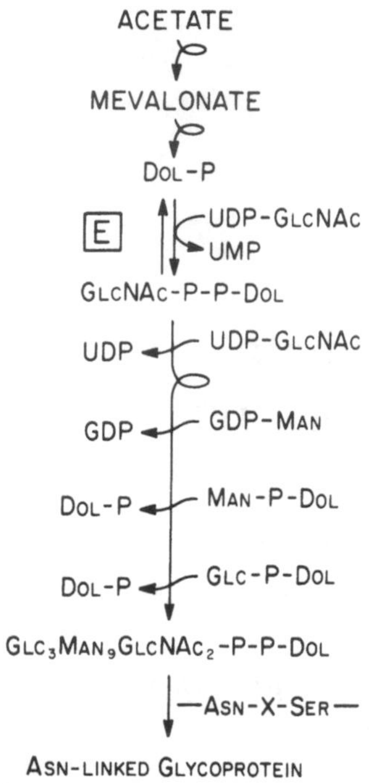

Figure 1 Reactions involved in synthesis of asparaginyl glycoproteins in eukaryotic cells. The target enzyme of tunicamycin action, *N*-acetylglucosaminyl phosphate transferase, is shown as a boxed "E." Arrows with loop indicate multiple steps. Abbreviations: Dol, Dolichol; GlcNAc, *N*-acetylglucosamine; P, phosphate; Man, mannose; Glc, glucose; UDP, uridine diphosphate; UMP, uridine monophosphate; GDP, guanosine diphosphate; Asn, asparagine; X, any amino acid; Ser, serine.

teins by binding tightly and inhibiting (apparent $K_i = 5 \times 10^{-8}$) the first enzyme that adds a sugar molecule to dolichyl phosphate—uridine diphosphate-*N*-acetyl D-glycosamine:dolichyl phosphate *N*-acetyl glucosamine-1 phosphate transferase (*N*-acetylglucosaminyl phosphate transferase, E in Fig. 1). This enzyme is unique among the enzymes synthesizing the oligosaccharide–lipid molecule, because it transfers a monosaccharide–phosphate unit rather than simply a monosaccharide.

The structure of TM suggests that it is a bisubstrate analogue, resembling UDP-*N*-acetylglucosamine and dolichol [2]; see Figure 1. It contains uracil, a fatty acid residue, and two glycosidically linked sugars. Different forms of TM vary in their fatty acid composition. The presence of the fatty acid moiety imparts its lipophilic nature and allows TM to partition into membranes [3]. TM has been used extensively to inhibit glycosylation [2] of both soluble glycoproteins, such as secreted proteins and lumenal lysosomal enzymes, and membrane-associated glycoproteins, such as receptors and viral envelope proteins.

III. ISOLATION OF TUNICAMYCIN-RESISTANT CELLS

An important property of TM is that it is cytotoxic to mammalian cells. We found that the LD_{50} of TM of wild-type Chinese hamster ovary (CHO) cells is 0.18 μg/ml. Only 1% of the cells survive in 0.28 μg TM/ml [4]. Exposure of mammalian cells to increasing concentrations of potent, specific inhibitors of growth-essential enzymes has been used to select cell lines that overproduce specific proteins [5]. We therefore decided to isolate CHO cells resistant to TM in order to obtain a cell line producing large amounts of *N*-acetylglucosaminyl phosphate transferase [4]. Unmutagenized wild-type CHO cells (1×10^6) were exposed to 0.9 μg TM/ml. The surviving cells (0.01%) were grown in this concentration of TM until the cells reached confluency (about 2×10^7 cells in a T-75 flask). A portion of these cells (5×10^5 cells) were then treated with 2.9 μg TM/ml and the surviving cells grown to confluency. This stepwise selection procedure was repeated with 9, 15, and finally 27 μg TM/ml. At each step only about 0.01% of the cells survived. Despite repeated attempts, a population of cells resistant to a higher level of TM (81 μg/ml) could not be isolated.

From the population of CHO cells resistant to 27 μg TM/ml, we also isolated, using microtiter dishes, clones of CHO cells resistant to 27 μg TM/ml. Four clones were isolated, all having essentially identical properties [6]. The LD_{50} for the TM-resistant population was 35 μg TM/ml, while the LD_{50} for 3E11 cells, one of the TM-resistant clones, was greater than 100 μg TM/ml.

IV. OVERPRODUCTION OF THE TARGET ENZYME

To determine the level of activity of the target enzyme *N*-acetylglucosaminyl phosphate transferase, membranes were prepared from four TM-resistant clones and assayed for transferase activity. Addition of detergent to the assay aids in the solubilization of substrates and membranes, but for cells grown in the presence of TM, added detergent also results in TM solubilization and therefore enzyme inhibition in vitro [4]. Because of the persistence of this lipophilic compound in cell membranes [3] for several generations, assays could be performed only after a TM reprieve. TM-resistant cells were cultured in the presence of TM except for the last 14 days before membrane preparation. All four TM-resistant clones displayed a 15-fold higher specific activity of the transferase than did wild-type cells [4,6]. Moreover, the enzyme produced by the TM-resistant cells was identical to the wild-type enzyme with regard to its sensitivity to TM.

The observed increase in *N*-acetylglucosaminyl phosphate transferase activity in TM-resistant cells appeared to be specific for that enzyme. Two other enzymes that synthesize monosaccharide-dolichols, mannosylphosphoryldolichol (Man-P-Dol) and glucosylphosphoryldolichol (Glc-P-Dol) synthases, had identical specific activities in membranes prepared from wild-type and TM-resistant cells [6]. Although TM has also been shown to inhibit Glc-P-Dol synthase [7], no increase in the specific activity of this enzyme was detected. TM has also been shown to block ganglioside biosynthesis in isolated Golgi vesicles, by inhibiting the transport of UDP-galactose into the vesicles [8]. The transport of UDP-galactose into Golgi vesicles prepared from wild-type and TM-resistant cells has not yet been compared.

V. MORPHOLOGY OF TM-RESISTANT CELLS

The TM-resistant cells had an altered morphology (larger and flatter), as shown by phase contrast microscopy, when compared to wild-type cells [4]. After release from the substratum, rounded TM-resistant cells had a 50% larger cell diameter and therefore a larger internal volume than wild-type cells [6]. However, wild-type and TM-resistant cells had similar internal morphologies when examined by electron microscopy [6]. Therefore, the overproduction of the target enzyme, a membrane protein [6], did not result in the elaboration of a crystalloid endoplasmic reticulum. This structure has been seen previously in UT-1 cells, which were isolated as compactin-resistant CHO cells and overproduced the endoplasmic reticular membrane enzyme, HMG-CoA reductase.

VI. GLYCOSYLATION PHENOTYPE OF TM-RESISTANT CELLS

We examined the rate of Asn-linked glycosylation in TM-resistant cells. In these studies, we compared wild-type cells with both TM-resistant cells cultured in the presence of TM (lacking transferase activity assayable in vitro) and TM-resistant cells grown in the absence of TM (showing a 15-fold increase in enzyme activity in the in vitro assay).

Interestingly, a 15-fold overproduction of the *N*-acetylglucosaminyl phosphate transferase did not increase the overall glycosylation rate [6,9]. In normal growth medium, the rate of incorporation of [2-^{3}H] (from 1 to 5 hours) into TCA-precipitable glycoproteins was comparable among wild-type cells, TM-resistant cells grown with the drug, and TM-resistant cells grown without the drug [9]. Early labeling experiments [6] showed small differences in the glycosylation rates but were complicated by potential differences in pool sizes among the different cell types.

Although the glycosylation rates were similar, the oligosaccharides transferred to the newly glycosylated proteins differed. This was determined by characterizing the radioactive products obtained during a 30-min incubation of cells with [2-^{3}H]mannose [6]. Wild-type cells and TM-resistant cells grown in the presence of TM transferred only oligosaccharides that were sensitive to *endo*-β-*N*-acetylglucosaminidase H to protein, indicating that they were high-mannose type, derived from transfer of $Glc_3Man_9GlcNAc_2$. On the other hand, TM-resistant cells grown without TM transferred only 50% high-mannose oligosaccharides; these cells also transferred significant amounts of a smaller $Glc_3Man_5GlcNAc_2$ oligosaccharide, which is resistant to hydrolysis by *endo*-β-*N*-acetylglucosaminidase H. Transfer of this truncated oligosaccharide is also observed in B4-2-1 cells, which lack Man-P-Dol synthase and therefore have reduced amounts of Man-P-Dol [1].

The levels of the monosaccharide-lipids, Man-P-Dol and Glc-P-Dol, and of Dol-P, and of the oligosaccharide lipid, oligosaccharide-P-P-dolichol, in the three cell types were also determined [9]. As expected from the nature of the oligosaccharide transferred to protein, the level of Man-P-Dol was lower in TM-resistant cells grown without TM than in either wild-type (2-fold) or TM-resistant cells grown with TM (10-fold). In addition, the levels of Glc-P-Dol and Dol-P were also lower in these cells. *N*-Acetylglucosaminylpyrophosphoryldolichol, the product of the target enzyme, was not detectable in any of these cell types. TM-resistant cells grown without TM had higher levels of oligosaccharide-lipid than the other cell types, and as on the protein, the structure of the predominant lipid-linked oligosaccharide was also different. The major oligosaccharide-lipid species in wild-type cells and TM-resistant cells grown in TM was $Glc_3Man_9GlcNAc_2$-P-P-Dol, while in TM-resistant cells grown without TM the major species was $Man_5GlcNAc_2$-P-P-Dol.

A summary of the glycosylation phenotype of TM-resistant cells grown without TM compared to wild-type and TM-resistant cells grown in TM is presented in Figure 2. Overproduction of active *N*-acetylglucosaminyl phosphate

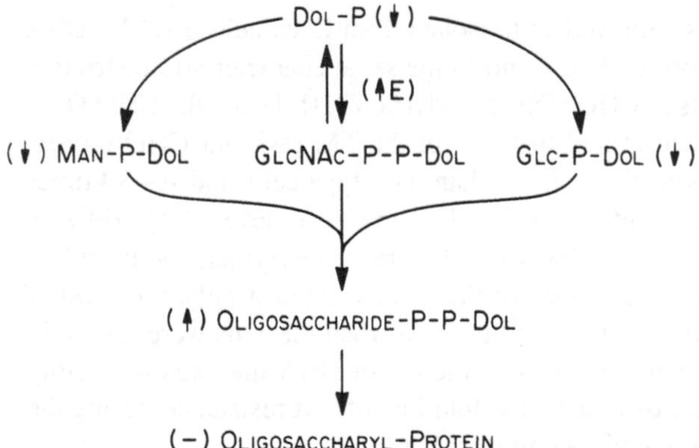

Figure 2 Glycosylation phenotype of TM-resistant cells grown in the absence of TM compared to either wild-type cells or TM-resistant cells grown in the presence of TM. The direction of the difference between the cells is indicated in parentheses. For abbreviations, see the legend of Figure 1.

transferase leads to an increase in the level of oligosaccharide-P-P-Dol at the expense of Man-P-Dol, Glc-P-Dol, and Dol-P. The nature of the predominant oligosaccharide-lipid differed from normal as a result of the lowered amount of Man-P-Dol, the substrate for the addition of the last four mannoses of the $Glc_3Man_9GlcNAc_2$ oligosaccharide [1]. It is not clear why an increase in the amount of oligosaccharide-lipid did not result in increased glycosylation. This failure is perhaps due to the alteration in the structure of the transferred oligosaccharide. This seems unlikely however, because transfer of this species in B4-2-1 cells does not result in a reduced level of glycosylation [1]. There may be other factors regulating glycosylation in these TM-resistant cells.

VII. AMPLIFICATION OF GENE

Gene amplification has been demonstrated as a mechanism of overproduction of target enzymes in drug-resistant cells. We examined the TM-resistant cells for the presence of homogeneously staining regions (HSRs) or double minute chromosomes, indicative of high levels of gene amplification. Neither was present in the TM-resistant cells, although the drug-resistant phenotype and the 15-fold increase in the activity of the target enzyme were stable when the cells were grown without drug. Karyotype analysis revealed that the TM-resistant cells did have chromosomal translocations, perhaps resulting from a translocation between chromosomes Z-7 and Z-5a [6], consistent with a moderate amount of gene amplification.

When the *ALG7* gene from yeast became available, we used it to determine that the CHO TM-resistant cells had three to four times more DNA hybridizing to *ALG7* than did wild-type cells [10]. The *ALG7* gene confers TM resistance to yeast and is presumably the gene for the yeast *N*-acetylglucosaminyl phosphate transferase.

The *ALG7* sequences also allowed us to isolate from CHO cells a 1.4 kb DNA fragment that hybridized both to *ALG7* and to the same characteristic restriction enzyme generated fragments of CHO DNA as did *ALG7* [11]. Using the CHO DNA fragment as a probe, we determined that one of the TM-resistant CHO clones, 3E11, had 12 times more hybridizing DNA than wild-type cells and 10–30 times more hybridizing RNA than wild-type cells. This increased level of hybridizing DNA and RNA compared closely to the level of increased enzymatic activity (15-fold) we had detected in the membranes of these cells. Interestingly, the level of hybridizing DNA, but not RNA, increased twofold when the cells were grown in the presence of TM. This may represent a fraction of DNA that was not stably amplified. We had previously detected a twofold loss of TM resistance during the first 5 days of culturing these cells without TM [6].

We examined the TM-resistant cells for amplification of two other genes: the gene for the P-glycoprotein known to be amplified in multidrug-resistant CHO cells and the gene for glycosylation site binding protein [11]. Neither of these sequences was amplified in our TM-resistant cells.

Two other laboratories have isolated TM-resistant cells [12,13]. Using a stepwise procedure, Lehrman and his coworkers [12] isolated CHO cells 80-fold resistant to TM that had 10-fold higher levels of the target enzyme in their membranes. From DNA and RNA blots probed with a portion of the CHO gene, it appeared that these TM-resistant CHO cells contained 50-fold more copies of the gene than wild-type cells and 40-fold more copies of the RNA. Kink and Chang, also using a procedure to gradually acclimate *Leishmania* cells to increasing concentrations of drug [13], isolated TM-resistant *Leishmania* cells. These TM-resistant cells had 30- to 100-fold more DNA hybridizing to the *ALG7* probe than did wild-type cells. Membranes from these cells had up to 15-fold higher levels of *N*-acetylglucosaminyl phosphate transferase.

VIII. ISOLATION OF cDNA ENCODING *N*-ACETYLGLUCOSAMINYL PHOSPHATE TRANSFERASE

Using the CHO transferase gene fragment as probe, we screened a cDNA library prepared from TM-resistant cells and isolated a cDNA containing a full-length open reading frame encoding a protein that is 43% identical to the yeast *N*-acetyl-glucosaminyl phosphate transferase [11]. Its idcntify as the CHO *N*-acetyl-glucosaminyl phosphate transferase is currently being demonstrated by expression studies.

The CHO protein encoded by the cDNA contains 408 amino acids and has a derived molecular weight of 44,900 D. Two highly homologous polypeptides obtained by purification of the enzyme from bovine mammary tissue by Vijay and coworkers [14] have similar molecular weights (46,000 and 50,000 D) as determined by sodium dodecyl sulfate polyacrylamide gel electrophoresis. As seen in Figure 3, this protein is very hydrophobic, consistent with its membrane localization [6] and its aggregation during gel filtration [14].

A few speculations concerning the topology and structure of this protein can be made based on the deduced amino acid sequence (see Fig. 4 and Ref. 11). The protein contains a putative dolichol binding domain (residues 67–79), based on comparison to a consensus sequence (Leu-Phe-Val-X-Phe-X-X-IIe-Pro-Phe-X-Phe-Tyr) derived from the sequences of three yeast glycosyltransferses that utilize dolichyl derivatives as substrates [15]. The protein does not contain either the KDEL sequence, a localization signal for soluble endoplasmic reticular proteins, or sequences that appear to be retention signals for transmembrane proteins in the endoplasmic reticulum. Nor was a functional signal sequence found near the amino terminus.

However, as seen in Figure 3, two well-defined potential membrane spanning regions are found. The charges on the amino acids surrounding the first potential membrane spanning region are compatible with an orientation of the amino-terminal sequence on the cytoplasmic side of the endoplasmic reticulum; the

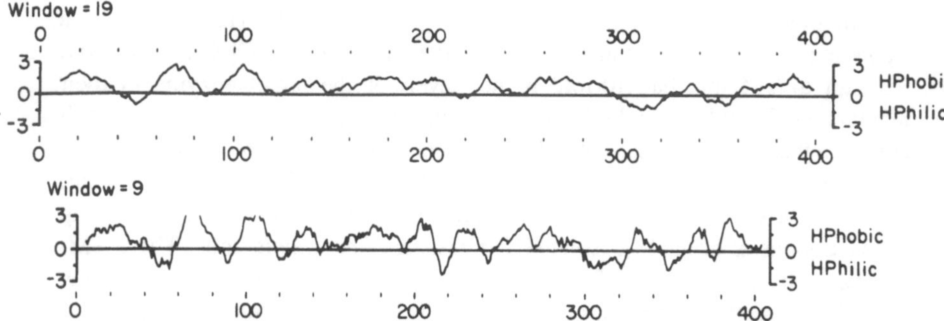

Figure 3 Kyte–Doolittle plots [11] of the hydrophobicity/hydrophilicity of the deduced amino acid sequence of the CHO *N*-acetylglucosaminyl phosphate transferase. For this analysis, the scanning windows were set for 19 (top curve) or for 9 (bottom curve).

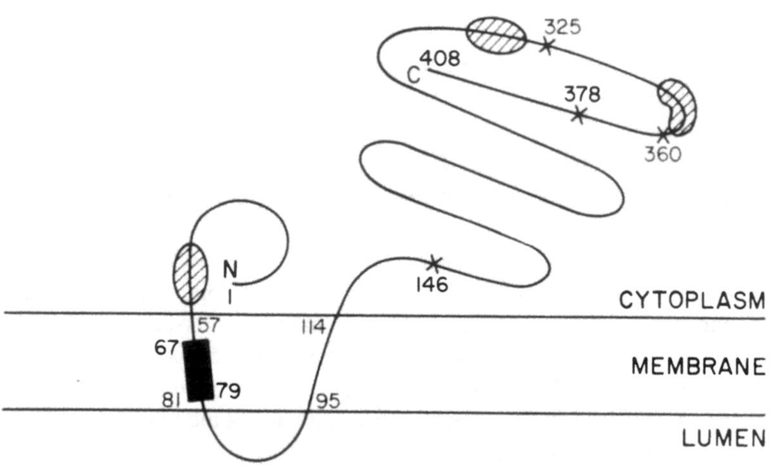

Figure 4 Model for the topography of the *N*-acetylglucosaminyl phosphate transferase. Examination of the deduced amino acid sequence [11] led to the components of this model. No functional signal sequence is found near the amino terminus. Two well-defined potential membrane spanning regions are located from residues 57 to 81 and 95 to 114 and are oriented based on the charges on the amino acids surrounding the first potential membrane spanning region. The orientation of the first membrane spanning region places the amino terminus of the protein on the cytoplasmic side of the endoplasmic reticulum, while the presence of a second membrane spanning region puts the carboxyl terminus on the cytoplasmic side as well. A putative dolichol binding domain (residues 67–79) is found in the first membrane spanning region. Four potential glycosylation sites (Asn 146, Asn 325, Asn 360, and Asn 378) were found. Finally, three hydrophilic regions (residues 45–56; 300–320; and 346–354) are indicated in this model by hatched circles based on the data shown in Figure 3. The overall hydrophobic nature of the protein suggests a tightly folded protein, consistent with its insensitivity to proteases [1].

presence of a second membrane spanning region also places the carboxyl-terminal sequence on the cytoplasmic side of the membrane (Fig. 4). The cytoplasmic regions of the protein appear to contain three hydrophilic areas (Fig. 3: residues 45–56, 300–320, and 346–354), which may participate in binding of the sugar nucleotide. Early biochemical experiments [1] to determine the topography of this protein were hampered by its insensitivity to protease in the absence or presence of detergent. The topography of the transferase has been controversial, although the synthesis of $Man_5GlcNAc_2$-P-P-Dol from $GlcNAc_2$-P-P-Dol is thought to occur on the cytoplasmic face of the endoplasmic reticulum [16]. The four potential Asn-X-Ser glycosylation sites in the protein (Fig. 4; residues 146, 325, 360, and 378) are also found in the cytoplasmic region and probably are not glycosylated, since according to this model that part of the protein would never interact with the membrane. This is consistent with Vijay's work on the purified protein [14], which does not bind to concanavalin A–Sepharose columns and is resistant to endoglycosidase treatments.

ACKNOWLEDGMENTS

The authors thank Dr. Adina Kaiden for her careful reading of this manuscript. This work was supported by a Public Health Service grant GM36570 from the National Institutes of Health, Bethesda, Maryland.

REFERENCES

1. Krag, S. S. Mechanisms and functional role of glycosylation in membrane protein synthesis. *Curr. Topics Membranes Transport*, *24*:181–249 (1985).
2. Elbein, A. D. Inhibitors of the biosynthesis and processing of N-linked oligosaccharide chains. *Annu. Rev. Biochem.* *56*:497–534 (1987).
3. Takatsuki, S., and Tamura, G. Preferential incorporation of tunicamycin, an antiviral antibiotic containing glucosamine, into the cell membranes. *J. Antibiot.* *25*:362–364 (1972).
4. Criscuolo, B. A., and Krag, S. S. Selection of tunicamycin-resistant Chinese hamster ovary cells with increased *N*-acetylglucosaminyl transferase activity. *J. Cell Biol.* *94*: 586–591 (1982).
5. Alt, F. W., Kellems, R. E., and Schimke, R. T. Synthesis and degradation of folate reductase in sensitive and methotrexate-resistant lines of S-180. *J. Biol. Chem.* *251*: 3063–3074 (1976).
6. Waldman, B. C., Oliver, C., and Krag, S. S. A clonal derivative of tunicamycin-resistant Chinese hamster ovary cells with increased *N*-acetylglucosamine-phosphate transferase activity has altered asparagine-linked glycosylation. *J. Cell. Physiol.* *131*: 302–317 (1987).
7. Elbein, A. D., Gafford, J., and Karz, M. S. Inhibition of lipid-linked saccharide synthesis: Comparison of tunicamycin, streptovirudin, and antibiotic 24010. *Arch. Biochem. Biophys.* *196*:311–319 (1979).

8. Yusuf, H. K. M., Pohlentz, G., Schwarzmann, G., and Sandhoff, K. Ganglioside biosynthesis in Golgi apparatus of rat liver. Stimulation by phosphatidyl glycerol and inhibition by tunicamycin. *Eur. J. Biochem.* *134*:47–54 (1983).
9. Rosenwald, A. G., Stoll, J., and Krag, S. S. Regulation of glycosylation. Three enzymes compete for a common pool of dolichyl phosphate in vivo. *J. Biol. Chem.* *265*:14544–14553 (1990).
10. Scocca, J. R., Hartog, K. O., and Krag, S. S. Evidence of gene amplification in tunicamycin-resistant Chinese hamster ovary cells. *Biochem. Biophys. Res. Commun.* *156*:1063–1069 (1988).
11. Scocca, J. R., and Krag, S. S. Sequence of a cDNA which specifies the uridine diphosphate-*N*-acetyl D-glucosamine: Dolichol phosphate *N*-acetyl glucosamine-1 phosphate transferase from Chinese hamster ovary cells. *J. Biol. Chem.* *265*:20621–20626 (1990).
12. Lehrman, M. A., Zhu, X., and Khounlo, S. Amplification and molecular cloning of the hamster tunicamycin-sensitive *N*-acetylglucosamine-1-phosphate transferase gene. The hamster and yeast enzymes share a common peptide sequence. *J. Biol. Chem.* *263*:19796–19803 (1988).
13. Kink, J. A., and Chang, K.-P. Tunicamycin-resistant *Leishmania mexicana amazonensis*: Expression of virulence associated with an increased activity of *N*-acetylglucosaminyltransferase and amplification of its presumptive gene. *Proc. Natl. Acad. Sci. USA*, *84*:1253–1257 (1987).
14. Shailubhai, K., Dong-Yu, B., Saxena, E. S., and Vijay, I. K. Purification and characterization of UDP-*N*-acetyl-D-glucosamine:dolichol phosphate *N*-acetyl-D-glucosamine-1-phosphate transferase involved in the biosynthesis of asparagine-linked glycoproteins in the mammary gland. *J. Biol. Chem.* *263*:15964–15972 (1988).
15. Albright, C. F., Orlean, P., and Robbins, P. W. A 13-amino acid peptide in three yeast glycosyltransferases may be involved in dolichol recognition. *Proc. Natl. Acad. Sci. USA*, *86*:7366–7369 (1989).
16. Hirschberg, C. B., and Snider, M. D. Topography of glycosylation in the rough endoplasmic reticulum and Golgi apparatus. *Annu. Rev. Biochem.* *56*:63–87 (1987).

19

Amplification of *CAD* Genes

George R. Stark *Imperial Cancer Research Fund, London, England*

I. ASPARTATE TRANSCARBAMYLASE AND PALA

Aspartate transcarbamylase catalyzes the formation of carbamyl-L-aspartate from carbamyl-P and L-aspartate, an early step in the synthesis of uridine 5′-phosphate (UMP) from simple precursors in virtually all organisms (Fig. 1). Detailed information on the mechanism of catalysis, accumulated over many years by studying the enzyme from *Escherichia coli* [1], led to the design of *N*-(phosphonacetyl)-L-aspartate (PALA), an analogue of the transition state (Fig. 2) intended to bind tightly to the active site [2]. PALA is a potent reversible inhibitor of *E. coli* aspartate transcarbamylase ($K_i \approx 10^{-8}$ M). The conformational changes induced when PALA binds to this allosteric enzyme have been studied by crystallography [3] and by several over methods [4].

As expected from the conservation of aspartate transcarbamylases in evolution [5], PALA is also a potent inhibitor of the mammalian enzymes (K_i about 10^{-9} M for the enzyme from Syrian hamsters) and of UMP biosynthesis in extracts of cells [6,7]. Crucially for the discussion to follow, PALA is able to enter intact cells in tissue culture [6] and in animals [8], where it inhibits the synthesis of UMP specifically: all toxic effects are reversed by providing uridine as an alternative precursor to UMP through the salvage pathway. PALA is chemically and metabolically stable, even in intact animals [8]. Starvation for pyrimidine nucleotides is cytotoxic and, even at an early stage, it was observed that colonies of PALA-resistant tissue culture cells grew out after most original cells had been killed by

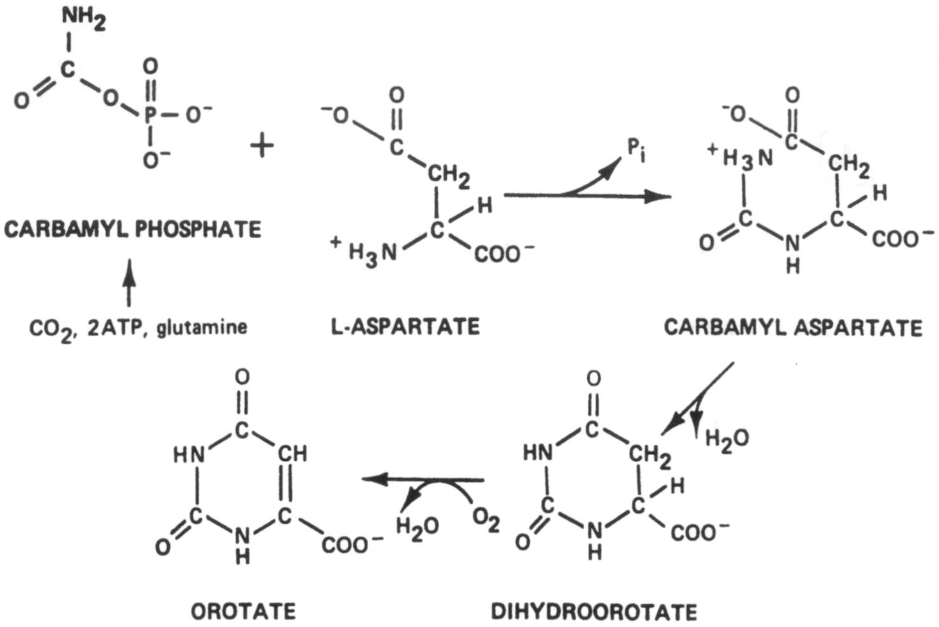

Figure 1 De novo pathway of UMP biosynthesis to orotate.

the drug and that these resistant cells had substantially higher levels of aspartate transcarbamylase activity [6].

The development of a new synthetic route for PALA [6] made large amounts of the compound available for extensive studies in cell culture and in animals. PALA has excellent activity against some mouse tumors, especially Lewis lung carcinoma, where, in contrast to all other agents tested, it is often curative [9].

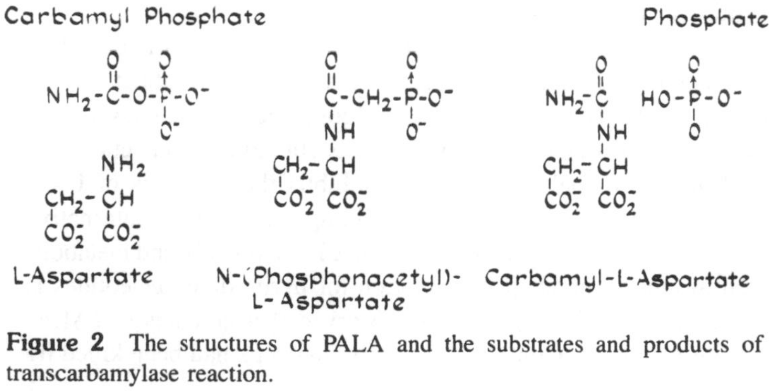

Figure 2 The structures of PALA and the substrates and products of the aspartate transcarbamylase reaction.

Although PALA has been tested extensively as a potential antitumor agent in man, the early promise of the mouse experiments has not been realized. Perhaps this disappointment is related to observations made in tissue culture [10] and in animals [11] that inhibition of cell growth by PALA was remarkably dependent on small differences in aspartate transcarbamylase activity, with about 20 times more PALA required to inhibit growth of cells containing only twice as much enzyme. Recent clinical trials have focused on the possible activity of PALA as a biochemical modulator, to enhance the cytotoxicity of other drugs through specific blockade of UMP synthesis [12].

II. CAD

Syrian hamster cells resistant to PALA contain not only more of the target aspartate transcarbamylase but also more of the two enzymes that flank it in the UMP biosynthetic pathway, despite the fact that these are not inhibited by PALA [10]. Purification to homogeneity and characterization of aspartate transcarbamylase from cells resistant to very high concentrations of PALA revealed that the enzyme was contained within a single polypeptide having a relative molecular mass of about *M*r 200,000 that comprised a multifunctional protein called CAD: carbamyl-*P*-synthetase, aspartate transcarbamylase, and dihydro-orotase [13].

Resistance to PALA was achieved in a stochastic process not dependent upon the presence of the drug, with a probability of about 2×10^{-5} per cell per division [10]. Highly resistant cell lines, selected in several steps of increasing PALA concentration and containing more than 100 times more CAD than wild-type cells, were stable in the absence of PALA [10]. Increased CAD in PALA-resistant cells was due to an increase in an mRNA about 8 kb long that was found in polysomes and could be translated into CAD protein in vitro [14]. A partial cDNA clone 2.3 kb long was then obtained [15]. It has been very difficult to obtain a full-length cDNA clone for CAD, for unknown reasons. The longest obtained thus far (6.5 kb) includes the complete aspartate transcarbamylase domain [16].

III. GENE AMPLIFICATION

The 2.3 kb cDNA clone was used [15] to demonstrate that the increase in aspartate transcarbamylase activity was accounted for by an approximately equal increase in the number of *CAD* genes per cell in every PALA-resistant clone examined (Fig. 3). This was the second example of gene amplification in mammalian cells, following the earlier demonstration that methotrexate resistance in mouse cells was due to amplification of the dihydrofolate reductase gene *DHFR* [17]. There was a strong contrast between the stability of *CAD* amplification in highly resistant Syrian hamster cells, where the amplified genes were chromosomal [18], and the instability of *DHFR* amplification in mouse cells [17], where the amplified genes

Figure 3 Amplified DNA in PALA-resistant cells. Genomic DNA was digested with *Eco*RI and a Southern transfer was probed with *CAD* cDNA. The relative level of CAD protein is shown above each track for three independent PALA-resistant clones. (From Ref. 15.)

were episomal. However, later work showed that chromosomally amplified *CAD* genes were also unstable when first- and second-step mutants, with low levels of amplification, were examined [19]. Cloned *CAD* cDNA was used to isolate genomic clones that spanned the entire gene, which is about 25 kb long [20]. About 40 introns were revealed by heteroduplex mapping, using electron microscopy. Two coamplified genes were discovered near the 5′ end of *CAD*, and their mRNAs were overproduced in proportion to the increase in gene copy number, making it clear that amplification was not restricted to the *CAD* gene itself.

IV. SIZE AND STRUCTURE OF AMPLIFIED DOMAINS CONTAINING *CAD*

A partial clone, 8.6 kb long and devoid of repetitive sequences, was isolated from *CAD* genomic DNA and used to analyze the locations of *CAD* in normal and PALA-resistant cells [18]. The sensitivity of the in situ hybridizations was increased greatly by using dextran sulfate, which also improves the sensitivity of Northern and Southern analyses [21]. The single copy *CAD* gene was mapped to the short arm of Syrian hamster chromosome B9. Four PALA-resistant clones, selected in several steps to high levels of resistance, contained about 100 amplified *CAD* genes within long regions of chromosomes, usually in an expanded domain that replaced the short arm of the same B9 chromosome [18]. From the length of the expanded domain and the gene copy number, it was possible to estimate that 500–1000 kb of DNA was coamplified with each copy of *CAD*, making it clear that the borders of the amplification event extended far beyond those of the *CAD* gene. Wahl et al. [22] found that rRNA genes were often a part of this coamplified DNA.

It seemed feasible to investigate the nature of these 500–1000 kb domains by cloning the coamplified DNA and then using such clones as probes of Southern transfers. To this end, Brison et al. [23] developed a general method for cloning any genomic DNA fragment increased in abundance by 10-fold or more, based on physical isolation of the fraction of sheared total genomic DNA that reassociates with appropriate kinetics. Probes isolated in this way from mutant and wild-type cells were depleted of repetitive sequences by annealing them to immobilized total genomic DNA and then used to differentially screen genomic libraries prepared from the DNA of PALA-resistant cells. The genomic clones thus isolated were used to demonstrate that different DNA sequences were coamplified with *CAD* in independent isolates of PALA-resistant cells [23]. This approach was carried further by Ardeshir et al. [24], who isolated sets of clones corresponding to 162 and 68 kb, respectively, of coamplified DNA from two highly resistant mutant cell lines. The cloned probes were then used on Southern transfers of DNAs from 12 different resistant cell lines, selected in one to five steps of increasing PALA concentration. Regions amplified in one mutant were often not amplified in another and, within a single mutant, different regions were amplified to different degrees. The results suggested that the amplified DNA was not comprised of identical tandemly arranged regions in any mutant and that major rearrangements of DNA probably accompanied some of the amplification events. Novel restriction fragments, formed by recombination and likely to link amplified regions to one another, were much more frequent in highly amplified DNA than in the moderately amplified DNA of first-step mutants. Legouy et al. [25] analyzed four such novel joints in detail and found that they were of different types and that they probably resulted from nonhomologous recombinations. Zieg et al. [26] examined newly amplified regions in first-step mutants by in situ hybridization with tritium-labeled probes and found that even when the degree of amplification was low, the *CAD* genes were located on abnormally long chromosome arms. However, the difficulty of banding the same chromosomes that hybridized to the *CAD* probe prevented their positive identification at this time.

The work was carried further by Giulotto et al. [27], who obtained from a single-step mutant a set of cosmid probes totaling 380 kb, corresponding to a

65 kb region surrounding the *CAD* gene, plus six additional coamplified regions that were not joined directly to each other or to *CAD*. These probes were applied to Southern transfers of genomic DNA derived from 33 independent single-step mutants, selected in low concentrations of PALA and having about three-fold amplifications of *CAD*. Again, the degree of coamplification of the outlying regions was not the same as that of *CAD* in many of the mutants. Only three single copy novel joints were found, each in a different mutant and each corresponding to a different probe. This extremely low frequency of novel joints, together with the result of Zieg et al. [26] that *CAD* probes hybridize in situ to unusually long chromosome arms in single-step mutants, led to the strong suggestion that the amount of DNA coamplified with each copy of *CAD* was much larger in single-step mutants (estimate > 10,000 kb per region) than in multiple-step mutants (estimate 500–1000 kb per region). The realization that such huge regions might be involved made their detailed analyses by the techniques of molecular biology seem rather unattractive.

Saito et al. [19] examined the evolution of amplified DNA in families of cell lines selected from first-step mutants with high concentrations of PALA. The large regions amplified initially were not reamplified intact. Instead, subregions were reamplified preferentially and some of the DNA amplified initially was lost, reducing the amount of coamplified DNA accompanying each copy of *CAD*. In other words, the concentration of *CAD* genes in amplified DNA increased with increasing copy number and resistance to PALA. After several steps of amplification, highly amplified condensed arrays were found that brought many *CAD* genes close together. In striking contrast to the stability of these arrays, also observed by Kempe et al. [10], the low-copy chromosomal arrays formed early were quite unstable in the absence of PALA. These results suggested that different mechanisms might be involved in the first step and in the later evolution of an already amplified array. It is relevant to note that Miele et al. [28] found that *CAD* gene amplification in Chinese hamster cells caused the elongated chromosomes to be unstable.

The indications of initial instability and of very large domains made it clear that it was imperative to examine new amplification events as soon as possible after they had occurred. The new technique of in situ hybridization using biotin-labeled probes and fluorescein-labeled avidin seemed ideal for the purpose, since individual cells could be examined at very early times. When this was done by Smith et al. [29], the results were dramatic. The amplified *CAD* genes of new colonies containing about 10^5 cells are present in multiple copies of very large regions of DNA, each tens of megabases long. Virtually every copy of *CAD* is seen as a well-resolved discrete spot (Fig. 4). By combining in situ hybridization with banding of the same chromosomes, it was revealed that the extra DNA is always chromosomal and usually is linked to the end of the short arm of one copy of chromosome B9, which retains one copy of *CAD* at its usual location. The other B9 homologue is

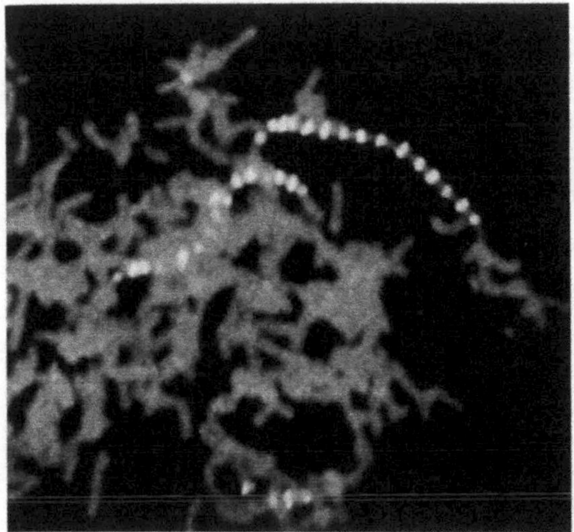

Figure 4 Example of a chromosome carrying about 15 amplified *CAD* genes present in a new colony (clone 20-1b) containing about 10^5 cells. The *CAD* genes were revealed by the binding of fluorescein-labeled avidin to biotin-labeled DNA probes after in situ hybridization. The chromosomes were stained the propidium iodide. (From. Ref. 29.)

present and unaltered. Individual cells within a single clone have highly variable (2–15) numbers of *CAD* genes. It is important to note that there is no selection for the high copy number of *CAD* genes observed in many of the cells, greatly in excess of the number of genes needed to achieve resistance to the low concentrations of PALA employed. The widely spaced *CAD* genes are usually interspersed with similarly spaced G-negative regions that often appear to be constricted (Fig. 5). This feature is not seen outside the newly amplified DNA or in wild-type cells. To account for all the results, it is proposed that all or most of the short arm is transferred from one B9 chromosome to another in the initial step of *CAD* amplification in Syrian hamster cells. In subsequent cell cycles, this initial duplication expands rapidly by unequal but homologous sister chromatid exchanges, to generate the heterogeneity of copy number seen in 10^5-cell clones. In this case, mechanisms involving participation of extrachromosomal DNA or overreplication of limited chromosomal regions seem to be ruled out by the data.

When resistant clones were examined much later, at the 10^{15}-cell stage, the extended arrays were not usually seen and the amplified *CAD* genes were found in more condensed structures (Fig. 6). These must be formed by events secondary to the ones that generate the initial structure and may even involve different mechanisms.

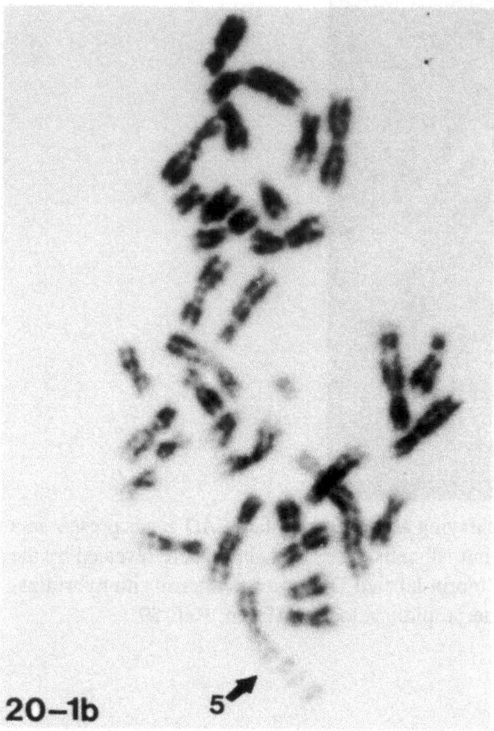

Figure 5 Example of a chromosome spread banded with Wright's stain after analysis by in situ hybridization, showing a G-negative region. The marker chromosome from clone 20-1b is labeled to show the number of separate copies of *CAD* genes observed and their approximate location. (From Ref. 29.)

Future analysis of *CAD* gene amplification will include a more detailed description of the changes in amplified DNA with time, beginning only a few generations after the initial event. Preliminary results with PALA-resistant cells observed only three or four generations after the first event of amplification are similar to those obtained with 10^5-cell colonies. DNA containing widely spaced *CAD* genes is added to the end of a normal B9 chromosome (K. Smith and G. R. Stark, unpublished data). The in situ technique can also be used to examine the nature of the initial event in amplificator cells [30] or in cells in which the rate of amplification is increased transiently in response to DNA damage or growth arrest [31]. It will also be interesting to reexamine the situation in primary cells, where it has not been possible to select stable PALA-resistant colonies [32,33], to see whether the early events revealed by in situ analysis in cell lines can be observed.

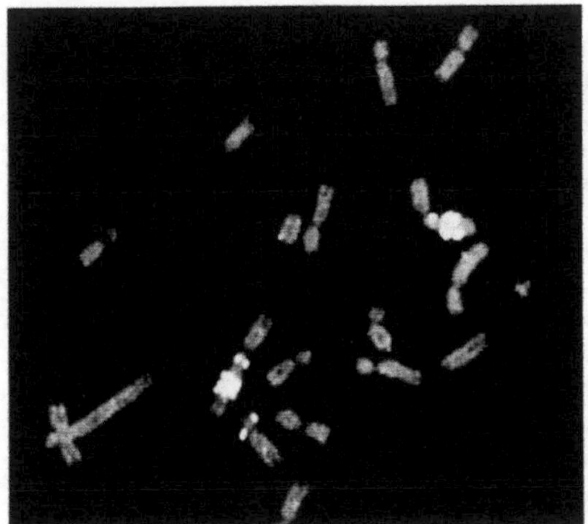

Figure 6 Example of a chromosome carrying amplified *CAD* genes in a PALA-resistant clone (15.3), examined when the total number of cells had reached about 10^{15} cells. (K. Smith and G. R. Stark, unpublished data.)

REFERENCES

1. Jacobson, G. R., and Stark, G. R. Aspartate transcarbamylase, in *The Enzymes* (P. D. Boyer, ed.), Academic Press, New York, 1973, pp. 225–297.
2. Collins, K. D., and Stark, G. R. Aspartate transcarbamylase. Interaction with the transition state analogue *N*-(phosphonacetyl)-L-aspartate. *J. Biol. Chem. 246*:6599–6605 (1971).
3. Kantrowitz, E. R., and Lipscomb, W. N. *Escherichia coli* aspartate transcarbamylase: The relation between structure and function. *Science, 241*:669–674 (1988).
4. Foote, J., and Schachman, H. K. Homotropic effects in aspartate transcarbamylase. *J. Mol. Biol. 186*:175–184 (1985).
5. Evans, D. R. CAD, a chimeric protein that initiates de novo pyrimidine biosynthesis in higher eukaryotes, in *Multidomain Proteins: Structure and Evolution* (D. G. Hardie and J. R. Coggin, eds.), Elsevier, New York, 1986, pp. 283–331.
6. Swyryd, E. A., Seaver, S. S., and Stark, G. R. *N*-(Phosphonacetyl)-L-aspartate, a potent transition state analog inhibitor of aspartate transcarbamylase, blocks proliferation of mammalian cells in culture. *J. Biol. Chem. 249*:6945–6950 (1974).
7. Hoogenraad, N. J. Reaction mechanism of aspartate transcarbamylase from mouse spleen. *Arch. Biochem. Biophys. 161*:76–82 (1974).
8. Yoshida, T., Stark, G. R., and Hoogenraad, N. J. Inhibition by *N*-(phosphonacetyl)-L-aspartate of aspartate transcarbamylase activity and drug-induced cell proliferation in mice. *J. Biol. Chem. 249*:6951–6955 (1974).
9. Johnson, R. K., Inouye, T., Goldin, A., and Stark, G. R. Antitumour activity of *N*-(phosphonacetyl)-L-aspartic acid, a transition-state inhibitor of aspartate transcarbamylase. *Cancer Res. 36*:2720–2725 (1976).

10. Kempe, T. D., Swyryd, E. A., Bruist, M., and Stark, G. R. Stable mutants of mammalian cells that overproduce the first three enzymes of pyrimidine nucleotide biosynthesis. *Cell, 9*:541–550 (1976).
11. Johnson, R. K., Swyryd, E. A., and Stark, G. R. Effects of *N*-(phosphonacetyl)-L-aspartate on murine tumors and normal tissues in vivo and in vitro and the relationship of sensitivity to rate of proliferation and level of aspartate transcarbamylase. *Cancer Res. 38*:371–378 (1978).
12. Leyland-Jones, B., and O'Dwyer, P. J. Biochemical modulation: Application of laboratory models to the clinic. *Cancer Treat. Rep. 70*:219–229 (1986).
13. Coleman, P. F., Suttle, D. P., and Stark, G. R. Purification from hamster cells of the multifunctional protein that initiates de novo synthesis of pyrimidine nucleotides. *J. Biol. Chem. 252*:6379–6385 (1977).
14. Padgett, R. A., Wahl, G. M., Coleman, P. F., and Stark, G. R. *N*-(Phosphonacetyl)-L-aspartate-resistant hamster cells overaccumulate a single mRNA coding for the multifunctional protein that catalyzes the first steps of UMP synthesis. *J. Biol. Chem. 254*:974–980 (1979).
15. Wahl, G. M., Padgett, R. A., and Stark, G. R. Gene amplification causes overproduction of the first three enzymes of UMP synthesis in *N*-(phosphonacetyl)-L-aspartate-resistant hamster cells. *J. Biol. Chem. 254*:8679–8689 (1979).
16. Shigesada, K., Stark, G. R., Maley, J. A., Niwsander, L. A., and Davidson, J. N. Construction of a cDNA to the hamster *CAD* gene and its application toward defining the domain for aspartate transcarbamylase. *Mol. Cell. Biol. 5*:1735–1742 (1985).
17. Alt, F. W., Kellems, R. E., Bertin, J. R., and Schimke, R. T. Selective multiplication of dihydrofolate reductase genes in methotrexate-resistant variants of cultured murine cells. *J. Biol. Chem. 253*:1357–1370 (1978).
18. Wahl, G. M., Vitto, L., Padgett, R. A., and Stark, G. R. Single-copy and amplified *CAD* genes in Syrian hamster chromosomes localized by a highly sensitive method for in situ hybridization. *Mol. Cell. Biol. 2*:308–319 (1982).
19. Saito, I., Groves, R., Giulotto, E., Rolfe, M., and Stark, G. R. Evolution and stability of chromosomal DNA coamplified with the *CAD* gene. *Mol. Cell. Biol. 9*:2445–2452 (1989).
20. Padgett, R. A., Wahl, G. M., and Stark, G. R. Structure for the gene for CAD, the multifunctional protein that initiates UMP synthesis in Syrian hamster cells. *Mol. Cell. Biol. 2*:293–301 (1982).
21. Wahl, G. M., Stern, M., and Stark, G. R. Efficient transfer of large DNA fragments from agarose gels to diazobenzyloxymethyl-paper and rapid hybridization by using dextran sulfate. *Proc. Natl. Acad. Sci. USA, 76*:3683–3687 (1979).
22. Wahl, G. M., Vitto, L., and Rubinitz, J. Co-amplification of rRNA genes with *CAD* genes in *N*-(phosphonacetyl)-L-aspartate-resistant Syrian hamster cells. *Mol. Cell. Biol. 3*:2066–2075 (1983).
23. Brison, O., Ardeshir, F., and Stark, G. R. General method for cloning amplified DNA by differential screening with genomic probes. *Mol. Cell. Biol. 2*:578–587 (1982).
24. Ardeshir, F., Giulotto, E., Zieg, J., Brison, O., Liao, W. S. L., and Stark, G. R. Structure of amplified DNA in different Syrian hamster cell lines resistant to *N*-(phosphonacetyl)-L-aspartate. *Mol. Cell. Biol. 3*:2076–2088 (1983).

25. Legouy, E., Fossar, N., Lhomond, G., and Brison, O. Structure of four amplified DNA novel joints. *Somatic Cell Mol. Genet. 15*:309–320 (1989).
26. Zieg, J., Clayton, C. E., Ardeshir, F., Giulotto, E., Swyryd, E. A., and Stark, G. R. Properties of single-step mutants of Syrian hamster cell lines resistant to *N*-(phosphonacetyl)-L-aspartate. *Mol. Cell. Biol. 3*:2089–2098 (1983).
27. Giulotto, E., Saito, I., and Stark, G. R. Structure of DNA formed in the first step of *CAD* gene amplification. *EMBO J. 5*:2115–2121 (1986).
28. Miele, M., Bonatti, S., Menichini, P., Ottaggio, L., and Abbondandolo, A. The presence of amplified regions affects the stability of chromosomes in drug-resistant Chinese hamster cells. *Mutat. Res. 219*:171–178 (1989).
29. Smith, K. A., Gorman, P. A., Stark, M. B., Groves, R. P., and Stark, G. R. Distinctive chromosomal structures are formed very early in the amplification of CAD genes in Syrian hamster cells. *Cell, 63*:1219–1227 (1990).
30. Giulotto, E., Knights, C., and Stark, G. R. Hamster cells with increased rates of DNA amplification, a new phenotype. *Cell, 48*:837–845 (1987).
31. Schimke, R. T. Gene amplification in cultured cells. *J. Biol. Chem. 263*:5989–5992 (1988).
32. Wright, J. A., Smith, H. S., Watt, F. M., Hancock, M. C., Hudson, D. L., and Stark, G. R. DNA amplification is rare in normal human cells. *Proc. Natl. Acad. Sci. USA*, *87*:1791–1795 (1990).
33. Tlsty, T. D. Normal diploid human and rodent cells lack a detectable frequency of gene amplification. *Proc. Natl. Acad. Sci. USA*, *87*:3132–3136 (1990).

20

A Reversible Selection System for UMP Synthase Gene Amplification and Deamplification

D. Parker Suttle *St. Jude Children's Research Hospital, Memphis, Tennessee*
Mazin B. Qumsiyeh *T. C. Thompson Children's Hospital, Chattanooga, Tennessee*

I. INTRODUCTION

With the selection system described in this chapter, it is possible to characterize both the generation of stably amplified DNA and the rapid loss of that amplified DNA from its intrachromasomal location. By alternating the selective drug pressure, it is possible to go through multiple cycles of amplification and deamplification in a single cell lineage. High resolution G-banding (trypsin-Giemsa) and in situ hybridization techniques were used to document the generation and loss of expanded chromosomal regions (ECRs) and to reveal the chromosome rearrangements that have accompanied amplification and deamplification events.

II. UMP SYNTHASE, INHIBITORS, AND REVERSE SELECTION

A. Bifunctional Nature of UMP Synthase

The last two steps of the de novo pyrimidine pathway are the addition of ribose 5′-monophosphate from 5′-phosphoribosyl-1-pyrophosphate to orotic acid to form orotidine 5′-monophosphate (OMP), followed by the decarboxylation of OMP to form uridine monophosphate (UMP) (Fig. 1). The two activities catalyzing these reactions, orotate phosphoribosyltransferase (OPRT) and orotidylate

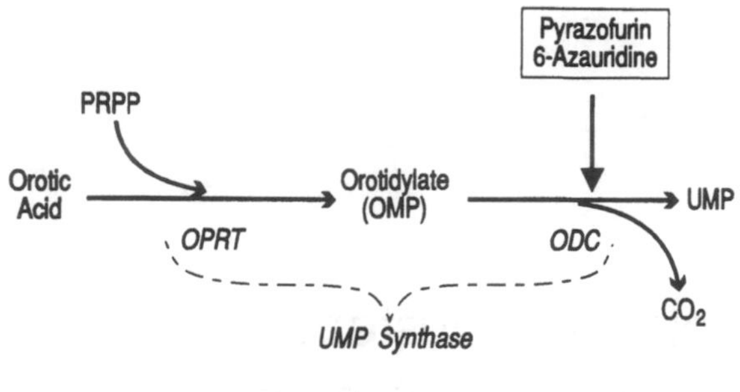

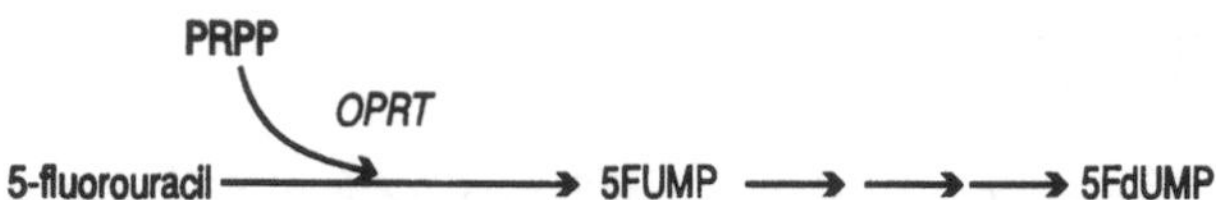

Figure 1 The last two steps of the de novo pyrimidine pathway catalyzed by UMP synthase. The monophosphate nucleotide forms of pyrazofurin and 6-azauridine act as inhibitors of the decarboxylase reaction. 5-Fluorouracil is metabolized by OPRT to 5′-fluorouridine monophosphate (5FUMP). Further metabolism of 5FUMP leads to the formation of 5′-fluorodeoxyuridine monophosphate, an inhibitor of thymidylate synthetase.

decarboxylase (ODC), are contained in a single polypeptide chain comprising the bifunctional protein UMP synthase [1]. Human UMP synthase contains 480 amino acids and has a molecular weight of 52,199 kDa [2]. The two activities are contained within distinct domains of the protein. OPRT activity is contained in the NH_2-terminal half of the protein, while the ODC activity is in the COOH-terminal portion [2].

B. Drug Selections for Amplification

Of the several drugs that can act as inhibitors of UMP synthase activity, pyrazofurin (PF) and 6-azauridine (6AUR) were chosen to select for amplification of the UMP synthase gene. Pyrazofurin (Eli Lilly, Indianapolis, IN), a C-nucleoside antibiotic isolated from *Streptomyces candidus*, has a molecular structure similar to OMP [3] and is a potent inhibitor of the decarboxylase activity of the UMP synthase protein. 6-Azauridine is also an inhibitor of the decarboxylase activity, but is about 40–50 times less cytotoxic. Following transport into the cell, the two drugs are metabolized to the monophosphate nucleotide form, which inhibits the decarboxylase activity of UMP synthase. PF is phosphorylated by adenosine kinase [4], while 6AUR phosphorylation is catalyzed by uridine kinase. When selecting cells for resistance, both drugs are added to the media in approximately equitoxic concentrations. This dual drug selection ensures that the development of resistance is due to UMP synthase gene amplification, since the simultaneous loss of activity for both the adenosine and uridine kinases would be a very rare event. As is the case for many of the selection systems that result in gene amplification, the development of relatively high levels of PF+6AUR resistance requires stepwise increases in the level of the drugs used for selection. With selection in increasing concentrations of PF+6AUR, resistant cells increase the levels of ODC activity secondary to UMP synthase gene amplification. With the increase in ODC, there is a coordinate increase in the activity of OPRT, since these activities are contained in the single UMP synthase protein.

C. Drug Selection for Deamplification

The initial step in the metabolism of the drug 5-fluorouracil (5FU) is catalyzed by OPRT (Fig. 1). Cells with increased levels of OPRT activity will metabolize more 5FU, and thus are subjected to higher levels of the toxic metabolite. Therefore, although increases in UMP synthase make the cells more resistant to PF+6AUR, increased UMP synthase activity also increases the sensitivity of the cells to the drug (5FU). This bifunctional nature of UMP synthase combines with the presence of drugs that are inhibitors of or activated by the enzyme to provide a unique reversible selection system for gene amplification. Cells that have increased levels of UMP synthase will be more resistant to PF+6AUR and more sensitive to 5FU, but cells that have low levels of UMP synthase will be more resistant to 5FU and more sensitive to PF+6AUR.

III. AMPLIFICATION AND DEAMPLIFICATION OF UMP SYNTHASE

The test of the reversible selection for UMP synthase activity was carried out using the Chinese hamster lung (CHL) cells, V79 (Fig. 2). Four steps of selection in increasing concentrations of PF+6AUR resulted in a cell line, PUC4-2, that was resistant to 40 μM PF + 1.0 mM 6AUR and had approximately 56 times the UMP synthase activity as the wild-type CHL cells [5]. The increased level of activity corresponded with an increased number of UMP synthase gene copies as determined by hybridization of a UMP synthase specific cDNA probe to quantitative slot blots of the wild-type and resistant cells. As judged by the level of 5FU that results in 50% growth inhibition with 96 hr of continuous drug exposure, the PUC4-2 cells are about 70 times more sensitive to the cytotoxic effects of 5FU than are the wild-type CHL cells [6]. This differential sensitivity to 5FU offers the possibility of selecting for cells that have undergone a deamplification event resulting in the loss of the amplified UMP synthase genes and a return to wild-type level of UMP synthase activity.

Before beginning selection for deamplification events, we grew out new cell lines from single cell isolates of the PUC4-2 cell line. This was done to ensure a

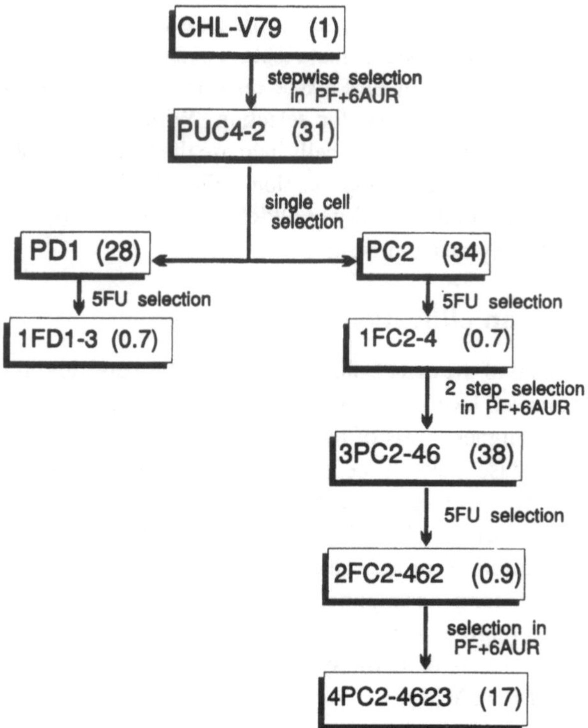

Figure 2 The lineage and selection steps involved in three cycles of amplification and deamplification. At each step either isolated clones or single cells were selected following growth in the indicated drug(s). The number of parentheses indicates the estimated number of UMP synthase gene copies in the cell lines normalized to the CHL-V79 parental cell line. The PC2 lineage of cells was isolated by single cell dilution in the absence of any drugs. The PD1 cells were grown from a single cell in the continuous presence of pyrazofurin and 6-azauridine. The concentration of drugs used for selection is given in the text.

homogeneous cell population containing amplified UMP synthase gene copies. More than 15 single cell isolates were grown into populations and all had levels of UMP synthase activity 25–50 times the wild-type CHL level. Starting at the point of single cell cloning, some cells were grown continuously in the presence of the selecting concentration of PF+6AUR (labeled PD) and others were removed from drug selection (labeled PC).

A. Stability of UMP Synthase Amplification

To determine whether deamplification was a spontaneous event in the amplified cell population, the UMP synthase activity of the cells was checked following growth of the cells in the absence of selective pressure. One single cell clone (PD-1) grown to a population in 40 μM PF + 1 mM 6AUR, and two single cell clones (PC2 and PC-10) grown in the absence of drugs were put in media without drug selection and monitored for 72 weeks by periodic determination of the level of UMP synthase. During that period there was no loss in the UMP synthase activity in these cells, indicating that the DNA amplification was stable.

B. Selection for Deamplification

We studied three PF+6AUR-resistant cell lines that overproduced UMP synthase secondary to a stable gene amplification of approximately 30-fold. These cell lines were subjected to selection in 5 μM 5FU. Individual colonies were isolated following 14–21 days in 5FU and grown to mass population in the absence of any drug selection (labeled 1FC or 1FD). Assays of activity and determination of the relative gene copy number by slot blot hybridization analysis showed that all the 5FU-selected cell lines had near wild type levels of UMP synthase activity and gene copies. The loss of the amplified UMP synthase genes in the 5FU-selected cells did not depend on previous growth of the cells in the absence of PF+6AUR. Revertants were isolated from both the PD cells, grown continuously in the presence of PF+6AUR, and the PC cells, grown from single cells in the absence of drugs. Cells that had lost their amplified UMP synthase genes would not be expected to survive in the PD cell population, which was maintained in 40 μM PF + 1.0 mM 6AUR until the start of the selection in 5FU.

C. Cycles of Amplification and Deamplification

Two of the deamplified cell lines, 1FD1-6 and 1FC2-4, were reselected in a low levels of PF+6AUR to determine whether reamplification of the UMP synthase gene would occur. Surviving colonies were isolated, grown into populations, and found to have increased UMP synthase activity. One cell line, 3PC2-46, was taken through two additional steps of increasing PF+6AUR selection (up to 50 μM PF + 1.25 mM 6AUR), and analysis showed UMP synthase activity and gene copy number similar to the PC2 cell line. Amplification was stable in the 3PC2-46 cell line, with no loss of enzyme activity detected after 37 weeks of cell growth in the absence of PF+6AUR. The 3PC2-46 cells were again subject to selection in 5 μM 5FU and one of the surviving colonies, 2FC2-462, was grown into a population. The UMP synthase activity of this line was again reduced to near wild type levels. Reselection of this line in 0.2 μM PF + 5 μM 6AUR resulted in the production of cells comprising a third cycle of amplification. The 4PC2-4623 cell line was established from a single colony and found to have about 17 times the level of UMP synthase activity as the V79 cells. Figure 2 summarizes the results of three cycles of amplification and deamplification in a single cell lineage of the PC2 cell line.

IV. CYTOGENETIC ANALYSIS OF AMPLIFICATION CYCLES

UMP synthase activity assays and determination of relative gene copy number are good indicators of gene amplification, but these data are not adequate to indicate the precise location or arrangement of the amplified DNA and do not provide information about the mechanism of loss of the amplified DNA. Detailed cytogenetic analysis was used to determine the chromosomal location of the amplified DNA and to assess chromosomal aberrations associated with the amplification and deamplification process.

A. Karyotype of Parental CHL V79 Cells

Normal Chinese hamster cells contain two copies of the UMP synthase gene, which maps on hamster chromosome 4p2 [7]. The V79 cells used in our selection system were highly derived and aneuploid, having acquired many rearrangements during evolution from a Chinese hamster lung biopsy [8]. Many chromosomes in this cell line are derivative chromosomes that have undergone at least one rearrangement. However, the present karyotype of V79 A3 is relatively stable, as evidenced by its similarities to karyotypes reported earlier by Thacker [8]. In the V79 cell line there are three marker chromosomes (M1–M3) that appear to be the result of a double reciprocal translocation involving both homologues of chromosomes 4 and a chromosome 5. A minimum of four breakpoints and the generation of an acrocentric chromosome containing material from 4p are involved. M1 is a der(4q;4p;4q), M2 is a der(4p;5q), and M3 is the small acrocentric chromosome probably containing a portion of 4p (Fig. 3). The rearrangements were established in the parental V79 cells long before any drug selection was initiated and thus are not directly related to the process of amplification. However, the two chromosomes M2 and M3 provided sites of gene amplification in the various selected cell lines.

B. Chromosome Amplification and Rearrangements

Initial amplification produced a cell line in which UMP synthase is stably amplified as a homogeneously staining region (HSR) on M3 [9]. Recent fluorescent in situ hybridization experiments (Qumsiyeh, Goohra, and Suttle) document that in addition to amplified domains on M3, single copy sites were retained on M1 and M2 in areas recognized cytogenetically as corresponding to chromosome region 4p2. This would suggest that the parental V79 cell line has copies of the gene on chromosomes M1, M2, and M3. These gene loci are substantiated by radioactive in situ hybridization analysis of V79 cells that showed peaks of silver grains on each of these marker chromosomes [7].

Amplification and deamplification in this system are accompanied by formation and deletion of expanded chromosomal regions (ECRs) carrying the amplified copies of the UMP synthase genes. Cells selected in the initial amplification step

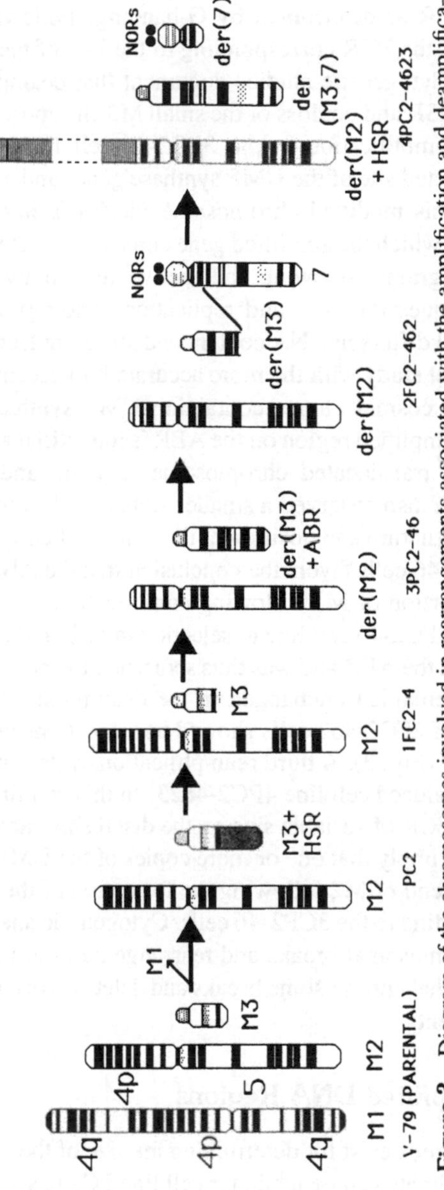

Figure 3 Diagram of the chromosomes involved in rearrangements associated with the amplification and deamplification of UMP synthase: NOR, nuclear organizing region. The M1 chromosome, which is constant throughout the cell lineage, is not depicted after the V79 cell. The location of the UMP synthase gene in each of the cell lines is detailed in the text.

(PC2) developed an HSR as determined by G-banding. Following selection in 5FU, there was a loss of the HSR corresponding to the loss of the UMP synthase gene copies (1FC2-4). Cytogenetic studies document that deamplification is the result of the loss of the HSR and not loss of the small M3 chromosome. Two linked events accompanied reamplification in the 3PC2-46 cell line: a chromosomal break in M2 at the predicted site of the UMP synthase gene, and a modification of the structure of M3. This modified chromosome, der(M3), has an abnormally banded region (ABR) in which the amplified gene copies are located, as evidenced by the distribution of grains following radioactive in situ hybridization [9]. Various banding techniques (G-, C-, and replication banding) suggest that the ABR in the 3PC2-46 cell differs in DNA content and structure from the HSR seen in the PC2 cells. A recent study with the more accurate fluorescent in situ hybridization technique using cosmid clones containing UMP synthase genomic sequences shows that the amplified region on the ABR is interstitial and is situated at the junction between a translocated chromosome segment and what was the terminal area on M3. We also observed a smaller signal on the broken end of M2 (Fig. 3). The alternate structure and position of the amplified UMP synthase DNA on (der)M3 in the 3PC2-46 cells favors the conclusion that the ABR originated by acquiring the deleted portion of M2 following amplification.

Upon subjecting the 3PC2-46 cell line to selection in 5FU and deamplification, the M3 chromosome lost the ABR and was thus shortened to about half its original length. The der (M2) remained unchanged in the deamplified cell line labeled 2FC2-462. A minority of 2FC2-462 cells showed translocations involving the der (M3) and chromosome 7 (Fig. 3). A third reamplification cycle starting from cells carrying the t(7;M3) produced cell line 4PC2-4623. In this cell line the amplified genes are located in an ECR of variable size on the distal short arm of the deleted M2 (Fig. 3). This would imply that one or more copies of the UMP synthase gene remained on the broken end of M2 following the formation of the (der)M3 in the reamplification steps leading to the 3CP2-46 cells. Cytogenetic analysis supported the hypotheses that chromosomal breaks and rearrangements are associated with gene amplification, and that chromosome breaks and deletions occur in connection with deamplification events.

C. Size of the Amplified DNA Regions

Although no direct measures exist for determining the size of the amplified region in these cell lines, an estimate can be made for cell line PC2 based on cytogenetic data. Using calculations as described by Nunberg et al. [10] based on the relative size of the ECR and the size of the hamster genome, we estimate the amplicon size for PC2 to be 1000–2000 kb. The 3PC2-46 cells are unique in that fluorescent in situ hybridization revealed amplified copies in a very localized area on the der(M3). Thus we would expect amplicon size to be much smaller in the 3PC2-46 cells. The variability in the size of the ECR in 4PC2-4623 makes it difficult to estimate the amplicon size.

V. CONCLUDING REMARKS

In this system, gene amplification occurs as stable expanded chromosome regions that occur at sites already bearing UMP synthase loci. These sites were disrupted by chromosome rearrangements during selection and amplification (e.g., breaks in M2 during selection of the 3PC2-46 cells), indicating a role for chromosome reorganization. The involvement of chromosome rearrangements in gene amplification has also been seen in earlier studies [11–13]. This study supports proposed mechanisms of gene amplification that do not require an extrachromosomal element as an intermediate step in the amplification process. We have also shown that deamplification can occur quickly by excision of expanded chromosome regions from stable intrachromosomal locations. The different ECR structures and fluorescent hybridization patterns observed indicate that different amplicon structures are possible even for the same gene in the same cell line. We have yet to understand the possible relationships of these different types of ECR structure to the initial mechanism of gene amplification or to latter condensation of the amplified regions. This and other questions about the functional and mechanical aspects of gene amplification must await further detailed study of these cell lines.

REFERENCES

1. Jones, M. E. Pyrimidine nucleotide biosynthesis in animals: Genes, enzymes, and regulation of UMP biosynthesis. *Annu. Rev. Biochem. 49*:253–279 (1980).
2. Suttle, D. P., Bugg, B. Y., Winkler, J. K., and Kanalas, J. J. Molecular cloning and nucleotide sequence for the complete coding region of human UMP synthase. *Proc. Natl. Acad. Sci. USA*, *85*:1754–1758 (1988).
3. Gutowski, E., Sweeney, M. J., Delong, D. C., Hamill, R. L., and Gerzon, K. Biochemistry and biological effects of the pyrazofurins (pyrazomycins): Initial clinical trial. *Ann. NY Acad. Sci. 255*:544–551 (1975).
4. Suttle, D. P., Harkrader, R. J., and Jackson, R. C. Pyrazofurin-resistant hepatoma cells deficient in adenosine kinase. *Eur. J. Cancer*, *17*:43–51 (1981).
5. Suttle, D. P., and Stark, G. R. Coordinate overproduction of orotate phosphoribosyltransferase and orotidine-5′-phosphate decarboxylase in hamster cells resistant to pyrazofurin and 6-azauridine. *J. Biol. Chem. 254*:4602–4607 (1979).
6. Suttle, D. P. A reversible selection system for UMP synthase gene amplification and deamplification. *Somatic Cell Mol. Genet. 15*:435–443 (1989).
7. Qumsiyeh, M. B., and Suttle, D. P. Localization of the adenosine deaminase, transferrin, and UMP synthetase genes on Chinese hamster chromosomes 4 and 6 by in situ hybridization. *J. Hered. 81*:111–116 (1990).
8. Thacker, J. The chromosomes of a V-79 Chinese hamster cell line and a mutant subline lacking *HPRT* activity. *Cytogenet. Cell Genet. 29*:16–25 (1981).

9. Qumsiyeh, M. B., and Suttle, D. P. Cytogenetic analysis of amplification and deamplification of UMP synthase genes in Chinese hamster cells. *Somatic Cell Mol. Genet. 15*:503–512 (1989).
10. Nunberg, J. N., Kaufman, R. J., Schimke, R. T., Urlaub, G., and Chasin, L.A. Amplified dihydrofolate reductase genes are localized to a homogeneously staining region of a single chromosome in a methotrexate resistant Chinese hamster ovary cell line. *Proc. Natl. Acad. Sci. USA, 75*:5553–5556 (1978).
11. Biedler, J. L., Melera, P. W., and Spengler, B. A. Specifically altered metaphase chromosomes in antifolate-resistant Chinese hamster cells that overproduce dihydrofolate reductase. *Cancer Genet. Cytogenet. 2*:47–60 (1980).
12. Flintoff, W. F., Livingston, E., Duff, C., and Worton, R. G. Moderate-level gene amplification in methotrexate-resistant Chinese hamster ovary cells is accompanied by chromosomal translocation at or near the site of the amplified *DHFR* gene. *Mol. Cell. Biol. 4*:69–76 (1984).
13. Stallings, R. L., Munk, A. C. Longmire, J. L., Hildebrand, C. E., and Crawford, B. D. Assignment of genes encoding metallothioneins I and II to Chinese hamster chromosome 3: Evidence for the role of chromosome rearrangement in gene amplification. *Mol. Cell. Biol. 4*:2932–2936 (1984).

21

Amplification of the Genes Encoding the R1 and R2 Subunits of Mammalian Ribonucleotide Reductase

Lars Thelander *University of Umeå, Umeå, Sweden*

I. INTRODUCTION

Cells with increased ribonucleotide reductase activity are unique among drug-resistant cell lines with amplified copies of specific genes because ribonucleotide reductase activity is dependent on *two* different gene products. These are the two nonidentical subunits of ribonucleotide reductase, proteins R1 and R2. These subunits are inactive alone, but together they form an active enzyme complex. The subunits are encoded by different genes located on different chromosomes. This arrangement poses many interesting and complicated solutions to drug resistance by gene amplification.

II. INTRODUCTION TO RIBONUCLEOTIDE REDUCTASE

Before division, a cell needs to duplicate its DNA. DNA synthesis requires deoxyribonucleotides, and these are made by a direct reduction of the corresponding ribonucleotides. The reaction, which is the first unique step leading to DNA synthesis, is catalyzed by the enzyme ribonucleotide reductase. In mammalian cells this enzyme consists of two nonidentical homodimeric subunits called proteins R1 and R2, which are inactive alone. Together, however, the subunits form an active $\alpha_2\beta_2$ enzyme complex [1–3].

Ribonucleotide reduction is controlled in a number of ways. There is only one enzyme catalyzing the reduction of all four ribonucleoside diphosphates, but both the substrate specificity and overall enzyme activity are allosterically controlled by nucleoside triphosphate effectors. This allosteric control ensures a balanced supply of the four deoxyribonucleotides needed for DNA synthesis, which is very important for error-free replication. Furthermore, enzyme activity is correlated to DNA synthesis showing maximal activity in S phase cells [1–3]. Finally, there also seems to be a close correlation between ribonucleotide reductase activity and DNA repair. This has been demonstrated in yeast, where DNA damage induces expression of ribonucleotide reductase mRNAs [4].

III. MAMMALIAN RIBONUCLEOTIDE REDUCTASE

A. Enzyme Structure

The R1 and R2 subunits of mammalian ribonucleotide reductase were purified to homogeneity from calf thymus and overproducing mouse cells, respectively [5,6]. Recently, pure recombinant mouse R2 subunit could be obtained in high yields from *Escherichia coli* carrying an expression plasmid [7]. The R1 subunit has a molecular weight of 170,000 and contains binding sites for the nucleoside triphosphate allosteric effectors as well as for the ribonucleoside diphosphate substrates (Fig. 1). It also contributes redox active dithiols to the active site, which is formed from both subunits.

The R2 subunit has a molecular weight of 90,000, and each of the two polypep-

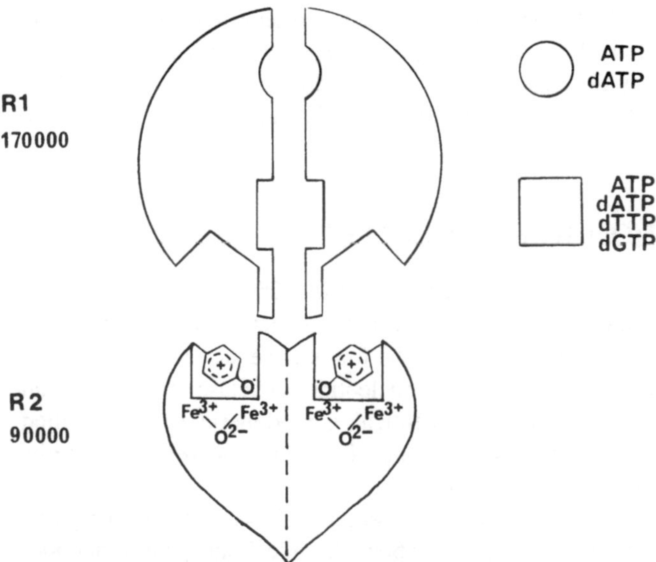

Figure 1 Mammalian ribonucleotide reductase.

tide chains contains a dimeric iron center (Fig. 1). Each center is formed by a pair of antiferromagnetically coupled high spin ferric ions mediated by a μ-oxo bridge [8]. During its formation, each center generates a stable tyrosyl free radical essential for activity [9]. The tyrosyl radical is the target for a number of antiproliferative drugs such as hydroxyurea, polyhydroxybenzohydroxamic acids, and α(*N*)-heterocyclic carboxaldehyde thiosemicarbazones [10,11]. Recently, the three-dimensional structure of the *E. coli* ribonucleotide reductase R2 subunit was solved to 2.2 Å resolution. The *E. coli* protein is very similar to the mammalian R2 [12]. In the bacterial protein, the main motif, an eight-helix bundle, forms a 140° angle at the polypeptide interface, resulting in a heart-shaped structure, and all residues involved in iron binding (2 His, 3 Glu, and 1 Asp) are conserved between *E. coli* and mouse. The X-ray structure also shows that the tyrosyl free radicals are buried in the protein 10 Å away from the surface.

B. Regulation of Enzyme Activity

In synchronously growing cell cultures, ribonucleotide reductase activity parallels the fraction of S phase cells [1–3]. However, the levels of the two subunits are differentially controlled during the cell cycle [13]. The R1 subunit, measured by a two-site monoclonal enzyme-linked immunoassay, showed almost constant levels all through the cell cycle. In contrast, active R2 subunit, estimated from direct electron paramagnetic resonance measurements of the tyrosyl free radical in frozen cells, showed a strictly S phase correlated expression. Activity measurements of the individual subunits, performed in the presence of an excess of the other subunit, showed that the R1 activity was constant during the cell cycle and that this subunit was present in excess. The R2 activity, like the R2 protein levels, was S phase specific. Therefore, ribonucleotide reductase activity during the cell cycle is controlled by de novo synthesis and breakdown of R2 subunit, which has a half-life of 3 hr. The S phase correlated increase in R2 activity is not the result of enzyme activation as shown by pulse-chase experiments using deuterated tyrosine [14]. The R1 subunit, with a half-life of more than 20 hr, is not cell cycle regulated but is present only in proliferating cells.

These results have been confirmed by analyzing the situation in individual cells using monoclonal antibodies and immunocytochemistry [15,16]. Independent of the cell cycle, both subunits were always found exclusively in the cytoplasm [15,17]. Therefore, deoxyribonucleotides must be synthesized in the cytoplasm and then transported into the nucleus.

C. Gene Structure and Localization

To study the regulation of the expression of the R1 and R2 subunits at the gene level, full-length cDNAs were cloned from an Okayama–Berg cDNA library made from hydroxyurea-resistant, R2-overproducing mouse cells [18]. The R1 cDNA contained 3064 bp encoding a polypeptide chain of molecular weight 90,000, consisting of 792 amino acid residues. The R2 cDNA contained 2113 bp encoding a polypeptide chain of 45 kD consisting of 390 amino acid residues. The cDNAs were used to map the ribonucleotide reductase genes on mouse and human chromosomes by in situ hybridization and Southern blots of interspecies cell hybrids [19,20]. Only one R1-related sequence was present in the mouse genome and this was localized to chromosome 7. In contrast, there were four R2-related sequences localized to chromosomes 4, 7, 12, and 13. The active R2 gene was assigned to chromosome 12 while the other R2-related sequences represent pseudogenes. In the human genome, the R1 gene was assigned to chromosome 11p15 and the active R2 gene to chromosome 2p24-p25. In addition, there were R2 pseudogenes on both arms of chromosome 1 and on the X chromosome. The ribonucleotide reductase genes were also mapped in hamster cells, where the R1 gene was assigned to chromosome 3 and the R2 gene to chromosome 7 [21].

Recently, the functional mouse R2 gene was cloned and sequenced [22]. The transcribed sequence of 5770 bp contains 10 exons separated by nine 95–917 bp introns. The promoter region is G+C-rich and contains TTTAAA as well as CCAAT and potential Sp1 binding sites. There are no obvious sequence similarities to other cell cycle regulated genes. Transfection of Balb/3T3 mouse cells with the complete genomic R2 clone allowed isolation of stable transformants showing 10-fold decreased sensitivity to hydroxyurea. This shows that the cloned R2 gene is the functional gene, and it establishes a conclusive link between hydroxyurea resistance and R2 expression in mammalian cells.

D. Expression of mRNA During the Cell Cycle

The levels of R1/R2 mRNA in resting cells are very low, and therefore a very sensitive antisense RNA, solution hybridization, RNase protection assay was developed, capable of detecting down to 1 molecule of R1/R2 mRNA per cell. The assay was used to measure the R1/R2 mRNA levels during the cell cycle in elutriated cells as well as in cells synchronized by isoleucine or serum starvation [23]. The levels of both transcripts were very low in G_0/G_1 cells, showed a marked increase as cells progressed into S phase, and then declined as cells passed into G_2 + M phase. Both transcripts increased in parallel slightly before the rise in S phase cells and reached the same levels. From these data it is possible to explain the relative lack of cell cycle variation in R1 protein levels observed earlier simply as a consequence of the long half-life of the R1 protein.

IV. HYDROXYUREA AND GENE AMPLIFICATION

A. Target of Drug

Inhibition of mammalian ribonucleotide reductases by hydroxyurea is an early observation. Studies using homogeneous *E. coli* reductase in vitro showed that the drug acts as a one-electron reductant and specifically scavenges the tyrosyl free radical of the R2 subunit, which is formed by a one-electron oxidation of a tyrosyl residue [24]. These studies were later extended to the mammalian ribonucleotide reductase [25,26]. To regenerate the tyrosyl radical in hydroxyurea-inactivated mouse R2 protein, addition of Fe(II) plus oxygen is sufficient, and this regeneration occurs spontaneously in living cells [14]. The difference between the *E. coli* and mammalian systems in this respect may be that hydroxyurea, in addition to scavenging the tyrosyl radical, can labilize the iron center in the mammalian system [27]. This may occur by reducing the Fe(III) to Fe(II) and then chelating out the iron. Each dimeric Fe(III) center of the mouse R2 subunit can generate a tyrosyl free radical during its formation [9]. The growth inhibitory effect of hydroxyurea on mammalian cells can be fully explained by the inhibition of the R2 subunit, since transformation of cells with the complete R2 gene makes them hydroxyurea resistant [22].

B. Hydroxyurea-Resistant Cell Lines and Gene Amplification

Hydroxyurea enters the cell by a diffusion process, and the drug has been used clinically in the treatment of solid tumors and leukemias and polycythemia vera, and in the control of proliferation of psoriasis [28]. Early experiments by Lewis and Wright suggested the involvement of ribonucleotide reductase in the development of resistance to hydroxyurea in mammalian cells [29]. Selection of resistance to hydroxyurea was also shown not to be associated with the multidrug-resistant phenotype [30].

The R1 and R2 subunits are present in nonstoichiometric amounts in most cells, with the R2 subunit always being limiting. To assist in the purification of the R2 subunit, cell lines with a resistance to hydroxyurea of 100-fold were selected by growing the parent cell line over a one-year period in the presence of stepwise increasing concentrations of drug [25,26]. The resistant cells had a 10-fold increase in reductase activity and a 30-fold increase in R2 activity, but the R1 protein was not overproduced. The tyrosyl free radical content was increased 20–40-fold compared to the parental cells.

Much of the information about the R2 subunit has been obtained by studying hydroxyurea-resistant, R2 overproducing cells, and extracts from them. The availability of the hydroxyurea-resistant mouse TA 3 cells allowed the first preparation of a homogeneous mammalian R2 subunit [6]. It also made it possible to clone first the mouse R2 cDNA and then to map and clone the functional gene encoding the mouse R2 subunit [18,22]. The resistant TA 3 mouse cells showed a stable fivefold amplification of the R2 gene on chromosome 12, while the R1 gene was still present as a single copy. The resistant cells showed a 50–100 fold excess of the R2 mRNA compared to the parental cells and a tenfold excess of R1 mRNA.

Subsequently, a large number of different hydroxyurea-resistant cell lines from hamster, mouse, rat, and human origin have been characterized and shown to contain increased ribonucleotide reductase activity, increased levels of R2 mRNA and protein, and in the majority of cases, an R2 gene amplification [31]. In contrast, there is no unambiguous demonstration of a hydroxyurea-resistant cell line containing the same levels of R2 subunit but with an altered sensitivity to the drug.

Most cell lines showing multiple copies of the R2 gene still only have one copy of the R1 gene. This may reflect the presence of the R1 subunit in excess in most cells and may indicate that the levels of the R2 subunit are limiting for enzyme activity. As the concentration of hydroxyurea is increased, however, the R1 subunit eventually starts to be limiting. The cells now contain high levels of mainly radical-lacking R2 molecules competing with the few radical-containing R2 molecules for protein R1 to form an active enzyme complex. To further increase the resistance to hydroxyurea, a cell would seem to have to increase the R1 subunit levels. This is actually demonstrated for a hydroxyurea-resistant hamster cell line, where stepwise selection for resistance against up to 1.3 mM drug resulted in R2 gene amplification only as seen earlier. However, in the next selection step using 8 mM hydroxyurea, a resistant cell line showing a fivefold amplification of the R1 gene in addition to the R2 gene amplification was obtained [32]. This was accompanied by increases in R1 mRNA and protein. Similar observations have been reported for two highly hydroxyurea-resistant mouse cell lines [33].

C. Mechanisms Behind the Development of Hydroxyurea Resistance

A series of clonally related hydroxyurea-resistant mouse cell lines, selected by a stepwise procedure for increasing drug resistance, were carefully investigated with respect to ribonucleotide reductase activity, protein, mRNA, and gene copy number [34]. Each step was followed by increased cellular resistance and ribonucleotide reductase activity. A very early event occurring during the first step in the selection was the amplification of the R2 gene, and this was accompanied by increased levels of R2 mRNA. Very little further amplification and increase in R2 mRNA occurred in the later selection steps, although cellular resistance and ribonucleotide reductase activity increased significantly in each step. This is explained by a posttranscriptional regulatory mechanism involving increased translational efficiency of protein R2 with no alternation in protein R2 half-life. By this mechanism, resistant cells were able to accumulate high levels of R2 protein. Up to 4 mM hydroxyurea concentration, there was no change in R1 mRNA levels, but cells selected for resistance to 5 mM drug showed an elevated R1 mRNA and protein level compared to wild-type cells. These changes occurred without R1 gene amplification.

The increased levels of R1 and R2 mRNA in resistant cells compared to the parental cell lines are due to increased rates of transcription of both genes [35]. The normal cell cycle regulation of mRNA expression was found even in the resistant cells, although at about 30 times higher level of R2 mRNA [23].

It is well documented that hydroxyurea can affect the levels of R1 and R2 proteins in resistant cells posttranscriptionally [35,36]. A drug-resistant mouse cell line showed a stable sixfold amplification in the R2 gene copy number and contained two- to threefold increased R1 protein levels and 50-fold increased R2 protein levels when grown out of drug. When grown in the presence of hydroxyurea, both the R1 and R2 protein levels were further increased, but this increase was not accompanied by increased mRNA levels. Careful measurements showed that the drug-induced increase in R1 and R2 protein levels in the resistant cells was a result of an increased rate of synthesis of both proteins combined with a slower turnover rate.

Reversion of hydroxyurea resistance in an unstable drug-resistant hamster cell line was accompanied by a decline in ribonucleotide reductase activity and R2 mRNA levels, as well as loss of R2 gene amplification [37].

An interesting observation is that in all hydroxyurea-resistant cell lines examined from human, hamster, rat, and mouse origin, there was an elevation in ferritin heavy and/or light mRNA levels with no change in gene copy number [27]. This alteration in ferritin gene expression indicates that ferritin may play a role in the establishment of hydroxyurea resistance. However, it should be remembered that hydroxyurea is a weak iron chelator.

D. Coamplification of the R2 Gene and the Gene Encoding Ornithine Decarboxylase

One cDNA isolated by differential hybridization from a cDNA library made from a hydroxyurea-resistant hamster cell line was found to code for ornithine decarboxylase. During the stepwise selection of cells resistant to increasing concentrations of hydroxyurea, the degree of R2 gene amplification corresponded to the degree of ornithine decarboxylase gene amplification, indicating coamplification of the two genes [38]. This observation was later extended to another independently isolated hamster cell line as well as to a multistep human myeloid leukemic cell line resistant to hydroxyurea [21]. In addition to the R2 and ornithine decarboxylase genes, all cells showed coamplification of genomic sequences homologous to p5-8 cDNA, which encodes a 55 kDa protein of unknown function. Reversion of hydroxyurea resistance in an unstable population of hamster cells was accompanied by a parallel decrease in the gene copy number of protein R2, ornithine decarboxylase, and p5-8, indicating that these three genes form part of a single amplicon on chromosome 7 in hamster cells or on chromosome 2 in human cells [39]. Most likely ornithine decarboxylase and the gene product of p5-8 play only a passive coamplification role in establishing the hydroxyurea-resistant phenotype, since this coamplification is not observed in mouse cells. Neither mouse cells selected for resistance to hydroxyurea nor mouse cells selected for resistance to α-difluoromethylornithine showing amplification of the R2 and ornithine decarboxylase genes, respectively, showed coamplification of the two genes (G. McClarty and O. Heby, personal communications).

ACKNOWLEDGMENT

This work was supported by grants from the Swedish Natural Science Research Council.

REFERENCES

1. Thelander, L., and Reichard, P. Reduction of ribonucleotides. *Annu. Rev. Biochem. 48*:133–158 (1979).
2. Lammers, M., and Follmann, H. The ribonucleotide reductases—A unique group of metalloenzymes essential for cell proliferation. *Structure Bonding*, *54*:27–91 (1983).
3. Reichard, P. Interactions between deoxyribonucleotide and DNA synthesis. *Annu. Rev. Biochem. 57*:349–374 (1988).
4. Elledge, S. J., and Davis, R. W. DNA damage induction of ribonucleotide reductase. *Mol. Cell. Biol. 9*:4932–4940 (1989).
5. Thelander, L., Eriksson, S., and Åkerman, M. Ribonucleotide reductase from calf thymus. *J. Biol. Chem. 255*:7426–7432 (1980).
6. Thelander, M., Gräslund, A., and Thelander, L. Subunit M2 of mammalian ribonucleotide reductase. *J. Biol. Chem. 260*:2737–2741 (1985).
7. Mann, G. J., Gräslund, A., Ochiai, E.-I., Ingemarson, R., and Thelander, L. Purification and characterization of recombinant mouse and herpes simplex virus ribonucleotide reductase R2 subunit. *Biochemistry*, *30*:1939–1947 (1991).
8. Gräslund, A., Sahlin, M., and Sjöberg, B.-M. The tyrosyl free radical in ribonucleotide reductase. *Environ. Health Perspect. 64*:139–149 (1985).
9. Ochiai, E.-I., Mann, G. J., Gräslund, A., and Thelander, L. Tyrosyl free radical formation in the small subunit of mouse ribonucleotide reductase. *J. Biol. Chem. 265*:15758–15761 (1990).
10. Kjøller Larsen, I., Sjöberg, B.-M., and Thelander, L. Characterization of the active site of ribonucleotide reductase of *Escherichia coli*, bacteriophage T4 and mammalian cells by inhibition studies with hydroxyurea analogues. *Eur. J. Biochem. 125*:75–81 (1982).
11. Thelander, L., and Gräslund, A. Mechanism of inhibition of mammalian ribonucleotide reductase by the iron chelate of 1-formylisoquinoline thiosemicarbazone. *J. Biol. Chem. 258*:4063–4066 (1983).
12. Nordlund, P., Sjöberg, B.-M., and Eklund, H. Three-dimensional structure of the free radical protein of ribonucleotide reductase. *Nature*, *345*:593–598 (1990).
13. Engström Y., Eriksson, S., Jildevik, I., Skog, S., Thelander, L., and Tribukait, B. Cell cycle-dependent expression of mammalian ribonucleotide reductase. *J. Biol. Chem. 260*:9114–9116 (1985).

14. Eriksson, S., Gräslund, A., Skog, S., Thelander, L., and Tribukait, B. Cell cycle-dependent regulation of mammalian ribonucleotide reductase. *J. Biol. Chem.* *259*:11695–11700 (1984).
15. Engström, Y., Rozell, B., Hansson, H.-A., Stemme, S., and Thelander, L. Localization of ribonucleotide reductase in mammalian cells. *EMBO J.* *3*:863–867 (1984).
16. Mann, G. J., Musgrove, E. A., Fox, R. M., and Thelander, L.Ribonucleotide reductase M1 subunit in cellular proliferation, quiescence, and differentiation. *Cancer Res.* *48*:5151–5156 (1988).
17. Engström Y., and Rozell, B. Immunocytochemical evidence for the cytoplasmic localization and differential expression during the cell cycle of the M1 and M2 subunits of mammalian ribonucleotide reductase. *EMBO J.* *7*:1615–1620 (1988).
18. Thelander, L., and Berg, P. Isolation and characterization of expressible cDNA clones encoding the M1 and M2 subunits of mouse ribonucleotide reductase. *Mol. Cell. Biol.* *6*:3433–3442 (1986).
19. Brissenden, J. E., Caras, I., Thelander, L., and Francke, U. The structural gene for the M1 subunit of ribonucleotide reductase maps to chromosome 11, band p15, in human and to chromosome 7 in mouse. *Exp. Cell Res.* *174*:302–308 (1988).
20. Yang-Feng, T. L., Barton, D. E., Thelander, L., Lewis, W. H., Srinivasan, P. R., and Francke, U. Ribonucleotide reductase M2 subunit sequences mapped to four different chromosomal sites in humans and mice: Functional locus identified by its amplification in hydroxyurea-resistant cell lines. *Genomics*, *1*:77–86 (1987).
21. Tonin, P. N., Stallings, R. L., Carman, M. D., Bertino, J. R., Wright, J. A., Srinivasan, P. R., and Lewis, W. H. Chromosomal assignment of amplified genes in hydroxyurea-resistant hamster cells. *Cytogenet. Cell Genet.* *45*:102–108 (1987).
22. Thelander, M., and Thelander, L. Molecular cloning and expression of the functional gene encoding the M2 subunit of mouse ribonucleotide reductase: A new dominant marker gene. *EMBO J.* *8*:2475–2479 (1989).
23. Björklund, S., Skog, S., Tribukait, B., and Thelander, L. S-phase-specific expression of mammalian ribonucleotide reductase R1 and R2 subunit mRNAs. *Biochemistry*, *29*:5452–5458 (1990).
24. Atkin, C. L., Thelander, L., Reichard, P., and Lang, G. Iron and free radical in ribonucleotide reductase. *J. Biol. Chem.* *248*:7464–7472 (1973).
25. Åkerblom, L., Ehrenberg, A., Gräslund, A., Lankinen, H., Reichard, P., and Thelander, L. Overproduction of the free radical of ribonucleotide reductase in hydroxyurea-resistant mouse fibroblast 3T6 cells. *Proc. Natl. Acad. Sci. USA*, *78*: 2159–2163 (1981).
26. Gräslund, A., Ehrenberg, A., and Thelander, L. Characterization of the free radical of mammalian ribonucleotide reductase. *J. Biol. Chem.* *257*:5711–5715 (1982).
27. McClarty, G. A., Chan, A. K., Choy, B. K., and Wright, J. A. Increased ferritin gene expression is associated with increased ribonucleotide reductase gene expression and the establishment of hydroxyurea resistance in mammalian cells. *J. Biol. Chem.* *265*: 7539–7547 (1990).
28. Wright, J. A., McClarty, G. A., Lewis, W. H., and Srinivasan, P. R. Hydroxyurea and related compounds, in *Drug Resistance in Mammalian Cells*, Vol. 1, (R. S. Gupta, ed.), CRC Press, Boca Raton, Florida, 1989.

29. Lewis, W. H., and Wright, J. A. Altered ribonucleotide reductase activity in mammalian tissue culture cells resistant to hydroxyurea. *Biochem. Biophys. Res. Commun. 60*:926–933 (1974).
30. Tagger, A. Y., and Wright, J. A. Molecular and cellular characterization of drug resistant hamster cell lines with alterations in ribonucleotide reductase. *Int. J. Cancer, 42*:760–766 (1988).
31. Wright, J. A., Alam, T. G., McClarty, G. A., Tagger, A. Y., and Thelander, L. Altered expression of ribonucleotide reductase and role of M2 gene amplification in hydroxyurea-resistant hamster, mouse, rat and human cell lines. *Somatic Cell Mol. Genet. 13*:155–165 (1987).
32. Cocking, J. M., Tonin, P. N., Stokoe, N. M., Wensing, E. J., Lewis, W. H., and Srinivasan, P. R. Gene for M1 subunit of ribonucleotide reductase is amplified in hydroxyurea-resistant hamster cells. *Somatic Cell Mol. Genet. 13*:221–233 (1987).
33. Hurta, R. A. R., and Wright, J. A. Amplification of the genes for both components of ribonucleotide reductase in hydroxyurea resistant mammalian cells. *Biochem. Biophys. Res. Commun. 167*:258–264 (1990).
34. Choy, B. K., McClarty, G. A., Chan, A. K., Thelander, L., and Wright, J. A. Molecular mechanisms of drug resistance involving ribonucleotide reductase: Hydroxyurea resistance in a series of clonally related mouse cell lines selected in the presence of increasing drug concentrations. *Cancer Res. 48*:2029–2035 (1988).
35. McClarty, G. A., Chan, A. K., Engström, Y., Wright, J. A., and Thelander, L. Elevated expression of M1 and M2 components and drug-induced posttranscriptional modulation of ribonucleotide reductase in a hydroxyurea-resistant mouse cell line. *Biochemistry, 26*:8004–8011 (1987).
36. McClarty, G. A., Chan, A. K., Choy, B. K., Thelander, L., and Wright, J. A. Molecular mechanisms responsible for the drug-induced posttranscriptional modulation of ribonucleotide reductase levels in a hydroxyurea-resistant mouse L cell line. *Biochemistry, 27*:7524–7531 (1988).
37. McClarty, G. A., Chan, A. K., Choy, B. K., and Wright, J. A. Reversion of hydroxyurea resistance, decline in ribonucleoside reductase activity, and loss of M2 gene amplification. *Biochem. Biophys. Res. Commun. 145*:1276–1282 (1987).
38. Srinivasan, P. R., Tonin, P. N., Wensing, E. J., and Lewis, W. H. The gene for ornithine decarboxylase is co-amplified in hydroxyurea-resistant hamster cells. *J. Biol. Chem. 262*:12871–12878 (1987).
39. McClarty, G. A., Tonin, P. N., Srinivasan, P. R., and Wright, J. A. Relationships between reversion of hydroxyurea resistance in hamster cells and the coamplification of ribonucleotide reductase M2 component, ornithine decarboxylase and P5-8 genes. *Biochem. Biophys. Res. Commun. 154*:975–981 (1988).

22

DNA Amplification in Histidinol-Resistant Chinese Hamster Ovary Cells

Florence W. L. Tsui *The Toronto Hospital, Western Division, Toronto, Ontario, Canada*

Louis Siminovitch *Mount Sinai Hospital, Toronto, Ontario, Canada*

I. INTRODUCTION

The proper functioning of aminoacyl–tRNA synthetases is critical to protein synthesis because this class of enzymes catalyzes the attachment of amino acids to the correct tRNAs. In bacteria, the amino alcohols methioninol, tyrosinol, isoleucinol, and leucinol competitively inhibit the activation of the corresponding amino acids. This information suggested an approach to the possible isolation in eukaryotic cells of aminoacyl–tRNA synthetase mutations which would facilitate the study of specific facets of protein synthesis in such cells. In principle, isolates resistant to these drugs could involve a variety of lesions in the genes responsible for the synthesis of the synthetases, including their amplification. The latter class is of particular interest because its members provide excellent starting material for cloning of the genes. The feasibility of obtaining amplification mutants of the synthetases was first demonstrated by Gerken and Arfin [1], who showed that cells resistant to borrelidin, an inhibitor of threonyl–tRNA synthetase, overproduce the target enzyme.

In our own laboratory, we were in fact able to isolate histidinol-resistant mutants of Chinese hamster ovary (CHO) cells that showed amplification of histidyl–tRNA synthetase (HRS). In this chapter we present our studies on (1) the isolation of the histidinol-resistant CHO cells, (2) the isolation of cDNA clones for HRS and demonstration that the histidinol-resistant CHO cells had amplified the gene for HRS, (3) structural organization of the chromosomal gene for HRS and identification of three neighboring genes that are coamplified in the histidinol-resistant CHO cells, and (4) preliminary examination of the nature of gene amplification in the histidinol-resistant CHO cells. Much of this work has been published and is reviewed only briefly here.

II. ISOLATION OF HISTIDINOL-RESISTANT CHINESE HAMSTER OVARY CELLS

Histidinol is a competitive inhibitor of histidyl–tRNA synthetase in animal cells. First-step mutants selected in the presence of histidinol were found to be about two to four times more resistant to the drug. Isolates with increased histidinol resistance were then obtained by growing the first-step mutants in gradually increasing concentrations of histidinol. As indicated by the survival curves (Fig. 1), the mutant cell lines were much more resistant than the parental cells with respect to plating efficiency. The former showed increases in their D10s of about 160–290-fold over the wild-type lines. As expected, these increases in resistance were paralleled by elevations in enzyme activity (14–29-fold increase), although the level of resistance was much greater than that of the enzyme. Using in vivo

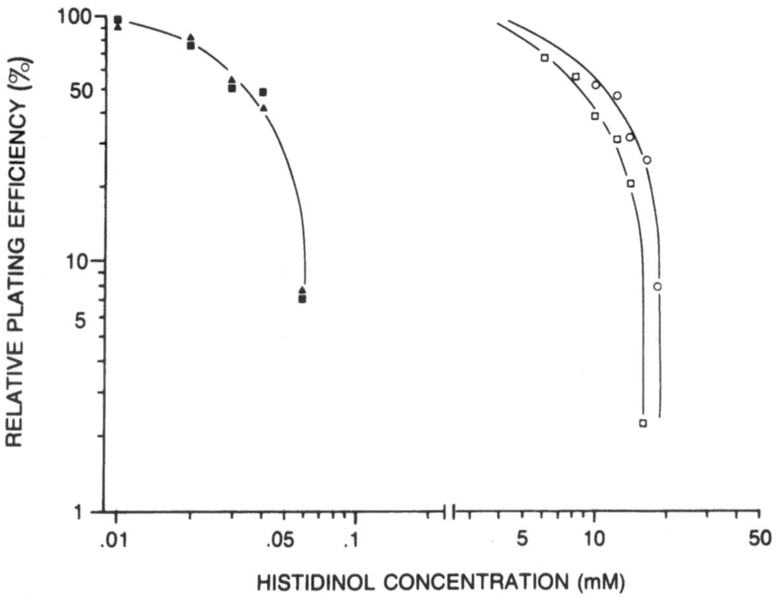

Figure 1 Survival curves of two wild-type CHO cell lines (▲, wt2; ■, wt5) and two histidinol-resistant cell lines (○, 2H2; □, 1H3) in the presence of different concentration of histidinol.

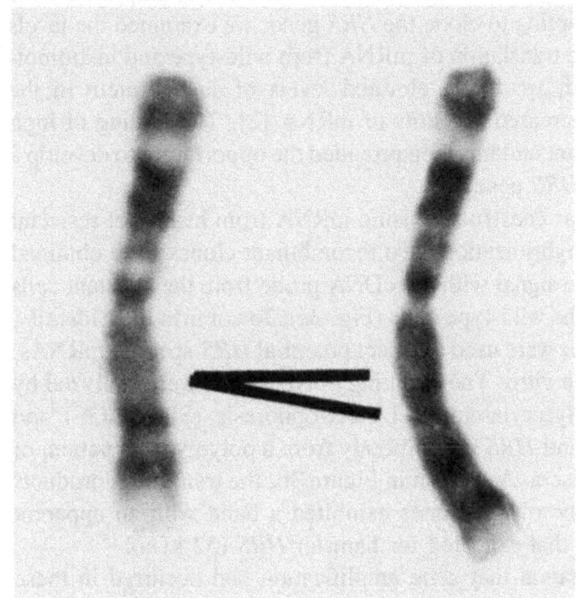

Figure 2 Comparison of the normal chromosome 2 (left) from the wild-type CHO cells and the abnormal chromosome 2 (right) from the histidinol-resistant CHO cells.

[^{35}S]methionine labeling and immunoprecipitation, we found that the resistant cells contained more HRS enzyme protein than wild-type cells [2]. No difference was observed in the K_m or K_i of the enzyme between parental and resistant cells, indicating that the observed resistance was not due to structural changes in the enzyme and more likely could be attributed to overproduction of the enzyme.

Further support for this hypothesis come from karyotypic analysis. As shown in Figure 2, we found an interstitial addition of chromosomal material in the long arm of chromosome 2 (right) in the resistant line. This is characteristic of amplified lines [3], essentially representing a major increase in chromosomal material.

III. ISOLATION OF cDNA CLONES FOR HISTIDYL–tRNA SYNTHETASE (HRS) AND DEMONSTRATION THAT THE HISTIDINOL-RESISTANT CHO CELLS HAD AMPLIFIED THE GENE FOR HRS

Although the foregoing data provided strong evidence that the histidinol-resistant mutants owed their resistance to amplification of the gene for HRS, unequivocal confirmation required that the gene be cloned and then shown to be amplified at the genomic level. Before attempting to clone the *HRS* gene, we examined the levels of HRS protein after in vitro translation of mRNA from wild-type and histidinol-resistant cells. As expected, we found elevated levels of such protein in the resistant cells, indicating increased amounts of mRNA [2]. The finding of high levels of mRNA in the resistant mutants then provided the opportunity to develop a strategy for cloning of the *HRS* gene.

A small cDNA library was constructed using mRNA from histidinol-resistant CHO cells. By differential hybridization, two recombinant clones were obtained that showed a strong positive signal with the cDNA probe from the resistant cells and little or no signal from the wild-type cells (Fig. 3a). To confirm their identity, clones isolated in this manner were used to select potential *HRS*-specific mRNAs, which were then translated in vitro. The resulting polypeptides were analyzed by sodium dodecyl sulfate polyacrylamide gel electrophoresis (SDS-PAGE) and immunoprecipitation using anti-*HRS* autoantibody from a polymyositis patient or rabbit anti-hamster *HRS* antisera. As shown in Figure 3b, the translation products immunoprecipitated from one of the clones exhibited a band with an apparent molecular weight similar to that expected for hamster *HRS* (52 kDa).

The conclusive demonstration that gene amplification had occurred in these histidinol-resistant cells was obtained by comparing the intensities of signals from a Southern blot containing serial dilutions of DNA from the resistant and wild-type cells, both electrophoresed in the same agarose gel [2]. As shown in Figure 4, compared to the wild-type cells, the copy number of the gene for HRS was approximately 30-fold higher in the resistant cells. Using the same *HRS* cDNA as a probe, we also compared the abundance of *HRS* mRNA in the resistant and wild-type cells by Northern blot analysis. We found that the level of the 2 kb *HRS* transcript was again about 30-fold higher, corresponding to the genomic DNA copy number, as well as the increased amount of HRS protein and enzyme activity. In summary, all these data provided very strong evidence that the isolated histidinol-resistant mutants contained amplified copies of the gene for HRS.

Figure 3 (a) Screening of recombinant cDNAs by differential hybridization: A, blot probed with $^{P}32$-labeled single-stranded cDNAs from wild- type CHO cells and B, blot probed with $^{P}32$-labeled single-stranded cDNAs from histidinol-resistant CHO cells. Arrows indicate the recombinant clone that showed a differential signal using the two probes mentioned. (b) SDS-PAGE analysis of in vitro translated products of mRNAs selected from histidinol-resistant CHO cells using the putative *HRS* recombinant isolated initially by differential hybridization: lane 1, translation products selected with control plasmid; lane 2, translation products selected with the putative *HRS* recombinant; lane 3, immunoprecipitate of translation products shown in lane 2 using anti-*HRS* autoantibodies containing polymyositis patient serum; and lane 4, immunoprecipitate of translation products shown in lane 2 using a polyclonal rabbit anti-hamster *HRS* antibodies.

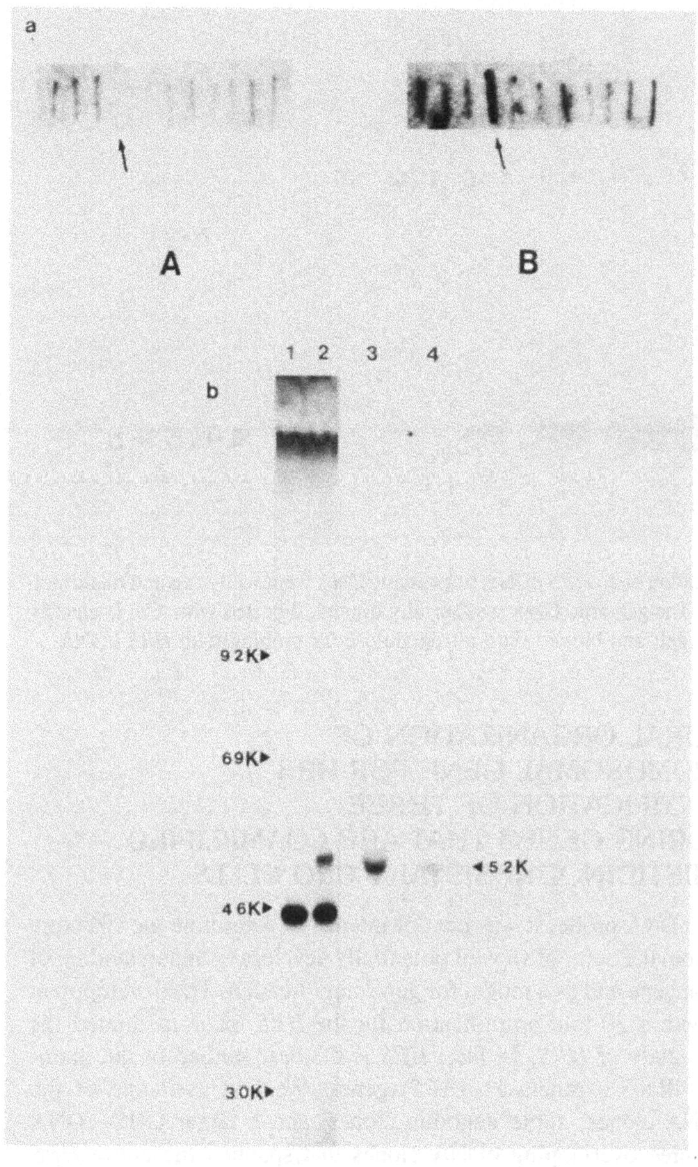
a
A
B
1 2 3 4
b
92K
69K
52K
46K
30K

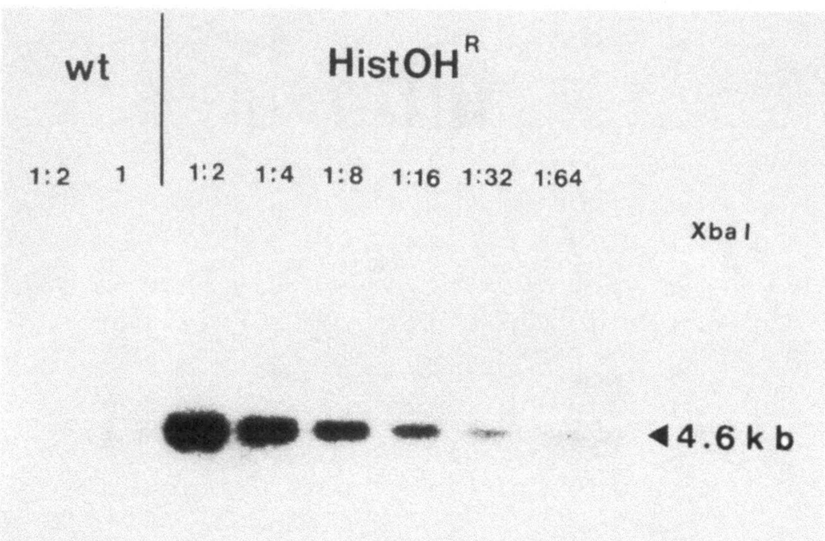

Figure 4 Hybridization of an *HRS* cDNA to genomic DNA from wild-type and histidinol-resistant CHO cells. The genomic DNA was serially diluted, digested with *Xba* I, electrophoresed in agarose gel, and blotted onto nitrocellulose for probing with *HRS* cDNA.

IV. STRUCTURAL ORGANIZATION OF THE CHROMOSOMAL GENE FOR HRS AND IDENTIFICATION OF THREE NEIGHBORING GENES THAT ARE COAMPLIFIED IN THE HISTIDINOL-RESISTANT CHO CELLS

Having obtained a cDNA probe, it was next of interest to determine the structure of the gene, both from the point of view of potentially developing understanding of the regulation of the gene and as a model for gene amplification. The development of the cell lines with a 30-fold amplification for the *HRS* locus facilitated the detailed molecular study of *HRS*. In fact, *HRS* is the best studied of the mammalian aminoacyl–tRNA synthetase (*AARS*) genes. We first used one of the original small cDNA clones, some genomic clones, and a larger CHO cDNA library to isolate three overlapping cDNA clones that spanned the entire *HRS* hamster gene [4]. We were then able to obtain a full-length human *HRS* cDNA from a human fibroblast library [4]. This clone was particularly useful in showing that the cDNAs we had isolated really coded for *HRS*, since it was possible to correct an *HRS*-defective, temperature-sensitive CHO mutant by transfection with the human cDNA (unpublished data). Having the hamster cDNAs in hand, we were in a position to study the structure of genomic *HRS*.

A. Structural Analysis of the Hamster Chromosomal Gene for HRS

Three overlapping phage clones encompassing 29 kb of the genomic DNA in which the chromosomal gene for HRS is located were first isolated and analyzed using conventional methods [4]. We found that the hamster genomic *HRS* gene contains 18 kb and 13 exons. The gene, like many other housekeeping genes, has multiple initiation sites [5]. Our studies on functional delineation of the hamster *HRS* promoter indicated that sequences containing -35 bases at the 5′ boundary are sufficient for some promoter activity. There appears to be a strong positive element in sequences between -61 and -122. Also, an inhibitory element seems to reside in sequences between -211 and -501 (unpublished results). Thus, at the molecular level, our present evidence indicates that transcription of the *HRS* gene can be regulated in a positive or inhibitory manner. Inhibition of transcription by upstream sequences has been reported in at least three other housekeeping gene promoters [6–8], but their functional significance is not clear.

Using RNase protection analysis, we also identified a 300 bp exon located only 36 bp away from the 5′-most start site of the *HRS* transcript. This exon hybridizes to a 3.5 kb transcript, which transcribes from a different strand of DNA in the 5′ region of the *HRS* gene. This divergent 3.5 kb transcript also has multiple transcription initiation sites [5]. The identity and function of the 3.5 kb transcript are not known. Divergent transcripts have been reported in other housekeeping enzyme gene loci, the most well studied example being the *DHFR* locus [9]. Although it appears that divergent transcripts might be a characteristic of most housekeeping enzyme gene loci, their significance is not understood. In respect to the *HRS* gene locus, we have a unique situation in which two overlapping but distinct promoters direct divergent transcription to two genes (*HRS* and an opposite strand gene of unknown function). Analysis of DNA sequences from the 5′ flanking region of a number of housekeeping genes (*HPRT*, *DHFR*, and malic enzyme) has revealed the presence of multiple copies of the Spl binding GC boxes. However, it seems unlikely that Spl plays any role in the regulation of *HRS* gene expression, since the presence of only one GGCGGG sequence located in the first exon of the gene is indicated by our DNA sequencing data of the 5′ region of the *HRS* chromosomal gene [5]. Future molecular analysis of this locus should shed light on some novel aspects of regulation of genes arranged as in *HRS*.

B. Identification of Coamplified Genes Located in the Vicinity of the Hamster *HRS* Locus

As indicated earlier, one of our objectives in carrying out the molecular analysis of the *HRS* was to develop further information about gene amplification. Thus, during the course of our studies we have been interested in identifying other gene loci that might be amplified with the *HRS* gene, as the amplification unit. As will be seen below, we have indeed shown the presence of two such genes.

1. Molecular Cloning of a Gene Located at the 3′ End of the HRS Locus

During the initial cloning of the *HRS* cDNA by differential hybridization, we obtained a recombinant clone that did not code for *HRS* but showed a strong positive signal with the cDNA probe from the resistant line and little or no signal with the cDNA probe from the wild-type cells. Hybrid selection and in vitro translation analysis indicated that this clone contained a cDNA specific to a 75 kDa polypeptide and that the 75 kDa polypeptide was also overexpressed [4]. From Southern and Northern blot analysis using this recombinant clone (pX75), we ascertained that the level of amplification of the genomic sequences and the corresponding 1.9 kb transcript was also about 30-fold. The polypeptide does not cross-react with antibody to histidyl–tRNA synthetase. However, results from in vitro experiments using [^{35}S]methionine labeling indicated that the 75kDa protein does not appear to accumulate inside the resistant cells in vivo, suggesting that it might be unstable or rapidly exported. Molecular studies using cDNAs covering the full length of the *HRS* gene and cDNA to the 75kDa polypeptide indicated that they were not related genes. We were able to obtain a phage clone (*HRS*λ1) from a CHO genomic library that hybridized to both the *HRS* cDNAs and the cDNA to the 75kDa polypeptide (pX75) [5]. Restriction enzyme mapping and Southern blot analysis then revealed that the exon corresponding to pX75 is located 9 kb at the 3′ end of the last exon (exon 13) of the chromosomal gene for HRS. Since we have found a number of unamplified DNA fragments with sequences homologous to pX75, this gene seems to be a member of a gene family.

2. Molecular Cloning of a Gene That Shares Some Homology with HRS and Is Located Within the Amplification Unit in the Histidinol-Resistant Cells

In work on the HRS genomic DNA, we had isolated from a CHO genomic library prepared in phage EMBL3 a unique *Sal*I-*Bam*HI fragment (from a phage clone called *HRS*λ2) [5]. In an attempt to identify *HRS* clones with more 5′ sequences, this genomic fragment was used as a probe to screen a second cDNA library made from mRNA of the resistant cells. In addition to identifying a few *HRS* cDNA clones that hybridized strongly to this genomic fragment, we isolated a cDNA clone (with an insert size of 1.4 kb) that hybridized weakly to the same probe. This cDNA clone hybridized to a 1.8 kb transcript. Preliminary DNA sequencing data of this clone (c19) has revealed a 38% homology to the *HRS* nucleotide sequence over a stretch of 100 bp in exons 9 and 10 of the *HRS* genomic DNA. The identity and function of this gene are not yet known. However, to our surprise, Southern and Northern blot analyses revealed that this cDNA clone (c19) located within the vicinity of the *HRS* locus was also amplified about 30-fold in the histidinol-

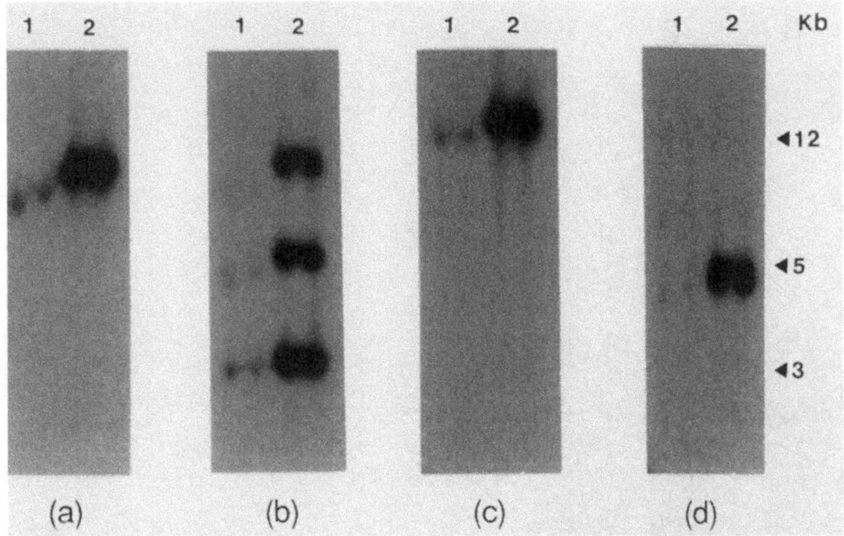

Figure 5 Southern blots of *Eco*RI-digested genomic DNA from wild type (lane 1) and histidinol-resistant (lane 2) CHO cells. (a) Blot probed with *HRS* cDNA containing the 5′ half of the *HRS* transcript. (b) Blot probed with a genomic fragment containing the first exon of the 3.5 kb opposite strand transcript. (c) Blot probed with an unknown cDNA (c19) with some homology to *HRS* sequences. (d) Blot probed with an unknown cDNA (X75) that coded for a 1.9 kb transcript.

resistant cells. Preliminary results using pulsed-field gel electrophoresis (PFGE) and Southern blot analysis have indicated that the two genes are probably less than 150 kb apart. It is not yet known whether this gene is situated at the 5′ or the 3′ end of the *HRS* gene.

In summary, we have cloned four different genes (including histidyl–tRNA synthetase) that are amplified in histidinol-resistant CHO cells. As shown in Figure 5, similar levels of amplification have been observed in all cases. These include the *HRS* gene itself, a divergent gene that codes for an unknown 3.5 kb transcript located only 36 bp upstream of the gene, a gene (pX75) that codes for an unknown 1.9 kb transcript situated 9 kb downstream of the gene for HRS, and another amplified gene (c19) with some homology to the gene for HRS.

V. NATURE OF GENE AMPLIFICATION IN THE HISTIDINOL-RESISTANT CHO CELLS

We have been unable to amplify the histidinol-resistant cells more than 30-fold in enzyme activity. This contrasts with the results in other amplification systems,

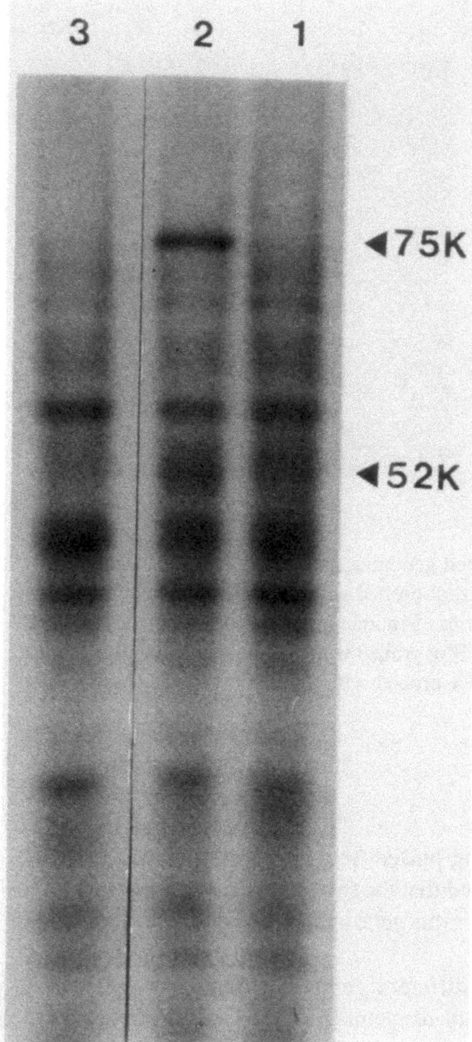

Figure 6 SDS-PAGE analysis of in vitro translated products of mRNAs isolated from wild-type cells (lane 1), histidinol-resistant cells (lane 2), and histidinol-resistant cells after growth in drug-free medium for 2 months (lane 3).

where increases exceeding 100-fold have been found. There are several possibilities that could account for this restriction:

1. Cells cannot survive with higher enzyme levels.
2. Since the promoter of the opposite strand gene overlaps with that of the *HRS* gene, expression of *HRS* may be affected by increased expression of this opposite strand gene.
3. Superlevels of the gene products of the coamplified genes could be toxic.

These hypotheses are currently being tested by transfection experiments.

When histidinol-resistant cells are taken off selection (i.e., by growing them in drug-free medium) for 2 months, one obtains cells that exhibit only a twofold increase in HRS enzyme activity. In vitro translation of mRNA from these cells followed by SDS-PAGE analysis has revealed the absence of amplified in vitro translatable gene products (Fig. 6), indicating that the cells have lost their amplified DNA. Experiments are now underway to examine this loss of the gene and presumably, the amplicon, at the molecular level.

The HRS system has already proved to be a very useful one from the point of view of development of understanding of the structure of a housekeeping gene and in respect to gene amplification. The gene itself is of interest not only because of its role in protein synthesis, but because HRS autoantibodies are a prominent feature of patients with polymyositis [10]. Much remains to be done, however, to obtain a comprehensive knowledge of this whole system. Such work is currently being pursued in our laboratory.

ACKNOWLEDGMENTS

This work was supported by grants from the Arthritis Society and the Medical Research Council of Canada awarded to F.W.L.T. and by grants from the National Cancer Institute of Canada awarded to L.S. We thank Dr. S. Mok for work on DNA sequencing of c19, Dr. H. W. Tsui for work on pulsed-field gel electrophoresis and Southern blotting, and Ms. B. Bessey for typing the manuscript.

REFERENCES

1. Gerken, S. C., and Arfin, S. M. Chinese hamster ovary cells resistant to borrelidin overproduce threonyl–tRNA synthetase. *J. Biol. Chem. 259*:11160–11161 (1984).
2. Tsui, F. W. L., Andrulis, I. A., Murialdo, H., and Siminovitch, L. Amplification of the gene for histidyl–tRNA synthetase in histidinol resistant Chinese hamster ovary cells. *Mol. Cell. Biol. 5*:2381–2388 (1985).
3. Cowell, J. K. Double minutes and homogenously staining regions: Gene amplification in mammalian cells. *Annu. Rev. Genet. 16*:21–59 (1982).
4. Tsui, F. W. L., and Siminovitch, L. Isolation, structure and expression of mammalian genes for histidyl–tRNA synthetase. *Nucleic Acid. Res. 15*:3349–3367 (1987).

5. Tsui, F. W. L., and Siminovitch, L. Structural analysis of the 5′ region of the chromosomal gene for hamster histidyl–tRNA synthetase. *Gene*, *61*:349–361 (1987).
6. Boyer, T. G., Krug, J. R., and Maquat, L. E. Transcriptional regulatory sequences of the housekeeping gene from human triosephosphate isomerase. *J. Biol. Chem. 264*: 5177–5187 (1989).
7. Morioka, H., Tennyson, G. E., and Nikodem, V. M. Structural and functional analysis of the rat malic enzyme gene promoter. *Mol. Cell. Biol. 8*:3542–3545 (1988).
8. Melton, D. W., McEwan, C., McKie, A. B., and Reid, A. M. Expression of the mouse *HPRT* gene: Deletional analysis of the promoter region of an X-chromosome-linked housekeeping gene. *Cell*, *44*:319–328 (1986).
9. Linton, J. P., Yen, J. J., Selby, E., Chen, Z., Chinsky, J. M., Liu, K., Kellems, R. E., and Crouse, G. F. Dual bidirectional promoters at the mouse *DHFR* locus: Cloning and characterization of two mRNA classes of divergently transcribed *Rep-1* gene. *Mol. Cell. Biol. 9*:3058–3072 (1989).
10. Ramsden, D. A., Chen, J., Miller, F. W., Misener, V., Bernstein, R. M., Siminovitch, K. A., and Tsui, F. W. L. Epitope mapping of the cloned human autoantigen, histidyl–tRNA synthetase: Analysis of the myositis-associated anti-Jo-1 autoimmune response. *J. Immunol. 143*:2267–2272 (1989).

23

Amplification of the P-Glycoprotein Gene Family and Multidrug Resistance

Lela Veinot *University of Toronto, Toronto, Ontario, Canada*

Victor Ling *The Ontario Cancer Institute and University of Toronto, Toronto, Ontario, Canada*

I. INTRODUCTION

A. The Multidrug Resistance Phenotype

Multidrug resistance (MDR) in human malignancies is likely to be a major cause of failure in cancer chemotherapy. Characterization of independently derived MDR cell lines clearly illustrates that MDR can be mediated by the overproduction of a 170 kDa glycosylated plasma membrane protein called P-glycoprotein (*Pgp*). This chapter focuses on the role of amplification of the P-glycoprotein gene family in the expression of MDR in a number of cultured mammalian cell lines. In-depth discussions of the biochemistry, genetics, and molecular biology of MDR can be found in several reviews [1–12].

MDR is characterized by simultaneous acquisition of resistance to structurally and functionally unrelated drugs following selection for resistance to a single cytotoxic drug. The drugs associated with MDR are predominantly alkaloids and antibiotics of plant and fungal origin. Acquired resistance to these compounds is thought to be achieved by the lowered intracellular drug concentration that is characteristic of MDR cell lines. The MDR phenotype can be completely or partially reversed by a diverse group of compounds termed "chemosensitizers," which are characterized by their ability to increase drug accumulation and retention in resistant cells. Compounds that act as chemosensitizers include nonionic detergents, local anesthetics, calcium channel blockers, calmodulin inhibitors, and steroids. MDR cells often display increased sensitivity (collateral sensitivity) to one or more of the chemosensitizers. The underlying mechanism of this cytotoxicity is not understood, although disruption of the plasma membrane has been postulated.

Pgp was first implicated in MDR during the characterization of a series of MDR Chinese hamster ovary (CHO) subclones, which had undergone stepwise selection for resistance to colchicine (CH^R). A high level of resistance to colchicine could be achieved only by a multiple stepwise selection, which implied that this phenotype was related to a gene dosage mechanism. The most striking feature of CH^RCHO cells was a significant increase in the concentration of *Pgp* in the plasma membrane; *Pgp* was present in very low levels in the AuxB1 parental cell line. The relative amount of *Pgp* correlated with the relative level of drug resistance in independent mutants and in revertants. The development of specific antibodies directed against *Pgp* was used for Western blot analysis to show that overexpression of *Pgp* was associated with MDR cell lines from different species (hamster, mouse, and human).

The use of monoclonal antibodies directed against *Pgp* led to the isolation of the first *Pgp* cDNA clones derived from the MDR CHO cell line, CH^RB30. *Pgp* genomic DNA and cDNA sequences were also cloned by taking advantage of MDR cells containing amplified sequences. This was undertaken using either the technique of in-gel renaturation or by differential hybridization of C_0t-fractionated genomic DNA and cDNA libraries (for review, see Ref. 1). Subsequently, full-length cDNA clones for *Pgp* have been isolated from hamster, mouse, and human sources. Transfection studies using either full-length *Pgp* cDNA or genomic DNA have demonstrated directly that overexpression of *Pgp* is sufficient to mediate MDR in cultured mammalian cells.

A model of Pgp as a membrane-associated, energy-dependent, drug-efflux pump has been postulated based on its predicted amino acid sequence. Figure 1 depicts the proposed structure of *Pgp* relative to the plasma membrane. The *Pgp* molecule (ca.1280 amino acids in length) is a duplicated molecule with each half containing six potential transmembrane regions and a putative ATP-binding region. The transmembrane regions likely form a pore or channel through the plasma membrane.

B. The P-Glycoprotein Gene Family

Pgp is a highly conserved molecule and has been described in species as diverse as yeast and man. Much of the proposed structure and function of *Pgp* is based on its remarkable amino acid similarity to a group of well-characterized bacterial transport proteins and especially to the 66 kDa *Escherichia coli* hemolysin secretion protein, HlyB. The high sequence similarity between mammalian P-glycoproteins, prokaryotic transport proteins, and other eukaryotic transport proteins suggests that they are all members of a supergene family. This family is characterized by an extensive sequence similarity involving about 200 amino acids that encompass a highly conserved ATP-binding domain.

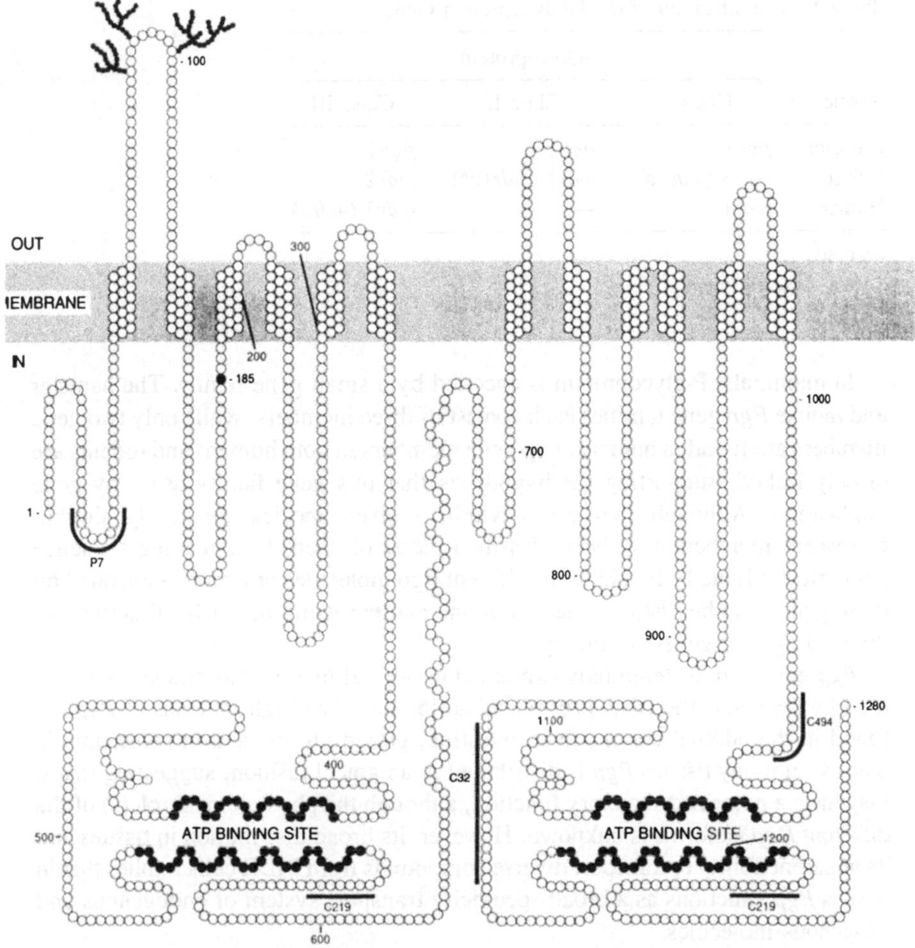

Figure 1 Proposed orientation of Pgp within the plasma membrane lipid bilayer. Each circle represents a single amino acid residue; the branched structures represent putative N-linked carbohydrates on the first extracellular domain. The proposed ATP-binding motifs are represented by solid circles. The positions of epitopes that have been mapped thus far include the monoclonal antibodies (C219, C494, C32) that were used to localize the ATP-binding regions to the cytoplasm and the polyclonal antibody (P7) raised against a synthetic peptide and are represented by the solid lines. Position 185 localizes a spontaneous mutation (valine is substituted for glycine) found in the human *mdr*1 gene and was isolated from an MDR cell line stepwise selected to resist colchicine. Transformants expressing the mutant gene showed an altered cross-resistance profile, indicating that position 185 may be involved in drug binding. (From Ref. 8; references not shown here can be found in Ref. 8.)

Table 1 Classification of the P-Glycoprotein Genes

Species	P-Glycoprotein Class I	Class II	Class III
Hamster	*pgp*1	*pgp*2	*pgp*3
Mouse	*mdr*3 (*mdr*1a)[a]	*mdr*1 (*mdr*1b[b])	*mdr*2
Human	*mdr*1	—	*mdr*3 (*mdr*2)

[a]Ref. 30.
Source: Adapted from Ref. 9.

In mammals, P-glycoprotein is encoded by a small gene family. The hamster and mouse *Pgp* gene families each consist of three members, while only two gene members are found in humans. *Pgp* gene members in both humans and rodents are closely linked, supporting the hypothesis that this gene family arose by gene duplication. Although *Pgp* genes within a given species are highly similar, equivalent members have been classified based on their 3′ noncoding sequence similarities (Table 1). In addition, different Pgp molecules or isoforms encoded by these genes can be distinguished from one another using monoclonal antibodies directed against gene-specific epitopes.

Pgp genes are differentially expressed in normal mammalian tissues as measured by gene-specific DNA probes and antibodies. The highest levels of *Pgp* are found in the adrenal cortex, liver, intestine, gravid uterus, and in some muscle tissues. In many tissues *Pgp* is distributed in an apical fashion, suggesting that it may have a polarized secretory function, although the physiological role(s) of the different *Pgp* isoforms is unknown. However, its broad distribution in tissues and its apparent ability to transport diverse compounds from MDR cells implies that in tissues *Pgp* functions as a broad-specificity transport system of endogenous and exogenous molecules.

Pgp is also found widespread in human cancers, including leukemias, lymphomas, sarcomas, and carcinomas. The clinical significance of this expression with respect to response to chemotherapy is currently under investigation. It remains to be established at what level(s) *Pgp* becomes a significant factor in a patient's response to chemotherapy.

II. P-GLYCOPROTEIN GENE AMPLIFICATION AND EXPRESSION

A. Karotypic Features of Amplification

Many MDR cell lines containing amplified copies of *Pgp* genes also contain double-minute (DM) chromosomes or homogeneously staining regions (HSRs).

Characterization of these karotypic abnormalities has been reviewed recently [2]. Amplified genes have been also identified in mammalian cells as part of submicroscopic extrachromosomal circular molecules or episomes. Ruiz et al. [13] have shown that episomes in the size range of 600–750 kb may contain the *mdrl* (class I; see Table 1) gene in the highly vinblastine-resistant KB-V1 human epidermoid cell line. KB-V1 cells, selected through multiple steps, contained more than a 100-fold increase in the *mdr*1 gene copy number per cell [14,15]. Although these cells also contained DMs, the number of DMs per cell was substantially less than the level of amplification of the *Pgp* gene [14]. Thus, it has been postulated that the episomes may be early products of amplification from which DMs are generated [13,16]. It has not been shown whether episomes exist in other MDR cell lines or are generated during the initial steps of a multistep selection regimen for drug resistance.

B. Characteristics of P-Glycoprotein Amplicons

An unanswered question is how the pattern of cross-resistance can differ from one MDR cell line to another and even for MDR cell lines selected from the same parental cell line using the same selecting drug. One hypothesis is that overexpression of more than one amplified gene contributes to the MDR phenotype. In this way, the pleiotropic nature of the MDR phenotype could be explained by differential amplification and/or expression of these genes. This hypothesis has been addressed by identifying genes that are coamplified in different MDR sublines and determining the correlation between gene amplification/expression and drug resistance. This section discusses the results of studies that have shown the following results.

1. The size of the amplicon encompassing *Pgp* genes can be large, in the order of megabases, but highly resistant cells tend to have shorter amplicons.
2. Genes flanking the *Pgp* gene family are frequently coamplified in MDR cells; however, there may not be a direct correlation between the amplification and expression of these genes and MDR.
3. Differential amplification of the *Pgp* gene family occurs frequently in highly resistant cells, suggesting that sequences near or within *Pgp* genes are involved in the amplification process itself.
4. *Pgp* genes usually are differentially expressed in MDR cells, suggesting that expression of amplified *Pgp* genes is under regulation.

1. *Amplicon Size and Coamplification*

The *Pgp* gene family as well as large stretches of flanking sequences are often amplified in MDR cells. Individual amplicons in the Chinese hamster CH^{R}C5 subline and its progenitor CH^{R}B3 were first shown by pulsed-field electrophoresis to span at least 1500 kb [17–19]. Subsequently, the mean amplicon size in B3$^+$

cells, a subclone of CH^{R}B3 containing a large HSR, was estimated to be in the order of 4.5–10.1 mb. This was determined using Southern blot hybridization and cytogenetic techniques [20]. *Pgp* amplicons as large as 625 kb, encompassing the entire *Pgp* family, have been found in mouse MDR cell lines [21]. As described in Section II.A, human *Pgp* amplicons as large as 750 kb have also been detected, although the two *Pgp* genes map within only 500 kb of each other [22].

The large size of amplicons in MDR cells indicated that other genes, in addition to *Pgp* genes, may be amplified and expressed in MDR cells. Indeed, six classes of cDNA clones representing genes overexpressed in the Chinese hamster CH^{R}C5 MDR cell line, but not in the AuxB1 drug-sensitive parental cell line, were isolated and shown to be linked to *Pgp* genes [17–19]. Three of these six classes (classes 2a, 2b, and 2c) corresponded to the *Pgp* gene family, while another (class 4) was shown to encode sorcin (V19), a cytosolic calcium binding protein homologous to calpain [22,23]. Genes represented by these cDNA clones were shown to be conserved and coamplified with *Pgp* genes in mouse and human MDR cell lines. Coamplification of these genes implied that they may contribute to drug resistance [23,24]. However, only *Pgp* genes have been found to be consistently amplified and overexpressed in independent MDR sublines lines [17,18,24]. In addition, cDNA transfection studies have shown that overproduction of Pgp is sufficient to confer cross-resistance and collateral sensitivity (for reviews, see Refs. 1, 9).

2. *Differential Amplification of P-Glycoprotein Genes*

During multiple steps of selection for increased drug resistance, the mean amplicon size decreases, resulting in DNA rearrangements, perhaps novel joints, and differential amplification of the *Pgp* gene family. Of the six cDNA classes described above, three of the linked genes, including two *Pgp* genes (classes 1, 2a and 2b, respectively), were not coamplified further during the selection of the CH^{R}C5 clone from its progenitor, CH^{R}B3 [17,19]. In the vincristine-resistant human neuroblastoma subline SH-SY5Y/VCR, the *mdr*1 (class I) gene copy number was 30-fold greater than that of the parental SH-SY5Y cell line, but no amplification of the *mdr*3 (class III) gene was detected [23]. The cross-resistance profile of the SH-SY5Y/VCR cells was similar to that of the vincristine-resistant human neuroblastoma cell line MC-1XC/VCR, which was shown to contain a 30-fold increase in the copy number of both the *mdr*1 and *mdr*3 genes. In two independent sublines of the SEWA murine TC13K cell line selected to resist colcemid, amplification of the *mdr*3 (class I) gene was detected, whereas the *mdr*1 (class II) and *mdr*2 (class III) genes were not amplified [24]. In another study, independent MDR cell lines, derived from macrophagelike murine J774.2 cells, selected stepwise to resist either colchicine or taxol, were found to contain equally amplified copies of the *mdr*1a (class I) and *mdr*1b (class II) genes, but amplification of the *mdr*2 (class III) gene was not detected [25,26].

In summary, these results illustrate that differential amplification of the *Pgp* gene family is common in highly resistant MDR cell lines and that amplification of the class III (Table 1) *Pgp* gene member occurs infrequently. Thus, one end of the *Pgp* amplicon often lies within the *Pgp* loci, excluding the class III *Pgp* gene from further amplification. It is unlikely that this pattern of differential amplification is coincidental. It is possible that specific sequences are located within the *Pgp* loci, such as a site of unequal crossing over or a replication origin that promotes DNA amplification and thus prevents further amplification of this gene member. It should be noted also that overexpression of the class III *Pgp* gene is not correlated with drug resistance. Transfection studies show that overexpression of the class III *Pgp* gene is not capable of conferring MDR onto transfected drug-sensitive cells (for reviews, see Refs. 1, 9). Thus, coamplification of this gene may be fortuitous because it happens to be tightly linked to other *Pgp* genes that can confer drug resistance.

It is interesting to note here that although it is possible to isolate rodent MDR cells that have undergone selective amplification of the class I *Pgp* gene without further amplification of the class II *Pgp* gene (see, e.g., Ref. 17), to date no MDR cell line has been described that has undergone amplification of the class II *Pgp* gene in the absence of class I *Pgp* gene amplification. This phenomenon is not easily explained, but it may be due to restrictions imposed by the amplification process itself. Alternatively, although less likely, there may be selective pressures acting that favor overexpression via gene amplification of the class I gene alone or of both the class I and II *Pgp* genes, but not overexpression of the class II *Pgp* gene alone. However, transfection studies have shown that overexpression of either of these two gene members alone is sufficient to confer MDR upon transfection into drug-sensitive cells.

3. *Amplification of Transfected P-Glycoprotein Genes*

Differential amplification of the *Pgp* gene family has been also observed in cells transfected with genomic DNA encoding *Pgp* (for review, see Ref. 2). For example, mouse LTA cells transfected with genomic DNA from the colchicine-resistant CH^RC5 hamster cell line and selected for colchicine resistance showed a cross-resistance profile that was similar to that of the donor cells [27,28]. Southern blot hybridization and Western blot analysis showed that only the hamster *pgp*1 (class I) gene was maintained in independent transfectants [27, and Deuchars and Ling, unpublished data]. The basis for the selective amplification of this transfected gene is not understood. It is possible that the drug used in the selection somehow affected the overexpression of this particular *Pgp* gene. Recently, Devault and Gros [29] showed that the cross-resistance profiles of cells transfected with mouse *mdr*1 (class II) cDNA was distinct from that seen for cells expressing murine *mdr*3 (class I) cDNA. Alternatively, it is possible that the transfected hamster *pgp*1 (class I) genomic sequence or flanking sequences contained an origin of replication that rendered it more amplifiable than the other transfected *Pgp* genes. It has been shown by several groups that transfected DNA can destabilize the chromosomal site at its site of integration and may play a role in its amplification (see, e.g., Ref. 30). It is interesting to note here that hamster genomic DNA from drug-sensitive AuxB1 cells was apparently incapable of manifesting the MDR phenotype in transfected LTA cells [27,28,31]. These results would suggest that amplified sequences encoding *Pgp* or perhaps flanking sequences may be modified in MDR cells in some way that favors its amplification upon transfection.

4. *Expression of Amplified P-Glycoprotein Genes*

Amplified *Pgp* genes are not all equally expressed. This suggests that regulatory sequences of the unamplified genome may continue to regulate the amplified sequences. Raymond et al. [21] showed that independent mouse fibroblast LTA MDR clones selected by a single-step regimen using colchicine underwent equal coamplification of the three *Pgp* genes. Among these sublines the level of *mdr*1 (class II) gene expression correlated with the *mdr*1 gene copy number, whereas the amplified *mdr*2 (class III) and *mdr*3 (class I) genes were underexpressed. On the other hand, the mouse P388 ADM-2 cell line selected to resist adriamycin contained equal copy numbers of the three *Pgp* genes, but overexpressed only the *mdr*3 (class I) gene, with no detectable increase in *mdr*1 (class II) and *mdr*2 (class III) gene expression [21]. Unlike rodent cells, amplified copies of the human class I (*mdr*1) *Pgp* gene are disproportionately overexpressed relative to the degree of gene amplification. For example, in the vincristine-resistant human neuroblastoma cell lines SH-SY5Y/VCR and MC-1XC/VCR described above, each was shown to contain a 30-fold increase in the copy number of the *mdr*1 (class I) and *mdr*3 (class III) genes, but the *mdr*1 gene was shown to be disproportionately overexpressed in both cell lines. Overexpression of the *mdr*1 gene may indicate that up-regulation of this gene may also occur prior to or in concert with gene amplification [23].

C. Hypoxia and P-Glycoprotein Gene Amplification

The frequency with which cells become resistant to drugs can be increased by prior exposure to treatments that perturb DNA replication such as transient hypoxia, UV irradiation, and the carcinogen *N*-acetoxy-2-acetylaminofluorene. It has been shown that during the recovery of cells to inhibitors of DNA synthesis, a significant proportion of the cells have > 4C DNA content. The frequency of selecting drug-resistant variants has been shown to be higher using this subpopulation compared to cells containing a normal haploid DNA content. How DNA overreplication occurs as a result of these treatments is not known. Rice et al. [32] subjected CHO cells to transient hypoxia prior to selecting for doxorubicin resistance. These cells manifested a 10-fold increase in the frequency of drug-resistant cells compared to a control population. Amplification of the *Pgp* genes accounted for the drug resistance in approximately 50% of the cell variants. These results supported the hypothesis that gene amplification in these cells may involve DNA overreplication plus a variety of recombinational events [33].

Results similar to those described above were obtained using murine KHT-LP1 cells selected in a single step to resist doxorubicin following exposure to hypoxia [34]. However, hypoxia-treated EMT6/Ro murine cells, containing overreplicated DNA, as indicated by their > 4C content of DNA, were found to be no more resistant to doxorubicin than the control population [34]. In the same study, DNA overreplication was not detected in three independent human tumor cell lines following transient hypoxia. Thus, it appears that the effects of hypoxia may be phenomena dependent on cell line and perhaps species, as well. The mechanisms that underlie these different results are unknown.

III. MULTIDRUG RESISTANCE IN THE ABSENCE OF GENE AMPLIFICATION

Amplification of *Pgp* genes is not a prerequisite for the MDR phenotype. There are now examples to illustrate that regulation of *Pgp* gene expression is complex and can be affected at various levels of the biosynthetic pathway. Human MDR cell lines selected to resist a low concentration of a drug frequently overproduce both *Pgp* and *Pgp* mRNA in the absence of gene amplification. As an example, SKOV3 human ovarian carcinoma mutants, selected to resist a low concentration of either vincristine or vinblastine, contained increased amounts of both *Pgp* and *Pgp* mRNA without a detectable increase in the *Pgp* gene copy number [35]. It was only upon further selection for increased drug resistance that an abrupt increase in the *Pgp* gene copy number was observed. Similar findings have been reported using other human cell lines selected for a low level of MDR (for review, see Refs. 1, 9). Overexpression of either the murine *mdr*1 (class II) or *mdr*3 (class I) genes has been described in MDR cell lines in the absence of a detectable level of amplification (see, e.g., Ref. 21). In another study, the actinomycin D resistant DC-3F/ADXC hamster subline showed only a five-fold increase in *Pgp* gene amplification relative to the DC-3F parental cells, but there was a 97-fold increase in the level of *Pgp* mRNA [36]. Thus, it is likely that changes at the level of transcription or perhaps mRNA stability can be altered in MDR cells in the absence of gene amplification.

The level of Pgp in drug-resistant cells can increase upon selection for variants that display increased resistance, but without a further increase in either the level of *Pgp* mRNA or its gene copy number. The human SK VCR 0.25 human ovarian carcinoma MDR subline, selected stepwise for vincristine resistance, was shown to contain increased amounts of *Pgp*, *Pgp* mRNA, and amplified copies of the native *Pgp* genes [35]. However, SK VCR 2, a subline derived from SK VCR 0.25 by selecting cells that grew in a eight-fold higher concentration of vincristine, displayed a marked increase in cross-resistance and also a three-fold higher level of *Pgp* without a commensurate change in its level of mRNA or gene copy number. This is the first example indicating that MDR in human cells may be mediated by *Pgp* at either the translational or posttranslational level.

MDR in rodent cells may also be mediated at the level of translation and/or posttranslational regulation of Pgp. Independent sublines of the Chinese hamster ovary cell line AuxB1 pretreated with multiple lethal doses of X-ray consistently expressed drug resistance, overproduced *Pgp*, without a detectable increase in the *Pgp* gene copy number or the level of the *Pgp* mRNA transcripts encoded by the three gene members [37]. These sublines exhibited many of the hallmarks of MDR, but unlike most drug-selected MDR cell lines, these and other sublines selected in a similar manner lacked significant cross-resistance to anthracyclines (for review, see Ref. 38). It has been postulated that one or more mutations in the Pgp molecule could affect the cross-resistance in these sublines. As described in Figure 1, at least one mutation in *Pgp* is known to affect the cross-resistance profile of MDR cells selected for colchicine resistance.

In summary, overexpression of *Pgp* genes is not always a function of gene dosage alone. Several studies suggest that the *Pgp* genes are regulated independently of one another. The molecular basis that favors the up-regulation of the *Pgp* genes as opposed to gene amplification is not understood. The role of tissue-specific factors that may regulate *Pgp* gene expression at the level of transcription, translation, or posttranslation should also be considered when studying the role of *Pgp* gene amplification in multidrug resistance.

IV. CONCLUSIONS

MDR mediated by P-glycoproteins in cultured mammalian cells is frequently the result of expression of amplified copies of the *Pgp* gene family. Studies suggest that in addition to gene amplification, other mechanisms acting on all levels of *Pgp* gene expression may operate to affect the level of Pgp and may work in concert with *Pgp* gene amplification. Thus, MDR and overexpression of *Pgp* appear to be the result of a complex process and not simply the result of increased gene dosage. The *Pgp* gene family has yet to be fully exploited as a model system to study the mechanism(s) of gene amplification. Further analysis of MDR cell lines is required to answer questions about the mechanism(s) of amplification of *Pgp* genes in drug-sensitive and highly resistant cells.

ACKNOWLEDGMENTS

The authors thank their colleagues, especially Kathy Deuchars and Roman Zastawny, at the Ontario Cancer Institute, for critical reading of this chapter.

Studies conducted in the authors' laboratory were supported by Public Health Service grant CA37130 from the U.S. National Institutes of Health and grants from the National Cancer Institute of Canada.

REFERENCES

1. Endicott, J. A., and Ling, V. The biochemistry of P-glycoprotein-mediated multidrug resistance. *Annu. Rev. Biochem. 58*:137–171 (1989).
2. Bradley, G., Juranka, P.F., and Ling, V. Mechanism of multidrug resistance. *Biochem. Biophys. Acta, 948*:87–128 (1988).
3. Riordan, J. R., and Ling, V. Genetic and biochemical characterization of multidrug resistance. *Pharmacol. Ther. 28*:51–75 (1985).
4. Gottesman, M. M., and Pastan, I. The multidrug transporter, a double-edged sword. *J. Biol. Chem. 263*:12163–12166 (1988).
5. Croop, J. M., Gros, P., and Housman, D. E. Genetics of multidrug resistance. *J. Clin. Invest. 81*:1303–1309 (1988).
6. Gerlach, J. H., Kartner, N., Bell, D. R., and Ling, V. Multidrug resistance. *Cancer Surv. 5*:25–46 (1986).
7. Gottesman, M. M., and Pastan, I. Resistance to multiple chemotherapeutic agents in human cancer cells. *Trends Pharmacol. Sci. 9*:54–58 (1988).
8. Juranka, P. F., Zastawny, R. L., and Ling, V. P-Glycoprotein: Multidrug-resistance and a superfamily of membrane-associated transport proteins. *FASEB J. 3*:2583–2592 (1989).
9. Veinot-Drebot, L., and Ling, V. The molecular biology of multidrug resistance, in *Molecular Biology of Cancer Genes* Vol. 14 (M. Sluyser, ed.), Ellis Horwood, Toronto, 1990, pp. 263–289.
10. Ling, V. Colchicine and colcemid in drug resistance in mammalian cells, in *Anti-cancer and Other Drugs* Vol. 7 (R. S. Gupta, ed.), CRC Press, Boca Raton, Florida, 1989, pp. 139–154.
11. Ling, V. Multidrug-resistant mutants, in *Molecular Cell Genetics* (M. M. Gottesman, ed.), Wiley, New York, 1985, pp. 773–787.
12. Ling, V. Genetic basis of drug resistance in mammalian cells, in *Drug and Hormone Resistance in Neoplasia*, Vol. 1 (N. Bruchovsky and J. H. Goldie, eds.), CRC Press, Boca Raton, Florida, 1982, pp. 1–19.
13. Ruiz, J. C., Choi, K., Van Hoff, D. D., Robinson, I. B., and Wahl, G. M. Autonomously replicating episomes contain *mdr*1 genes in a multidrug-resistant human cell line. *Mol. Cell. Biol. 9*:109–115 (1989).
14. Fogo, A. T., Whang-Peng, J., Gottesman, M. M., and Pastan, I. Amplification of DNA sequences in human multidrug-resistant KB carcinoma cells. *Proc. Natl. Acad. Sci. USA, 82*:7661–7665 (1985).
15. Shen, D.-W., Cardarelli, C., Hwang, J., Cornwell, M., Richert, N., Ishii, S., Pastan, I., and Gottesman, M. M. Multiple drug-resistant human KB carcinoma cells independently selected for high-level resistance to colchicine, adriamycin, or vinblastine show changes in expression of specific proteins. *J. Biol. Chem. 261*:7762–7770 (1986).

16. Wahl, G. M. The importance of circular DNA in mammalian gene amplification. *Cancer Res. 49*:1333–1340 (1989).
17. de Bruijn, M. H. L. Van der Bliek, A. M., Biedler, J. L., and Borst, P. Differential amplification and disproportionate expression of five genes in three multidrug-resistant Chinese hamster lung cell lines. *Mol. Cell. Biol. 6*:4717–4722 (1986).
18. Van der Bliek, A. M., Van der Velde-Koerts, T., Ling, V., and Borst, P. Overexpression and amplification of five genes in a multidrug-resistant Chinese hamster ovary cell line. *Mol. Cell. Biol. 6*:1671–1678 (1986).
19. Jongsma, A. P. M., Spengler, B. A., Van der Bliek, A. M., Borst, P., and Biedler, J. L. Chromosomal localization of three genes coamplified in the multidrug-resistant CH^RC5 Chinese hamster ovary cell line. *Cancer Res. 47*:2875–2878 (1987).
20. Jongsma, A. P. M., Riethorst, A., and Aten, J. A. Determination of the DNA content of a six-gene amplicon in the multidrug-resistant Chinese hamster cell line CH^RB3 by flow karotyping. *Cancer Res. 50*:2803–2807 (1990).
21. Raymond, M., Rose, E., Housman, D. E., and Gros, P. Physical mapping, amplification, and overexpression of the mouse *mdr* gene family in multidrug-resistant cells. *Mol. Cell. Biol. 10*:1642–1651 (1990).
22. Van der Bliek, A. M., Meyers, M. B., Biedler, J. L., Hes, E., and Borst, P. A 22-kd protein (sorcin/V19) encoded by an amplified gene in multidrug-resistant cells, is homologous to the calcium-binding light chain of calpain. *EMBO J. 5*:3201–3208 (1986).
23. Van der Bliek, A. M., Baas, F., Van der Velde-Koerts, T., Biedler, J. L., Meyers, M. B., Ozols, R. F., Hamilton, T. C., Joenje, H., and Borst, P. Genes amplified and overexpressed in human multidrug-resistant cell lines. *Cancer Res. 48*:5927–5932 (1988).
24. Stahl, F., Martinsson, T., Dahlof, B., and Levan, G. Amplification and overexpression of genes in SEWA murine multidrug-resistant cells and mapping of the P-glycoprotein genes to chromosome 5 in mouse. *Hereditas, 108*:251–258 (1988).
25. Horwitz, S. B., Goei, S., Greenberger, L., Lothstein, L., Mellado, W., Samar, S. N., Yang, C.-P. H., and Zeheb, R. Multidrug resistance in the mouse macrophage-like cell line J774.2, in *Mechanisms of Drug Resistance in Neoplastic Cells* (P. V. Wooley and K. D. Tew, eds.), Academic Press, New York, 1988, pp. 223–242.
26. Hsu, S.-H., Lothstein, L., and Horwitz, S. B. Differential overexpression of three *mdr* gene family members in multidrug-resistant J774.2 mouse cells. *J. Biol. Chem. 264*:12053–12062 (1989).
27. Deuchars, K. L., Du, R.-P., Naik, M., Evernden-Porelle, D., Kartner, N., Van der Bliek, A. M., and Ling, V. Expression of hamster P-glycoprotein and multidrug resistance in DNA-mediated transformants of mouse LTA cells. *Mol. Cell. Biol. 7*: 718–724.
28. Dabenham, P. G., Kartner, N., Siminovitch, L., Riordan, J. R., and Ling, V. DNA-mediated transfer of multiple drug resistance and plasma membrane glycoprotein expression. *Mol. Cell. Biol. 2*:881–889 (1982).
29. Devault, A., and Gros, P. Two members of the mouse *mdr* gene family confer multidrug resistance with overlapping but distinct drug specificities. *Mol. Cell. Biol. 10*:1652–1663 (1990).

30. Ruiz, J. C., and Wahl, G. M. Chromosomal destabilization during gene amplification. *Mol. Cell. Biol. 10*:3056–3066 (1990).
31. Robertson, S. M., Ling, V., and Stanners, C. P. Co-amplification of double minute chromosomes, and cell surface P-glycoprotein in DNA mediated transformants of mouse cells. *Mol. Cell. Biol. 4*:500–506 (1984).
32. Rice, G. C., Ling, V., and Schimke, R. T. Frequencies of independent and simultaneous selection of Chinese hamster cells for methotrexate and doxorubicin (adriamycin) resistance. *Proc. Natl. Acad. Sci. USA*, *84*:9261–9264 (1987).
33. Schimke, R. T., Sherwood, S. W., Hill, A. B., and Johnston, R. N. Overreplication and recombination of DNA in higher eukaryotes: Potential consequences and biological implications. *Proc. Natl. Acad. Sci. USA*, *83*:2157–2161 (1986).
34. Luk, C. K., Veinot-Drebot, L., Tjan, E., and Tannock, I. F. Effect of transient hypoxia on sensitivity to doxorubicin in human and murine cell lines. *J. Natl. Cancer Inst. 82*:684–692 (1990).
35. Bradley, G., Naik, M., and Ling, V. P-Glycoprotein expression in multidrug-resistant human ovarian carcinoma cell lines. *Cancer Res. 49*:2790–2796 (1989).
36. Scotto, K. W., Biedler, J. L., and Melera, P.W. Amplification and expression of genes associated with multidrug resistance in mammalian cells. *Science*, *232*:751–755 (1986).
37. Hill, B. T., Deuchars, K., Hosking, L. K., Ling, V., and Whelan, R. D. H. Overexpression of P-glycoprotein in mammalian tumor cell lines after fractionated X-irradiation in vitro. *J. Natl. Cancer Inst. 82*:607–612 (1990).
38. Hill, B. T. In vitro drug–radiation interactions using fractionated X-irradiation regimens, in *Antitumor Drug Radiation Interactions* Vol. 12 (B. T. Hill and A. S. Bellamy, eds.), CRC Press, Boca Raton, Florida, 1990, pp. 207–224.

24

Glutamine Synthetase Gene Amplification in Chinese Hamster Ovary Cells

Richard H. Wilson *University of Glasgow, Glasgow, Scotland*

I. GLUTAMINE SYNTHETASE: THE ENZYME

The enzyme glutamine synthetase (GS) [L-glutamate:ammonia ligase (ADP-forming), EC 6.3.1.2] catalyzes the transfer of an ammonia molecule to glutamic acid to form glutamine, the corresponding amide, using the hydrolysis of ATP to drive the reaction. The glutamine thus formed can be regarded as a stable source of nontoxic ammonia in cells, and it acts as an amino donor in a wide variety of biosynthetic reactions. In mammals amino groups from glutamine are found in purines, pyrimidines, amino sugars, NAD, and carbamyl phosphate (for arginine and urea synthesis). Glutamine hydrolyzes spontaneously under physiological conditions with a half-life of a few days. In vivo, glutaminases hydrolyze glutamine to re-form glutamate and ammonia for further involvement in intermediary metabolism.

In bacteria, fungi, and plants, which can assimilate inorganic nitrogen, glutamine synthesis is usually the first stage in the assimilation of ammonium into organic nitrogen compounds, whereas in animals, which exist in a relative surplus of nitrogen, the primary function of GS in intermediary metabolism is to regulate the relative fluxes of carbon and nitrogen [1]. In mammals, plasma concentrations of glutamine, at about 0.5mM, are higher than that of any other amino acid.

One of the conundrums of glutamine metabolism in mammals is the observation that glutamine, not glucose, or even an intermediate between glutamine and the tricarboxylic acid cycle (glutamate or 2-oxoglutarate), is the major respiratory substrate for many cell types (see, e.g., Ref. 1). This leads to a cycle of glutamine synthesis and export by the liver and a return transport of alanine and other amino acids from peripheral tissues. Another consequence of this respiratory function of glutamine is that most, but not all, cultured cells have an absolute requirement for glutamine, in spite of possessing adequate levels of glutamine synthetase. In mammals, GS is found at low "housekeeping" levels ($\approx 0.01\%$ of total protein) in all cell types, with a few specialized cell types expressing levels approximately 100-fold higher. High levels of GS are found in a subset of hepatocytes, glial cells, kidney proximal tubule cells, and maturing adipocytes. Only hepatocytes surrounding the efferent venules express high levels of GS, presumably regulating circulating plasma levels of glutamine. Expression in glial cells would relate to the functions of these cells in minimizing the levels of ammonium, and possibly glutamate, in the neurones they contact. Whereas the expression of GS in the kidney could be involved in pH regulation, it is uncertain why a cell synthesizing fats should require high levels of an ammonia transfer enzyme unless this is a consequence of the respiratory element of glutamine metabolism mentioned earlier.

Sequences from a variety of GS proteins from prokaryotes and eukaryotes can be classified into two related types, a "prokaryotic" type existing as a 12-mer of approximately 480 amino acids (see, e.g., Ref. 2) and a "eukaryotic" type existing as an 8-mer of about 360 amino acids (see, e.g., Refs. 3–5). A crystal structure for a prokaryote type GS has been solved by X-ray diffraction [2]. Some species of bacteria, including endosymbiotic *Rhizobia*, may possess GSs of both types. Among vertebrates there appears to be only one expressed *GS* gene, but plants may express up to four different *GS* genes, with tissue-specific GSs and targeting of one GS type to chloroplasts.

Methionine sulfoximine (Msx) is a highly effective inhibitor of all GS types [6]. It functions by acting as a pseudosubstrate, becoming phosphorylated at the S-linked N atom. The phosphorylated Msx then stays effectively irreversibly bound to the GS active site. In mammals large doses of Msx act as a convulsant; indeed Msx was first discovered as the psychoactive component of overbleached flour made into dog biscuits that induced a form of "canine hysteria." Phosphinothricin, derived from a streptomycete antibiotic, is another potent GS inhibitor with commercial applications as a herbicide.

II. GLUTAMINE SYNTHETASE GENES

The Chinese hamster *GS* gene may be taken as typical of vertebrate *GS* genes characterized to date (see, e.g., Ref. 4). It extends some 10,450 bp from cap site to final poly(A) site with seven exons and six introns, including an intron in the 5′ untranslated region of the mRNA (Fig. 1). The alfalfa gene selected for by gene amplification [5] has an intron structure almost completely different, although

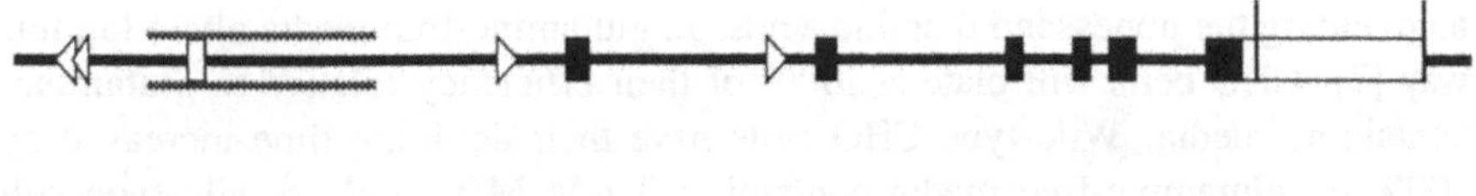

Figure 1 The Chinese hamster glutamine synthetase gene. Exons are open boxes (untranslated) or solid boxes (translated), transcription being from left to right. Polyadenylation sites are represented by vertical bars, *Alu* sequences by open triangles with their points at the poly (A) ends. The CpG-undermethylated island is delineated by bars.

exon 5 of the vertebrate gene exactly corresponds to exon 7 of the alfalfa gene. I am uncertain as to the significance of these observations for competing theories of the long-term evolution of spliced genes. All vertebrate *GS* genes appear to produce mRNAs polyadenylated at two different sites. The rarer *GS* mRNA generated from a weak poly(A) site has a short (≈ 100 base), and the major *GS* mRNA a long (≈ 1500 base), 3′ untranslated region to the poly (A) tail. No differential role has been ascribed to these alternate mRNAs.

The multiplicity of possibly conflicting controls required by the *GS* gene may well be reflected in the structure of its promoter, in that most cell types express *GS* in a housekeeping manner, while a few cell types of diverse embryological origin show high levels of expression. The Chinese hamster *GS* gene has a large undermethylated CpG island, with a high G+C:A+T base ratio, around its promoter and extending 1.6 kbp downstream. This island contains a profusion of restriction sites for infrequent cutters of mammalian DNA with two *Nae* I, three *Nar* I, one *Sal* I, three *Sma* I, two *Sst* II and one *Xho* I sites, all of which are unmethylated in both wild-type and *GS*-gene-amplified cells. This structure would presumably reflect the low level "housekeeping" nature of *GS* promoter expression in most cell types, including the Chinese hamster ovary (CHO) cell. However, the *GS* promoter possesses motifs characteristic of high level expression genes such as a CCAAT box and a TATA box (even if the latter is an AATA box in the rodent *GS* genes). As little as 400 bp of the Chinese hamster gene upstream of the cap site are required for the maximal response of chloramphenicol acetyltransferase constructs in transient expression studies in transfected CHO cells.

III. GLUTAMINE SYNTHETASE GENE AMPLIFICATION

Although *GS* is a suitable target gene to select for gene amplification in cultured mammalian cells, normal cell culture media contain considerable amounts of glutamine, which would interfere with selection for increased GS levels. Consequently media that are effectively glutamine-free must be prepared. This is achieved by using dialyzed serum, omitting the glutamine from the medium, and augmenting the nonessential amino acids. In glutamine-free media altered in this way [7], CHO cells will plate at 100% of their efficiency relative to glutamine-containing media. Wild-type CHO cells have their doubling time increased by 100% on glutamine-free media containing 3 μM Msx, and no wild-type cell colonies can establish themselves on glutamine-free media containing more than 5 μM Msx. However glutamine is capable of completely reversing the inhibitory effects of 20 mM Msx, showing the exquisite specificity of Msx for the inhibition of glutamine synthesis without side effects on other components of cellular metabolism. The only consequence of Msx administration to cultured mammalian cells is thus to induce a phenocopy state of glutamine auxotrophy, which can be reversed either by the administration or by the increased synthesis of glutamine.

Selection for resistance to Msx on glutamine-free media will produce colonies of cells that are found, when tested, to contain 10 or so copies of the *GS* gene. From these cells further variants of cells resistant to higher concentrations of Msx can be produced by stepwise selection, eventually producing cells resistant to millimolar levels of Msx. These highly resistant cells may contain a 1000-fold amplification of the *GS* gene and produce up to 30% of their soluble protein as GS [7]. In situ hybridization of metaphase spreads of one such cell line (C4M), using a 3.5 kb region of the *GS* gene as a probe, reveals that the amplified *GS* genes are clustered on a small acrocentric chromosome that is not present on the original wild-type CHO cells (Fig. 2). It is not yet known where the single copy *GS* genes map in the wild-type CHO genome (cf Ref. 8). In interphase cells, this cluster of amplified *GS* genes is localized to a small region of the nucleus, even though the total length of the amplified region is at least 70 Mbp.

The morphology of these cells with high levels of *GS* gene expression is slightly abnormal, tending to a more rounded "fried-egg" form. Whether this is a consequence of having to organize their cytoskeletons around a mass of glutamine synthetase enzyme molecules is not known.

No kinetic differences are apparent when the GS of the gene-amplified cells is studied. It would appear to be difficult to produce a GS enzyme that is insensitive to Msx, presumably because Msx or its phosphorylated form is a close mimic of a transition state of the catalytic process. Hence most examples of Msx resistance in mammalian cells are due to GS overproduction. It is possible that at high levels of *GS* gene amplification, however, certain transcription factors have become limiting as they are titrated out onto the amplified *GS* promoters, and it may be possible that the highest levels of gene amplification lead to increased levels of GS production only if there is an overproduction of one or more transcription factors.

Cells expressing the ultimate levels of *GS* expression have to be maintained on Msx-containing media. When transferred down to inhibitor-free media, most cells stop growing and die, with colonies developing from a small percentage of cells. These cells have lost their amplified *GS* genes, usually by chromosome loss (e.g., Fig. 2) and are almost as sensitive to Msx as the original wild-type cells.

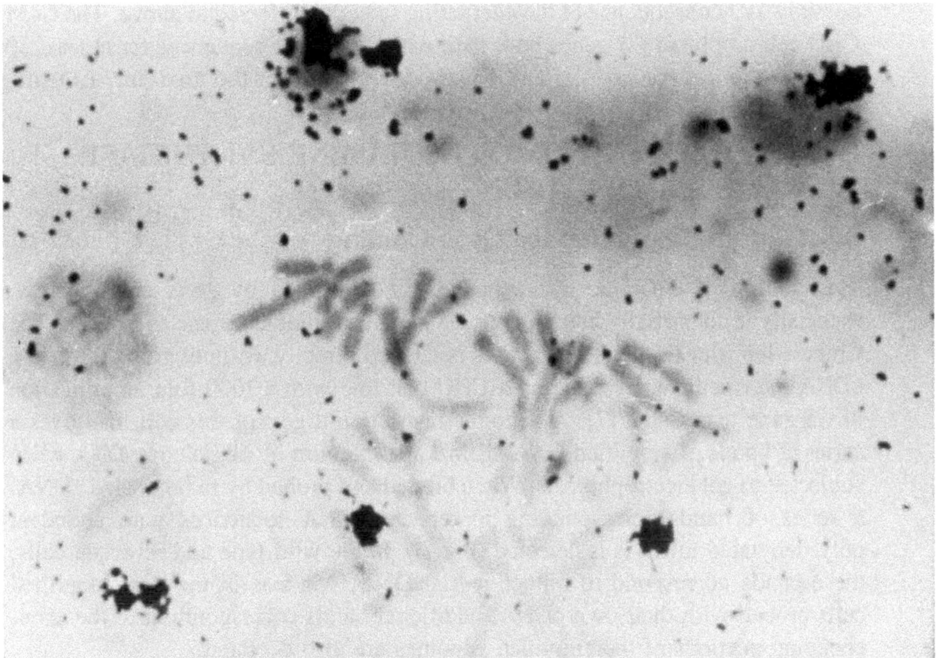

Figure 2 In situ hybridization of colchicine-arrested, *GS*-gene-amplified C4M CHO cells with a tritium labeled *GS* gene probe (data of B. E. Hayward). A mitotic spread of a quasi-tetraploid cell can be seen with two clusters of amplified *GS* genes obscuring the small acrocentric chromosomes carrying them. Three interphase nuclei can be seen, with amplified *GS* genes clustered in one location in the nucleus together with one interphase nucleus with no amplified *GS* genes.

Presumably the expression of levels of uninhibited GS many times those found in normal cells would seriously imbalance intermediary metabolism and lead to cell sickness, if not cell death. As a consequence of the possibility that forward and reverse selection have been inadvertently applied during the development of *GS*-gene-amplified cell lines, the structure of the *GS* gene amplification unit in these cells may have been subject to considerable selection for its ability to respond to this bidirectional selection.

Pulsed-field gel electrophoresis of restriction digests of DNA from the most highly gene-amplified cell line show a repeating structure with repeat units of 70 or 140 kbp. Mapping experiments to date have been unable to confirm the relative orientation of these units, although the results are most compatible with a head-to-head, tail-to-tail orientation of the units [9]. If these results are confirmed, then these *GS* amplicons should be among the smallest of mammalian amplicons, possibly as a consequence of the alternating selection referred to above. The C4M CHO cell amplified *GS* genes have their native flanking sequences for at least 20 kbp upstream of the cap site and 10 kbp downstream of the final poly (A) site.

IV. THE EXPLOITATION OF GLUTAMINE SYNTHETASE GENE AMPLIFICATION

A. Glutamine Synthetase Gene Cloning

The cloning of a *GS* gene can be greatly facilitated by gene amplification, especially if no partially homologous *GS* gene is available for use as a probe. The Chinese hamster *GS* gene was able to be cloned directly, without going through a cDNA intermediate, by utilizing a CHO cell line with a 1000-fold amplification of *GS* gene expression [7]. A genomic restriction digest of this cell line gives a series of bands of amplified DNA against a continuum of single copy DNA when subjected to gel electrophoresis. When blotted and probed by radiolabeled cDNA, a series of bands corresponding to repeated DNA sequences with abundant polyadenylated mRNAs is detected (Fig. 3). In the wild-type and revertant cells, these bands correspond to mitochondrial DNA, whereas in the gene-amplified cells probed with their own cDNA, additional bands corresponding to the exon-encoding portions of the amplified genomes are also detected.

Once a single restriction fragment that apparently contained all the *GS* exons detectable had been identified, it was possible to cut this fragment from the gel and clone it. Subsequent studies using cDNA probes that extended further in the 5′ direction showed that additional exons could be detected upstream. The promoter region of the Chinese hamster *GS* gene proved refractory to cloning this way, apparently as a result of poison sequences within the first intron, with a yield of positive clones only 0.2% of that expected given the proportion of amplified DNA in the bands isolated.

The cDNA probing technique for the detection of amplified coding DNA can be made more sensitive by quenching the interfering mitochondrial DNA signal by using cloned mtDNA, but the objective of being able to recognize and clone any newly amplified coding region of the genome, such as in primary amplificants or tumor cells, appears to be beyond the sensitivity of this technique.

Gene amplification was also used to facilitate the cloning of an alfalfa *GS* gene by selecting for cell culture variants of alfalfa cells resistant to the GS-inhibiting herbicide, phosphinothricin [10].

B. Gene Expression by Coamplification with Glutamine Synthetase Genes

The ease of selection of *GS* gene amplification in certain cell types has led to its exploitation in the expression of foreign genes in mammalian cells. GS has some

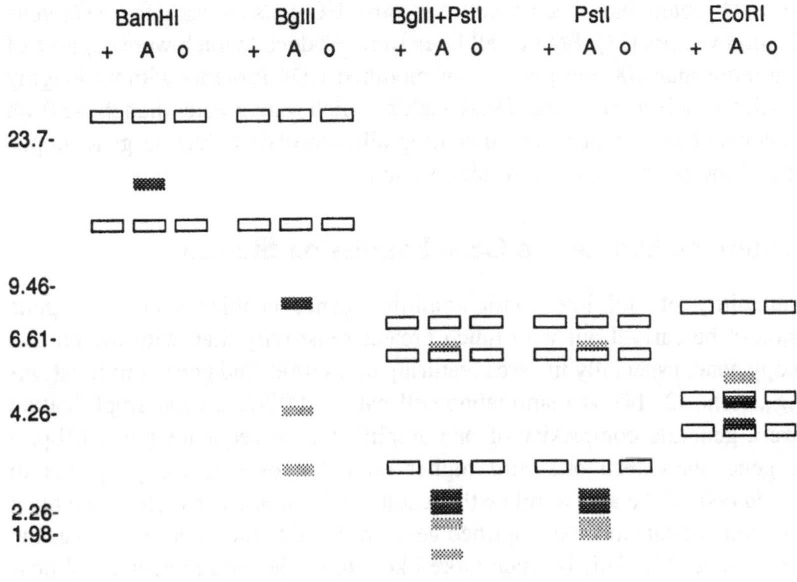

Figure 3 A summary of the banding patterns found by probing Southern blots of restriction digests of genomic DNA from wild-type KG1 (+), *GS*-gene-amplified C4M (A), and revertant C4O (o) CHO cells with radioactive cDNA prepared from C4M cells. Open boxes are mitochondrial DNA bands, heavily shaded bands are *GS* gene bands found in the original experiments [7], lightly shaded bands are those found subsequently.

advantages over other enzymes for the selection of gene amplification. GS is a fairly stable enzyme in the cellular environment and is not expressed in a cell-cycle-dependent manner, unlike several enzymes involved in nucleic acid metabolism. Consequently the incoming gene may be selected for against a background of the native gene expression because of the constancy of the latter. Second, expression of *GS* may be selected for over a factor of 10^3 by using Msx selection, Msx inhibition being highly specific for glutamine synthesis, as stated above. These two observations make it possible to manufacture *GS* gene amplification vectors for foreign gene expression in cells that are wild type for the *GS* gene [11,12]. One such vector uses the SV40 late promoter to drive expression of a *GS* minigene at a low level (but high enough to enable selection above the background of endogenous gene expression), while the human cytomegalovirus promoter/enhancer drives high level expression of an inserted cDNA construct. In addition, the *GS*-gene-amplified cells produced for gene expression purposes may have growth advantages insofar as they are capable of growth on cheaper media that do not contain the chemically unstable glutamine. Among other applications, soluble forms of CD4 protein have been produced from CHO cells by using these *GS* gene amplification vectors [13]. Stable cell lines were produced which were capable of secreting more than 100 mg per liter of modified CD4 proteins without lengthy amplification or selection stages. These yields, which were greater than those from other systems of gene expression, including dihydrofolate reductase gene amplification and the Baculovirus expression system.

C. Glutamine Synthetase Gene Expression Studies

The availability of cell lines with amplified genes enables studies on gene expression to be carried out with much greater sensitivity than with the normal single copy gene, especially in cases featuring up to 1000-fold gene amplifications (see, e.g., Refs. 12, 14). A mammalian cell with a 1000-fold gene amplification will have a genomic complexity of one amplified gene sequence per 3 Mbp, a relative gene abundance one-third higher than that of single copy genes in *Escherichia coli*. To be able to relate the results to the single copy gene, one must be able to demonstrate that the amplified genes respond to the same controls as the single copy gene [14]. This is much more likely to be the case in gene amplification, where a region of up to several hundred kilobases of DNA flanking the selected gene may be preserved intact. Although one hopes that the processes of gene amplification can maintain the epigenetic status of the gene with the same fidelity as normal replication, the long-range interactions with the appropriate dominant control regions, matrix attachment sites, and distant enhancer complexes are more likely to be maintained than with alternate forms of gene copy number increase that utilize a naked DNA intermediate.

One of the best studies of the in vivo status of promoter chromatin used a hybrid construct where the mouse metallothionein I promoter drove the expression of a mouse dihydrofolate reductase minigene [15]. Amplification of this construct could be selected for by methotrexate resistance. Although the construct used only about 1775 bp of the metallothionein promoter, which had to be assembled into chromatin from naked DNA in the absence of epigenetic or chromosomal positional information, the resulting gene-amplified promoters showed convincing binding of factors to DNA sites and changes in binding profiles in response to induction by zinc when base alkylation sensitivity to dimethyl sulfate (DMS) was studied.

Similar techniques have been used to study the amplified *GS* genes in CHO cells. Cells were briefly exposed to DMS and single strand breaks induced in the DNA at sites of purine alkylation (7-methyl guanine and, to a lesser extent, 3-methyl adenine). An end-labeled oligonucleotide was used to prime upstream extension of these DNA fragments using multiple cycles of *Taq* DNA polymerase to produce labeled DNA fragments terminated by the sites of alkylation in the chromatin DNA. These fragments could then be separated on a standard DNA sequencing gel and visualized by autoradiography.

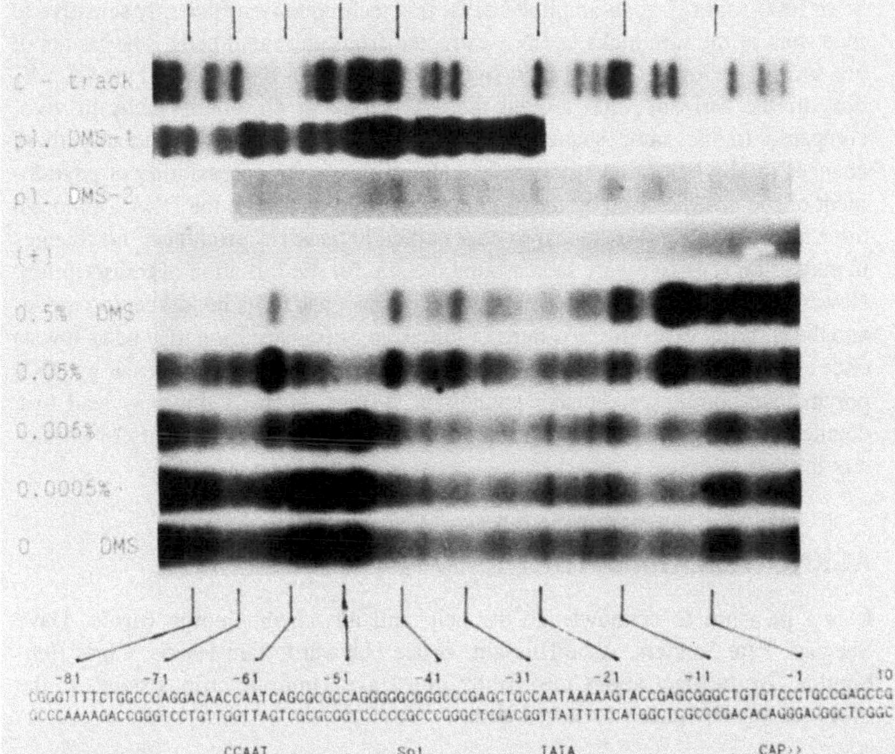

Figure 4 In vivo studies of *GS* promoter chromatin status in *GS*-gene-amplified C4M CHO cells (data of D. L. Thom). A ^{32}P-labeled oligonucleotide complementary to bases −32 to −51 was used to prime upstream extension off DNA templates using multiple cycles of *Taq* DNA polymerase activity. Key: C-track, a standard T7 DNA polymerase sequencing reaction terminated with dideoxycytidine; pl. DMS-1 and pl. DMS-2, two exposures of an extension reaction using DMS-treated plasmid DNA; (+), extension reaction using DNA from wild-type KG1 cells treated with 0.5% DMS; 0 to 0.5% DMS, series of extension reactions on DNA extracted from *GS*-gene-amplified C4M CHO cells treated with the stated concentration of DMS. An alignment with the DNA sequence of the Chinese hamster *GS* promoter is shown. Note: 1. Spontaneous DNA nicking occurs at a high rate at the 3′ end of the CCAAT box and at the 5′ end of the Spl site; nicking is also enhanced for the bases around the site for the initiation of transcription. 2. Methylation (relative to naked plasmid DNA) of guanine −42 in the Spl site and guanine −32 5′ of the TATA box is enhanced. 3. One (or both) of adenines −65 and −66 (just 5′ of the CCAAT box) has vastly increased methylation sensitivity. 4. The 5′ end of the Spl site and the bases immediately 5′ appear to be protected against methylation.

In 1000-fold *GS*-gene-amplified cells, this technique is sufficiently sensitive to map sites of the rare nicks in DNA extracted from untreated cells. The results of one such experiment (Fig. 4) show hypermethylation of bases around the CCAAT box, in the Spl site, and around the TATA box in chromatin DNA in vivo, compared to the same sequences in naked plasmid DNA. Presumably these changes in alkylation sensitivity reflect alterations to the accessibility or environment of the bases caused by the binding of protein factors to the DNA. Although three factor-binding sites appear to be occupied in these *GS* promoters, no changes in alkylation sensitivity are seen around the site for the initiation of transcription. However, given the low level of expression of this gene in its housekeeping mode, and the stability of its mRNA, rates of initiation of transcription may be as low as once every 10 min or so, leading to a low occupancy of the initiation site by RNA polymerase complexes. Further studies on systems such as this may lead to a detailed mapping of the molecular interactions involved in the activity of promoters in vivo.

ACKNOWLEDGMENTS

It is a pleasure to acknowledge the help and advice of George Birnie, Dave Sherratt, Pete Sanders, Ashe Hussain, Bruce Hayward, Tim Harris, Chris Bebbington, Andy MacCabe, Craig Gibbs, and Dave Thom at various stages in the evolution of the studies reviewed here.

REFERENCES

1. Krebs, H. Glutamine metabolism in the animal body, in *Glutamine Metabolism, Enzymology and Regulation* (J. Mora and R. Palacios, eds.), Academic Press, New York, 1980, pp. 319–329.
2. Almassy, R. J., Janson, C. A., Hamlin, R., Xuong, N.-H., and Eisenberg, D. Novel subunit–subunit interactions in the structure of glutamine synthetase. *Nature*, *323*: 304–309 (1986).
3. Hayward, B. E., Hussain, A., Wilson, R. H., Lyons, A., Woodcock, V., McIntosh, B., and Harris, T. J. R. The cloning and nucleotide sequence of cDNA for an amplified glutamine synthetase gene from the Chinese hamster. *Nucleic Acids Res.* *14*:999–1008 (1986).
4. Kuo, C. F., and Darnell, J. E., Jr. Mouse glutamine synthetase is encoded by a single gene that can be expressed in a localized fashion. *J. Mol. Biol.* *208*:45–56 (1989).
5. Tischer, E., DasSarma, S., and Goodman, H. M. Nucleotide sequence of an alfalfa glutamine synthetase gene. *Mol. Gen. Genet.* *203*:221–229 (1986).
6. Ronzio, R. A., Rowe, W. B., and Meister, A. Studies on the mechanism of inhibition of glutamine synthetase by methionine sulfoximine. *Biochemistry*, *8*:1066–1075 (1969).

7. Sanders, P. G., and Wilson, R. H. Amplification and cloning of the Chinese hamster glutamine synthetase gene. *EMBO J. 3*:65–71 (1984).
8. Trask, B. J., and Hamlin, J. L. Early dihydrofolate reductase gene amplification events in CHO cells usually occur on the same chromosome arm as the original locus. *Genes Dev. 3*:1913–1925 (1989).
9. Ford, M., and Fried, M. Large inverted duplications are associated with gene amplification. *Cell, 45*:425–430 (1986).
10. Donn, G., Tischer, E., Smith, J. A., and Goodman, H. M. Herbicide-resistant alfalfa cells: An example of gene amplification in plants. *J. Mol. Appl. Genet. 2*:621–635 (1984).
11. Bebbington, C. R., and Hentschel, C. C. G. The use of vectors based on gene amplification for the expression of cloned genes in mammalian cells, in *DNA Cloning: A Practical Approach*, Vol. 3 (D. M. Glover, ed.), IRL Press, Oxford, 1987, pp. 163–188.
12. Cockett, M. I., Bebbington, C. R., and Yarranton, G. T. High level expression of tissue inhibitor of metalloproteinases in Chinese hamster ovary cells using glutamine synthetase gene amplification. *Bio/Technology, 8*:662–667 (1990).
13. Davis, S. J., Ward, H. A., Puklavec, M. J., Willis, A. C., Williams, A. F., and Barclay, A. N. High level expression in Chinese hamster ovary cells of soluble forms of CD4 T lymphocyte glycoprotein including glycosylation variants. *J. Biol. Chem. 265*:10410–10418 (1990).
14. Bhandari, B., Wilson, R. H., and Miller, R. E. Insulin and dexamethasone stimulate transcription of an amplified glutamine synthetase gene in CHO cells. *Mol. Endocrinol. 1*:403–407 (1987).
15. Mueller, P. R., Salser, S. J., and Wold, B. Constitutive and metal-inducible protein: DNA interactions at the mouse metallothionein I promoter examined by in vivo and in vitro footprinting. *Genes Dev. 2*:412–427 (1988).

III

Amplification of Transferred Genes

25

Amplification and Expression of Transfected Genes in Mammalian Cells

Randal J. Kaufman *Genetics Institute, Cambridge, Massachusetts*

I. INTRODUCTION

The ability to engineer mammalian cells to produce specific proteins has had tremendous impact on our understanding of cell biology and gene expression, as well as protein structure and function; in addition, proteins that were previously available in only limited quantity can now be obtained in large quantities. The ability to select for stable integration of plasmid DNA into the host chromosome has permitted the generation of stably transfected cell lines that indefinitely express the desired gene product. High level expression of transfected DNA can be obtained by selection for amplification of the transfected DNA. This strategy has become frequently used to establish cell lines that express high levels of therapeutically useful proteins. Advantages of using mammalian cells for the expression of foreign genes include the following: (1) the expressed proteins can be readily synthesized and secreted into the growth medium; (2) protein folding and disulfide bond formation are usually like these processes in the natural protein, and therefore, the proteins are usually produced in a functional form and resistant to degradation; (3) glycosylation, both aparagine- and serine/threonine-linked, generally occurs at normal positions; (4) many other posttranslational modifications can occur; and (5) multimeric proteins can be correctly assembled. Although a number of different host–vector systems have been described, significant success has been obtained by cotransfection of Chinese hamster ovary cells (CHO) with the desired gene and an amplifiable selectable genetic marker such as dihydrofolate reductase (DHFR). In this case it is possible to select cells that have amplified copies of the *DHFR* gene by growth in the presence of methotrexate (MTX), a tight-binding inhibitor of DHFR. Among the advantages of using CHO cells for the production of therapeutically useful proteins are the following: (1) amplified genes are integrated into the host chromosome and thus are stably maintained even in the absence of continued drug selection, (2) a variety of proteins have been properly expressed and secreted at high levels in CHO cells, (3) the cells grow well in the absence of serum, (4) the cells can grow either attached or in suspension, (5) the cells have been scaled up to greater than 5000 liters, and (6) CHO cells are being used to produce several proteins that are approved for clinical use (e.g., tissue plasminogen activator and erythropoietin). This chapter reviews genetic engineering of mammalian cells using amplification of transfected genes, improved vectors for selection and coamplification of transfected genes, and methods to overcome factors limiting expression of transfected genes.

II. AMPLIFICATION OF TRANSFECTED GENES IN MAMMALIAN CELLS

A. The Utility of Gene Amplification to Improve Expression

Stepwise selection for growth of cultured animal cells in progressively increasing concentrations of MTX results in cells with increased levels of DHFR as a consequence of a proportional increase in the *DHFR* gene copy number [1]. Most, if not all, genes become amplified at some frequency (approximately 10^{-4}, although this number can vary extensively) (for review, see Ref. 2). Under appropriate selection conditions, which favor the growth of cells harboring amplification of a particular gene, a population of cells that contain the amplified gene will outgrow the general population. In the absence of drug selection, the amplified gene is most frequently lost.

Since the degree of gene amplification, in most cases, is proportional to the level of gene expression, it offers a convenient means to increase expression of any particular gene. Although the copy number of a wide variety of genes can be amplified as a consequence of applying appropriate selective pressures, for many genes direct selection methods are not available. In these cases it is possible to introduce the gene of interest with a selectable and amplifiable marker gene into the cell and subsequently select for amplification of the marker gene to generate cells that have coamplified the desired gene. Table 1 lists the selection systems for amplifiable genes that have demonstrated utility in transfection experiments. *DHFR* is the most widely used amplifiable marker gene for this purpose.

Several general features are characteristic of amplification of both endogenous genes and transfected genes in cultured mammalian cells. First, gene amplification is usually obtained after stepwise selection for increasing resistance. Second, the

Table 1 Amplifiable Selectable Markers for Mammalian Cells

Selection	Gene	Ref.
Methotrexate	Dihydrofolate reductase	20
Cadmium	Metallothionein	54
PALA	*CAD*	19
Adenosine, alanosine, and 2′-deoxycoformycin	Adenosine deaminase	55
Adenine, azaserine, and coformycin	Adenylate deaminase	56
6-Azauridine, pyrazofuran	UMP synthetase	57
Mycophenolic acid	Xanthine-guanine-phosphoribosyltransferase	58
Multiple drugs[a]	P-Glycoprotein 170	59
Methionine sulfoximine	Glutamine synthetase	60
β-Aspartyl hydroxamate or albizziin	Asparagine synthetase	61
α-Difluoromethyl-ornithine	Ornithine decarboxylase	62

[a]Adriamycin, vincristine, colchicine, actinomycin D, puromycin, cytocholasin B, emetine, maytansine, Baker's antifolate.

amplified DNA displays variable stability, and both size and structure of the amplified unit are variable. Finally, the stability, size, and structure of the amplified genes may change with time. These characteristics suggest that multiple mechanisms are likely responsible for gene amplification and stabilization.

B. Amplification upon Stepwise Selection

Gene amplification generally occurs upon stepwise selection for resistance to increasing concentrations of the selective agent. When larger selection steps are employed, mechanisms other than gene amplification are observed. For example, single-step, high level resistance to MTX may result in cells with an altered, MTX-resistant DHFR enzyme [3,4] or in cells that contain altered MTX transport properties [5,6]. Since gene amplification requires DNA replication and cell division, optimal amplification occurs when cells are subject to severe, but not absolute, growth-limiting conditions. In some cases resistant colonies may appear but fail to continue growing, whereas in other cases only after prolonged selection does the growth rate eventually return to that of the original parental line. These observations suggest that multiple genetic changes occur from the time a cell first acquires resistance to the time it is established as a well-proliferating, highly drug-resistant clonal cell line.

Many variables can influence the rate of gene amplification. A number of approaches have been taken to increase the frequency of gene amplification. Perturbation of DNA synthesis by treatment with hydroxyurea [7], aphidicolin [7],

UV gamma irradiation [8], hypoxia [9], carcinogens [10], and arsenate [11] can enhance the frequency of specific gene amplification 10–100-fold. Agents that promote cell growth, such as phorbol esters or insulin, can also increase the frequency of gene amplification [12]. Interestingly, cells resistant to one selective agent as a result of gene amplification exhibit increased frequencies for amplification of other genes [13]. This "amplificator" phenotype may be one particular property of cells selected for stepwise increasing levels of drug resistance. For example, this phenotype is associated with an altered DNA repair capacity [14]. Alternatively, it may result from a general disruption of DNA replication which occurs upon drug treatment [15]. The rate of amplification is higher in tumorigenic cells than in their nontumorigenic counterparts [16–18]. Chromosomal position of a transfected gene can also dramatically (> 100-fold) affect its frequency of amplification [19–21]. The role of cis-acting elements that affect gene amplification frequency is under investigation. A 1.45 kb middle repetitive element (termed HSAG) has been shown to increase the frequency of gene amplification more than 100-fold [22–24]. A 370 bp element from the nontranscribed spacer region of the murine ribosomal gene locus can also increase the copy number of a selectable marker when introduced into recipient cells [25]. Interestingly, both elements have AT-rich segments and contain inverted repeats. These features have been observed in hot spots for recombination in the region of the adenylate deaminase gene [26] and at novel joints produced by gene amplification [27,28]. Interestingly, these elements have also been observed at the *DHFR* origin of replication [29]. These cis-acting elements may provide origins of replication or recombinagenic sites for gene amplification.

C. Variable Stability, Size, and Structure of Amplified Genes

When drug-resistant cells are propagated in the absence of the selective agent, the amplified genes may be maintained or lost. Amplified genes in newly selected resistant cell lines are generally unstable. As cells are propagated for increasing periods of time in the presence of the selection agent, the amplified genes, whether transfected or endogenous, exhibit increased stability [30–32]. In some cases, the degree of stability of the amplified genes in the absence of drug selection may correlate with the localization of the amplified genes and the complexity of the associated karyotypic alterations. Amplified genes localized to double minute chromosomes are usually lost upon propagation in the absence of selection [32]. Double minute chromosomes are small paired extrachromosomal elements that lack centromeric function and thus segregate randomly at mitosis. Most evidence suggests that double minute chromosomes are circular DNA structures [33]. In contrast, stably amplified genes are usually integrated into the chromosome and frequently are associated with expanded chromosomal regions termed homogeneously staining regions (HSRs) [34]. This term reflects the lack of banding in these regions after Giemsa–trypsin banding procedures.

Amplified endogenous [35] as well as amplified transfected heterologous [36] *DHFR* genes in MTX-resistant mouse fibroblasts are usually associated with double minute chromosomes. In contrast, amplified endogenous [34] and transfected heterologous [37] *DHFR* genes in MTX-resistant Chinese hamster ovary cells are frequently associated with expanded chromosomal regions. In a few cases, the amplified transfected DNA may be present with little cytological perturbation [37]. This may reflect a smaller size of the amplified unit within the chromosome. CHO cells selected in the presence of MTX for amplification of heterologous *DHFR* genes have generally proven to be very stable upon propagation in the absence of MTX [37]. Instability associated with chromosomal amplified genes may correlate with more complex chromosomal alterations. For example, cells selected for amplified genes may become tetraploid, a characteristic possibly associated with increased instability. In other cases, transfected and amplified DNA may be observed in HSRs associated with extremely large or dicentric chromosomes (Fig. 1). This observation led to a proposal that bridge–breakage–fusion cycles could be important in generating chromosome aberrations and perhaps duplication and instability in DNA transfer experiments. Recombination of exogenous DNA with a chromosome might displace or disturb function of a telomere, thus rendering the chromosome susceptible to instability. After chromosome replication, both sister chromatids would contain abnormal termini, which might spontaneously fuse to form chromosomal loops or circles, and eventually dicentric chromosomes. Repeated cycles would generate duplicated inverted chromosomal regions, and stabilization could result from acquisition of telomere function. Chromosomal inversions and translocation have been observed in DNA transfection experiments [37–40]. Indeed, inverted duplications have been associated with gene amplification in several cases [41–43]. An early change detected upon amplification of the endogenous *DHFR* locus was a chromosome break with subsequent deletion [44]. These findings suggest that chromosome breakage may be an initial event in the amplification process. In contrast, studies on aspartate transcarbamylase (*CAD*) gene amplification upon selection for resistance to *N*-(phosphonacetyl)-L-aspartate (PALA) suggest that the initial event may yield tandem duplications of very large regions of DNA (10 mb), possibly by sister chromatid exchange [45].

The size of the amplified unit estimated by cytogenetic analysis is highly variable and may range from fewer than 100 to more than 500 kbp [34,41]. During propagation of cells in culture, the amplified DNA may undergo changes that involve (1) extrachromosomal circles becoming larger through tandem duplications [46,47], (2) the possibility that the positions of novel recombination break points between amplified units will change [47,48], or (3) the propagation of

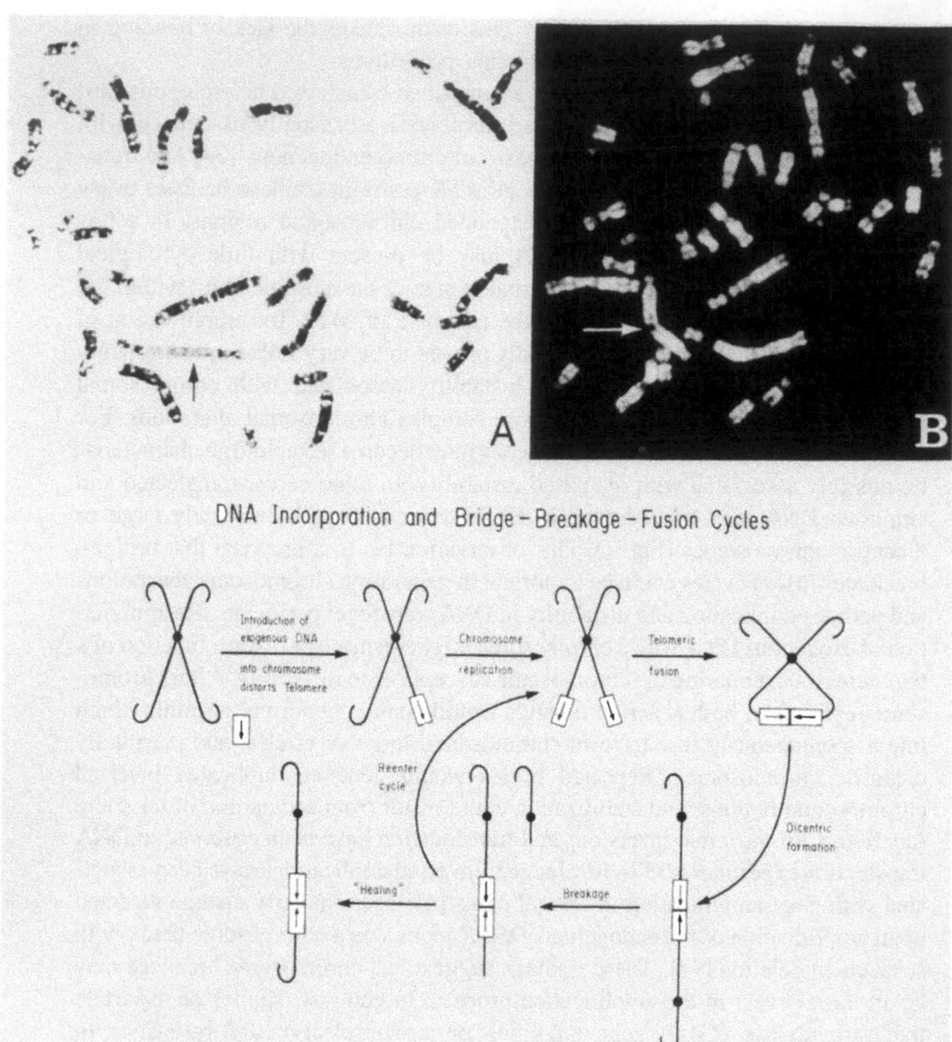

Figure 1 DNA integration and bridge–breakage–fusion cycles. (A) Metaphase spread of a cell containing transfected and amplified *DHFR* genes. The chromosomes are stained with Giemsa–trypsin and show dicentric chromosome containing an HSR. (B) Metaphase spread of a cell containing transfected and amplified *DHFR* genes in a broken dicentric chromosome. The spread was stained with quinacrine. (C) Model for bridge–breakage–fusion cycle in generating gene duplication.

mutations from one amplified unit to other units within an amplified array, possibly by gene conversion [49].

Insight into potential mechanisms that mediate stabilization of amplified genes have resulted from an analysis of the amplification of a transfected *CAD* gene [21,42]. In this case, the integrated *CAD* gene first is deleted and then undergoes replication as an extrachromosomal element. The extrachromosomal intermediate may increase in size by forming multiples of itself to eventually create double minute chromosomes. The extrachromosomal sequence may eventually integrate to form a homogeneously staining chromosomal region.

III. IMPROVED VECTORS FOR COAMPLIFICATION

A. A Historical Perspective

Most methods of DNA transfer yield DNA incorporation into 5–50% of the cells in the population. These cells express the DNA transiently for several days to several weeks. However, the DNA is eventually lost from the cell population unless some selection procedure is used to isolate cells that have stably integrated the foreign DNA into their genome. The limiting event for obtaining stable DNA transfer is the frequency of DNA integration into the chromosome of the cell, not the efficiency by which cells uptake DNA. Different cell lines as well as different transfection methods have been reviewed [50] which yield dramatically different efficiencies with respect to the frequency of stable integration, as well as to the amount of foreign DNA incorporated. The ability to select for incorporation of one gene by selection for a second cotransfected gene has been termed cotransformation [51]. In cotransformation, separate DNA molecules become ligated together inside the cell and subsequently cointegrate as a unit via nonhomologous recombination into the host chromosome. If DNA molecules are linked together within the chromosome, selection for amplification of one of the molecules usually results in coamplification of the linked DNA molecule. Since different cell lines and DNA transfection methods exhibit different potentials for cotransformation, approaches have been taken to obviate the requirement for linkage of the two separate vectors within the cell. First, separate DNA molecules may be ligated in vitro prior to transfection. By varying the ratio of input DNA molecules, it is possible to optimize the copy number of the desired expression vector relative to the selection marker [52]. As an alternative approach, vectors have been constructed that contain both the selection gene and a transcription unit for the desired gene within the same plasmid. For example, pRK1-4 [53] contains a *DHFR* transcription unit under control of the SV40 early promoter and cloning sites for insertion of cDNA clones to be transcribed from the adenovirus major later promoter. Even with the selection gene linked to the product gene, however, there is potential to amplify only the selectable marker gene without the product gene.

B. Generation of Efficient Dicistronic mRNA Expression Vectors

A more direct approach to link expression of the selectable amplifiable marker with the product gene is through the use of dicistronic mRNA expression vectors. Dicistronic mRNA expression vectors express the selection gene from the same transcription unit as the product gene. Because dicistronic mRNAs can translate two nonoverlapping open reading frames (ORFs) in mammalian cells [63], it is possible to select for the low level of internal translation initiation from an ORF contained within the 3′ end of the mRNA. The utilization of dicistronic mRNA expression vectors containing the desired gene in the 5′ position and a selectable gene in the 3′ position has proved to be a successful approach to the derivation of cells that express high levels of the gene product in the 5′ position [64,65]. In these vectors, translation of the desired ORF in the 5′ position within the mRNA is efficient while translation of the selectable marker within the 3′ position is very inefficient. Thus, cells require higher levels of the dicistronic mRNA to survive selection and synthesize high levels of the desired product encoded by the 5′ ORF.

Although dicistronic mRNA expression vectors have proved to be useful, their general utility is limited for several reasons. First, the frequency of translation initiation at the 3′ ORF is very low: approximately 1% of the frequency of initiation at the 5′ ORF [64,66]. Thus, in some cases it is not possible to produce adequate levels of the selection marker product to survive selection. Second, insertion of specific RNA sequences or AUG codons upstream may dramatically alter the efficiency of initiation at the downstream ORF. It has been reported that a minimal distance between the termination codon of the upstream ORF and the initiation codon of the downstream ORF is 100 bases [66]. Finally, experience with dicistronic *DHFR* amplification vectors has shown that deletions and/or rearrangements within the 5′ ORF frequently occur, presumably to allow more efficient translation of the selectable marker encoded by the 3′ ORF. These deletions and rearrangements may be monitored by Northern blot hybridization analysis. For example, independent clones were selected for *DHFR* expression using the pMT21 vector, which contains a cDNA encoding the α subunit of the translation initiation factor 2 (eIF-2α) [67]. This vector is designed to express a dicistronic mRNA that encodes eIF-2α in the 5′ position and *DHFR* in the 3′ position. Nineteen independent clones were derived by transfection of *DHFR*-deficient CHO cells and selection for growth initially in the absence of nucleosides, and further amplified by selection for increased *DHFR* expression by growth in 0.02 μM MTX. RNA isolated from these clones was subjected to Northern blot analysis to identify the presence of multiple mRNA species that hybridize to a *DHFR* probe (see below: Fig. 3, bottom). Frequently two hybridizing species are detected. The higher molecular weight species corresponds to the correctly sized dicistronic mRNA and also hybridizes to an eIF-2α probe (not shown). The multiple lower molecular weight mRNA species does not hybridize to an eIF-2α probe and likely represents mRNA derived from a deleted or rearranged vector. The deletions that occur generate *DHFR* mRNAs that differ slightly in mobility between the different clones. It is difficult to know whether these deletions occur as a result of selection against high level expression of the ORF in the 5′ position, or whether deletion improved DHFR translation to survive the MTX selection.

The limitations of the dicistronic expression vectors can be overcome by the use on an internal ribosome binding site to promote efficient internal initiation within the mRNA. The 5′ untranslated region from picornaviruses such as poliovirus [68] and encephalomyocarditis (EMC) virus [69] can promote ribosome binding at internal sites within an mRNA. In constructs that contain two ORFs, insertion of the picornavirus leader upstream from the second ORF results in efficient translation of that ORF independently of the presence of the first ORF. A new set of mammalian cell expression vectors have been generated that utilize the 5′ untranslated region from EMC virus to promote efficient internal translation initiation of selectable markers encoding *DHFR* [70] (pED, Fig. 2), a methotrexate-resistant *DHFR* [70]; (pED-*mtx*^{r}), neomycin phosphotransferase [71] (pEMC-*neo*), and adenosine deaminase (Michnick and Kaufman, unpublished; pEA). The use of pED is generally limited to *DHFR*-deficient cells. pEMC-*neo*, pEA, and pED-*mtx*^{r} may be used as dominant selectable markers in multiple types of cells. These vectors permit the rapid derivation of stably transfected cell lines that express high levels of the desired product. Figure 3 (top) shows Northern blot analysis of clones and several pools of transformants selected for *DHFR* expression using the pED vector containing the eIF-2α cDNA insert. In contrast to analysis of clones isolated using the pMT21 vector (Fig. 3, bottom), all clones as well as several pools of transformants derived from the pED vector express primarily a single mRNA of the expected size. The same mRNA is also detected by hybridization to an eIF-2α probe (not shown). Because there is no selective advantage for deletion of the ORF contained in the 5′ position, these vectors do not undergo deletion upon selection for further increases in expression. If deletion within the 5′ ORF occurs, it is a good indication that the gene may be detrimental to cell growth. The EMC series of vectors will be generally useful to rapidly obtain high level amplification and expression of heterologous genes in mammalian cells.

IV. USE OF GENE AMPLIFICATION FOR GENETIC MANIPULATION OF MAMMALIAN CELLS TO IMPROVE CELLULAR PRODUCTIVITY

A. Rate-Limiting Steps in Expression of Heterologous Genes

Many genes can be transfected, amplified, and expressed at high level in mammalian cells [72–78]. To identify potential limitations for expression of an

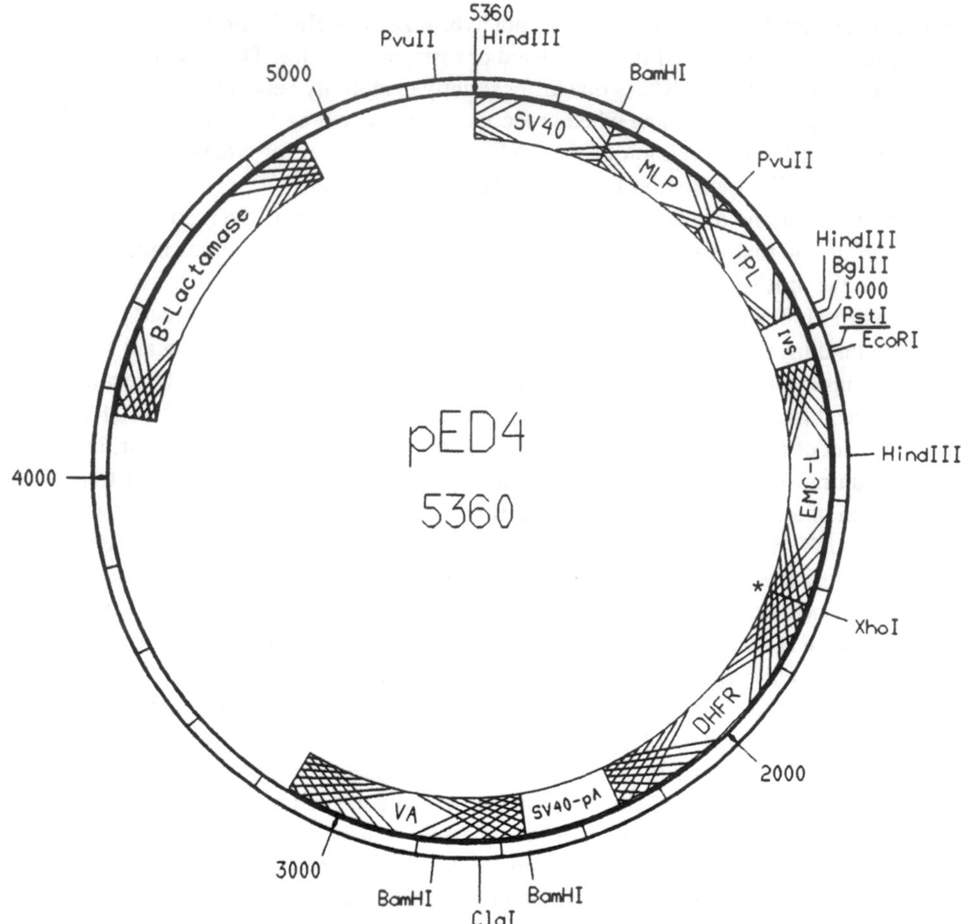

Figure 2 Map of dicistronic expression vector pED4. The components of the 5360 bp pED4 in the puc18 background are depicted as follows: SV40, origin of replication and enhancer element; MLP, the adenovirus major late promoter; TPL, the tripartite leader from adenovirus late mRNAs; IVS, an intervening sequence; *Pst* I and *Eco*RI, unique cloning sites; EMC-L, the 5′ untranslated region from EMC virus; DHFR, a murine *DHFR* coding region; SV40-pA, the SV40 early polyadenylation signal; VA, the adenovirus *VA* I RNA gene, and β-lactamase, a selectable gene for propagation in *E. coli*.

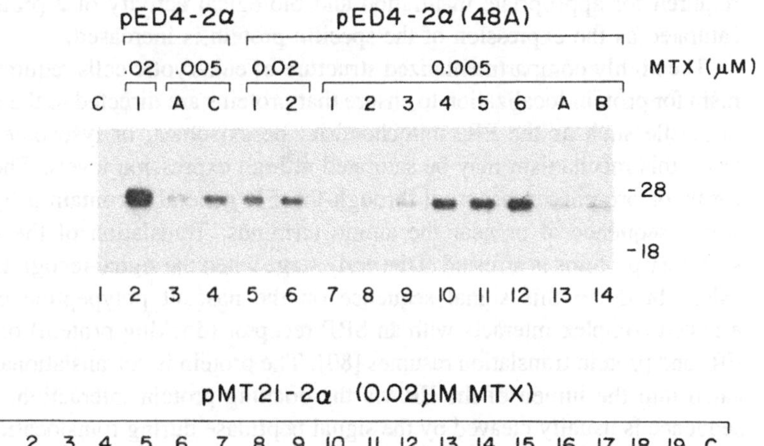

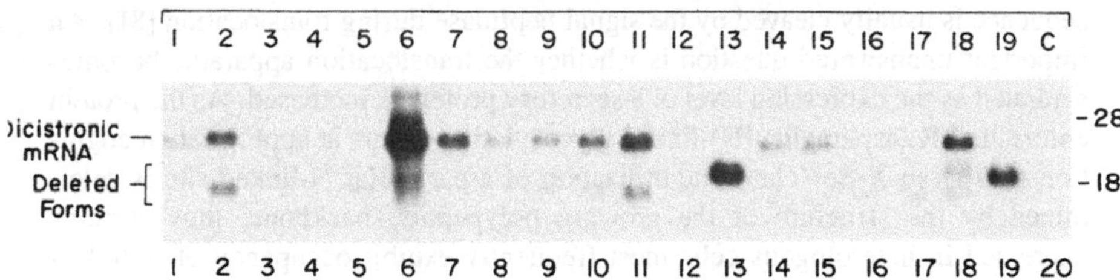

Figure 3 eIF-2α Expression from pMT21 and pED4 vectors in stably transformed cells. *DHFR*-deficient CHO cells were electroporated with plasmid DNA and pools (indicated by letters above lanes) or clones (indicated by numbers above the lanes) were selected for resistance to the concentration of MTX indicated. Total cell RNA was isolated and analyzed by Northern blot hybridization to a *DHFR* probe. Lane 1 in the top and lane 20 in the bottom represent RNA from control untransfected CHO cells.

amplified heterologous gene, it is essential to consider all aspects of gene expression. Gene expression may be limited at the level of transcription, processing of precursor mRNAs, mRNA transport to the cytoplasm, mRNA stability and translational efficiency, and protein stability. In many cases, however, protein expression is limited. In general, the limitation to high level expression of any heterologous gene in a mammalian cell host is very protein specific. Proteins that transit the secretory apparatus are subject to a variety of additional steps that may regulate the appropriate secretion of the mature polypeptide. Experience from studying the expression of a wide variety of proteins that transit the secretory apparatus demonstrates that the rate-limiting step in secretion is transport from the endoplasmic reticulum (ER) to the Golgi apparatus [79]. It is now known that expression of heterologous secretory proteins from transfected genes introduced into heterologous cells may also be limited as a result of inefficient transport from the ER to Golgi. In addition, a wide variety of posttranslational steps that are required for appropriate maturation and biological activity of a protein may be saturated as the expression of the specific protein is increased.

The highly compartmentalized structure of eukaryotic cells requires a mechanism for protein localization to ensure that proteins are directed to the appropriate organelle such as the ER, mitochondria, peroxisomes, or lysosomes. In many cases this mechanism may be saturated at high expression levels. The precursor forms of proteins transported through the ER generally contain a hydrophobic signal sequence at or near the amino terminus. Translation of the mRNAs of secretory proteins is arrested at the early stage when the signal recognition particle (SRP) binds to this signal sequence on the nascent polypeptide chain. This arrested complex interacts with an SRP receptor (docking protein) on the rough ER, and protein translation resumes [80]. The protein is cotranslationally translocated into the lumen of the ER via the docking protein interaction. The signal sequence is usually cleaved by the signal peptidase during translocation [81]. An important unanswered question is whether the translocation apparatus becomes saturated as the expression level of a secretory protein is increased. As the protein enters the ER, asparagine (N)-linked glycosylation occurs at appropriate recognition sites (Asn-X-Ser/Thr). The utilization of a particular N-linked site is determined by the structure of the growing polypeptide backbone; thus proteins expressed in heterologous cells most frequently exhibit occupancy of N-linked sites very similar to that of the native polypeptide [82]. As the protein is translocated into the lumen of the ER, it undergoes a complex series of reactions that result in attaining its final appropriate conformation (Fig. 4). These steps include protein folding [83], disulfide bond formation [84], oligomerization of

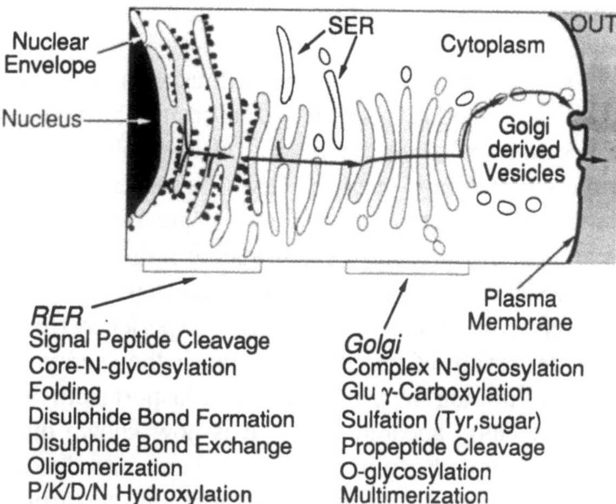

Figure 4 Posttranslational modification within the secretory pathway.

multisubunit proteins [83], hydroxylation of specific amino acids (e.g., proline, aspartic acid, asparagine, lysine) [85], trimming of glucose and mannose residues from the core N-linked oligosaccharides [82], and fatty acid acylation [86]. The culmination of these steps results in the vesicle-mediated transport of the polypeptide from the ER to the *cis*-Golgi compartment [87,88]. As the protein transits through the Golgi complex, further modifications occur, including addition of *N*-acetylglucosamine, galactose, and sialic acid to N-linked carbohydrate [82]; serine and threonine *O*-linked glycosylation; sulfation of tyrosine [89] and carbohydrate residues; phosphorylation; cleavage of propeptides [90]; and perhaps γ-carboxylation of vitamin-K-dependent proteins [91]. When the protein reaches the *trans*-Golgi network, there exist three different pathways [92]: (1) membrane-bound and secreted proteins are constitutively transported to the cell surface, (2) proteins may enter the regulated secretory pathway and be stored in vesicles for release upon stimulation, or (3) proteins may be targeted to lysosomal structures. All cells contain a constitutive secretory pathway. Only some cells have the capacity to store proteins in specialized granules and release them upon stimulation. This regulated pathway is required to mediate correct processing of proinsulin to insulin and proopiomelanocortin to ACTH. Thus, certain proteins require this specialized secretory pathway to mediate appropriate maturation of the precursor polypeptides.

B. The Rate of Protein Folding in the Endoplasmic Reticulum and the Role of Immunoglobulin Binding Protein (BiP)

Different proteins exhibit dramatically different rates of transit through the ER. It is not known what structural features within a protein are responsible for the individual rates of transit. However, most results suggest that the rate by which a protein attains its appropriate conformation is a crucial limiting factor in the rate of exit from the ER [93,94]. Experiments with mutant proteins demonstrate that altered protein structures may result in improper folding. An improperly folded protein may exhibit altered solubility properties, resulting in its aggregation and retention within the ER. An important unanswered question is: What role do proteins within the ER play to facilitate protein folding? The most extensively studied protein in this class is the glucose-regulated protein of 78 kDa (GRP78), or also known as immunoglobulin binding protein (BiP). BiP is a lumenal protein of the ER homologous to hsp70 [95]. BiP is a member of the glucose-regulated protein family whose expression is induced by glucose starvation, treatment with inhibitors of N-linked glycosylation, or the presence of misfolded proteins within the ER [96–98]. BiP has been shown to bind ATP, and it has an ATPase activity [99].

The potential importance of BiP in secretion is assumed because misfolded, improperly glycosylated, or incompletely folded or assembled proteins are detected in association with BiP [100–103]. Stable association correlates with retention in the ER. This stable association may serve to target secretion incompetent proteins for degradation. Other proteins may be detected in a transient association with BiP [102,104–106]. In this case, interaction with BiP may be part of the normal pathway leading to secretion. For example, vesicular stomatitis virus glycoprotein (VSV G) exhibits a transient association with BiP [104]. However, it is only the forms of VSV G that are intermediates in the disulfide bonding process that are detected in association with BiP. After appropriate disulfide bonds form, the BiP-associated VSV G is released and then transits to the Golgi compartment.

The role of BiP association in secretion in CHO cells was studied by expressing high levels of proteins that display different degrees of association with BiP [102,105]. The proteins evaluated were tissue plasminogen activator, human coagulation factor VIII, a B domain deletion mutant of factor VIII (VIII-LA), human von Willebrand factor (vWF), and macrophage colony stimulating factor (M-CSF) (Table 2). Wild-type tissue plasminogen activator (tPA) expressed at high levels by coamplification with *DHFR* in CHO cells is detected only in a slight transient association with BiP. In contrast, similar amplification and expression of unglycosylated tPA (tPA-3X), prepared by mutagenesis of asparagines at the three consensus N-linked sites, results in a greater proportion of unglycosylated tPA in both stable and transient association with BiP [102]. Factor VIII and vWF were coexpressed in the same cell using two independent amplifiable markers. Factor

Table 2 BiP Association and Effect on Secretion of Increased and Reduced BiP Levels in CHO Cells[a]

		Effect of altered BiP level[b]	
Protein	BiP Association	Increased BiP	Reduced BiP
tPA-3X	Stable/transient	−−	++
WT FVIII	Stable/transient	n.d.	++
VIII-LA	Transient	−−−	++
vWF	Slight transient	−	n.d.
M-CSF	Nondetectable	None	n.d.

[a]The relative increase (+) or decrease (−) in secretion is depicted for different proteins that exhibit different extents of association with BiP. tPA-3X represents either wild-type tPA expressed in the presence of tunicamycin to block N-linked glycosylation or also a mutant with the three potential N-linked glycosylation sites mutagenized to glutamines. VIII-LA represents a factor VIII harboring an 880 amino acid deletion within the heavily glycosylated B domain.
[b]n.d., not determined.

VIII was coamplified with *DHFR* and vWF was coamplified with adenosine deaminase. In these factor VIII and vWF coproducing cells, the vWF serves to stabilize factor VIII upon secretion into the conditioned medium [105]. In these cells, factor VIII exhibits both transient and stable association with BiP, whereas vWF is detected only in a slight transient association with BiP [102]. The factor VIII B domain deletion mutant VIII-LA was also coamplified with *DHFR* in CHO cells. Expression of this mutant is fivefold greater than wild-type factor VIII, and this correlates with a less stable, more transient association with BiP. It is speculated that the lesser association of VIII-LA results from an increased ability for it to attain an appropriate conformation for release from BiP compared to wild-type factor VIII. M-CSF was coamplified with *DHFR* and it cannot be detected in association with BiP.

The effect of expression level of the BiP-associated protein was studied by evaluating secretion and BiP association as the expression levels of wild-type and mutant tPA were increased through increasing degrees of coamplification with *DHFR* [102]. Cells were obtained that expressed increasing amounts of the wild-type or mutant tPA. Increased synthesis of wild-type tPA correlated with increased secretion. In contrast, as the synthesis of unglycosylated tPA-3X was increased, a greater percentage of the newly synthesized tPA-3X was retained in a stable complex with BiP within the cell. Thus, increased expression of a secretory protein found in association with BiP resulted in a reduced efficiency of secretion and increased association with BiP.

The role of BiP in protein secretion was studied by altering the level of BiP and monitoring its effect on secretion of the different proteins that display different degrees of BiP association. To elicit a reduction of BiP within the cell, an antisense RNA strategy was employed [107]. Coamplification of an antisense BiP expression vector with adenosine deaminase yielded a three- to fourfold reduction in the steady state levels of BiP mRNA in a cell line expressing unglycosylated mutant tPA-3X. In these cells secretion of unglycosylated tPA-3X was improved several-fold by reduction of BiP. The secreted tPA exhibited a specific activity similar to the protein secreted in the absence of antisense BiP mRNA. Thus, it did not appear that unfolded tPA-3X was being secreted by reducing BiP level. Most dramatically, in the cells with reduced BiP, the newly synthesized tPA exhibited a lesser and transient association with BiP. In addition, expression of factor VIII in cells similarly reduced in BiP level resulted in a two- to fourfold increase in factor VIII secreted into the growth medium. These results (summarized in Table 2) both demonstrate that a reduction in BiP level can improve secretion of an associated protein and implicate BiP as playing a role in preventing secretion of associated proteins.

To increase the level of BiP, cells were selected for coamplification with *DHFR* and a BiP expression plasmid [108]. Surprisingly, BiP overexpression in these cells interfered with the transcriptional induction of all glucose-regulated proteins mediated by treatment with tunicamycin or the calcium ionophore A23187. This suggests that the overexpressed BiP can relieve the stress induced by these treatments, or that BiP is directly involved in signaling transcriptional induction of the glucose-regulated protein genes. In these cells the synthesis and secretion of factor VIII-LA, vWF, and M-CSF was studied. The results demonstrated that an increase in the steady state level of BiP extended the association of vWF and VIII-LA with BiP and prevented their secretion. In contrast, no effect on M-CSF expression was observed (Table 2). These results suggest that BiP serves as a retention mechanism for the selective retention of specific proteins in the ER. From these experiments, there is no indication that BiP serves to facilitate protein folding.

Is it possible to increase expression of a BiP-associated protein to a point at which newly synthesized protein will bypass association with BiP? This question was addressed by obtaining high level expression of factor VIII and vWF through coamplification with *DHFR* and adenosine deaminase as described above. After amplification of the factor VIII and vWF genes for expression at a high level, the factor VIII and vWF mRNA levels were further increased by addition of sodium butyrate to the conditioned medium for 24 hr. The factor VIII and vWF expression vectors can be transcriptionally induced by addition of sodium butyrate to the conditioned medium [109]. The factor VIII mRNA was induced 10-fold and the vWF mRNA was induced 4-fold. Analysis of the primary translation products demonstrated that factor VIII and vWF protein synthesis was elevated proportionally to the increase in mRNA. Although increased levels of vWF synthesis correlated with increased levels of vWF secreted into the conditioned medium, the majority of newly synthesized factor VIII was retained within the cell and was not secreted. Within the same cell, significant association of factor VIII with BiP did not significantly affect vWF secretion or vWF association with BiP. This suggests that factor VIII and vWF transit independently through the secretory pathway and that a block to one protein does not interfere with others. Most dramatically, sodium butyrate induction of transcription from the factor VIII gene in CHO cells elicited an 80-fold induction of BiP mRNA. In contrast, the induction of vWF alone did not dramatically induce BiP mRNA [109]. These results suggest that an increase in the expression of a BiP-associated protein cannot circumvent BiP association.

Since induction of factor VIII synthesis resulted in a dramatic BiP induction, it appeared that factor VIII synthesis may result in a stress response within the ER. Upon transmission electron microscopic observation, the CHO cells induced to synthesize factor VIII exhibited a dilated ER juxtaposed to many mitochondria (Fig. 5B) compared to control cells not treated with sodium butyrate (Fig. 5A). The presence of mitochondria in close proximity to the ER suggested that factor VIII synthesis and secretion may require high levels of ATP.

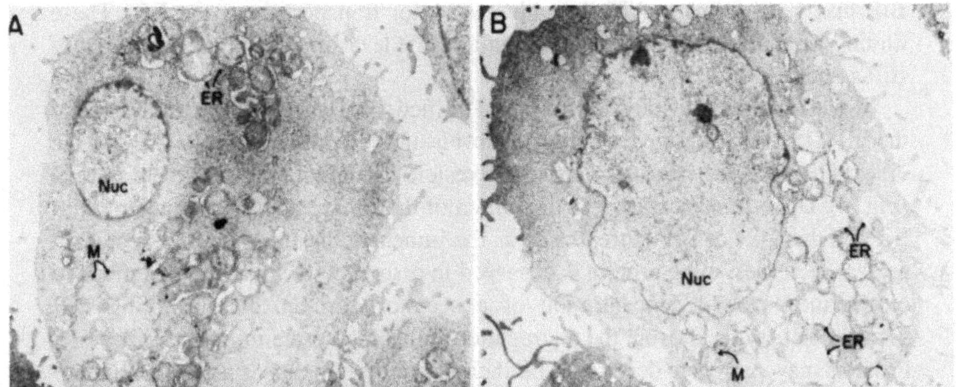

Figure 5 Electron microscopy of CHO cells induced to synthesize factor VIII. CHO cells in the absence (A) and presence (B) of induced factor VIII secretion were prepared for electron microscopy.

C. Some Proteins Require ATP-Dependent Release from BiP for Secretion

The importance for ATP in dissociation of protein-BiP complexes has been shown in vitro [110]. Complexes of BiP with immunoglobulin heavy chain immunoglobulin [95] or factor VIII (A. Dorner, unpublished observations) exhibit an ATP-dependent release from BiP upon incubation in vitro. The importance of ATP availability for the secretion of BiP-associated proteins in vivo was studied by examining the effect of depleting intracellular ATP levels through treatment of CHO cells with carbonylcyanide *m*-chlorophenylhydrazone (CCCP), an uncoupler of oxidative phosphorylation that reversibly inhibits transport from the ER [111–113]. Cells coexpressing factor VIII and vWF were pulse labeled with [^{35}S]-methionine and then chased in the presence of various concentrations of CCCP [114]. The results demonstrated that secretions of factor VIII and vWF synthesized in the same cells display different sensitivities to inhibition by CCCP. Low concentrations of CCCP (10 μM) inhibited secretion of factor VIII but not vWF coexpressed in the same cells. The finding that secretion of factor VIII but not vWF was inhibited at low CCCP concentration precludes a general disruption of secretion. Further studies demonstrated that this inhibition correlated with a block to dissociation of the BiP/factor VIII complex. Much higher concentrations of CCCP (200 μM) were required to block secretion of vWF in these cells. In a similar manner, secretion of unglycosylated tPA was also inhibited by low levels of CCCP compared to wild-type tPA [114]. These results indicate that secretion of BiP-associated proteins may be more sensitive to perturbation of intracellular ATP levels by CCCP treatment than unassociated proteins. In vivo dissociation from

BiP may be a primary ATP-dependent step for transport from the ER. These findings suggest that increased intracellular ATP levels may facilitate secretion of BiP-associated proteins.

Two ATP-requiring steps have been identified for transport through and out of the ER [115,116]: one step requires high levels of ATP for release from BiP and the other is likely at the stage of transitional vesicle formation upon exit from the ER (Fig. 6). Upon translocation into the lumen of the ER, proteins may interact with BiP. As the level of BiP increases or as the lumenal concentration of a secretory protein increases, more protein is observed in association with BiP. The influence of the intralumenal concentration of a secretory protein on promoting BiP association may result from the aggregation state of the protein transiting the ER. Proteins may aggregate depending on their solubility properties and intralumenal concentration. These aggregates may more avidly bind to BiP. Conversely, increased levels of BiP may promote association and aggregation of proteins transiting the ER which are incompletely folded or assembled. BiP-associated protein requires ATP hydrolysis for dissociation. If ATP is limiting, proteins associated with BiP are trapped and blocked in their transit through the ER. Proteins associated stably with BiP are likely destined for degradation. Analysis of the kinetics of appropriate disulfide bond formation and BiP release of VSV G protein suggest that folding and disulfide bond formation must precede release

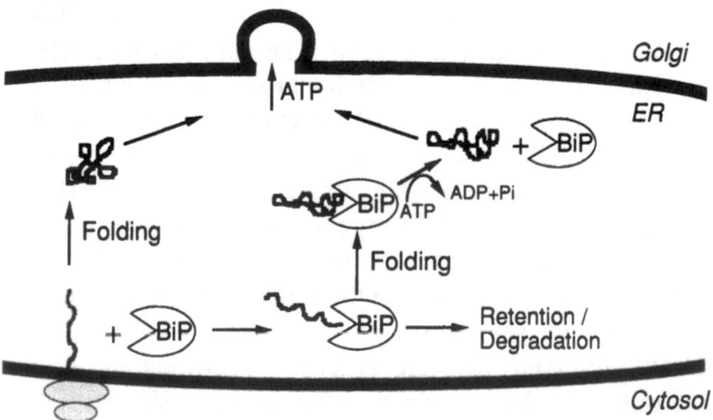

Figure 6 ATP-requiring steps for transit of secretory proteins from the endoplasmic reticulum: processes of protein translocation into the lumen of the ER, protein folding, and exit from the ER. ATP-dependent release from BiP is required for transport of proteins that exhibit significant association with BiP. It is not known whether protein folding also requires ATP or is facilitated by BiP. Vesicle-mediated transport from the ER is a second ATP-requiring step. Proteins found in significant association with BiP may be destined for degradation.

from BiP [104]. However, it is not known whether ATP hydrolysis mediated by BiP actually assists in protein folding. Another lumenal protein, protein disulfide isomerase (PDI), has been shown to assist the formation of appropriate disulfide bond [117]. It is not known whether all proteins exhibit BiP association because it is not possible to detect association of some proteins under the conditions used. Whether some proteins can transit the ER independent of BiP association is still an unanswered question.

D. Posttranslational Modifications That Limit Secretion

Most proteins synthesized in heterologous cells frequently exhibit the same modifications that occur on the native protein. However, in a number of situations the posttranslational modification may be inefficient. For example, the coagulation factor IX is a serine protease that exhibits γ-carboxylation of 12 glutamic acid residues within the amino-terminal region of the protein. These carboxyl groups are transferred by a vitamin-K-dependent reaction, and their presence is required for factor IX activity [85]. Expression of factor IX in CHO cells has resulted in high levels of factor IX antigen, but negligible factor IX activity possibly attributable to saturation of the carboxylase activity.

More significant is the observation that as expression of propolypeptides increases, saturation of the propeptide cleaving enzyme may occur [119–123]. Proteolytic processing of proproteins is an event that occurs late in both the regulated and constitutive secretory pathways of many different cell types. Expression of heterologous proteins in foreign cells has shown that the propeptide is usually cleaved correctly. However, cleavage is often incomplete. The inefficiency of cleavage may represent saturation of the endoproteolytic cleaving activity of the cell which is a result of large amounts of protein expressed. Alternatively, a portion of the expressed proprotein may bypass the compartment containing the endogenous protease, or its rate of transit through that compartment may be too rapid for complete cleavage to occur. One approach to the circumvention of this limitation to high level expression is to overexpress a propeptide processing enzyme in order to improve processing of proproteins expressed in heterologous cells.

The best characterized proprotein processing enzyme is the Kex2 endoprotease, which is responsible for appropriate processing of prepro-killer toxin and prepro-α-factor in *Saccharomyces cerevisiae* [124,125]. Kex2 cleaves after paired basic amino acid residues Lys-Arg and Arg-Arg (126,127]. It can accurately process proinsulin and proalbumin expressed in yeast [128,129] and can process proopiomelanocortin (POMC) upon transfection into mammalian cells [130]. Based on sequence homology to Kex2, a cDNA encoding a human homologue of Kex2 was isolated from a cDNA library prepared from the human HepG2 cell line

[131]. The deduced amino acid sequence demonstrated that the protein contained the conserved residues for the subtilysin type of serine protease. In addition, the protein contained a transmembrane domain, likely responsible for its presumed localization in the *trans*-Golgi. This protease has been given the acronym PACE (Paired basic Amino acid Cleaving Enzyme). Overexpression of PACE in mammalian cells can enhance the processing of the propeptide of human von Willebrand factor [131]. Thus, the expression of PACE in mammalian cells should allow the production of more completely processed protein products. In a similar manner, the identification and overexpression through gene amplification of other potentially rate-limiting enzymes involved in posttranslational modifications offers a solution to the inefficiency observed for these processes.

V. CONCLUSION

The ability to isolate and express genes has made available for pharmaceutical use unlimited quantities of proteins that previously could be isolated only in small quantities. Since many of the proteins of considerable interest contain multiple posttranslational modifications, most success at producing biologically active forms has come from expression of the desired gene in mammalian host cells. The use of gene amplification to increase expression level has provided a vital tool for obtaining high levels of expressed protein. However, there are still many obstacles to the eventual use of these proteins for pharmaceutical purposes. The limitations lie primarily in the production of the proteins in a cost-effective manner and the evaluation the efficacy of these molecules in carefully designed clinical studies. Future developments to improve production of these complex proteins will involve modification of the host cell machinery to properly and more efficiently process and secrete these molecules. Gene amplification offers an important approach to the expression of desired genes to improve properties of the host cell. These improvements are presently being complemented by advances in biochemical engineering that enable us to grow mammalian cells in very large volumes and at very high densities with reduced serum requirements. As a consequence, the cost for production of gram quantities of a protein from mammalian cells is approaching the cost for proteins derived from microbial systems, with all the advantages that mammalian systems afford.

ACKNOWLEDGMENTS

I gratefully thank my colleagues Andrew Dorner, Monique Davies, Debra Pittman, Louise Wasley, David Israel, Clive Wood, Patty Murtha-Riel, and Donna Michnick for their many contributions to the work described here. I thank Joanne Janeiro for assistance in preparation of this chapter.

REFERENCES

1. Alt, F. W., Kellems, R. E., Bertino, J. R., and Schimke, R. T. Selective multiplication of dihydrofolate reductase genes in methotrexate-resistant variants of cultured murine cells. *J. Biol. Chem.* *253*:1357–1370 (1978).
2. Schimke, R. T. Gene amplification in cultured cells. *J. Biol. Chem.* *263*:5989–5992 (1988).
3. Flintoff, W. F., Davidson, S. V., and Siminovitch, L., Isolation and partial characterization of three methotrexate-resistant phenotypes from Chinese hamster ovary cells. *Somatic Cell Genet.* *2*:245–261 (1976).
4. Haber, D. A., Beverley, S. M., and Schimke, R. T. Properties of an altered dihydrofolate reductase encoded by amplified genes in cultured mouse fibroblasts. *J. Biol. Chem.* *256*:9501–9510 (1981).
5. Sirotnik, F. M., Kurita, S., and Hutchison, D. J. On the nature of a transport alteration determining resistance to amethopterin in the L1210 leukemia. *Cancer Res.* *28*:75–80 (1968).
6. Assaraf, Y., and Schimke, R. T. Identification of methotrexate transport deficiency in mammalian cells using fluoresceinated methotrexate and flow cytometry. *Proc. Natl. Acad. Sci. USA*, *84*:7154–7158 (1987).
7. Hoy, C. A., Rice, G. C., Kovacs, M., and Schimke, R. T. Overreplication of DNA in S phase Chinese hamster ovary cells after DNA synthesis inhibition. *J. Biol. Chem.* *262*:11927–11934 (1987).
8. Tlsty, T. D., Brown, P. C., and Schimke, R. T. UV radiation facilitates methotrexate resistance and amplification of the dihydrofolate reductase gene in cultured 3T6 mouse cells. *Mol. Cell. Biol.* *4*:1050–1056 (1984).
9. Rice, G. C., Hoy, C., and Schimke, R. T. Transient hypoxia enhances the frequency of dihydrofolate reductase gene amplification in Chinese hamster ovary cells. *Proc. Natl. Acad. Sci. USA*, *83*:5978–5982 (1986).
10. Lavi, S., Carcinogen-mediated amplification of viral DNA sequences in simian virus 40-transformed Chinese hamster embryo cells. *Proc. Natl. Acad. Sci. USA*, *78*: 6144–6148 (1981).
11. Lee, T. C., Lamb, P. W., Tanaka, N., Gilmer, T. N., and Barrett, J. C. Induction of gene amplification by arsenic. *Science*, *241*:79–81 (1988).
12. Varshavsky, A. Phorbol ester dramatically increases incidence of methotrexate-resistant mouse cells: Possible mechanisms and relevance to tumor promotion. *Cell*, *25*:561–572 (1981).
13. Giulotto, E., Knights, C., and Stark, G. R. Hamster cells with increased rates of DNA amplification, a new phenotype. *Cell*, *48*:837–845 (1987).
14. Giulotto, E., Bertoni, L., Attolini, C., Rainaldi, G., and Anglana, M. BHK cell lines with increased rates of gene amplification are hypersensitive to ultraviolet light. *Proc. Natl. Acad. Sci. USA*, *88*:3484–3488 (1991).
15. Johnston, R. N., Feder, J., Hill, A. B., Sherwood, S. W., and Schimke, R. T. Transient inhibition of DNA synthesis results in increased dihydrofolate reductase synthesis and subsequent increased DNA content per cell. *Mol. Cell. Biol.* *6*:3373–3381 (1986).

16. Sager, R., Gadi, I. K., Stephens, L., and Grabowy, C. T. Gene amplification: An example of accelerated evolution in tumorigenic cells. *Proc. Natl. Acad. Sci. USA*, *82*:7015–7019 (1985).
17. Tlsty, T. D., Margolin, B. H., and Lum, K. Differences in the rates of gene amplification in nontumorigenic and tumorigenic cell lines as measured by Luria–Debruck fluctuation analysis. *Proc. Natl. Acad. Sci. USA*, *86*:9441–9445 (1989).
18. Otto, E., McCord, S., and Tlsty, T. D. Increased incidence of *CAD* gene amplification in tumorigenic rate lines as an indicator of genomic instability of neoplasmic cells. *J. Biol. Chem. 264*:3390–3396 (1989).
19. Wahl, G. M., Robert de Saint Vincent, B., and DeRose, M. L. Effect of chromosomal position on amplification of transfected genes in animal cells. *Nature*, *307*: 516–520 (1984).
20. Kaufman, R. J., and Sharp, P. A., Amplification and expression of sequences cotransfected with a modular dihydrofolate reductase cDNA gene. *J. Mol. Biol. 159*: 601–621 (1982).
21. Ruiz, J. C., and Wahl, G. M. Formation of an inverted duplication can be an initial step in gene amplification. *Mol. Cell. Biol. 8*:4302–4313 (1988).
22. McArthur, J. G., and Stanners, C. P. A genetic element that increases the frequency of gene amplification. *J. Biol. Chem. 266*:6000–6005 (1991).
23. Beitel, L. K., McArthur, J. G., and Stanners, C. P. Sequence requirements for the stimulation of gene amplification by a mammalian genomic element. *Gene 102*:149–156 (1991).
24. McArthur, J. G., Beitel, L. K., Chamberlain, J., and Stanners, C. P. Elements which stimulate gene amplification in mammalian cells—Role of recombinogenic sequences and transcriptional activation. *Nucleic Acids Res. 19*:2477–2484 (1991).
25. Wegner, M., Schwender, S., Dinkl, E., and Grummt, F. Interaction of a protein with a palindromic sequence from murine rDNA increases the occurrence of amplification-dependent transformation in mouse cells. *J. Biol. Chem. 265*:13925–13932 (1990).
26. Hyrien, O., Debatisse, M., Buttin, G., and Robert de Saint Vincent, B. The multicopy appearance of a large inverted duplication and the sequence at the inversion joint suggest a new model for gene amplification. *EMBO J. 7*:407–417 (1988).
27. Passananti, C., Davies, B., Ford, M., and Fried, M. Structure of an inverted duplication formed as a first step in a gene amplification event: Implications for a model of gene amplification. *EMBO J. 6*:1697–1703 (1987).
28. Legouy, E., Fossar, N., Lhomond, G., and Brison, O. Structure of four amplified DNA novel joints. *Somatic Cell Mol. Genet. 15*:309–320 (1989).
29. Burhans, W. C., Selegue, J. E., and Heintz, N. H. Isolation of the origin of replication associated with the amplified Chinese hamster dihydrofolate reductase domain. *Proc. Natl. Acad. Sci. USA*, *83*:7790–7794 (1986).
30. Kaufman, R. J., and Schimke, R. T. Loss and stabilization of amplified dihydrofolate reductase genes in cultured animal cells resistant to methotrexate. *Mol. Cell. Biol. 1*:1069–1076 (1981).
31. Kaufman, R. J., Brown, P. C., and Schimke, R. T. Loss and stabilization of amplified

dihydrofolate reductase genes in cultured animal cells resistant to methotrexate. III. Loss and stabilization of amplified dihydrofolate reductase in mouse sarcoma S-180 cell lines. *Mol. Cell. Biol. 1*:1084–1093 (1981).

32. Kaufman, R. J., Brown, P. C., and Schimke, R. T. Amplified dihydrofolate reductase genes in unstably methotrexate-resistant cells are associated with double minute chromosomes. *Proc. Natl. Acad. Sci. USA*, *76*:5669–5673 (1979).
33. Hamkalo, B., Farnham, P. J., Johnston, R., and Schimke, R. T. Ultrastructural features of minute chromosomes in a methotrexate-resistant mouse 3T3 cell line. *Proc. Natl. Acad. Sci. USA*, *82*:1126–1130 (1985).
34. Nunberg, J. H., Kaufman, R. J., Schimke, R. T., Urlaub, G., and Chasin, L. A. Amplified dihydrofolate reductase genes are localized to a homogenously staining region of a single chromosome in a methotrexate-resistant Chinese hamster ovary cell line. *Proc. Natl. Acad. Sci. USA*, *75*:5553–5556 (1978).
35. Brown, P. C., Beverly, S. M., and Schimke, R. T. Relationship of amplified dihydrofolate reductase genes in double minute chromosomes in unstably resistant mouse fibroblast cell lines. *Mol. Cell. Biol. 1*:1077–1083 (1981).
36. Murray, M. J., Kaufman, R. J., Latt, S. A., and Weinberg, R. A. Construction and use of a dominant, selectable marker: A Harvey sarcoma virus–dihydrofolate reductase chimera. *Mol. Cell. Biol. 3*:32–43 (1983).
37. Kaufman, R. J., Sharp, P. A., and Latt, S. A. The evolution of chromosomal regions containing transfected and amplified dihydrofolate reductase sequences. *Mol. Cell. Biol. 3*:699–711 (1983).
38. Robins, D. M., Axel, R., and Henderson, A. S. Chromosome structure and DNA sequence alterations associated with mutation of transformed genes. *J. Mol. Appl. Genet. 1*:191–203 (1981).
39. Fendrock, B., Destremps, M., Kaufman, R. J., and Latt, S. A. Cytological, flow cytometric, and molecular analysis of the rapid evolution of mammalian chromosomes containing highly amplified DNA sequences. *Histochemistry*, *84*:121–130 (1986).
40. Andrulis, J. L., and Siminovitch, L. Amplification of the gene for asparagine synthetase in "Gene Amplification" (R. T. Schimke, ed.), Cold Spring Harbor Laboratory, Cold Spring Harbor, New York, 1982, pp. 75–80.
41. Looney, J. E., and Hamlin, J. L. Isolation of the amplified dihydrofolate reductase domain from methotrexate-resistant Chinese hamster ovary cells. *Mol. Cell. Biol. 7*: 569–577 (1987).
42. Ruiz, J. C., and Wahl, G. M. Formation of an inverted duplication can be an initial step in gene amplification. *Mol. Cell. Biol. 8*:4302–4313 (1988).
43. Ford, M., and Fried, M. Large inverted duplications are associated with gene amplification. *Cell*, *45*:425–430 (1986).
44. Windle, B., Draper, B. W., Yin, Y., O'Gorman, S., and Wahl, G. M. A central role for chromosome breakage in gene amplification, deletion formation, and amplicon integration. *Genes Dev. 5*:160–174 (1991).
45. Smith, K. A., Gorman, P. A., Stark, M. B., Groves, R. P., and Stark, G. R. Distinctive chromosomal structure are formed very early in the amplification of *CAD* genes in Syrian hamster cells. *Cell*, *63*:1219–1227 (1990).

46. Carroll, S. M., DeRose, M. L., Gaudray, P., and Moor, C. M. Double minute chromosomes can be produced from precursors derived from a chromosomal deletion. *Mol. Cell. Biol. 8*:1525–1533 (1988).
47. Federspeil, N. A., Beverley, S. M., Schilling, J. W., and Schimke, R. T. Novel DNA rearrangements are associated with dihydrofolate reductase gene amplification. *J. Biol. Chem. 259*:9127–9140 (1984).
48. Giulotto, E., Saito, I., and Stark, G. R. Structure of DNA formed in the first step of *CAD* gene amplification. *EMBO J. 5*:2115–2121 (1986).
49. Roberts, J. M., and Axel, R. Gene amplification and gene correction in somatic cells. *Cell, 29*:109–119 (1982).
50. Keown, W. A., Campbell, C. R., and Kucherlapati, R. S. Methods for introducing DNA into mammalian cells. *Methods Enzymol. 185*:527–537 (1990).
51. Wigler, M., Sweet, R., Sim, G. K., Wold, B., Pellicer, A., Lacy, E., Maniatis, T., Silverstein, S., and Axel, R. Transformation of mammalian cells with genes from prokaryotes and eukaryotes. *Cell, 16*:777–785 (1979).
52. Takeshita, S., Tezuka, K.-I., Takahashi, M., Honkawa, H., Matsuo, A., Matsuishi, T., and Hashimoto-Gotoh, T. Tandem gene amplification in vitro for rapid and efficient expression in animal cells. *Gene, 71*:9–18 (1988).
53. Kaufman, R. J. Selection and coamplification of heterologous genes in mammalian cells. *Methods Enzymol. 185*:487–511 (1990).
54. Beach, L. R., and Palmiter, R. D. Amplification of the metallothionein-I in cadmium-resistant mouse cells. *Proc. Natl. Acad. Sci. USA, 78*:2110–2114 (1981).
55. Kaufman, R. J., Murtha, P., Ingolia, D. E., Yeung, C.-Y., and Kellems, R. E. Selection and amplification of heterologous genes encoding adenosine deaminase in mammalian cells. *Proc. Natl. Acad. Sci. USA, 83*:3136–3140 (1986).
56. Debatisse, M., Hyrien, O., Petit-Koskas, E., Robert de Saint Vincent, B., and Buttin, G. Secretion and rearrangement of coamplified genes in different lineages of mutant cells that overproduce adenylate deaminase. *Mol. Cell. Biol. 6*:1776–1781 (1986).
57. Kanalas, J. J., and Suttle, D. P. Amplification of the UMP synthase gene and enzyme overproduction in pyrazofurin-resistant rat hepatoma cells. *J. Biol. Chem. 259*:1848–1853 (1984).
58. Chapman, A. B., Costello, M. A., Lee, R., and Ringold, G. M. Amplification and hormone-regulated expression of a mouse mammary tumor virus-*Eco*gpt fusion plasmid in mouse 3T6 cells. *Mol. Cell. Biol. 3*:1421–1429 (1983).
59. Kane, S. E., Reinhard, D. H., Fordis, C. M., Pastan, I., and Gottesman, M. M. A new vector using human multidrug resistant gene as a selectable marker enables overexpression of foreign genes in eukaryotic cells. *Gene, 84*:439–446 (1989).
60. Cockett, M. I., Bebbington, C. R., and Yarranton, G. T. High level expression of tissue inhibitor of metalloproteinases in Chinese hamster ovary cells using glutamine synthetase gene amplification. *Biotechnology, 8*:662–667 (1990).
61. Cartier, M., Chang, M., and Stanners, C. Use of the *Escherichia coli* gene for asparagine synthetase as a selective marker for a shuttle vector capable of dominant transfection and amplification in animal cells. *Mol. Cell. Biol. 7*:1623–1628 (1987).
62. Chiang, T.-R., and McConlogue, L. Amplification of heterologous ornithine decarboxylase in Chinese hamster ovary cells. *Mol. Cell. Biol. 8*:764–769 (1988).

63. Kozak, M. Bifunctional messenger RNAs in eukaryotes. *Cell, 47*:481–483 (1986).
64. Kaufman, R. J., Murtha, P., and Davies, M. V. Translation efficiency of polycistronic mRNAs and their utilization to express heterologous genes in mammalian cells. *EMBO J. 6*:187–193 (1987).
65. Balland, A., Faure, T., Carvallo, D., Cordier, P., Ulrich, P., Fournet, B., De La Salle, H., and Lecocq, J.-P. Characterization of two differently processed forms of human recombinant factor IX synthesized in CHO cells transformed with a polycistronic vector. *Eur. J. Biochem. 172*:565–572 (1988).
66. Kozak, M. Effects of intercistronic length on the efficiency of reinitiation of eukaryotic ribosomes. *Mol. Cell. Biol. 7*:3438–3445 (1987).
67. Kaufman, R. J., Davies, M. V., Pathak, V., and Hershey, J. W. B. The phosphorylation state of eukaryotic initiation factor 2 alters translational efficiency of specific mRNAs. *Mol. Cell. Biol. 9*:946–958 (1989).
68. Pelletier, J., and Sonenberg, N. Internal binding of ribosomes to 5′ noncoding region of a eukaryotic mRNA: Translation of poliovirus. *Nature, 334*:320–325 (1988).
69. Jang, S. K., Davies, M. V., Kaufman, R. J., and Wimmer, E. Initiation of protein synthesis by internal entry of ribosomes into the 5′ nontranslated region of encephalomyocarditis virus RNA in vivo. *J. Virol. 63*:1651–1660 (1989).
70. Kaufman, R. J., Davies, M. V., Wasley, L. C., and Michnick, D. An internal ribosome entry site from EMC virus improves utility of vectors for stable expression of foreign genes in mammalian cells. *Nucleic Acids Res. 16*:4485–4490 (1991).
71. Wood, C. R., Morris, G. E., Alderman, E. M., Fouser, L., and Kaufman, R. J. An internal ribosome binding site can be used to select for homologous recombinants at an immunoglobulin heavy chain locus. *Proc. Natl. Acad. Sci. USA 88*:8006–8010 (1991).
72. Kaetzel, D. M., Browne, J. K., Wondisford, F., Nett, T. M., Thomason, A. R., and Nilson, J. H. Expression of biologically active bovine luteinizing hormone in Chinese hamster ovary cells. *Proc. Natl. Acad. Sci. USA, 82*:7280–7283 (1985).
73. Scahill, S. J., Devos, R., van der Heyden, J., and Fiers, W. Expression and characterization of the product of a human immune interferon cDNA gene in Chinese hamster ovary cells. *Proc. Natl. Acad. Sci. USA, 80*:4654–4658 (1983).
74. Gentry, L. E., Webb, N. R., Lim, G. I., Brunner, A. M., Ranchalis, J. E., Twardzik, D. R., Lioubin, M. N., Marquardt, H., and Purchio, A. F. Type 1 transforming growth factor beta: Amplified expression and secretion of mature and precursor polypeptides in Chinese hamster ovary cells. *Mol. Cell. Biol. 7*:3418–3427 (1987).
75. Kaufman, R. J., Wasley, L., Spiliotes, L., Gossels, S., Latt, S., Larsen, G., and Kay, R. Coamplification and coexpression of human tissue-type plasminogen activator and murine dihydrofolate reductase in Chinese hamster ovary cells. *Mol. Cell. Biol. 51*:1750–1759 (1985).
76. Bonthran, D. T., Handin, R. I., Kaufman, R. J., Wasley, L.C., Orr, E. C., Mitsock, L. M., Ewenstein, B., Loscalzo, J., Ginsberg, D., and Orkin, S. H. Pre-pro-von Willebrand factor: Structure and expression in heterologous cells. *Nature, 324*:270–273 (1986).
77. Wasley, L. C., Atha, D. H., Bauer, K. A., and Kaufman, R. J. Expression and characterization of human antithrombin III synthesized in mammalian cells. *J. Biol. Chem. 262*:14766–14772 (1987).

78. Oprian, D., Kaufman, R. J., Khorana, G. High level expression of rhodopsin in African green monkey kidney cells. *Proc. Natl. Acad. Sci. USA*, *84*::8874–8878 (1987).
79. Lodish, H. F., Kong, N., Snider, M., Strous, G. J. A. M. Hepatoma secretory proteins migrate from rough endoplasmic reticulum to Golgi at characteristic rates. *Nature*, *304*:80–83 (1983).
80. Bernstein, D. H., Poritz, M. A., Strub, K., Hoben, P. J., Brenner, S., and Walter, P. Model for signal sequence recognition from amino-acid sequence of 54 K subunit of signal recognition particle. *Nature*, *340*:482–486 (1989).
81. Blobel, G., and Dobberstein, B. Transfer of protein across membranes. I. Presence of proteolytically processed and unprocessed nascent immunoglobulin light chains on membrane-bound ribosomes of murine myeloma. *J. Cell Biol.* *67*:835–851 (1975).
82. Kornfeld, R., and Kornfeld, S. Assembly of asparagine-linked oligosaccharides. *Annu. Rev. Biochem.* *54*:631–669 (1985).
83. Hurtley, S. M., and Helenius, A. Protein oligomerization in the endoplasmic reticulum. *Annu. Rev. Biochem.* *57*:277–307 (1989).
84. Freedman, R. B., Bulleid, N. J., Hawkins, H. C., and Paver, J. L. Role of protein disulphide-isomerase in the expression of native proteins. *Biochem. Soc. Symp.* *55*: 167–192 (1989).
85. Friedman, P. A., and Przysiecki, C. T. Vitamin K dependent carboxylation. *Int. J. Biochem.* *19*:1–7 (1987).
86. Towler, D. A., Gordon, J. I., Adams, S. P., and Glaser, L. The biology and enzymology of eukaryotic protein acylation. *Annu. Rev. Biochem.* *57*:69–99 (1988).
87. Clary, D. O., Griff, I. C., and Rothman, J. E. SNAPs, a family of NSF attachment proteins involved in intracellular membrane fusion in animals and yeast. *Cell*, *61*: 709–721 (1990).
88. Kaiser, C. A., and Schekman, R. Distinct sets of SEC genes govern transport vesicle formation and fusion early in the secretory pathway. *Cell*, *61*:723–733 (1990).
89. Huttner, W. B., and Baeuerle, P. A. Protein sulfation on tyrosine, in *Modern Cell Biology*, Vol. 6 (Birgit H. Satiri, ed.), Alan R. Liss, New York, 1988, pp. 97–140.
90. Thomas, G., Thorne, B. A., and Hruby, D. E. Gene transfer techniques to study neuropeptide processing. *Annu. Rev. Physiol.* *50*:323–332 (1988).
91. Furie, B., and Furie, B. C. Molecular basis of vitamin K-dependent g-carboxylation. *Blood*, *75*:1753–1762 (1990).
92. Burgess, T. L., and Kelley, R. B. Constitutive and regulated secretion of proteins. *Annu. Rev. Cell Biol.* *106*:629–639 (1987).
93. Rothman, J. E. Protein sorting by selective retention in the endoplasmic reticulum and Golgi stack. *Cell*, *50*:1–23 (1987).
94. Pelham, H. R. B. Control of protein exit from the endoplasmic reticulum. *Annu. Rev. Cell. Biol.* *5*:1–23 (1989).
95. Munro, S. and Pelham, H. R. B. An hsp70-like protein in the ER: Identity with the 78 kD glucose-regulated protein and immunoglobulin heavy chain binding protein. *Cell*, *46*:291–300 (1986).
96. Chang, S. C. Wooden, S. K., Nakaki, T., Kim, T. K., Lin, A. Y., Kung, L.,

Attenello, J. W., and Lee, A. S. Rat gene encoding the 78-kDa glucose-regulated protein GRP78: Its regulatory sequences and the effect of protein glycosylation on its expression. *Proc. Natl. Acad. Sci. USA*, *84*:680–684 (1987).

97. Kozutsumi, Y., Segal, M., Normington, K., Gething, M.-J., and Sambrook, J. The presence of malfolded proteins in the endoplasmic reticulum signals the induction of glucose-regulated proteins. *Nature*, *332*:462–464 (1988).
98. Watowich, S. S., and Morimoto, R. I. Complex regulation of heat shock- and glucose-responsive genes in human cells. *Mol. Cell. Biol.* *8*:393–405 (1988).
99. Kassenbrock, C. K., and Kelly, R. B. Interaction of heavy chain binding protein (BiP/GRP78) with adenine nucleotides. *EMBO J.* *8*:1461–1467 (1989).
100. Bole, D. G., Hendershot, L. M., and Kearney, J. F. Posttranslational association of immunoglobulin heavy chain binding protein with nascent heavy chains in non-secreting and secreting hybridomas. *J. Cell Biol.* *102*:1558–1566 (1986).
101. Hurtley, S. M., Bole, D. G., Hoover-Litty, H., Helenius, A., and Copeland, C. S. Interactions of misfolded influenza virus hemagglutinin with binding protein (BiP). *J. Cell. Biol.* *108*:2117–2126 (1989).
102. Dorner, A. J., Bole, D. G., and Kaufman, R. J. The relationship of N-linked glycosylation and heavy chain binding protein association with the secretion of glycoproteins. *J. Cell Biol.* *105*:2665–2674 (1987).
103. Singh, I., Doms, R. W., Wagner, K. R., and Helenius, A. Intracellular transport of soluble and membrane-bound glycoproteins: Folding, assembly, and secretion of anchor-free influenza hemagglutinin. *EMBO J.* *9*:631–639 (1990).
104. Machamer, C. E., Doms, R. W., Bole, D. G., Helenius, A., and Rose, J. K. Heavy-chain binding-protein recognizes incompletely disulphide-bonded forms of vesicular stomatitis virus G protein. *J. Biol. Chem.* *265*:6879–6883 (1990).
105. Kaufman, R. J., Wasley, L. C., Davies, M. V., Wise, R. J., Israel, D. I., and Dorner, A. J. Effect of von Willebrand factor coexpression on the synthesis and secretion of factor VIII in Chinese hamster ovary cells. *Mol. Cell. Biol.* *9*:1233–1242 (1989).
106. Ng, D. T. W., Randall, R. E., and Lamb, R. A. Intracellular maturation and transport of the SV5 type II glycoprotein hemagglutinin-neuraminidase: Specific and transient association with GRP78-BiP in the endoplasmic reticulum and extensive internalization from the cell surface. *J Cell Biol.* *109*:3273–3289 (1989).
107. Dorner, A. J., Krane, M. A. G., and Kaufman, R. J. Reduction of endogenous GRP78 levels improves secretion of a heterologous protein in CHO cells. *Mol. Cell. Biol.* *8*:4063–4070 (1988).
108. Dorner, A. J., Wasley, L. C., and Kaufman, R. J. *EMBO J.* *11*:1563–1571 (1992).
109. Dorner, A. J., Wasley, L. C., and Kaufman, R. J. Increased synthesis of secreted proteins induces expression of glucose-regulated proteins in butyrate-treated Chinese hamster ovary cells. *J. Biol. Chem.* *264*:20602–20607 (1989).
110. Flynn, G. C., Chappell, T. G., and Rothman, J. E. Peptide binding and release by proteins implicated as catalysts of protein assembly. *Science*, *245*:385–390 (1989).
111. Jamieson, J. D., and Palade, G. E. Intracellular transport of secretory proteins in the pancreatic exocrine cell. IV. Metabolic requirements. *J. Cell Biol.* *39*:589–603 (1968).
112. Heytler, P. G. Uncoupling of oxidative phosphorylation by carbonylcyanamide

phenyhydrazones. I. Some characteristics of m-Cl-CCP action on mitochondria and chloroplasts. *Biochemistry*, *2*:357–361 (1963).

113. Argon, Y., Burkhardt, J. K., Leeds, J. M., and Milstein, C. Two steps in the intracellular transport of IgD are sensitive to energy depletion. *J. Immunol.* *142*:554–561 (1989).

114. Dorner, A. J., Wasley, L. C., and Kaufman, R. J. Protein dissociation from GRP78 and secretion is blocked by depletion of cellular ATP levels. *Proc. Natl. Acad. Sci. USA* *87*:7429–7432 (1990).

115. Balch, W. E., Elliott, M. M., and Keller, D. S. ATP-coupled transport of vesicular stomatitis virus G protein between the endoplasmic reticulum and the Golgi. *J. Biol. Chem.* *261*:14681–14689 (1986).

116. Balch, W. E., and Keller, D. S. ATP-coupled transport of vesicular stomatitis virus G protein: Functional boundaries of secretory compartments. *J. Biol. Chem.* *261*: 14690–14696 (1986).

117. Bulleid, N. J., and Freedman, R. B. Defective co-translational formation of disulphide bonds in protein disulphide-isomerase-deficient microsomes. *Nature*, *335*:649–651 (1988).

118. Kaufman, R. J., Wasley, L. C., Furie, B. C., Furie, B., and Shoemaker, C. B. Expression, purification, and characterization of recombinant γ-carboxylated factor IX synthesized in Chinese hamster ovary cells. *J. Biol. Chem.* *261*:9622–9628 (1986).

119. Warren, T. G., and Shields, D. Expression of preprosomatostatin in heterologous cells: Biosynthesis, posttranslational processing, and secretion of mature somatostatin. *Cell*, *39*:547–555 (1984).

120. Hellerman, J. G., Cone, R. C., Potts, J. T., Rich, A., Mulligan, R. C., and Kronenberg, H. M. Secretion of human parathyroid hormone from rat pituitary cells infected with a recombinant retrovirus encoding preproparathyroid hormone. *Proc. Natl. Acad. Sci. USA*, *81*:5340–5344 (1984).

121. Gentry, L. E., Webb, N. R., Lim, G. J., Brunner, A. M., Ranchalis, J. E., Twardzik, D. R., Lioubin, M. N., Marquardt, H., and Purchio, A. F. Type 1 transforming growth factor β: Amplified expression and secretion of mature and precursor polypeptides in Chinese hamster ovary cells. *Mol. Cell. Biol.* *7*:3418–3427 (1987).

122. Foster, D. C., Sprecher, C. A., Holly, R. D., Gambee, J. E., Walker, K. M., and Kumar, A. A. Endoproteolytic processing of the dibasic cleavage site in the human protein C precursor in transfected mammalian cells: Effects of sequence alterations on efficiency of cleavage. *Biochemistry*, *29*:347–354 (1990).

123. Derian, C. K., VanDusen, W., Przysiecki, C. T., Walsh, P. N., Berkner, K. L., Kaufman, R. J., and Friedman, P. A. Inhibitors of 2-ketoglutarate-dependent dioxygenases block aspartyl β-hydroxylation of recombinant human factor IX in several mammalian expression systems. *J. Biol. Chem.* *264*:6615–6618 (1989).

124. Mizuno, K., Nakamura, T., Oshima, T., Tanaka, S. and Matsuo, H. Yeast *KEX2* gene encodes an endopeptidase homologous to subtilisin-like serine proteases. *Biochem. Biophys. Res. Commun.* *156*:246–254 (1989).

125. Fuller, R. S., Brake, A. J., and Thorner, J. Yeast prohormone processing enzyme (*KEX2* gene product) is a Ca^{2+}-dependent serine protease. *Proc. Natl. Acad. Sci. USA*, *86*:1434–1438 (1989).

126. Julius, D., Brake, A., Blair, L., Kunisawa, R., and Thorner, J. Isolation of the putative structural gene for the lysine-arginine-cleaving endopeptidase required for processing of yeast prepro-alpha-factor. *Cell*, *37*:1075–1089 (1984).
127. Julius, D., Schekman, R., and Thorner, J. Glycosylation and processing of prepro-alpha-factor through the yeast secretory pathway. *Cell*, *36*:309–318 (1984).
128. Thim, L., Hansen, M. T., Norris, K., Hoegh, I., Boel, E., Forstrom, J., Ammerer, G., and Fiil, N. P. Secretion and processing of insulin precursors in yeast. *Proc. Natl. Acad. Sci. USA*, *83*:6766–6770 (1986).
129. Sleep, D., Belfield, G. P., and Goodey, A. R. The secretion of human serum albumin from the yeast *Saccharomyces cerevisiae* using five different leader sequences. *Bio/Technology*, *8*:42–48 (1990).
130. Thomas, G., Thorne, B. A., Thomas, L., Allen, R. G., Hruby, D. E., Fuller, R., and Thorner, J. Yeast *KEX2* endopeptidase correctly cleaves a neuroendocrine prohormone in mammalian cells. *Science*, *241*:226–230 (1988).
131. Wise, R. J., Barr, P. J., Wong, P. A., Bathurst, I. C., Brake, A. J., and Kaufman, R. J. Expression of a cDNA encoding a human proprotein processing enzyme: Correct cleavage of the von Willebrand factor precursor at a paired basic amino acid site. *Proc. Natl. Acad. Sci. USA*, *87*:9378–9382 (1990).

26

Consequences of Coamplification of Adenosine Deaminase and Nonselected Genes in a Heterodiploid Host

Sheila P. Little, Edward M. Johnstone, and Karen Ward* *Lilly Research Laboratories, Indianapolis, Indiana*

I. INTRODUCTION

Biochemical selection of a dominant genetic marker has proved to be a powerful tool to identify rare cell transformants containing cloned genes of interest. Cultured somatic cells containing integrated copies of foreign DNA and a biochemical or phenotypic marker have been isolated, and some have been shown to represent a stable change of genotype. Dominant selectable markers such as *Escherichia coli* guanosine phosphoribosyl transferase (*gpt*) and Tn5 aminoglycoside phosphotransferase (*neo*) have been used successfully to create stable transformants in Chinese hamster ovary (CHO) cells [1]. CHO cells have a stable, near-diploid karyotype, yet they appear to be extensively haploid or functionally hemizygous for a large number of genes. It is quite fortunate that early vector-mediated gene transfer was tested in this cell line.

The use of dominant selectable markers that are amplifiable in mammalian cells (dihydrofolate reductase, *DHFR*; adenosine deaminase, *ADA*), extensively reviewed by Kaufman [2], requires the inheritance of multiple copies of a foreign or recombinant gene. Researchers have frequently used diploid hosts: CHO cells and mouse L cells for gene amplification studies. CHO cells show some chromosomal rearrangements, but the karyotype remains stable. Transfected and amplified *DHFR* genes in mouse fibroblasts are usually associated with double minute

*No longer affiliated with Lilly Research Laboratories.

chromosomes. These extrachromosomal elements may integrate into the diploid genome and form homogeneously staining (chromosomal) regions (HSRs) [3]. HSRs are associated with stably amplified genes.

Usually the stability of the recombinant or foreign gene is considered to be important after a transformant has been isolated that is producing recombinant proteins for functional studies or therapeutic applications. The drug of selection is removed and stability is assessed. Here, instability can be attributed to the chromosomal location of the recombinant DNA, the size of the amplified region, and ploidy. In rat hepatoma cells where the endogenous adenosine deaminase has been amplified using deoxycoformycin (dCF), adenosine deaminase levels were reduced when one of the two HSRs was lost [4]. The question arises as to what strategy should be used if one cannot achieve any increase in nonselected, cotransfected genes early in the amplification protocol while selection is maintained.

In this chapter, we describe the coamplification of a nonselected gene, human protein C (*HPC*), an anticoagulant [5], with adenosine deaminase (ADA) in a heterodiploid (293 cell) host. We show that increases in selectable drug dose demonstrate a predictable increase of a coamplified gene in 293 cells. Yet, the number of clones that could be amplified by ADA selection was small in this heterodiploid host, and the level of expression in isolated clones decreased over time while under selection. We suggest that the unpredictable chromosome distribution that occurs in heterodiploid hosts might explain some of our findings.

II. ADENOSINE DEAMINASE AS SELECTABLE AND AMPLIFIABLE MARKER

A. Expression Vectors

Adenosine deaminase was used as a dominant selectable marker and amplifiable gene in our attempt to cotransfer recombinant genes. The selection of transformants is based on the ability to amplify *ADA* in the presence of *de novo* purine synthesis inhibitors and to detoxify high levels of adenosine [6]. The procedure consists of the addition of alanosine, aspartic acid analogue to block *de novo* purine biosynthesis, uridine 1.1 mM adenosine in the presence of a potent inhibitor of ADA, deoxycoformycin. The selection process is referred to as 11 AAU.

Several expression vectors were prepared to select for and amplify adenosine deaminase (Fig. 1, top left). The murine adenosine deaminase cDNA was ligated to a plasmid vector pEJ/VA containing the adenovirus *VA* I and *VA* II genes (pADA/VA). The cDNA for human protein C was cloned into pADA/VA for cotransfer of *ADA* and *HPC*. Plasmids were transfected into human embryonic kidney (293) cells using a calcium phosphate DNA precipitation method [7].

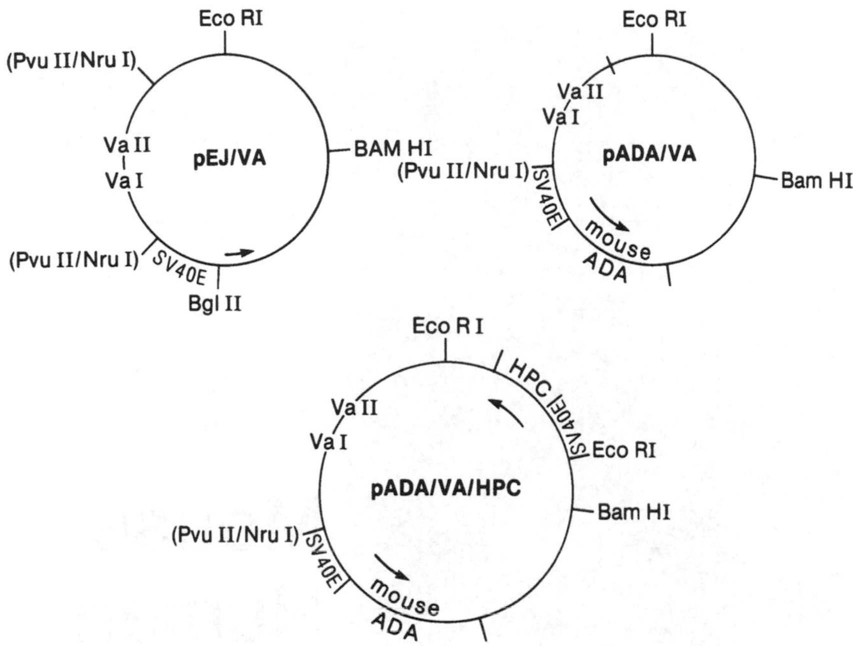

Figure 1 Schematic diagram of the expression vectors used in the gene transfer experiments discussed here. Plasmid pEJ/VA was constructed using a PSV2 β-globin deleted globin vector containing a *Bg*/III cloning site and the *VA* I and VA II genes (9833–11338 bp) of adenovirus 2 and a 321 bp *Sal* I to *Nru* I fragment of pBR322. Into this vector was blunt-end ligated the *Nco* I *Eco*RV mouse cDNA at the *Bg*/II site (pADA/VA). Finally, into the *Eco*RI site was cloned a cassette encoding the human protein C cDNA (Little et al., unpublished). (Reprinted with permission from Ref. 13.)

B. The Host

Because 293 cells had not been selected as a host in cotransfection experiments using adenosine deaminase amplification, we sought to assess the initial ADA level. ADA levels were measured using a modification of a commercially available kit (Authentikit, Innovative Chemistry, Inc., Marshfield, MA). Assay conditions consisted of 250 mM $NaPO_4$, pH 6.4, 50 mM adenosine, 15 mM 3-(4,5-dimethylthiazol-2-yl)-2,5-diphenyltetrazolium bromide (MTT), 0.2 mM phenazine methosulfate (PMS), 0.5 mM flavin adenine dinucleotide, 1 unit/ml xanthine oxidase, and 0.1 unit/ml purine nucleoside phosphorylase. The reaction is initiated with the addition of 10 μl of cytoplasmic or tissue extract; after 10 min at 37°C the absorbance at 565 nM is recorded. An aliquot of the cellular extract can also be

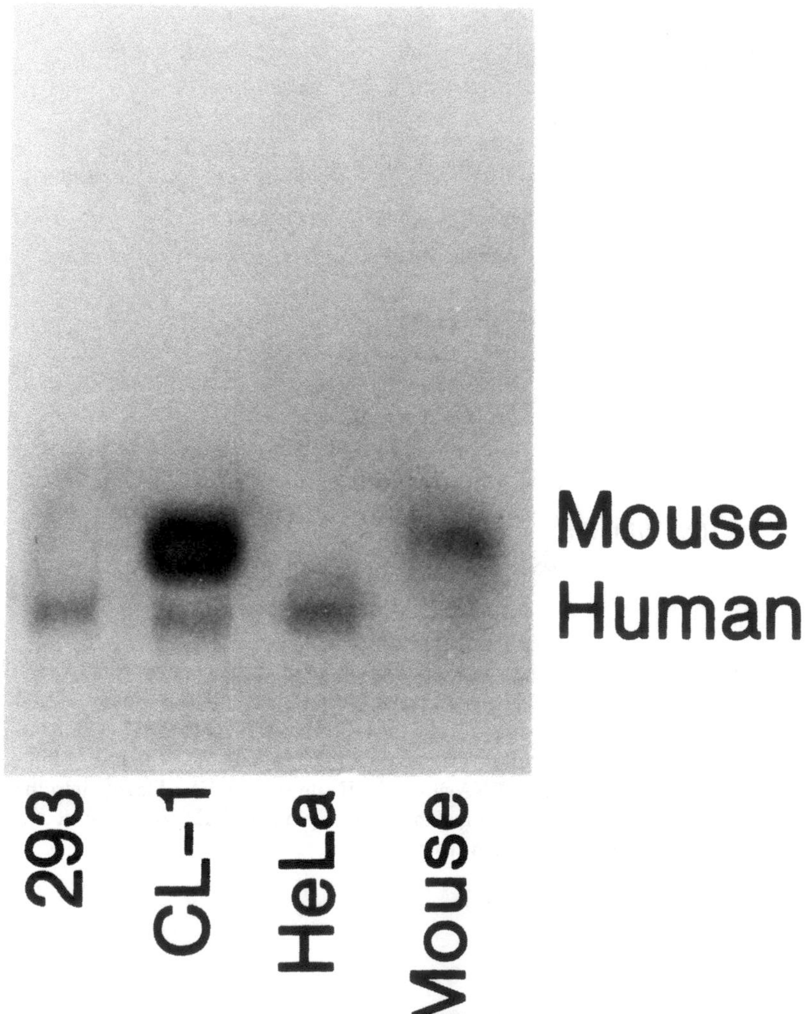

Figure 2 Analysis of ADA activity in a selected transformant following *ADA* gene transfer. Human kidney embryonic cells, 293, were selected in 11AAU and 3 nM deoxycoformycin following transfer of the ADA expression vector pADA/VA/hPC. Homogenates from 293 cells and a selected transformant (C1-1) were electrophoretically fractionated on agarose gels and the presence of mouse or human ADA activity determined by histochemical analysis. Extracts from a mouse and a human (HeLa) cell line served as controls to indicate the electrophoretic mobility of mouse and human ADA activity. (Reprinted with permission from Ref. 13.)

Table 1 Adenosine Deaminase: A Dominant Selectable Marker[a]

		Deoxycoformycin concentration (nM)			
Cell line	Plasmid	0.3	1	3	10
293	pADA/VA/HPC	OG	6×10^{-5}	5×10^{-6}	$<10^{-6}$
	Mock	OG	5×10^{-6}[b]	$<10^{-6}$	$<10^{-6}$

[a]The indicated cells were transfected with either pADA/HPC or by calcium phosphate and were subcultured 48 hr later into 11AAU selection medium at various concentrations of deoxycoformycin. Colonies were stained and counted 12 days after subculturing into selective medium. The frequency of colony formation expressed as the number of colonies relative to the number of cells plated into selective medium (5×10^{-6} cells) is shown. "Mock" refers to similarly treated cells that received no DNA; OG, overgrown.
[b]Clones not viable upon replating.
Source: Reprinted with permission from Ref. 13.

analyzed on preformed agarose gels (Fig. 2). For the relative adenosine deaminase level in 293 cells, see below (Table 2).

Upon transfection by pADA/VA/HPC of 293 cells, we were able to titrate ADA activity in the range of 1–3 nM dCF (Table 1). Because the concentration of dCF required to select for cells expressing the exogenous *ADA* gene varies as a function of the endogenous level of ADA, there is an advantage to start with a cell line that has a low level of ADA to provide conditions of less physiological stress to the host during selection and greater amplification potential.

We prepared an expression vector containing a portion of *ADA* cDNA sequence in an antisense orientation to try to reduce further the level of human *ADA* in 293 cells (Fig. 3). Housekeeping genes such as *DHFR* and *ADA* can be greater than 90% similar at the nucleotide level (e.g., mouse and human *ADA* [8]), however the 5′ and 3′ noncoding regions are not as similar (Fig. 3a). We, therefore, constructed a plasmid vector to express the 5′ noncoding region of the human adenosine deaminase gene under the transcriptional control of SV40 early promoter (Fig. 3b) in an antisense orientation. Clones were selected for resistance to 100 μg/ml hygromycin (Calbiochem) [9]. Table 2 shows the relative ADA activity of 293 cell clones that were isolated. It is significant that we could reduce endogenous ADA activity by 50%. The ADA level of a pool of transformants as well as the average level of six clones suggested a reduction by only 25% of the endogenous ADA activity. Perhaps this is an indication of the minimal level of ADA needed or that our antisense construct was not as efficient. Nevertheless, an attempt to reduce the endogenous ADA activity in order to reduce the initial levels of deoxycoformycin may be beneficial in some cell lines using this technology.

(a) Mouse CCGGGAACACGCTCGGAACCATG
Human CCGGGGCGCACGAGGGCACCATG

(b)

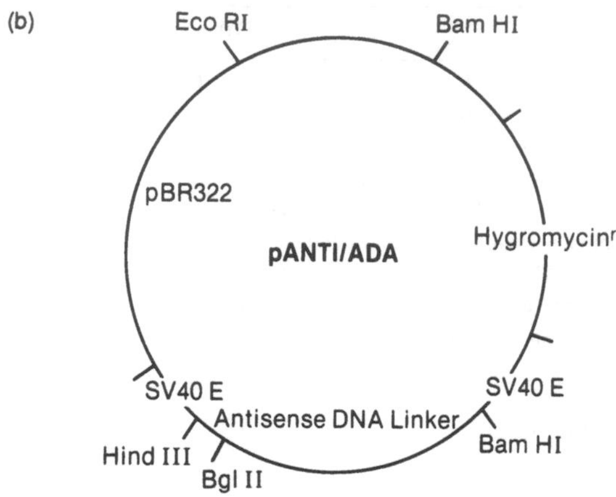

Figure 3 (a) Comparison of nucleotide sequence 18 bp 5′ to ATG of mouse and human adenosine deaminase. (b) Schematic diagram of pANTI/ADA. Plasmid pANTI/ADA consists of a *Hin*dIII-*Bam*HI fragment from pSV2*CAT* (ATCC no. 37155), an antisense linker representing the human sequence (a) a *Bgl*II-*Bam*HI fragment containing the SV40 small *t* intron and polyadenylation signal, and a *Bam*HI fragment containing a hygromycin resistance [9] expression cassette (Little et al., unpublished).

III. COAMPLIFICATION OF NONSELECTED GENE

Human embryonic kidney (293) transformants containing pADA/VA/HPC were selected in increasing concentrations of dCF. At each level of dCF, 24 clones were isolated and assayed for HPC antigen. HPC antigen levels for three independent subclones are shown in Figure 4. One clone, CL-1, selected at 3 nM dCF, was found to express a 30-fold increase in mouse adenosine deaminase over the

Table 2 Relative ADA Activity in Antisense Clones

Cell	ADA activity[a]	Cell	ADA activity[a]
293 parental	0.622	Clone 4	0.490
Clone 1	0.560	Clone 5	0.378
Clone 2	0.334	Clone 6	0.452
Clone 3	0.580		

[a]Relative ADA activity is expressed as absorbance at 565 nM per milligram of protein in cell lysates prepared according to the manufacturer of Authentikit. Absorbance at 565 nM of 0.1 equals approximately 100 international units of ADA/liter.

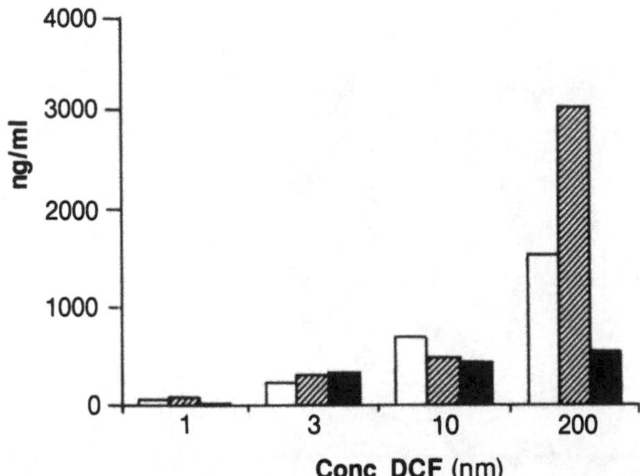

Figure 4 Cotransfer and coamplification of human protein C using ADA/HPC expression vectors. Plasmid pADA/VA/HPC was introduced into human kidney embryonic cell line 293, and transformants were selected in 11AAU and deoxycoformycin as shown in Table 1. Transformants were analyzed for the presence of mouse and human *ADA*, as shown in Figure II, and for human protein C production. Three transformants expressing protein C were selected in the continued presence of 11AAU for higher levels of deoxycoformycin resistance. The relationship between deoxycoformycin resistance and human protein C antigen production is shown. (Reprinted with permission from Ref. 13.)

endogenous hamster ADA level (Fig. 2). We could not further amplify the level of HPC of CL-1 with stepwise increases in dCF. The highest expressing clones selected at 10 nM dCF (350–377 ng/ml HPC) were pooled and new clones were isolated at 50 nM dCF (253–340 ng/ml HPC), hence no increase in HPC expression was apparent. Clones isolated at 200 nM dCF expressed HPC in the range of 318–3000 ng/ml HPC. This observation suggested that some clones, although infrequently isolated, may have undergone amplification of the non-selected marker. In general, it was difficult to identify high expressing clones after stepwise increases in dCF concentration. This observation is consistent with that of Walls et al. [10] when *DHFR* gene amplification was used to coamplify *HPC* in 293 cells. In addition, after several passages, the level of HPC antigen fluctuated even while under selection.

IV. KARYOTYPE ANALYSIS

The cell line 293 is an adenovirus 5 transformed primary human embryonic kidney cell and has a greater than 2 n (46) karyotype (59–62 chromosomes) (Fig. 5). Clone CL-1 and parental 293 cells were subjected to a limited karyotype analysis. Both cultures were extremely variable with respect to chromosome number. Both the parental and CL-1 cells appeared to have an abnormal chromo-

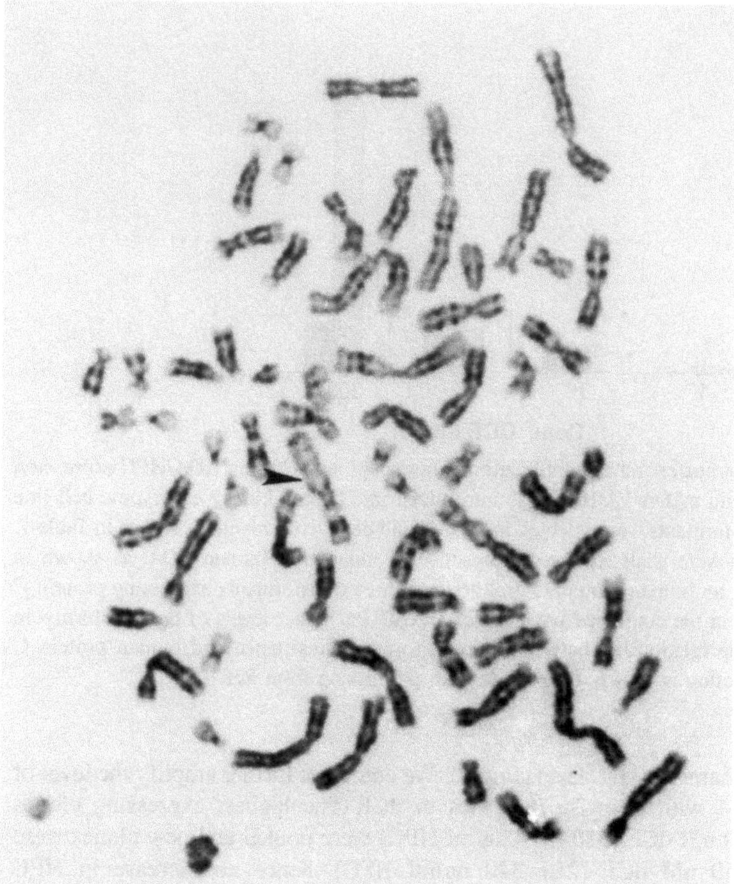

Figure 5 Giemsa–trypsin banded chromosome preparations of 293 cells. Arrow shows HSR-like region.

some 11 with an added segment on 11p which resembled an HSR, but the origin of this segment was not obvious. Thus, 293 cells were capable of maintaining HSR-like material. There were no outstanding differences to suggest that there was any new region of amplification after treatment with 11AAU/dCF.

V. CONSEQUENCES OF *ADA* GENE AMPLIFICATION IN MAMMALIAN CELLS

CHO cells have been the traditional host for gene expression using DHFR, and because of their apparent diploid karyotype and propensity to form HSRs, and so on, they appear to be the ideal host for *ADA* gene amplification and coexpression research [11,12]. If it is necessary for the recombinant gene to be expressed in a cell line other than a CHO cell derivative, it might be useful to obtain a karyotype analysis of the new host. The presence of HSRs of double minute chromosomes, determination of ploidy, and documentation of cell phenotype during gene amplification will be useful to predict its utility in future gene amplification research.

Our preliminary analysis of 293 cells confirms that this cell line is heterodiploid. Gene dosage changes can be the result of having cell lines that are heterodiploid. These cells may have several copies of a chromosome containing multiple integrations of the exogenous amplifiable gene. Therefore, a change in expression level of the nonselected gene may be partially the result of gene amplification or chromosomal distribution. Heterodiploid hosts show great variability with respect to chromosome number even in cloned isolates. Perhaps we could target our amplification units to a stable region of a chromosome, such as a preexisting HSR. Nevertheless, the maintenance of amplifiable genes in a heterodiploid host appears to be a complex problem and worthy of future research.

ACKNOWLEDGMENTS

We thank Teresa Gygi, Maria Poore, Lillian Wang, and Donald McClure for helpful discussions. We are grateful to Dr. Catherine Palmer, Indiana University Medical School, for preliminary cytogenetic analysis. We thank Barbara Fogleman for typing this manuscript.

REFERENCES

1. Howard, B. H., and McCormick, M. Vector-mediated gene transfer, in *Molecular Cell Genetics* (M. M. Gottesman, ed.), Wiley, New York, 1985, pp. 211–233.
2. Kaufman, R. J. Selection and co-amplification of heterologous genes in mammalian cells. *Methods Enzymol. 185*:537–566 (1990).
3. Murray, M. J., Kaufman, R. J., Latt, S. A., and Weinberg, R. A. Construction and use of a dominant, selectable marker: A Harvey sarcoma virus–dihydrofolate reductase chimera. *Mol. Cell Biol. 3*:32–43 (1983).
4. Rowland, P., Chiang, J., Jargiello-Jarrett, P., and Hoffee, P. A. Chromosome anomalies associated with amplification of the adenosine deaminase gene (*ADA*) in rat hepatoma cells. *Cytogenet. Cell Genet. 41*:136–144 (1986).
5. Beckman, R. J., Schmidt, R. J., Santerre, R. F., Plutzky, J., Crabtree, G. R., and Long, G. L. The structure and evolution of a 461 amino acid human protein C precursor and its messenger RNA, based upon the DNA sequence of cloned human liver cDNAs. *Nucleic Acid Res. 13*:5233–5247 (1985).
6. Kaufman, R. J., Wasley, L. C., Spiliotes, A. J., Gossels, S. D., Latt, S. A., Larsen, G. R., and Kay, R. M. Coamplification and coexpression of human tissue-type plas

minogen activator and murine dihydrofolate reductase sequences in Chinese hamster ovary cells. *Mol. Cell Biol.* *5*:1750–1759 (1985).

7. Graham, F. L., and van der Eb, A. J. A new technique for the assay of infectivity of human adenovirus 5 DNA. *Virology*, *52*:456–467 (1973).
8. Yeung, C.-Y., Ingolia, D. E., Roth, D. B., Shoemaker, C., Al-Ubaidi, M. R., Yen, J.-Y., Ching, C., Bobonis, C., Kaufman, R. J., and Kellems, R. E. Identification of functional murine adenosine deaminase cDNA clones by complementation in *Escherichia coli*. *J. Biol. Chem.* *260*:10299–10307 (1985).
9. Grinnell, B. W., Berg, D. T., Walls, J., and Yan, S. B. Trans-activated expression of fully gamma-carboxylated recombinant human protein C, and antithrombotic factor. *Bio/Technology*, *5*:1189–1192 (1987).
10. Walls, J. D., Berg, D. T., Yan, S. B., and Grinnell, B. W. Amplification of multicistronic plasmids in human 293 cell line and secretion of correctly processed recombinant human protein C. *Gene*, *81*:139–149 (1989).
11. Kaufman, R. J., Murtha, P., Ingolia, D. E., Yeung, C.-Y., and Kellems, R. E. Selection and amplification of heterologous genes encoding adenosine deaminase in mammalian cells. *Proc. Natl. Acad. Sci. USA*, *83*:3136–3140 (1986).
12. Kaufman, R. J., Wasley, L. C., Davies, M. V., Wise, R. J., Israel, D. I., and Dorner, A. J. Effect of von Willebrand factor coexpression on the synthesis and secretion of factor VIII in Chinese hamster ovary cell. *Mol. Cell Biol.* *9*:1233–1242 (1989).
13. Kellems, R. E., Johnstone, E. M., Ward, K. E., and Little, S. P. Adenosine deaminase: A dominant amplifiable genetic marker for use in mammalian cells, in *Genetics and Molecular Biology of Industrial Microorganisms* (C. L. Hershberger, S. W. Queener, and G. Hegeman, eds.), 1989.

27

Production of Proteins by Mammalian Cells Using Dominant Selectable and Amplifiable Markers

Mitchell E. Reff *IDEC Pharmaceuticals Corporation, La Jolla, California*

David S. Pfarr *SmithKline Beecham Pharmaceuticals, King of Prussia, Pennsylvania*

I. INTRODUCTION

When one desires to produce a protein that is most likely to be successfully synthesized in a mammalian cell, the recommended process depends on several different criteria, most importantly, the amount of the protein that is required. Microgram to milligram quantities of either cytoplasmic or secreted proteins can most efficiently be produced by transient expression using virus-derived vectors that produce a lytic infection in mammalian cells. These include vectors containing the SV40 origin in monkey kidney COS cells or vectors containing most of the polyoma virus in mouse cells. Milligram quantities of protein or cellular substrates expressing a specific gene can be made by creating stable mammalian cell lines. This is usually accomplished by introduction of the desired gene along with a dominant selectable marker gene into a mammalian cell, followed by selection for the marker gene.

Gene amplification is required to produce more than 10 mg of protein (or a cell substrate with abnormally high numbers of a specific cell surface receptor).

Just as important as the type of system (i.e., transient vs. stable) used for expression is the strategy of selection for high expression. High expression can be achieved rapidly by a combination of clonal selection and extensive screening. Isolated clones with stable high expression can be amplified and maximal levels of expression of most genes can be achieved in one or two amplification steps if clonal selection and extensive screening are done at each step. The practical result of this

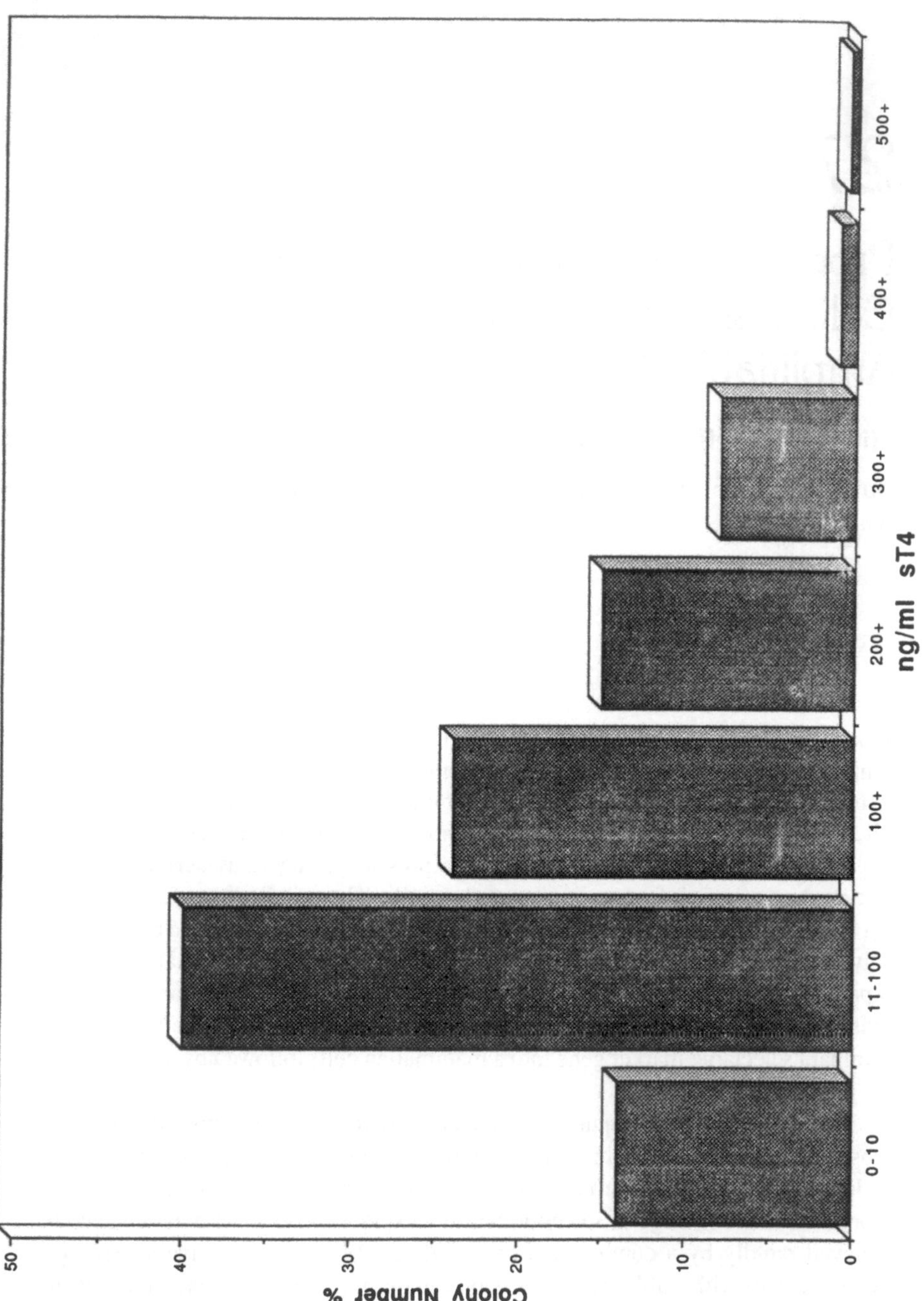
Colony Number %
0
10
20
30
40
50
0-10
11-100
100+
200+
300+
400+
500+
ng/ml sT4

strategy is the isolation of stable production cell lines in low levels of toxic drugs (i.e., 10 nM methotrexate), with low copy numbers of the expression genes (i.e., < 10 copies) and saturated expression levels of secretory glycoprotein (i.e., > 20 pg/cell/day or > 20 mg/liter/day) in a period of 8 weeks.

II. GENERAL ASPECTS OF HIGH LEVEL, STABLE EXPRESSION, AND AMPLIFICATION

A. Position Effects

The introduction of a dominant selectable marker gene into a mammalian cell followed by selection for stable expression of the marker gene results in cell clones at a frequency of about 10^{-3}–10^{-4} of the cells into which the DNA was introduced. A higher percentage of stable clones can be achieved by several different methods, but this is not necessary for isolating clones with high level expression. The introduced marker gene can be shown to be randomly integrated into cellular DNA. The cellular DNA into which the marker gene is integrated may or may not be part of a stable chromosome of the cell; other authors in this book deal with this observation. Homologous integration of exogenous DNA can be achieved at a much lower frequency than random integration in mammalian cells, but it has not yet been used extensively for high level expression.

Expression of a nonselected colinked piece of DNA (expression gene) from the stable marker cell clones varies over several orders of magnitude and fits a Poisson distribution around an average clonal expression level (Fig. 1). We have observed this phenomenon of variable expression, called a position effect, in several different cell types, and it has been reported by other investigators [1]. Position effects are not related to the copy number of the expression gene, but rather to the region of the genome into which the DNA is integrated. Position effects are dominant over whatever regulatory signals are used to express the colinked piece of DNA, although the regulatory signals do relate to the average clonal expression level. A simple corollary to this observation is that the more random integration colonies are screened, the higher the probability of finding an individual clone with very high expression. The methodology that we prefer for isolation of cell clones expressing a gene at high levels is described later in detail.

Other investigators have identified specific genetic elements that alleviate the position effect phenomena, but the levels of expression that they achieve do not appear to be any higher than the highest levels of expression achieved by random

Figure 1 Histogram of soluble CD4 (ST4) expression from individual G418-resistant CHO colonies transfected with the CD4 gene and a linked G418 resistance (neomycin phosphotransferase) gene; data from 228 colonies.

integration and clonal selection [2,3]. However, use of one of these position-independent elements may decrease the number of individual clones that need to be screened to identify one with high expression.

B. Vector Effects That Contribute to High Expression

We have found several factors that contribute to the high level coexpression of the nonselected gene in the majority of selected clones created by random integration. The most obvious of these are the genetic regulatory elements (promoter–enhancer, polyadenylation region) used to express the nonselected gene, which determine the average clonal level of expression. The genetic elements we use for expression are discussed in detail in the next section on expression vectors.

A less obvious factor is physical linkage between the selected and nonselected genes. Our finding in both the R1610 Chinese hamster fibroblast and the P3x63-Ag8.653 (P3) mouse myeloma cell line is that the nonselected gene is twice as likely to be expressed if it is physically linked than if it is cointroduced on another plasmid. These data are summarized in Table 1. All our stable expression experiments have been performed with linear vectors so that random breakage of the DNA prior to integration into the genome is not a factor. In addition, we have data in the R1610 cell line to show that linear DNA remains linear for several days following introduction of DNA into the cell, and it does not recircularize, or concatamerize (Fig. 2). These data are in agreement with our Southern data on transfected and selected, but nonamplified, mammalian clones derived from both the Chinese hamster ovary (CHO) and the P3 mouse myeloma cell lines, which

Table 1 Linkage Experiments

Plasmid and/or DNA fragments[a]	Total G418R colonies assayed	Expressing colonies assayed by colony screen	Expression (%)
Hamster cells			
TN	60	22	37
T + N	118	11	9
		Expressing colonies assayed by enzyme assay	
Mouse myeloma cells			
TN	176	38	22
T + N	185	22	12

[a]Cells were transfected with individual plasmids containing the tPA gene (t) and the neomycin phosphotransferase gene (N) or with a single plasmid containing both genes. G418-resistant colonies were assayed for expression of tPA using a S2251 assay [4] or a modified Western blot (colony screen).

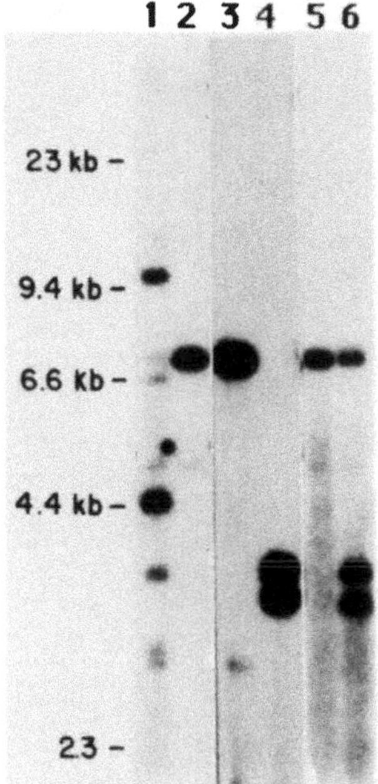

Figure 2 Southern blot analysis DNA isolated from R1610 cells 2 days after transfection with the plasmid DSP1. Probe was nick-translated DSP1. The plasmid DNA was made linear by digestion with *Bam*HI prior to transfection. There is no *Bgl*II site in DSP1. *Eco*RV digests the plasmic once. A *Bam*HI, *Eco*RV double-digest of DSP1 produces two fragments of about 3 kb each. Lanes 1 and 2 are marker lanes. Lane 1, 25 pg DSP1 supercoil mixed with 1 μg R1610 DNA digested with *Bgl*II; lane 2, 25 pg DSP1 *Bam*HI linear mixed with 1 μg R1610 DNA digested with *Bgl*II; lanes 3 and 4, total DNAs isolated from R1610 cells transiently transfected with *Bam*HI-digested DSPI, with lane 3 displaying 1 μg of *Bgl*II-digested DNA and lane 4, 1 μg of *Bgl*II- and *Eco*RV-digested DNA; lanes 5 and 6, DNAs isolated from purified nuclei of R1610 cells transfected with *Bam*HI-digested DSP1, with lane 5 displaying 1 μg of *Bgl*II digested DNA and lane 6, 1 μg of *Bgl*II- and *Eco*RV-digested DNA.

show copy number of 1 for the transfected DNA in the majority of the cell clones (Fig. 3).

The third factor that contributes to high level clonal expression is the "strength" of the selected marker gene. Strength here refers to the relative efficiency of the genetic regulatory elements surrounding the marker gene, which contribute to the production of marker protein in the cell. Our data suggest an inverse correlation between the strength of the selected marker gene and the average clonal expression of the nonselected gene. A "weaker" selectable marker leads to a shift in the average clonal expression of the nonselected gene in the

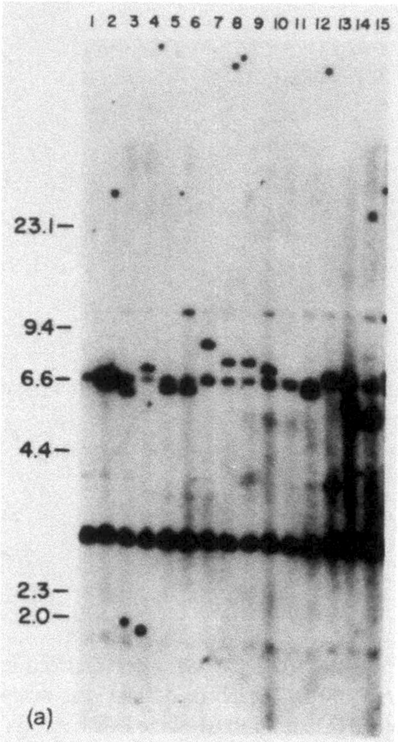

Figure 3a Southern blot analysis of DNA isolated from individual G418-resistant mouse myeloma colonies transfected with the tPA gene and a linked G418 resistance gene contained in the plasma TND (digested with *Bam*HI prior to the transfection). The DNA has been digested with *Dra*I. Probe is apiece of the tPA gene. Lane 1 is the parent cell line. Lane 2 is the parent cell line + 20 pg of transfected plasmid digested with *Bam*HI. Lanes 3–15 are DNAs from individual G418-resistant clones.

direction of higher level expression, indicating that a "weak" selectable marker selects for positive position effects. For example, we have observed an increase in average clonal expression of the nonselected gene using selectable markers made weaker by moving their enhancer-dependent promoter elements further from powerful enhancers on the same vector DNA, without altering any genetic parameters of the nonselected gene or any other parameters of the selectable marker (Fig. 4).

III. EXPRESSION VECTORS FOR MAMMALIAN CELLS

The expression vector TND contains three independent mammalian transcription cassettes [4]. These cassettes can be separated from the plasmid used to grow the

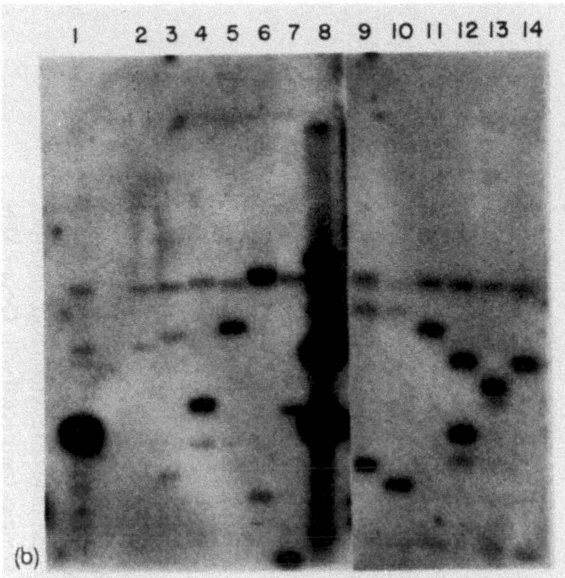

Figure 3b Southern blot analysis of individual G418-resistant CHO colonies transfected with a G418-resistant gene. Lane 1 is the parent cell line + 30 pg of vector. Lane 2 is the parent cell line. Lanes 3–4 are individual clones. All DNA was digested with *Eco*RI, which digests the linear piece of DNA transfected into the cell into a small and a large piece. The probe is identical to the piece of DNA transfected into the cell.

DNA in bacteria by digestion with the restriction endonuclease *Bam*HI. We have found that separating the bacterial plasmid sequences from the mammalian expression sequences prior to introduction of the DNA into mammalian cells increases the efficiency of isolating stable mammalian cell clones containing the transfected DNA [5].

Although position effects are dominant when looking at expression of individual colonies derived from randomly integrated DNA, the average level of expression obtained depends on the regulatory signal surrounding the expression gene. We have used strong non-cell-type, specific viral promoter–enhancers in our vector construction to drive transcription of the expression gene. For example, the vector TND uses the Rous long terminal repeat (LTR) promoter to initiate transcription of the expression gene tissue plasminogen activator (tPA) and a combination of Rous enhancer elements with SV40 enhancer elements to increase the levels of expression [6]. The piece of DNA in TND that contains the SV40

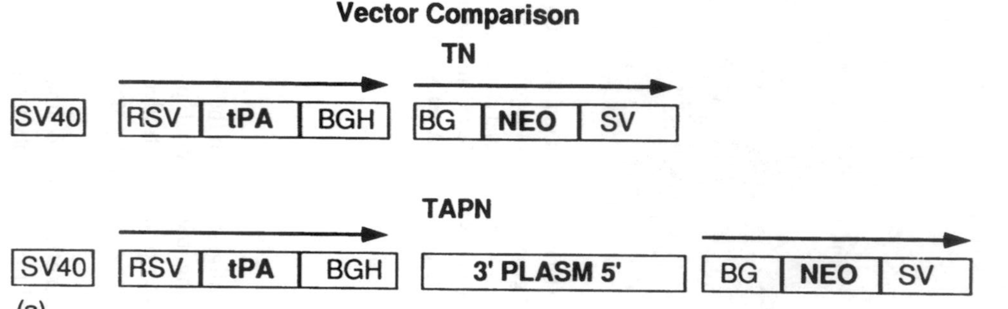

Figure 4 (a) Schematic representation of two plasmids TN and TAPN transfected into CHO cells: SV40, origin of replication and enhancer region; RSV, Rouse sarcoma virus LTR; BGH, bovine growth hormone polyadenylation region; BG, mouse β-globin major promoter; SV, SV40 early polyadenylation region; 3′ plasm 5′, plasminogen coding region in reverse orientation. (b) Histogram of tPA expression from individual G418-resistant CHO colonies transfected with TN or TAPN.

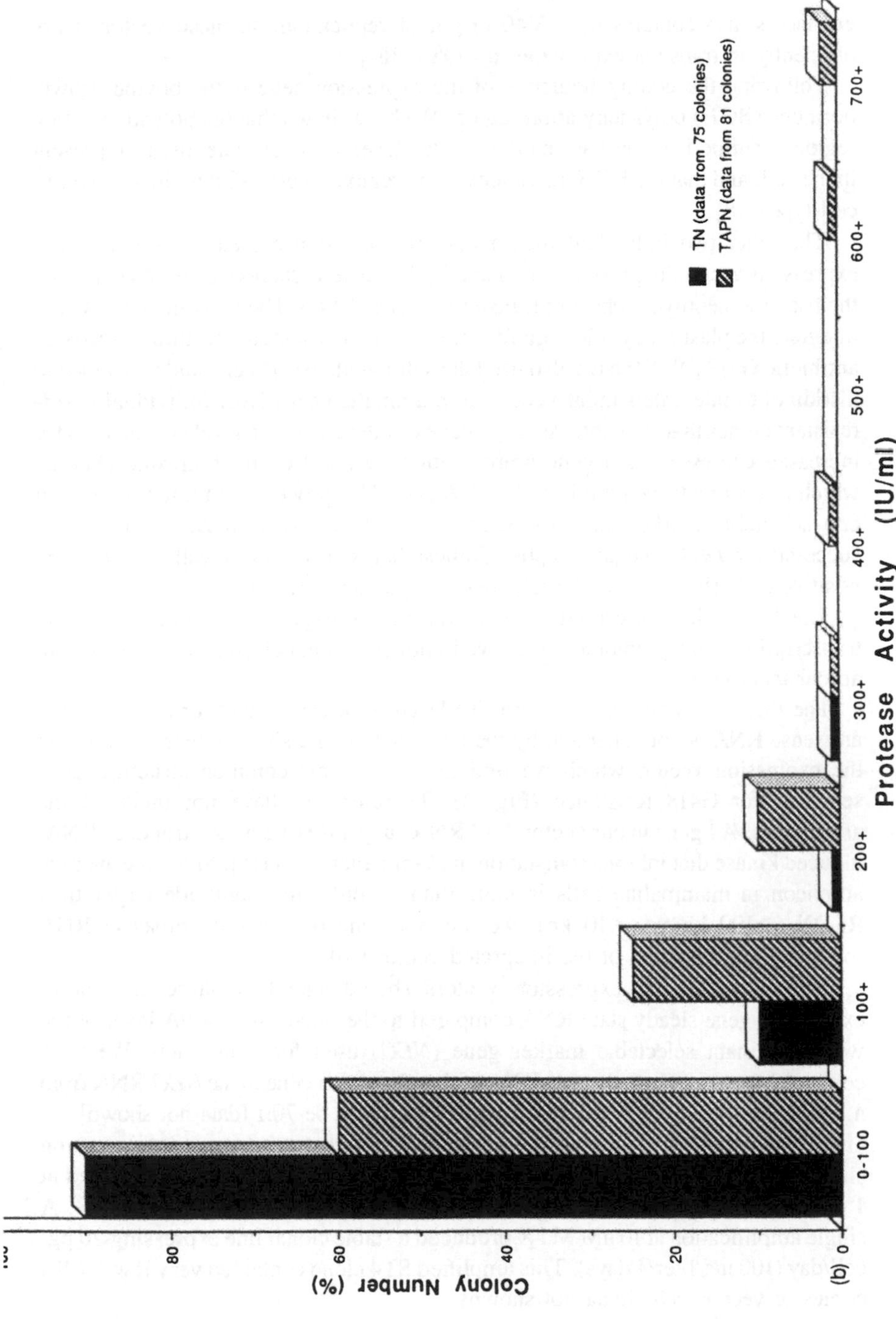
TN (data from 75 colonies)
TAPN (data from 81 colonies)
Colony Number (%)
80
60
40
20
0
0-100
100+
200+
300+
400+
500+
600+
700+
Protease Activity (IU/ml)
(b)

enhancers also contains the SV40 origin of replication, so these vectors work efficiently in transient expression in COS cells [7].

Following the coding sequence of the expression gene is the bovine growth hormone (BGH) polyadenylation region. We have shown that the polyadenylation region of chimeric genes has an effect on the level of steady state message present in the cell, and that the BGH polyadenylation region is very efficient in a variety of cell types. [8].

The other two individual mammalian transcription cassettes present in our expression vectors express the mouse dihydrofolate reductase gene (*DHFR*) and the bacterial neomycin phosphotransferase gene (*NEO*). The *NEO* gene is used to integrate the plasmid by selecting for cells that are resistant to the aminoglycoside antibiotic G418 [9]. We have also used the salmonella *His* D gene and resistance to histidinol to integrate similar vectors in mammalian cells [10]. Individual G418-resistant clones that are expressing the gene of interest at high level are selected for increased expression and gene amplification by the drug methotrexate (MTX), which is a competitive inhibitor of *DHFR* [11]. This procedure of integration with one selectable marker gene followed by amplification with *DHFR* has been successful in *DHFR*-negative cells (Chinese hamster ovary) as well as in *DHFR*-positive cells (bovine endothelial, mouse myeloma) [4,12,13].

The *NEO* selection cassette utilizes the mouse β-globin promoter to initiate transcription. This promoter is inactive in nonerythroid cells unless it is linked to an enhancer [14].

The three cassettes in the vector TND are transcriptionally oriented so that antisense RNA is not generated by the introduction of a single integrated copy of the expression vector, which we find to be the most common structure after selection for G418 resistance (Fig. 3). Therefore we have not included the adenovirus *VA* 1 gene in our vector. *VA* 1 RNA suppresses a double-stranded, RNA-induced kinase that inhibits translation in mammalian cells [15]. Since the average amplicon in mammalian cells is more than an order of magnitude larger than RLDN (>100 kb vs. <10 kb), we are not concerned about antisense RNA following amplification of the integrated vector [16].

The design of our expression vectors should lead to a large amount of expression gene steady state RNA compared to the steady state RNA level of the weak dominant selectable marker gene (*NEO*) used for integration. We have compared the ratio of steady state RNA of the expression gene to the *NEO* RNA from a high expressing nonamplified clone and found it to be 70:1 [data not shown].

We have expressed and amplified several different secretory proteins using similar vector systems. As an example, soluble CD4 (ST4) has been expressed at 1 pg/cell/day by several nonamplified clonal CHO cell lines selected in G418. A single amplification at 10 nM MTX produced a stable clonal line expressing 20 pg/cell/day (100 mg/liter/4 days). This amplified ST4 clone contained very low (< 10) copies of vector DNA [data not shown].

IV. SPECIFIC METHODOLOGY FOR ISOLATION OF MAMMALIAN CELL CLONES EXPRESSING A GENE AT HIGH LEVELS

Cells can be seeded directly following introduction of DNA into 96-well dishes, and then selected for a marker gene. We have found that waiting for 48 rather than 24 hr prior to beginning selection gives the maximum number of individual colonies. Under conditions in which only one-third of the wells in a 96-well dish achieve growth, the majority of wells are statistically clonal. This technique is easier than plating adherent cells into dishes and using cloning cylinders, and it also simplifies evaluation of individual clonal expression greatly. In addition, it is a useful strategy for nonadherent cells such as hybridoma fusion partners. Our observation has been that if several hundred clones are screened for expression (e.g., ten 96-well dishes per gene construct, each containing 25 colonies) several clones expression in the range of 1–3 pg/cell/day of a secretory glycoprotein usually can be obtained. This process takes about 3 weeks, and these cell lines, which usually contain a single gene copy of the expression gene, make enough of the protein for most research efforts.

If a higher level of expression is desired, a group of high level expressing clones are individually selected for gene amplification. The majority of high expression level clones selected for resistance to G418 can have their expression increased and their copy number amplified by a second selection in MTX. This method of clonal amplification works for *DHFR*-negative cell lines such as the CHO lines DG44 and DUXB11 as well as *DHFR*-positive cell lines such as mouse myeloma cell line P3, although fewer of the mouse clones respond to MTX by increasing expression. We usually attempt to amplify the 5 highest clones arising from the initial selection when using CHO and the 10 highest clones when using P3. The cells are plated in 96-well dishes at several levels of MTX, and resistant colonies are screened for expression. This process takes about 3–4 weeks, and the cell clones, which usually contain multiple copies of the expression gene, often are saturated for expression of secretory glycoproteins (> 20 pg/cell/day).

ACKNOWLEDGMENTS

The authors thank John J. Trill, Deborah Jiampetti, and Kelly McLeod, who performed much of the work on which this chapter is based.

REFERENCES

1. Yoshimura, F. K., and Chaffin, K. Different activities of viral enhancer elements before and after stable integration of transfected DNAs. *Mol. Cell. Biol.* 7:1296–1299 (1987).

2. Grosveld, F., Blom van Assendelft, G., Greaves, D. R., and Kollias, G. Position-independent, high-level expression of the human β-globin gene in transgenic mice. *Cell, 51*:975–985 (1987).
3. Phi-Van, L., von Kries, J. P., Ostertag, W., and Strätling, W. H. The chicken lysozyme 5′ matrix attachment region increases transcription from a heterologous promoter in heterologous cells and dampens position effects on the expression of transfected genes. *Mol. Cell. Biol. 10*:2302–2307 (1990).
4. Connors, R. W., Sweet, R. W., Noveral, J. P., Pfarr, D. S., Trill, J. J., Shebuski, R. J., Berkowitz, B. A., Williams, D., Franklin, S., and Reff, M. E. DHFR coamplification of t-PA in DHFR$^+$ bovine endothelial cells: In vitro characterization of the purified serine protease. *DNA*, *7*:651–661 (1988).
5. Trill, J., Pfarr, D., and Reff, M. Efficiency of transfection of linear DNA with and without pBR sequences. *ASBC, Fed. Proc. 463*:2040 (1987).
6. Gorman, C. M., Merlino, G. T., Willingham, M. C., Pastan, I., and Howard, B. H. The Rous sarcoma virus long terminal repeat is a strong promoter when introduced into a variety of eukaryotic cells by DNA-mediated transfection. *Proc. Natl. Acad. Sci. USA*, *79*:6777–6781 (1982).
7. Gluzman, Y. SV40-transformed simian cells support the replication of early SV40 mutants. *Cell, 23*:175–182 (1981).
8. Pfarr, D. S., Rieser, L. A., Woychik, R. P., Rottman, F. M., Rosenberg, M., and Reff, M. E. Differential effects of polyadenylation regions on gene expression in mammalian cells. *DNA*, *5*:115–122 (1986).
9. Colbere-Garapin, F., Horodniceanu, F., Kourilsky, P., and Garapin, A.-C. A new dominant hybrid selective marker for higher eukaryotic cells. *J. Mol. Biol. 150*:1–14 (1981).
10. Hartman, S. C., and Mulligan, R. C. Two dominant-acting selectable markers for gene transfer studies in mammalian cells. *Proc. Natl. Acad. Sci. USA*, *85*:8047–8051 (1988).
11. Kaufman, R. J., Wasley, L. C., Spiliotes, A. J., Grossels, S. O., Latt, S. A., Larsen, G. R., and Kay, R. M. Coamplification and coexpression of human tissue-type plasminogen activator and murine dihydrofolate reductase sequences in Chinese hamster ovary cells. *Mol. Cell. Biol. 5*:1750–1759 (1985).
12. Trill, J. J., Fong, K.-L., Shebuski, R. J., McDevitt, P., Rosa, M. D., Johanson, K. Williams, D., Boyle, K. E., Sellers, T. S. and Reff, M. E. Expression and characterization of finger protease (FP); a mutant tissue-type plasminogen activator (t-PA) with improved pharmacokinetics. *Fibrinolysis*, *4*:131–140 (1990).
13. Piérard, L., García Quintant, L., Reff, M. E., and Bollen, A. Production in eukaryotic cells and characterization of four hybrids of tissue-type and urokinase-type plasminogen activators. *DNA*, *8*:321–328 (1989).
14. Berg, P. E., Yu, J. K., Popovic, Z., Schumperli, D., Johansen, H., Rosenberg, M., and Anderson, W. F. Differential activation of the mouse β-globin promoter by enhancers. *Mol. Cell. Biol. 3*:1246–1254 (1983).
15. Kitajewski, J., Schneider, R. J., Safer, B., Munemitsu, S. M., Samuel, C. E., Thimmappaya, B., and Shenk, T. Adenovirus *VA* I RNA antagonizes the antiviral

action of interferon by preventing activation of the interferon-induced eIF-2α kinase. *Cell*, *45*:195–200 (1986).

16. Looney, J. E., Ma, C., Leu, T.-H., Flintoff, W. F., Troutman, W. B., and Hamlin, J. L. The dihydrofolate reductase amplicons in different methotrexate-resistant Chinese hamster cell lines share at least a 273-kilobase core sequence, but the amplicons in some cell lines are much larger and are remarkably uniform in structure. *Mol. Cell. Biol.* *8*:5268–5279 (1988).

IV

Gene Amplification and Cancer

28

Oncogene Amplification: Analysis of *myc* Oncoproteins

Kari Alitalo, Tomi P. Mäkelä, Kalle Saksela, and Päivi J. Koskinen *University of Helsinki, Helsinki, Finland*

Harri Hirvonen *University of Turku, Turku, Finland*

I. INTRODUCTION

The *myc* genes encode short-lived nuclear phosphoproteins, with putative roles in the regulation of normal cell growth and differentiation during embryogenesis and adult life (for a review, see Refs. 1,2). The *myc* genes are well conserved between species as divergent evolutionarily as fruit fly and man. Deregulated expression of *myc* genes has been linked to malignant growth in a large number of studies on human tumors and cultured cancer cell lines. Similar findings have also been obtained in transgenic animals expressing certain activated *myc* alleles. The c-*myc* gene was originally identified as the cellular counterpart of the v-*myc* gene of transforming avian retroviruses [3–5]. The N-*myc* [6] and L-*myc* [7] genes were thereafter cloned as c-*myc* homologous amplified oncogenes from human tumors. Members of the *myc* gene family have been found activated in several human tumors as a result of DNA rearrangements, such as chromosomal translocations or gene amplification [8–10]. The molecular mechanisms by which the *myc* proteins function to exert their effects on cellular growth are, however, not well understood.

The three *myc* proteins show approximately 50% overall identity and 75% similarity with each other in computer-aided sequence comparison. The degree of homology between the different *myc* proteins is unequally distributed, as these proteins share several domains that are almost identical. When overexpressed, all *myc* genes complement mutant c-*ras* oncogenes in the transformation of rat primary embryonic cells in culture and transform Rat-1A cells without the assistance of other oncogenes. Stimulation of cellular *myc* expression levels or changes in the posttranslational modification of *myc* proteins can be observed in cells following treatment by many types of growth promoting stimuli. These features suggest that the *myc* proteins participate in the final steps of mitogenic signal transduction.

The amino acid sequence of the c-*myc* protein is shown in Figure 1, with some defined polypeptide domains and the casein kinase II (CK II) phosphorylation

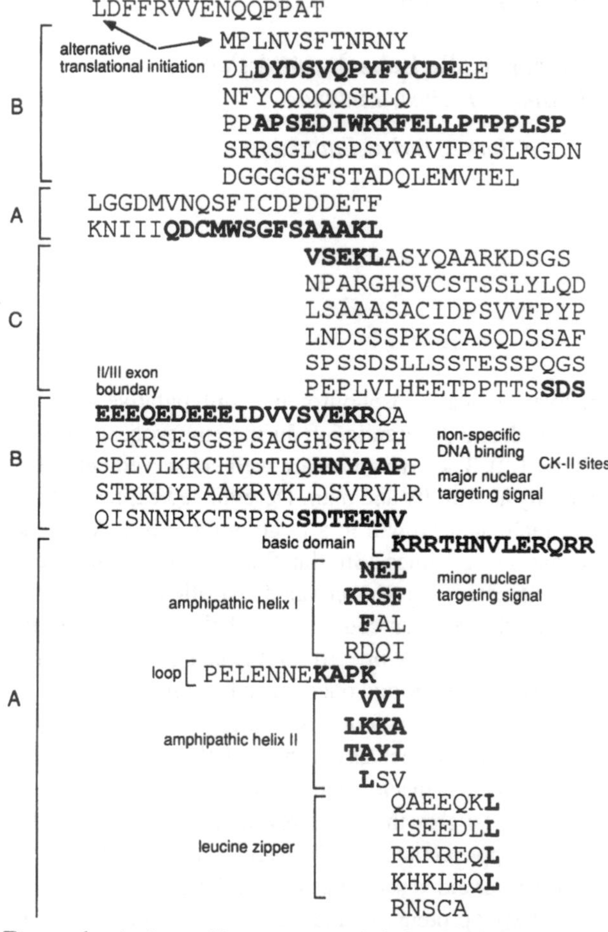

Figure 1 Amino acid sequence and defined polypeptide domains of the human c-*myc* protein: A, B and C refer to segments of decreasing importance for biological activity of the c-*myc* protein in the c-*ras* oncogene cotransformation assay of rat embryonic fibroblasts.

sites. According to recent studies, the *myc* proteins act as transcription factors involved in activation and/or repression of target genes (reviewed in Ref. 11). This hypothesis is strongly supported by the recently revealed structural similarities between the different *myc* proteins and a number of transcription factors. These include (1) demonstration of helix–loop–helix (bHLH) [12], and leucine zipper (bZIP) [13] dimerization and basic DNA-binding domains in the C-termini of the *myc* proteins, (2) binding of a bacterially produced C-terminal c-*myc* polypeptide [14] or a full-length c-*myc* protein in association with *max*, another bHLH-bZip protein [15], to a specific DNA sequence, (3) binding of a chimeric bHLH-transcription factor containing the putative c-*myc* DNA-binding domain to this same sequence, and the ability of this construct to compete with the wild-type c-*myc* gene in a transformation assay [16], and (4) potential of the N-terminus of the c-*myc* protein to activate transcription of a reporter gene via a heterologous promoter when supplied with the appropriate DNA-binding domain [17]. Furthermore, the N-termini of all *myc* proteins have acidic and proline-rich regions, which define trans-activating domains in many characterized transcription factors.

II. MECHANISMS OF c-*myc* REGULATION

Several differences exist between the expression of the three *myc* genes and their oncogenic potential. The c-*myc* gene is expressed in a wide variety of cells and tissues from two major promoters [18], and its mRNA is subject to a rapid regulation by transcriptional and posttranscriptional mechanisms in response to a variety of growth factors [19]. While the c-*myc* expression is enhanced during transition of cells from the G_0 to the G_1 phase, the levels of the c-*myc* mRNA and protein synthesis stay constant throughout the cell cycle. The cellular levels of c-*myc* mRNA are controlled by multiple mechanisms. The activity of c-*myc* promoters, the elongation of the nascent RNA chains, as well as the stability of the c-*myc* mRNA are all under complex cellular regulation (for a review, see Ref. 20). In contrast, the N-*myc* and L-*myc* genes are expressed in a restricted set of tissues, predominantly in primitive embryonic ones [21,22]. These differences are at least partly reflected by the oncogenic spectra of the three genes: activations of c-*myc* are commonly seen in a wide variety of tumors in various species (reviewed in Refs. 9, 23, 24), whereas the N-*myc* and L-*myc* genes are activated in a limited number of tumors [2,10,25].

The activity of the c-*myc* promoter region is controlled by cooperation of positively and negatively acting factors. Apart from several cis-acting regulatory regions that have been mapped relatively close to the c-*myc* promoters, a site 2.2 kb upstream of the P1 promoter of c-*myc* may also be involved [26]. The c-*myc* protein has been found to bind to this site and thereby stimulate its own expression in an autoregulatory way. Furthermore, c-*myc* protein can induce DNA replication through this same site. However, our electrophoretic mobility shift assays have indicated that a factor present in nuclear extracts from HeLa cells binds to this DNA fragment, but the binding does not correlate with the presence of the c-*myc* protein [27]. Further investigations are thus needed to characterize this factor, and to examine its role in normal and deregulated expression of the c-*myc* gene.

III. COEXPRESSION OF *myc* GENES DURING NEURAL DIFFERENTIATION

The developing human fetal brain is an interesting model for studying the developmental expression of the *myc* genes, and in particular their coupling to cell proliferation and differentiation in vivo. The distinct zonal distribution of proliferating, migrating, and differentiating cells in the developing fetal brain is unique among human tissues: the proliferating neuroepithelial precursor cells reside in the periventricular inner layer adjacent to the ventricle, while only postmitotic cells are present in the intermediate and the cortical layers. Upon cessation of cell

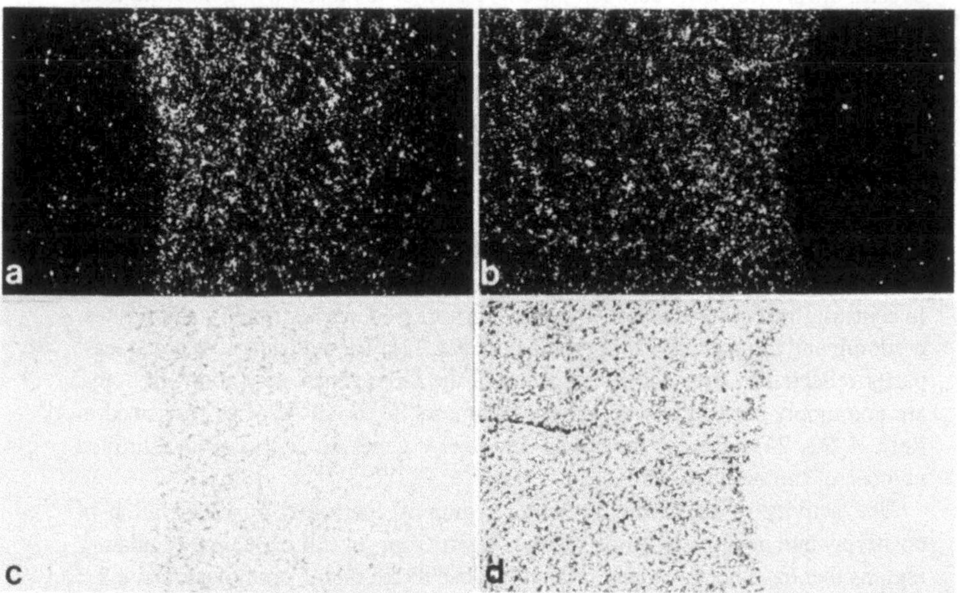

Figure 2 In situ hybridization analysis of L-*myc* expression in the fetal central nervous system: (a) and (c) views of cortical plate (postmitotic cells); (b) and (d) periventricular layers of human fetal cerebral sections hybridized with the L-*myc* probe. Dark field images (a, b) and bright (c, d).

division in the mitotic layers, the cells migrate through the intermediate layers to their future locations in the cortical plate. The cortical plate therefore grows by addition of new cells to it, not by cell division. This unique developmental organization of the fetal brain allows the assessment of the coupling of the *myc* gene expression to cell proliferation and differentiation in vivo.

We have analyzed the expression of *myc* genes in a developing human fetal central nervous system during the second trimester, using Northern hybridization, RNAse protection, and in situ hybridization [22]. Interestingly c-*myc*, N-*myc*, and L-*myc* were found to be expressed not only in the mitotic cells, but also in the postmitotic layers (Fig. 2). Our data suggest a role for all three *myc* genes in the normal differentiation and maturation of the human central nervous system. L-*myc* was found to be more widely expressed than previously thought; we detected L-*myc* mRNA in several fetal tissues such as in the skeletal and cardiac muscle, spleen, thymus, and skin, which contained the highest levels of L-*myc* mRNA [22]. We have also detected expression of L-*myc* and N-*myc* mRNAs by RNAse protection analyses in some acute leukemias and leukemic cell lines, as well as in normal adult bone marrow [28]. These results suggest that N-*myc* and L-*myc* may have a role in regulating the growth and differentiation of hematopoietic cells.

IV. *myc* POLYPEPTIDES

The c-*myc* gene encodes two polypeptide chains, which have distinct amino termini. The translation of the longer of the two c-*myc* polypeptides is initiated from a CTG codon in the first exon of c-*myc*, and removal or mutation of the first exon in Burkitt's lymphomas correlates with suppression of the synthesis of the larger protein [29]. Thus the unusual type of translational initiation might make c-*myc* prone to mutations that affect its protein product and may contribute to the oncogenic activation of c-*myc*. Our results show that the N-*myc* gene (but not L-*myc*; see below), also encodes two distinct polypeptides, which are initiated at two ATG codons only 24 bp apart in the N-*myc* second exon [30]. When expressed separately, these polypeptides behave as nuclear phosphoproteins with short half-lives similar to that of the normal N-*myc* protein. The close spacing of the N-*myc* translation initiation codons suggest a possible explanation for the relative resistance of the N-*myc* gene to insertional mutagenesis and chromosomal translocations, which commonly activate the c-*myc* gene and mutagenize its protein product, whose corresponding polypeptide initiation codons are located in separate exons 1.7 kb apart.

The *myc* proteins are diffusely distributed throughout the cytoplasm of mitotic cells, while in interphase cells they localize in the nuclei in a granular pattern excluding nucleoli [31]. However, overexpression of either the c-*myc* or v-*myc* protein results in accumulation of these proteins in amorphous, often phase-dense

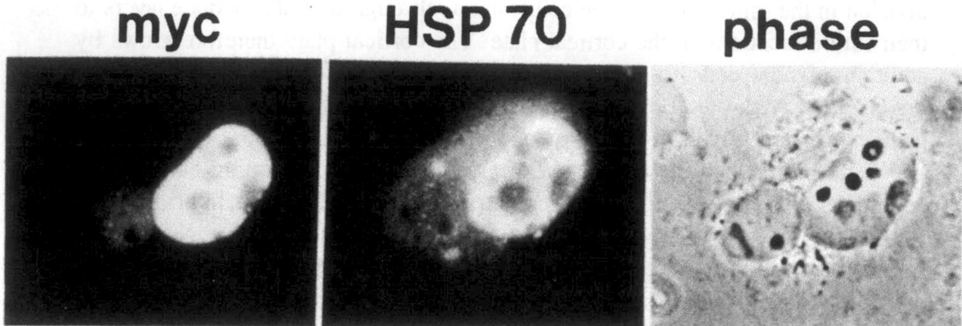

Figure 3 Nuclear colocalization of HSP70 and c-*myc* proteins induced by overexpression of c-*myc*. Panels represent COS cells that were transiently transfected with a c-*myc* expression vector and were double-stained with anti-c-*myc* and anti-HSP70 antibodies, as indicated.

nuclear structures in codistribution with the HSP70 heat shock protein (Fig. 3) [32]. Instead, the closely related N-*myc* protein does not accumulate in the nuclei or induce nuclear translocation of HSP70, suggesting differences in the post-translational regulation of the *myc* proteins. Studies with chimeric *myc* proteins and c-*myc* deletion mutants revealed that polypeptide sequences encoded by the second exon of c-*myc* are involved in both accumulation of the protein and in its colocalization with HSP70. Through this interaction HSP70 may regulate the availability of functional c-*myc* protein in the nuclei. In turn, c-*myc* may interfere with the cellular heat shock response by preventing HSP70 from binding to its normal nuclear targets during stress, which thus could result in increased cellular thermosensitivity.

V. PROTEIN KINASE C DEPENDENT PHOSPHORYLATION OF THE L-*myc* PROTEIN

The DNA-binding and/or trans-activation potential of several transcription factors is regulated by phosphorylation (recently reviewed in Ref. 33). Among the oncogene-encoded transcription factors, phosphorylation has been implicated as a regulatory mechanism for c-*fos*, c-*jun*, c-*myb*, c-*erb*-A, and c-*ets*-1 proteins. The c-*myc* protein is phosphorylated in two distinct regions by casein kinase II (CK-II) [34], but the role of this phosphorylation in regulation of c-*myc* remains to be studied.

We have shown elsewhere that the L-*myc* protein migrates as three distinct forms of 60, 64, and 66 kDa studied by sodium dodecyl sulfate polyacrylamide gel electrophoresis (SDS-PAGE) [35]. This heterogeneity arises from differential phosphorylation of a single polypeptide. The relative ratio of the different forms of L-*myc* protein is rapidly shifted in favor of the 66 kDa form when cells are treated with serum or chemical activators of protein kinase C, or by inhibition of serine/threonine protein phosphatases with okadaic acid. In vitro mutagenesis and phosphoamino acid analyses define the N-terminal serine residues 38 and 42 of L-*myc* as critical targets for the PK-C-dependent phosphorylation. This is the exclusive site of phosphorylation in the N-terminal third of the L-*myc*, and it is phosphorylated in vitro by glycogen synthetase kinase 3β (GSK-3β) [36]. Although GSK-3 was originally described as a cytoplasmic kinase involved in glycogen metabolism, the phosphorylation of nuclear oncoproteins by GSK-3 (c-*jun*, c-*myb*, L-*myc*) has raised the possibility that GSK-3 is involved in transcriptional regulation [37]. Potential phosphorylation sites similar to those of L-*myc* Ser 38 and 42 are present in c-*myc* and N-*myc* in a highly conserved region thought to represent a transcriptional activator domain. The N-terminal regions of c-*myc* [17] and L-*myc* (K. Saksela, unpublished) have been shown to function in transcriptional activation. Thus, it is tempting to speculate that the PK-C induced phosphorylation of serine residues 38 and 42 of L-*myc* could control the activity of this putative transcription factor as depicted in Figure 4. We suggest that the PK-C dependent phosphorylation is involved in rapid regulation of the functions of the L-*myc* protein. It appears likely that the kinase(s) that phosphorylate(s) L-*myc* residues 38 and 42 in a PK-C dependent manner is common to all *myc* proteins, and could be either GSK-3 or MAP kinase.

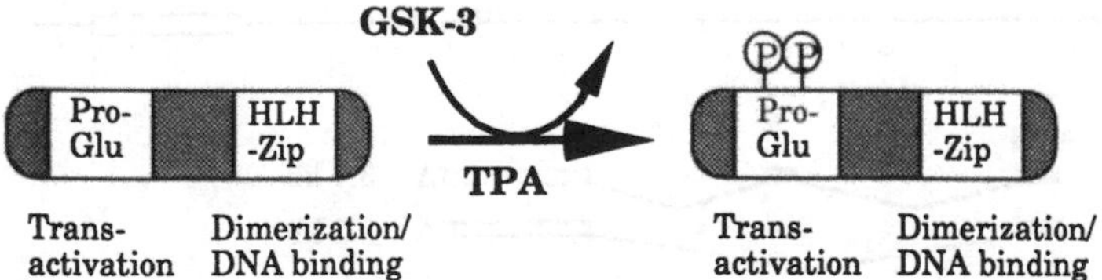

Figure 4 A putative model for the activation of L-*myc* as a result of protein kinase C induced phosphorylation. The proline-glutamine-rich trans-activation domain and the helix–loop–helix leucine zipper domain of the L-*myc* protein are shown schematically. Activation of protein kinase C by TPA causes a rapid phosphorylation of two glycogen synthetase kinase 3 (GSK-3) target sites in the L-*myc* trans-activator domain. This phosphorylation, which can be visualized as a mobility shift in SDS-PAGE, may cause a conformational and functional change also in vivo.

VI. A FUSION PROTEIN FORMED BY L-*myc* AND A NOVEL GENE IN SMALL CELL LUNG CANCER

With the exception of most Burkitt's lymphomas, where synthesis of one of the two normal c-*myc* polypeptides is abrupted [29], activation of the *myc* genes in human tumors has not been found to involve structural changes in the proteins encoded by the *myc* genes. DNA amplification of the L-*myc* gene is commonly detected in small cell lung cancer (SCLC) [2], but L-*myc* has not been implicated in other tumor types. We found an unusually large 72–77 kDa L-*myc* protein in our immunoprecipitation analyses of small cell lung carcinoma cell lines GLC28 and CORL47. Northern blotting analysis indicated that both cell lines expressed abnormal transcripts of 3.7 and 3.2 kb. Analysis of L-*myc* hybridizing cDNA clones from a GLC28 cell library revealed two distinct classes of cDNA, which had identical novel sequences immediately 5′ of either exon II or exon III. This novel sequence represented an upstream exon: a DNA rearrangement had placed a normally unrelated first exon upstream of L-*myc* in these SCLC cell lines (Fig. 5). Hybridization results indicated that the novel 5′ cDNA sequence represents a transcribed single locus in the human genome, but sequence analysis did not reveal a significant degree of homology with sequences in the databases searched. Thus it was concluded that the cDNA probe represents a new transcribed gene, which we have named *rlf* (for **r**earranged **L**-*myc* **f**usion) [38].

The *rlf*-L-*myc* rearrangements could not have been previously detected in Southern analyses using L-*myc* probes because of the distance of the breakpoints from L-*myc*. However, the rearrangements are readily detectable with the *rlf* probe because the breakpoints are less than 7 kbp from the *rlf* exon. We have also

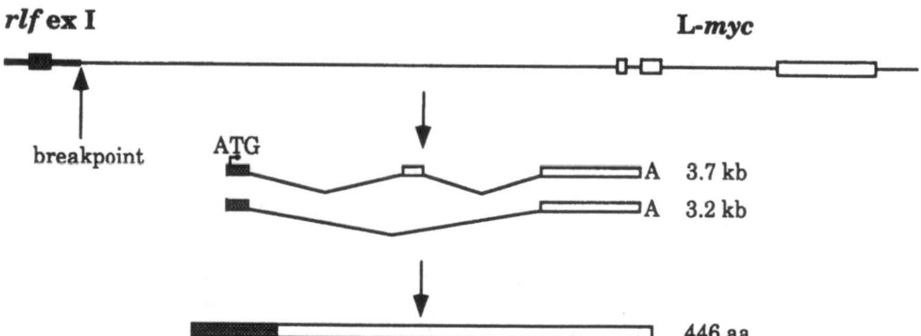

Figure 5 The *rlf*-L-*myc* gene fusion. Rearrangements joining exon I of a novel locus named *rlf* upstream of L-*myc* have been found in several SCLC cell lines and tumors. The rearrangement causes the formation of two chimeric alternatively spliced transcripts as shown. The 3.7 kb transcript encodes a 446 aa fusion protein with 79 aa of *rlf* fused to the L-*myc* protein.

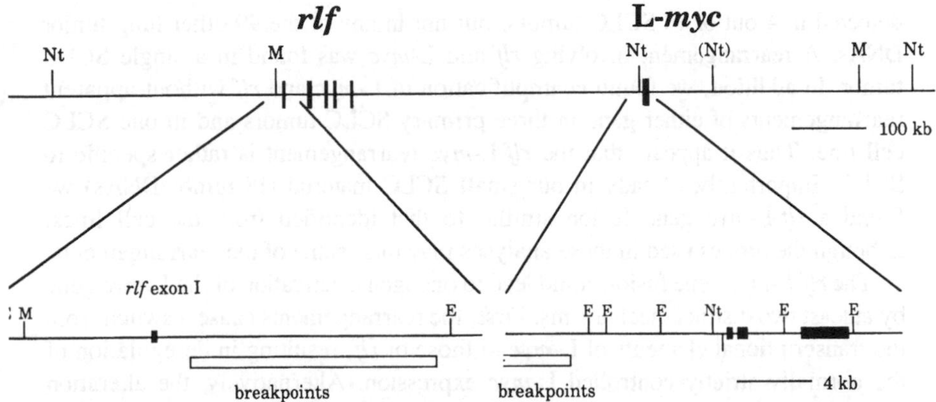

Figure 6 Germ line configuration of *rlf* and L-*myc*. The normal *rlf* locus, located approximately 480 kb upstream of the L-*myc* gene in chromosome band 1p32, consists of at least five exons shown schematically (not in scale) in the pulsed-field gel electrophoresis map in the top part of the figure. The regions surrounding *rlf* exon I and L-*myc* and the regions to which the characterized breakpoints have been mapped are shown in greater detail. Nt, *Not* I; M, *Mlu*I; E, *Eco*RI.

characterized the DNA rearrangement leading to the *rlf*-L-*myc* fusion protein. Our results demonstrate that normal *rlf* and L-*myc* genes are separated by about 480 kb of genomic DNA. Thus, the *rlf*-L-*myc* gene fusions are due to similar but not identical intrachromosomal rearrangements at 1p32 (Fig. 6) [39].

The presence of independent genetic lesions that cause the formation of identical chimeric *rlf*-L-*myc* proteins suggests a role for the fusion protein in the development of these tumors. In both cell lines used in this study the *rlf*-L-*myc* fusion gene has also been involved in a DNA amplification. Since hybridization analyses using an *rlf* probe show only a single rearranged, amplified band in restriction mapping, it can be concluded that the genomic amplifications in these cell lines must have occurred after the *rlf*-L-*myc* fusion. It is not known whether gene amplification has occurred soon after the rearrangement, or whether it represents a later event associated with the progression of these SCLCs. If the *rlf*-L-*myc* rearrangements were associated with DNA amplification, they would resemble the amplification and rearrangement described for the c-*myc* and *pvt* genes in the COLO320DM cell line, where rearrangement of *pvt* exon I 5′ of c-*myc* exon II has apparently also occurred prior to gene amplification.

To explore the incidence of the *rlf*-L-*myc* fusion, we analyzed a total of 117 primary lung tumors of various types with two different restriction enzymes to maximize the probability of detecting rearrangements, which can occur in a relatively large region of DNA [40]. DNA amplifications of the L-*myc* gene were detected in 4 out of 18 SCLC tumors, but not in any of the 99 other lung tumor DNAs. A rearrangement involving *rlf* and L-*myc* was found in a single SCLC tumor. In addition, we found coamplification of L-*myc* and *rlf* without apparent rearrangements of either gene in three primary SCLC tumors and in one SCLC cell line. Thus it appears that the *rlf*-L-*myc* rearrangement is rather specific to SCLC. Importantly, already in our small SCLC material (18 tumor DNAs) we found a *rlf*-L-*myc* gene fusion similar to that identified from the cell lines, although the probes used in these analyses may miss some of the rearrangements.

The *rlf*-L-*myc* gene fusion could lead to oncogenic activation of the L-*myc* gene by at least two distinct mechanisms. First, the rearrangements cause a switch from the transcriptional elements of L-*myc* to those of *rlf*, resulting in deregulation of the normally strictly controlled L-*myc* expression. Alternatively, the alteration of the amino terminus of L-*myc* could affect the activity or substrate specificity of the L-*myc* protein. The *rlf*-L-*myc* fusion protein represents the first example of a chimeric *myc* oncoprotein in human cancer. This may represent a novel activation mechanism for the *myc* family of proto-oncogenes.

REFERENCES

1. DePinho, R. A., Hatton, K. S., Morgenbesser, S. D., and Torres, R. The *myc* family of oncoproteins: Structure, biochemistry, and activities, in *Growth Regulation and Carcinogenesis*, Vol. I (W. R. Paukovits, ed.), CRC Press, Boca Raton, Florida, 1991, pp. 175–203.
2. Saksela, K. *myc* genes and their deregulation in lung cancer. *J. Cell Biochem*. *42*:153–180 (1990).
3. Sheiness, D., and Bishop, J. M. DNA and RNA from uninfected vertebrate cells contain nucleotide sequences related to the putative transforming gene of avian myelomacytosis. *J. Virol*. *31*:514–521 (1979).
4. Alitalo, K., Bishop, J. M., Smith, D. H., Chen, E. Y., Colby, W. W., and Levinson, A. D. Nucleotide sequence of the v-*myc* oncogene of avian retrovirus MC29. *Proc. Natl. Acad. Sci. USA*, *80*:100–104 (1983).
5. Colby, W. W., Chen, E. Y., Smith, D. H., and Levinson, A. D. Identification and nucleotide sequence of a human locus homologous to the v-*myc* oncogene of avian myelocytomatosis virus MC29. *Nature*, *301*:722–725 (1983).
6. Schwab, M., Alitalo, K., Klempnauer, K.-H., Varmus, H. E., Bishop, J. M., Gilbert, F., Brodeur, G., Goldstein, M., and Trent, J. Amplified DNA with limited homology to the *myc* cellular oncogene is shared by human neuroblastoma cell lines and a neuroblastoma tumor. *Nature*, *305*:245–248 (1983).
7. Nau, M. M., Brooks, B. J., Battey, J., Sausville, E., Gazdar, A. F., Kirsch, I. R., McBride, O. W., Bertness, V., Hollis, G. F., and Minna, J. D. L-*myc*, a new *myc*-related gene amplified and expressed in human small cell lung cancer. *Nature*, *318*: 69–73 (1985).
8. Klein, G., and Klein, E. Evolution of tumours and the impact of molecular oncology. *Nature*, *315*:190–195 (1985).

9. Cory, S. Activation of cellular oncogenes in hemopoietic cells by chromosome translocation. *Adv. Cancer Res. 47*:189–234 (1986).
10. Alitalo, K., and Schwab, M. Oncogene amplification in tumor cells. *Adv. Cancer Res. 47*:235–281 (1986).
11. Lüscher, B., and Eisenman, R. N. New light on Myc and Myb. Part I. Myc. *Genes Dev. 4*:2025–2035 (1990).
12. Murre, C., McCaw, P. S., and Baltimore, D., A new DNA binding and dimerization motif in immunoglobulin enhancer binding, daughterless, MyoD, and *myc* proteins. *Cell, 56*:777–783 (1989).
13. Landschulz, W. H., Johnson, P. F., and McKnight, S. L. The leucine zipper: A hypothetical structure common to a new class of DNA binding proteins. *Science, 240*: 1759–1764 (1988).
14. Blackwell, T. K., Kretzner, L., Blackwood, E. M., Eisenman, R. N., and Weintraub, H. Sequence-specific DNA binding by the c-*myc* protein. *Science, 250*:1149–1151 (1990).
15. Blackwood, E. M., and Eisenman, R. N. Max: A helix–loop–helix zipper protein that forms a sequence-specific DNA-binding complex with *myc*. Science, *251*: 1211–1217 (1991).
16. Prendergast, G. C., and Ziff, E. Methylation-sensitive sequence-specific DNA binding by the c-*myc* basic region. *Science, 251*:186–189 (1991).
17. Kato, G. J., Barrett, J., Villa-Garcia, M., and Dang, C. V. An amino-terminal c-*myc* domain required for neoplastic transformation activates transcription. *Mol. Cell. Biol. 10*:5914–5920 (1990).
18. Battey, J., Moulding, C., Taub, R., Murphy, W., Stewart, T., Potter, H., Lenoir, G., and Leder, P. The human c-*myc* oncogene: Structural consequences of translocation into the IgH locus in Burkitt lymphoma. *Cell, 34*: 779–787 (1983).
19. Kelly, K., Cochran, B. H., Stiles, C. D., and Leder, P. Cell-specific regulation of the c-*myc* gene by lymphocyte mitogens and platelet-derived growth factor. *Cell, 35*: 603–610 (1983).
20. Spencer, C. A., and Groudine, M. Control of c-*myc* regulation in normal and neoplastic cells. *Adv. Cancer Res. 56*:1–48 (1991).
21. Zimmerman, K. A., Yancopoulos, G. D., Collum, R. G., Smith, R. K., Kohl, N. E., Denis, K. A., Nau, M. M., Witte, O. N., Toran-Allerand, D., Gee, C. E., Minna, J. D., and Alt, F. W. Differential expression of the *myc* family genes during murine development. *Nature, 319*:780–783 (1986).
22. Hirvonen, H., Mäkelä, T. P., Sandberg, M., Kalimo, H., Vuorio, E., and Alitalo, K. Expression of the *myc* proto-oncogenes in developing human fetal brain. *Oncogene, 5*:1787–1797 (1990).
23. Bishop, J. M. Molecular themes in oncogenesis. *Cell, 64*:235–248 (1991).
24. Varmus, H. E. Viruses, genes, and cancer. I. The discovery of cellular oncogenes and their role in neoplasia. *Cancer, 55*:2324–2328 (1985).
25. Fourel, G., Trepo, C., Bougueleret, L., Henglein, B., Ponzetto, A., Tiollais, P., and Buendia, M. A. Frequent activation of N-*myc* genes by hepadnavirus insertion in woodchuck liver tumours: See comments. *Nature, 347*:294–298 (1990).
26. Ariga, H., Imamura, Y., and Iguchi-Ariga, S. M. DNA replication origin and

transcriptional enhancer in c-*myc* gene share the c-*myc* protein binding sequences. *EMBO J. 8*:4273–4279 (1989).
27. Saksela, K., Koskinen, P., and Alitalo, K. Binding of a nuclear factor to the upstream region of the c-*myc* gene. *Oncogene Res. 6*:73–76 (1991).
28. Hirvonen, H., Hukkanen, V., Salmi, T. T., Mäkelä, T. P., Pelliniemi, T.-T., and Alitalo, R. Expression of L-*myc* and N-*myc* proto-oncogenes in a subset of human leukemias and leukemia cell lines. *Blood 78*:3012–3020 (1991).
29. Hann, S. R., King, M. W., Bentley, D. L., Anderson, C. W., and Eisenman, R. N., A non-AUG translational initiation in c-*myc* exon I generates an N-terminally distinct protein whose synthesis is disrupted in Burkitt's lymphomas. *Cell, 52*:185–195 (1988).
30. Mäkelä, T. P., Saksela, K., and Alitalo, K. Two N-*myc* polypeptides with distinct amino termini encoded by the second and third exons of the gene. *Mol. Cell Biol. 9*: 1545–1552 (1989).
31. Winqvist, R., Saksela, K., and Alitalo, K. The *myc* proteins are not associated with chromatin in mitotic cells. *EMBO J. 3*:2947–2950 (1984).
32. Koskinen, P., Sistonen, L., Evan, G., Morimoto, R., and Alitalo, K. Nuclear colocalization of cellular and viral *myc* proteins with HSP70 in *myc*-overexpressing cells. *J. Virol. 65*:842–851 (1991).
33. Bohmann, D. Transcription factor phosphorylation: A link between signal transduction and the regulation of gene expression. *Cancer Cells, 2*:337–344 (1990).
34. Lüscher, B., Kuenzel, E. A., Krebs, E. G., and Eisenman, R. N. *Myc* oncoproteins are phosphorylated by casein kinase II. *EMBO J. 8*:1111–1119 (1989).
35. Saksela, K., Mäkelä, T. P., Evan, G., and Alitalo, K. Rapid phosphorylation of the L-*myc* protein induced by phorbol ester tumor promoters and serum. *EMBO J. 8*:149–157 (1989).
36. Saksela, K., Mäkelä, T. P., Hughes, K., Woodgett, J. R., and Alitalo, K. Activation of PKC increases phosphorylation of the L-*myc* trans-activator domain at a GSK-3 target site. *Oncogene 7*:347–353 (1992).
37. Woodgett, J. R. A common denominator linking glycogen metabolism, nuclear oncogenes and development. *Trends Biochem. Sci. 16*:177–181 (1991).
38. Mäkelä, T. P., Saksela, K., Evan, G., and Alitalo, K. A fusion protein formed by L-*myc* and a novel gene in SCLC. *EMBO J. 10*:1331–1335 (1991).
39. Mäkelä, T. P., Kere, J., Winqvist, R., and Alitalo, K. Intrachromosomal rearrangements fusing *rlf* and L-*myc* in SCLC. *Mol. Cell. Biol. 11*:4015–4021 (1991).
40. Mäkelä, T. P., Borrello, M. G., Shiraishi, M., and Alitalo, K. Rearrangement and coamplification of *rlf* and L-*myc* in primary SCLC tumors. *Oncogene 7*:405–409 (1992).

29

Biological and Clinical Significance of *MYCN* Amplification in Human Neuroblastomas

Garrett M. Brodeur *Washington University School of Medicine, St. Louis, Missouri*

I. CYTOGENETIC ABNORMALITIES AND TUMOR CELL PLOIDY IN NEUROBLASTOMAS

Three types of cytogenetic abnormality have been identified that are characteristic of human neuroblastomas (Fig. 1): deletion of the short arm of chromosome 1, double minute chromatin bodies (dmins), and homogeneously staining regions (HSRs). While the chromosome 1 deletion may represent loss of a neuroblastoma suppressor gene, the latter two abnormalities are known to be cytogenetic manifestations of gene amplification. The majority of neuroblastomas have near-diploid karyotypes, but a substantial number are hyperdiploid or near-triploid. The relationship between the ploidy of the tumor cells and the specific chromosome abnormalities described above has become clarified, and the molecular and clinical significance of these findings will be reviewed.

A. Deletions and Partial Monosomy for Chromosome 1p

Deletion of the short arm of chromosome 1 (Fig. 1a), resulting in partial 1p monosomy, has emerged as the most consistent cytogenetic abnormality in primary human neuroblastomas and tumor-derived cell lines (Table 1) [1–3]. This was the only numerical or structural cytogenetic abnormality that occurred with statistically increased frequency in near-diploid neuroblastomas ($p < 0.001$) [3]. However, it is notable that a substantial number of tumors karyotyped recently had modal chromosome numbers in the triploid range [4,5]. In contrast to the near-

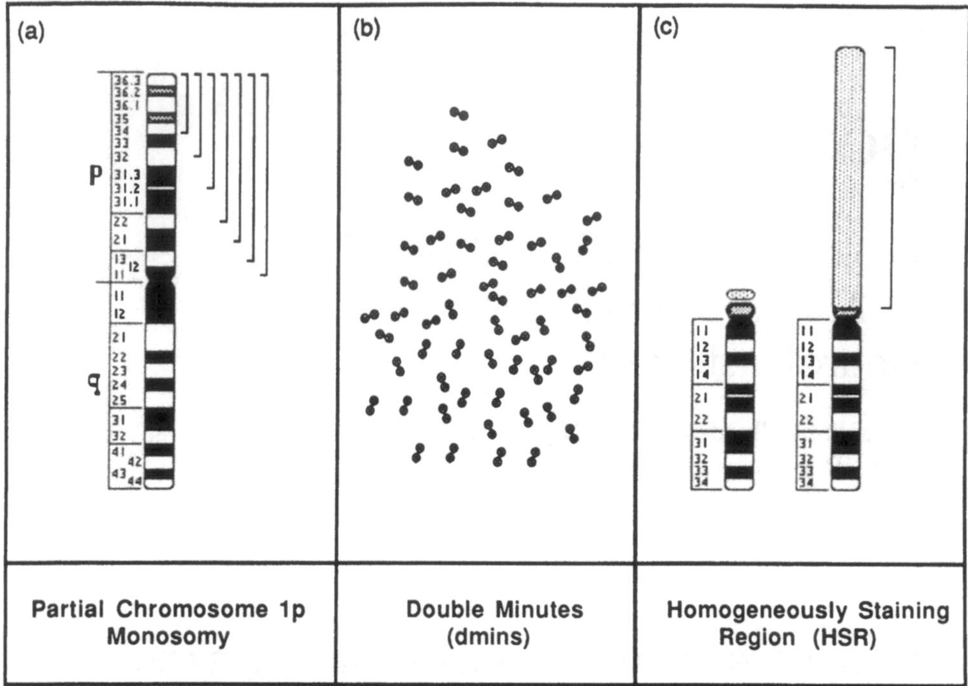

Figure 1 Common cytogenetic abnormalities in human neuroblastomas: diagrammatic representations of the three most common cytogenetic abnormalities seen in human neuroblastomas. (a) Deletions of the short arm of chromosome 1: brackets indicate that the region deleted in different tumors is variable in terms of its proximal break point, but the distal short arm appears to be deleted in all cases, resulting in partial 1p monosomy. (b) Extrachromosomal double-minute chromatin bodies (dmins). Dmins are seen on about 30% of primary neuroblastomas and a larger number of neuroblastoma cell lines. These are a cytogenetic manifestation of gene amplification, and in all cases analyzed so far, the amplification unit included the *MYCN* oncogene. (c) Homogeneously staining region (HSR). A representative HSR on the short arm of chromosome 13 is shown, but HSRs in neuroblastomas have been identified on at least 18 of the 24 different human chromosomes [24]. HSRs are a cytogenetic manifestation of gene amplification in which the amplified sequences are chromosomally integrated. All neuroblastomas with HSRs studied to date have had *MYCN* amplification.

diploid tumors, the near-triploid tumors rarely if ever had deletion or rearrangements of the short arm of chromosome 1, or cytogenetic manifestations of gene amplification.

The region of chromosome 1 most commonly deleted is between 1p32 and 1pter (Fig. 1a). However, recent molecular studies have localized the region that is most consistently deleted to subbands of chromosome 1p36 [6,7]. This deletion is

Table 1 Selected Chromosome Abnormalities in Near-Diploid Neuroblastomas

Tissue source	Del 1p	Double minute chromosomes	Homogeneously staining regions
Primary tumors	23/33 (70%)	15/33 (45%)	1/33 (3%)
Cell lines	23/27 (85%)	15/27 (56%)	14/27 (52%)
Total	46/60 (77%)	30/60 (50%)	15/60 (25%)

Source: Modified from Ref. 1.

thought to represent the loss of a putative neuroblastoma suppressor gene [1–3]. We postulate that loss or inactivation of one or both copies of a gene (or genes) at this site is critical for the development or progression of neuroblastoma. Alternatively, it is possible that this genetic lesion occurs only in a subset of neuroblastomas (see below).

B. Dmins, HSRs, and Gene Amplification

A substantial number of neuroblastomas have either extrachromosomal dmins, chromosomally integrated HSRs, or both in subpopulations of cells (Fig. 1b, 1c). These two abnormalities are cytogenetic manifestations of gene amplification. Dmins are the predominant form of amplified DNA in primary tumors, but dmins and HSRs are found with about equal frequency in neuroblastoma cell lines [1,3] (Table 1). Indeed, dmins and/or HSRs occur in about 90% of neuroblastoma cell lines. Evidence suggests that there is selection in vitro for cell lines derived from tumors that have preexisting dmins or HSRs, and there is no evidence to date that these abnormalities develop with time in culture, at least in neuroblastomas. It is unclear why HSRs are a more common form of amplified DNA in established cell lines than they are in primary tumors.

C. Tumor Cell Ploidy

Flow cytometric analysis of the DNA content (ploidy) of human neuroblastoma cells was first reported in a series of 35 infants [8]. In this analysis, abnormally high DNA content was associated with lower stages of tumor and a good response to initial chemotherapy, whereas subjects with a "diploid" DNA content were likely to have metastatic disease and a poor response to chemotherapy. Although the latter group of tumors had a "normal" DNA content, they likely had subtle genetic abnormalities that were not detectable by flow cytometry, such as dmins or chromosome 1 deletions. Subsequent studies have confirmed the prognostic importance of flow cytometric measurement of DNA content of neuroblastoma cells [9]. This modality provides a complementary approach to the cytogenetic and molecular analysis of human neuroblastomas, and it appears to provide important information in predicting response to therapy and outcome, at least in subsets of patients.

II. MOLECULAR ANALYSIS AND SIGNIFICANCE OF ONCOGENE AMPLIFICATION IN NEUROBLASTOMAS

A. Identification of *MYCN* Amplification

Although cytogenetic analysis of human neuroblastomas frequently has revealed dmins or HSRs in primary tumors and cell lines [1,3], the nature of the amplified sequences was not known until recently. Initially, evidence for amplification of genes associated with drug resistance was sought, but none was found. However, a study was undertaken to determine whether a proto-oncogene was amplified in a panel of neuroblastoma cell lines. An oncogene related to the viral oncogene v-*myc*, but distinct from *MYC*, was amplified in eight of the nine neuroblastoma cell lines tested [10]. The amplified *MYCN* sequence was mapped to the HSRs on different chromosomes in neuroblastoma cell lines, and the normal single copy locus was mapped to the distal short arm of chromosome 2 [11]. Thus, the *MYCN* locus was amplified in neuroblastomas regardless of whether they had dmins or an HSR, and regardless of the chromosomal location of the HSR.

B. Clinical Significance of *MYCN* Amplification

In collaboration with others, we began a study of primary tumors from untreated patients to determine whether *MYCN* amplification occurred. In the initial analysis of 63 primary tumors, amplification ranging from 3- to 300-fold per haploid genome was found in 24 tumors (38%) [12]. All cases with *MYCN* amplification in this initial study came from patients with advanced stages of disease (stages III and IV by the Evans staging system). Next, the progression-free survival of 89 patients was analyzed according to the stage of disease and *MYCN* copy number [13]. *MYCN* amplification clearly was associated with rapid tumor progression and a poor outcome, independent of the stage of the tumor. Two out of 16 patients with less advanced disease (stage II) had amplification. Although in general patients with stage II have a good prognosis, both patients with *MYCN* amplification had rapid tumor progression, compared to only one of the remaining 14 patients.

These studies were extended to more than 800 patients with neuroblastoma enrolled in protocols of the Children's Cancer Study Group (CCSG) and the Pediatric Oncology Group (POG) [1,2]. Examples of *MYCN* amplification seen in some of the primary tumors are shown in Figure 2. In general, the same pattern seen in our earlier studies has been borne out by the larger study (Table 2). It is now clear that, among patients with less advanced stages of disease traditionally associated with a good prognosis, a minority (5–10%) have tumors with *MYCN* amplification [1,2]. Our data indicate that virtually all these patients are destined

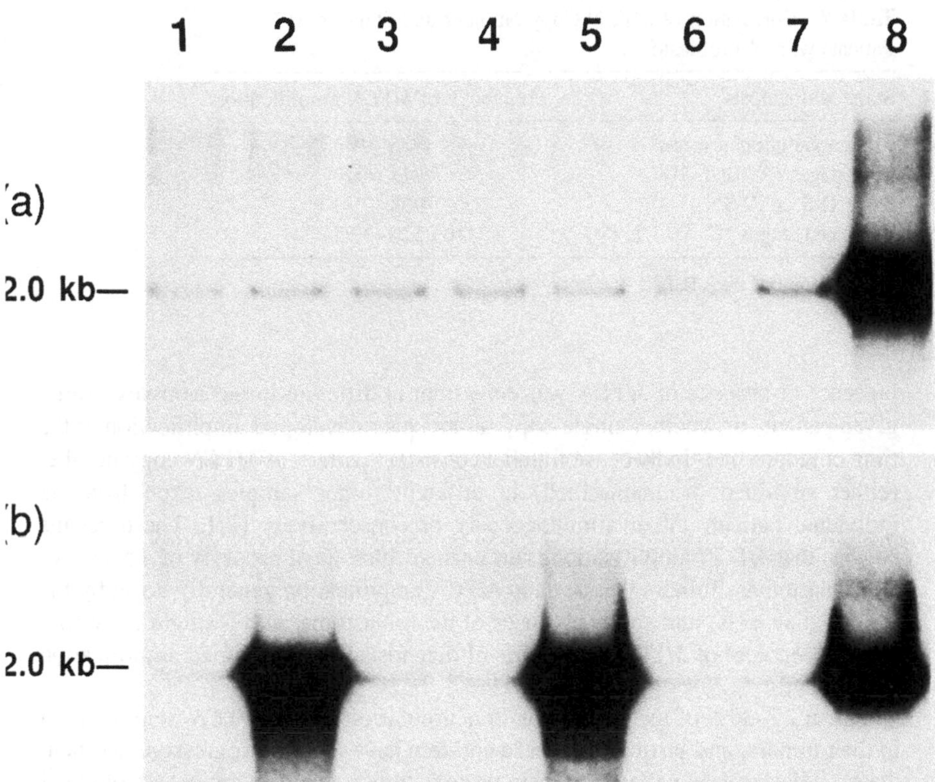

Figure 2 Southern blot showing *MYCN* amplification. Equal amounts of DNA were digested to completion with the restriction enzyme *Eco*RI, electrophoresed, blotted, and hybridized to a radioactive probe for the *MYCN* oncogene. In both rows, lane 1 represents DNA from a normal lymphoblastoid cell line as a single copy control, and lane 8 represents DNA from the NGP cell line, with 150 copies of *MYCN* per haploid genome, as determined by serial dilution and laser densitometry. (a) Lanes 2–7 represent six neuroblastomas with a single copy of *MYCN* per haploid genome. (b) Lanes 2 and 5 show examples of tumors with *MYCN* amplification, whereas the other tumors have the normal single copy signal.

to have rapid tumor progression and a poor outcome, similar to their counterparts with advanced stages of disease. More than 30% of patients with more advanced tumor stages had *MYCN* amplification, and they also had an expectedly poor outcome. Our findings that *MYCN* amplification is associated with a poor outcome regardless of the clinical stage of tumor is supported by preliminary studies from Japan and Europe [14–16].

We analyzed the *MYCN* copy number in multiple simultaneous or consecutive samples of neuroblastoma tissue from 60 patients [17] to determine whether the

Table 2 Correlation of *MYCN* Copy Number and Stage in 817 Patients with Neuroblastoma

Stage at diagnosis	Frequency of *MYCN* amplification
Benign ganglioneuromas	0/18 (0%)
Low stages (A, B; I, II)	9/213 (4%)
Stage D-S or IV-S	4/58 (7%)
Advanced stages (C, D: III, IV)	165/528 (32%)

Source: Modified from Ref. 1.

presence or absence of *MYCN* was consistent in different tumor samples from a given patient, or whether single copy tumors ever developed amplification at the time of recurrence. Indeed, we found a consistent pattern of *MYCN* copy number (either amplified or unamplified) in different tumor samples taken from an individual patient, either simultaneously or consecutively [17]. These results suggest that *MYCN* amplification is an intrinsic biological property of a subset of neuroblastomas. Tumors that develop *MYCN* amplification generally do so by the time of diagnosis, and so far no cases of neuroblastoma with a single copy (per haploid genome) of *MYCN* at the time of diagnosis have developed amplification subsequently.

About 25–30% of the children with neuroblastoma have *MYCN* amplification in their tumors, and virtually all these children have rapidly progressive and fatal disease. However, in patients with single copy tumors, there is not yet a biological marker or explanation why half these patients do not survive. A general correlation has been demonstrated between *MYCN* copy number and expression, but a high level of *MYCN* expression in single copy tumors does not appear to identify a subset with a particularly poor outcome [16,18–20]. It is still possible that activation of *MYCN* by mechanisms other than amplification or overexpression may play an important role. In addition, it is likely that either activation of other oncogenes, deletion of chromosome 1p, or other genetic lesions may contribute to the poor clinical outcome in these patients. We have sought evidence for amplification or overexpression of other oncogenes, including *MYC*, *MYCL*, *RASN*, *RASH*, *RASK*, *EGFR1*, *HER2*, *SIS*, *SRC*, *MYB*, *FOS*, and *ETS* in neuroblastomas, but none was found [1]. Thus, in the subset of the patients lacking *MYCN* amplification, there is no consistent evidence to date correlating activation of an oncogene other than *MYCN* with poor outcome.

C. The Association Between Chromosome 1p Deletion and *MYCN* Amplification

The presence of *MYCN* amplification in human neuroblastomas has been shown to correlate strongly with advanced clinical stage and poor prognosis [1,12,13,

15,16,18–20]. Our recent studies [6] showed a very strong correlation between *MYCN* amplification and chromosome 1p LOH ($p < 0.001$), indicating that LOH was common in patients with amplification (8 out of 9 patients, or 89%). Indeed, the correlation between these two abnormalities in neuroblastoma cell lines is almost 100% [1,3]. Both *MYCN* amplification and deletion of chromosome 1p (as detected by cytogenetic analysis) [4,5,21] appear to be strongly correlated with a poor clinical outcome and with each other, but it is not yet clear whether they are independent prognostic variables. Nevertheless, they appear to characterize a genetically distinct subset of very aggressive neuroblastomas.

Our data showing the consistency of *MYCN* copy number over time [17] would suggest that *MYCN* amplification is an intrinsic biological property of a subset of tumors, so it must occur relatively early in these cases. Since cases with *MYCN* amplification represent a subset of patients with chromosome 1p deletion, we suspect that the 1p deletion may precede the development of amplification. Recent studies of the cytogenetic features of tumors identified as a result of mass screening of infants for neuroblastomas in Japan [22] suggest that the majority of patients identified have lower stages of disease, and virtually all the tumors are in the hyperdiploid or triploid range.

Flow cytometric analyses of the DNA content of neuroblastomas, as well as cytogenetic studies of modal chromosome number, have associated hyperdiploid and triploid karyotypes with lower stages of disease and favorable outcomes [4,5,8,9,22,23]. Therefore, the results of the screening study have suggested at least two possibilities: either (1) all neuroblastomas begin as tumors with a more favorable genotype and phenotype, and some evolve into more aggressive tumors with adverse genetic features; or (2) there are at least two different subsets of neuroblastoma, and the more favorable group presents earlier and therefore is the predominant group detected by screening. Our data, combined with those discussed above, are more consistent with the latter explanation [1,2,6,21].

III. STRUCTURAL ANALYSIS OF THE *MYCN* AMPLIFIED DOMAIN IN NEUROBLASTOMAS

Estimates of the size of the amplified domain around the *MYCN* proto-oncogene in neuroblastomas have ranged from 300 to 3000 kb, based on physical, chemical, and electrophoretic measurements of the amplified DNA [24]. However, all these approaches to the mapping of the size of the amplified domain have been indirect. An attempt has been made to clone and map the amplified domain around *MYCN* in a representative neuroblastoma cell line [25,26]. A region spanning 140 kb around the *MYCN* locus was mapped with cosmid and λ clones, and additional amplified clones were identified that were not contained in the 140 kb contiguous region. The entire 140 kb contiguous locus was amplified in a panel of 12 primary neuroblastomas with *MYCN* amplification, whereas the noncontiguous fragments were amplified in subsets of them [26]. These data indicate that although each tumor had a relatively unique pattern of amplified DNA fragments, there was considerable similarity among the amplification units of different tumors.

Amler and Schwab [27] have analyzed the amplified domain of a series of neuroblastoma cell lines with *MYCN* amplification, most in the form of chromosomally integrated HSRs. They analyzed the amplified domain by pulsed-field gel electrophoresis and hybridization with DNA probes that represent the 5′ and 3′ ends of the *MYCN* gene. They confirmed the heterogeneity of size of the amplified domain seen in different neuroblastomas demonstrated by earlier studies [25,26,28,29]. They also concluded that most amplified regions of DNA consisted of multiple tandem arrays of DNA segments ranging in size from 100 to 700 kb, and that *MYCN* was at or near the center of the amplified units. Nonetheless, their data were most consistent with both tandem and inverted repeats, and rearrangements were more commonly found in the cell lines with higher *MYCN* copy number (>50–100 copies per haploid genome).

We have begun a structural analysis of the amplified domain in human neuroblastomas in order to determine the size and heterogeneity of this region in different tumors and cell lines, as well as its clinical implications. Because of the large size of the domain, we have used the yeast artificial chromosome (YAC) cloning vector system [30]. To date, 16 YACs have been identified, which contain segments of the amplified domain from a representative neuroblastoma cell line, and the YACs can be arranged in a contiguous linear map of ≥ 1 Mb [30]. All the YAC clones are consistent with a single, linear map of the region, with few if any rearrangements identified thus far. Our data also indicate that the core of the domain amplified in different tumors is at least 500 kb. The very large size suggests that there may be other genes near *MYCN* whose expression is important in mediating the aggressive phenotype associated with *MYCN* amplification. Alternatively, the nearest origin(s) of replication may be a considerable distance from *MYCN*, requiring a large domain to be stably replicated as an extrachromosomal element.

Our data show greater uniformity of the amplified domain within a given tumor and are more consistent with the deletion model for the initial event in the development of gene amplification, as suggested by Wahl and others [31–33]. This is also supported by recent data documenting excision of *MYCN* from one of the two chromosome 2 homologues [34]. The deletion model suggests that a large region containing the selectable marker and an origin of replication is deleted and circularizes, forming an extrachromosomal episome (Fig. 3). This episome segregates randomly during cell division, but cells with more copies of the marker gene have a selective advantage and accumulate to a certain stable average number. Although there would inevitably be heterogeneity in the number of episomes per cell, the average number in a population should be relatively stable. Larger episomes or episome multimers would be visible with the light microscope and called dmins. As a rare event, the episomes would integrate into a chromosome at

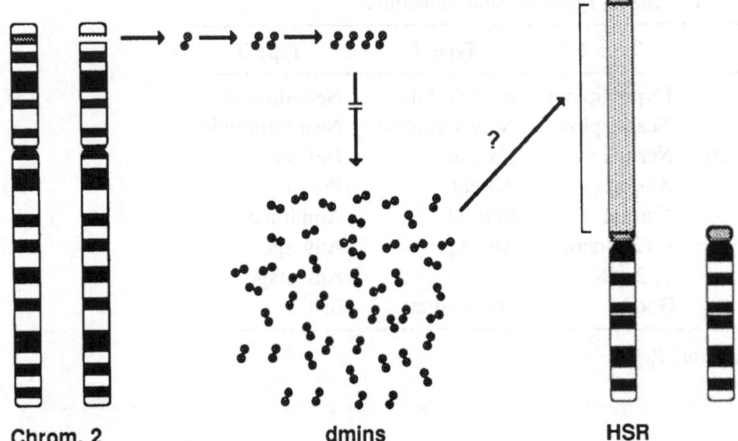

Figure 3 Hypothetical process of *MYCN* amplification in human neuroblastomas. The location of the normal *MYCN* gene is shown by the shaded band, 2p24 on the left. The current data suggest that a large region of DNA containing the *MYCN* gene is deleted from one homologue of chromosome 2 and forms an extrachromosomal element or episome, which is probably circular. It must contain an origin of replication but lacks a centromere or kinetocore, so it does not segregate evenly between daughter cells during mitosis. However, since it seems to provide a growth advantage in vivo and in vitro, the episomes accumulate in a subset of cells but remain unstable and therefore heterogeneously distributed among the cells in the population. If these episomes are large enough to be seen in the light microscope, they are called dmins. As a rare event, particularly in cells in vitro with preexisting *MYCN* amplification, the dmins linearly integrate into a chromosome at a seemingly random site.

an apparently random site forming an HSR (Fig. 3). Because of the secondary recombinational events, the structure of the amplified domains from HSRs would likely be more rearranged and heterogeneous. This might explain the apparent discrepancies in results of amplified domain analysis seen between various laboratories.

IV. SUMMARY AND CONCLUSIONS

Patterns are emerging, based on cytogenetic, molecular, and flow cytometric analysis, which suggest that neuroblastomas may be assigned to three genetically distinct groups (Table 3). The first comprises those with hyperdiploid or triploid modal karyotypes (or compatible DNA content by flow cytometry). Deletion of chromosome lp, dmins, HSRs, or molecular evidence of *MYCN* amplification are rarely seen. These patients are more likely to be infants with localized disease, and

Table 3 Clinical/Genetic Types of Neuroblastoma

Feature	Type 1	Type 2	Type 3
Ploidy	Hyperdiploid Near-triploid	Near-diploid Near-tetraploid	Near-diploid Near-tetraploid
Chromosome 1p	Normal	Normal	Deleted
Dmins, HSR	Absent	Absent	Present
MYCN copy	Normal	Normal	Amplified
Age	<12 months	Any age	Any age
Stage	1, 2, 4S	3, 4	Any stage
Outcome	Good	Intermediate	Bad

Source: Modified from Ref. 2.

they generally have a very favorable prognosis. The second group consists of tumors that are generally near-diploid or tetraploid modal chromosome number or DNA content. These tumors lack chromosome 1p deletion, dmins, HSRs, or *MYCN* amplification, and the patients are more likely to be more than 1 year old and to have advanced stages of disease; while their overall outcome is generally poor, their tumors respond at least initially to treatment, and they may survive for several years before they succumb to their disease. The third group consists of tumors that are also generally near-diploid or tetraploid, with chromosome 1p deletion (LOH), with or without dmins or HSRs (i.e., *MYCN* amplification). These patients also are more likely to be more than 1 year old and to have advanced stages of disease. In contrast to the second group, they generally respond to treatment only transiently, if at all; they experience rapid progression and die within months to a year or so. It remains to be determined whether tumors in one group ever evolve or "progress" into a less unfavorable group, but current evidence would suggest that they are genetically distinct.

The molecular and cytogenetic analysis of human neuroblastomas promises to make available a great deal of information that would be difficult to obtain otherwise. First, these studies may permit the localization of one or more neuroblastoma predisposition loci. This would make possible the identification of hereditary cases, so that family counseling, as well as prenatal diagnosis of affected individuals in informative families, could be provided. Second, genetic markers provide a more objective means to classify tumors that may appear similar histologically. Third, genetic analysis by karyotype, flow cytometry, and/or molecular analysis (for oncogene activation or allelic loss) provides information that has prognostic significance and can direct the most appropriate choice of treatment. Finally, genetic analysis of tumor-specific rearrangements may permit the identification of specific genetic lesions and biochemical pathways on which future therapeutic approaches may be focused.

ACKNOWLEDGMENTS

Some of the material in this chapter has been published previously [1,2]. This work was supported in part by National Institutes of Health grants RO1-CA-39771, KO4-CA-01027, and U10-CA-05887, the Children's United Research Effort, and the Fern Waldman Memorial Fund for Research in Childhood Cancer.

REFERENCES

1. Brodeur, G. M., and Fong, C. T. Molecular biology and genetics of human neuroblastoma. *Cancer Genet. Cytogenet. 41*:153–174 (1989).
2. Brodeur, G. M. Neuroblastoma—Clinical applications of molecular parameters. *Brain Pathol. 1*:47–54 (1990).
3. Brodeur, G. M., Green, A. A., Hayes, F. A., Williams, K. J., Williams, D. L., and Tsiatis, A. A. Cytogenetic features of human neuroblastomas and cell lines. *Cancer Res. 41*:4678–4686 (1981).
4. Kaneko, Y., Kanada, N., Maseki, N., Sakurai, M., Tsuchida, Y., Takeda, T., Okabe, I., and Sakurai, M. Different karyotypic patterns in early and advanced stage neuroblastomas. *Cancer Res. 47*:311–318 (1987).
5. Christiansen, H., and Lampert, F. Tumour karyotype discriminates between good and bad prognostic outcome in neuroblastoma. *Br. J. Cancer, 57*:121–126 (1988).
6. Fong, C. T., Dracopoli, N. C., White, P. S., Merrill, P. T., Griffith, R. C., Housman, D. E., and Brodeur, G. M. Loss of heterozygosity for chromosome 1p in human neuroblastomas: Correlation with N-*myc* amplification. *Proc. Natl. Acad. Sci. USA, 86*:3753–3757 (1989).
7. Weith, A., Martinsson, T., Cziepluch, C., Bruderlein, S., Amler, L. C., and Berthold, F. Neuroblastoma consensus deletion maps to 1p36.1-2. *Genes Chromosomes Cancer, 1*:159–166 (1989).
8. Look, A. T., Hayes, F. A., Nitschke, R., McWilliams, N. B., and Green, A. A. Cellular DNA content as predictor of response to chemotherapy in infants with unresectable neuroblastoma. *New Engl. J. Med. 311*:231–235 (1984).
9. Look, A. T., Hayes, F. A., Shuster, J. J., Douglass, E. C., Castleberry, R. P., and Brodeur, G. M. Clinical implications of tumor cell ploidy and N-*myc* gene amplification in childhood neuroblastoma. *J. Clin. Oncol. 9*:581–591 (1991).
10. Schwab, M., Alitalo, K., Klempnauer, K. H., Varmus, H. E., Bishop, J. M., Gilbert, F., Brodeur, G., Goldstein, M., and Trent, J. M. Amplified DNA with limited homology to *myc* cellular oncogene is shared by human neuroblastoma cell lines and a neuroblastoma tumour. *Nature, 305*:245–248 (1983).
11. Schwab, M., Varmus, H. E., Bishop, J. M., Grzeschik, K. H., Naylor, S. L., Sakaguchi, A. Y., Brodeur, G., and Trent, J. Chromosome localization in normal human cells and neuroblastomas of a gene related to c-*myc*. *Nature, 308*:288–291 (1984).
12. Brodeur, G. M., Seeger, R. C., Schwab, M., Varmus, H. E., and Bishop, J. M. Amplification of N-*myc* in untreated human neuroblastomas correlates with advanced disease stage. *Science, 224*:1121–1124 (1984).

13. Seeger, R. C., Brodeur, G. M., Sather, H., Dalton, A., Siegel, S. E., Wong, K. Y., and Hammond, D. Association of multiple copies of the N-*myc* oncogene with rapid progression of neuroblastomas. *New Engl. J. Med. 313*:1111–1116 (1985).
14. Tsuda, T., Obara, M., Hirano, H., Gotoh, S., Kubomura, S., Higashi, K., Kuroiwa, A., Nakagawara, A., Nagahara, N., and Shimizu, K. Analysis of N-*myc* amplification in relation to disease stage and histologic types in human neuroblastomas. *Cancer, 60*:820–826 (1987).
15. Nakagawara, A., Ikeda, K., Tsuda, T., Higashi, K., and Okabe, T. Amplification of N-*myc* oncogene in stage II and IVS neuroblastomas may be a prognostic indicator. *J. Pediatr. Surg. 22*:415–418 (1987).
16. Bartram, C. R., and Berthold, F. Amplification and expression of the N-*myc* gene in neuroblastoma. *Eur. J. Pediatr. 146*:162–165 (1987).
17. Brodeur, G. M., Hayes, F. A., Green, A. A., Casper, J. T., Wasson, J., Wallach, S., and Seeger, R. C. Consistent N-*myc* copy number in simultaneous or consecutive neuroblastoma samples from sixty individual patients. *Cancer Res. 47*:4248–4253 (1987).
18. Seeger, R. C., Wada, R., Brodeur, G. M., Moss, T. J., Bjork, R. L., Sousa, L., and Slamon, D. J. Expression of N-*myc* by neuroblastomas with one or multiple copies of the oncogene. *Progr. Clin. Biol. Res. 271*:41–49 (1988).
19. Nisen, P. D., Waber, P. G., Rich, M. A., Pierce, S., Garvin, J. R. J., Gilbert, F., and Lanskowsky, P. N-*myc* oncogene RNA expression in neuroblastoma. *J. Natl. Cancer Inst. 80*:1633–1637 (1988).
20. Slavc, I., Ellenbogen, R., Jung, W.-H., Vawter, G. F., Kretschmar, C., Grier, H., and Korf, B. R. *myc* gene amplification and expression in primary human neuroblastoma. *Cancer Res. 50*:1459–1463 (1990).
21. Hayashi, Y., Kanda, N., Inaba, T., Hanada, R., Nagahara, N., Muchi, H., and Yamamoto, K. Cytogenetic findings and prognosis in neuroblastoma with emphasis on marker chromosome 1. *Cancer, 63*:126–132 (1989).
22. Hayashi, Y., Inabada, T., Hanada, R., and Yamamoto, K. Chromosome findings and prognosis in 15 patients with neuroblastoma found by VMA mass screening. *J. Pediatr. 112*:67–71 (1988).
23. Franke, F., Rudolph, B., Christiansen, H., Harbott, J., and Lampert, F. Tumour karyotype may be important in the prognosis of human neuroblastoma. *J. Cancer Res. Clin. Oncol. 111*:266–272 (1986).
24. Brodeur, G. M., and Seeger, R. C. Gene amplification in human neuroblastomas: Basic mechanisms and clinical implications. *Cancer Genet. Cytogenet. 19*:101–111 (1986).
25. Kinzler, K. W., Zehnbauer, B. A., Brodeur, G. M., Seeger, R. C., Trent, J. M., Meltzer, P. S., and Vogelstein, B. Amplification units containing human N-*myc* and c-*myc* genes. *Proc. Natl. Acad. Sci. USA, 83*:1031–1035 (1986).
26. Zehnbauer, B. A., Small, D., Brodeur, G. M., Seeger, R., and Vogelstein, B. Characterization of N-*myc* amplification units in human neuroblastoma cells. *Mol. Cell. Biol. 8*:522–530 (1988).
27. Amler, L. C., and Schwab, M. Amplified N-*myc* in human neuroblastoma cells is often arranged as clustered tandem repeats of differently recombined DNA. *Mol. Cell. Biol. 9*:4903–4913 (1989).

28. Shiloh, Y., Shipley, J., Brodeur, G. M., Bruns, G., Korf, B., Donlon, T., Schreck, R. R., Seeger, R., Sakai, K., and Latt, S. A. Differential amplification, assembly and relocation of multiple DNA sequences in human neuroblastomas and neuroblastoma cell lines. *Proc. Natl. Acad. Sci. USA*, *82*:3761–3765 (1985).
29. Shiloh, Y., Korf, B., Kohl, N. E., Sakai, K., Brodeur, G. M., Harris, P., Kanda, N., Seeger, R. C., Alt, F., and Latt, S. A. Amplification and rearrangement of DNA sequences from the chromosomal region 2p24 in human neuroblastomas. *Cancer Res.* *46*:5297–5301 (1986).
30. Schneider, S. S., Zehnbauer, B. A., Vogelstein, B., and Brodeur, G. M. Yeast artificial chromosome (YAC) vector cloning of the *MYCN* amplified domain in human neuroblastomas. *Progr. Clin. Biol. Res.* *366*:71–76 (1991).
31. Carroll, S., DeRose, M., Gaudray, P., Moore, C., VanDevanter, D. N., Von Hoff, D., and Wahl, G. Double minute chromosomes can be produced from precursors derived from a chromosomal deletion. *Mol. Cell. Biol.* *8*:1525–1533 (1988).
32. Stark, G. R., Debatisse, M., Giulotto, E., and Wahl, G. M. Recent progress in understanding mechanisms of mammalian DNA amplification. *Cell*, *57*:901–908 (1989).
33. Wahl, G. M. The importance of circular DNA in mammalian gene amplification. *Cancer Res.* *49*:1333–1340 (1989).
34. Hunt, J. D., Valentine, M., and Tereba, A. Excision of N-*myc* from chromosome 2 in human neuroblastoma cells containing amplified N-*myc* sequences. *Mol. Cell. Biol.* *10*:823–829 (1990).

30

Amplification of the c-*abl* Oncogene in Chronic Myelogenous Leukemia

Steven J. Collins and Mark T. Groudine *Fred Hutchinson Cancer Research Center and University of Washington Medical School, Seattle, Washington*

I. INTRODUCTION

The human c-*abl* cellular oncogene was originally identified by virtue of its homology to v-*abl*, the transforming gene of the Abelson murine leukemia virus (Ab-MuLV). This virus is a recombinant of the Moloney murine leukemia virus genome and the normal mouse cellular c-*abl* gene [1]. Ab-MuLV is an acutely oncogeneic retrovirus that induces pre-B-cell leukemia approximately 2–3 weeks after injections into susceptible mouse strains. Moreover v-*abl* is also capable of transforming rodent fibroblast cell lines such as NIH3T3 cells in vitro. The mechanism of transformation of cells by v-*abl* is unclear; however, the tyrosine kinase activity demonstrated in v-*abl* appears to be essential for inducing malignant transformation. Although the c-*abl* cellular oncogene is expressed in a variety of different normal tissues [2], its normal physiological role remains unclear.

Abnormalities in the structure and expression of the human c-*abl* cellular oncogene have been associated with Philadelphia chromosome positive chronic myelogenous leukemia (CML). This form of leukemia is a clonal disorder involving a pluripotent stem cell from which granulocytes, erythrocytes, platelets, monocytes, macrophages, and B lymphocytes arise [3]. CML has distinct clinical phases, including a relatively benign chronic phase, characterized by hyperplasia of functionally mature granulocytes, and a blast crisis phase, usually lethal, in which immature blasts predominate. An important genetic event in the clonal development of CML involves the acquisition of the Philadelphia chromosome (Ph), which is present in more than 90% of cases of CML. This aberrant chromosome marker is generated by a reciprocal translocation between chromosomes 9 and 22 in which the c-*abl* oncogene is translocated from the distal end of the q arm of chromosome 9 to a relatively restricted 5–6 kb region on chromosome 22, termed the break point cluster region (*bcr*) [4]. This region lies within the coding sequences of a relatively large (> 50 kb of genomic DNA) gene that has been termed the *bcr* gene [5]. This translocation results in a fusion gene that expresses an 8 kb *bcr-abl* transcript that encodes the aberrant *bcr-abl* fusion protein product (P210) observed in CML cells [6–8]. Although the physiological consequences of the generation of this aberrant fusion protein are unclear, the p210 *bcr-abl* fusion product has enhanced tyrosine kinase activity compared with the normal p145 c-*abl* product [8]. Moreover abnormalities in the structure and expression of the c-*abl* cellular oncogene have not been described in any other type of human malignancy other than CML and Ph positive acute lymphatic leukemia.

In this chapter we discuss various experimental observations related to the possible role of amplification of the *bcr-abl* fusion gene in the pathogenesis of CML and, in particular, in the progression of CML from chronic phase to blast crisis.

II. AMPLIFICATION OF *bcr-abl* IN K-562 AND OTHER CML CELLS

K-562 is a continuously proliferating human hematopoietic cell line that was derived from the malignant pleural effusion of a patient with CML blast crisis. We have noted elsewhere that the K-562 cell line harbors a four- to eightfold genomic amplification of the *bcr-abl* fusion gene that is associated with enhanced expression of the 8 kb *bcr-abl* fusion transcript [9]. Moreover, the λ immunoglobulin constant region locus is also amplified some four- to eightfold in this cell line, and in situ hybridization suggests that both the c-*abl* and λ immunoglobulin constant region genes are tandemly amplified on an aberrant marker chromosome in K-562 [10]. These observations suggest that both the *bcr-abl* fusion gene and the λ immunoglobulin constant region locus reside on the same amplified unit in K-562. Recent long-range mapping studies using pulsed-field gel electrophoresis indicate that the λ immunoglobulin constant region locus resides at least 200 kb but no more than 1200 kb upstream of the *bcr* locus [11]. Thus the amplified unit (amplicon) harboring *bcr-abl* and the λ locus appears to be quite large in these cells. Interestingly while K-562 cells exhibit enhanced expression of the *bcr-abl* fusion transcript, this cell line does not display any transcripts related to the λ immunoglobulin constant region genes.

Genomic amplification of specific cellular oncogenes has been noted in human malignancies of a number of different types. In general the presence of such gene amplification correlates with the progression of malignancy. For example, genomic amplification of the N-*myc* cellular oncogene is present in approximately 40% of cases of childhood neuroblastoma and correlates with advanced anatomic stage of the disease (stages III and IV) and with poor prognosis [12]. This genomic N-*myc* amplification appears to be associated with tumor progression rather than tumor initiation, since early stage tumors are histologically malignant but rarely exhibit N-*myc* genomic amplification. Similarly, genomic amplification of the *neu* (c-*erbB*-2) cellular oncogene is present in approximately 30% of primary human breast cancers and may correlate with advanced disease and a relatively poor prognosis [13]. Given this apparent correlation between genomic cellular oncogene amplification and progression of malignancy, we wished to determine whether the four- to eightfold genomic amplification of *bcr-abl* that we had noted in the K-562 CML blast crisis cell line frequently occurs during the progression of CML from chronic phase to blast crisis.

To address this question, we looked for evidence of genomic amplification of the *bcr-abl* fusion gene in patients with CML at various stages of the disease. We analyzed leukemic cell samples from more than 50 patients with CML—both chronic phase and blast crisis—by Southern blot analysis of restriction endonuclease digested leukemic DNA hybridized with an *abl*-specific probe. We noted the band intensity of the *abl*-specific bands compared with control globin-specific bands. While some CML blast crisis samples harboring multiple Ph chromosomes displayed two- to threefold genomic amplification of *bcr-abl* (see below), in these samples the level of *bcr-abl* gene amplification was directly related to the number of Ph chromosomes per metaphase. These results indicate that a single copy of the *bcr-abl* fusion gene, rather than multiple copies, was present on each of the multiple Ph chromosomes. In none of these fresh patient CML DNA samples was observed the four- to eightfold degree of tandem genomic *bcr-abl* amplification noted in K-562 cells. Thus K-562 is unusual among CML cells in exhibiting this four- to eightfold *bcr-abl* genomic amplification, and genomic amplification of *bcr-abl* appears to occur relatively infrequently in CML.

III. ACQUISITION OF MULTIPLE PHILADELPHIA CHROMOSOMES IN CML

The evolution of CML from chronic phase to blast crisis is commonly associated with karyotypic changes, the most frequent of which is the development of two or more Ph chromosomes. This acquisition of multiple Ph chromosomes is noted in approximately 50% of blast crisis CML samples. Do the additional Ph chromosomes that frequently arise during the progression of CML from chronic phase to blast crisis result from a duplication of a preexisting Ph chromosome, or do they result from a new 9:22 translocation event? Is the acquisition of the additional Ph chromosome(s) important in the evolution of CML from chronic phase to blast crisis? We addressed the first question utilizing a Southern blot analysis of genomic digests of CML samples probed with a *bcr* fragment. The break point cluster region is an approximately 5–6 kb region on chromosome 22, which lies within the coding region of the *bcr* gene. In most cases of CML, the break point on the Ph chromosome lies within the break point cluster region, thus a Southern blot analysis of CML DNA utilizing molecular probes within and flanking this region will detect an aberrant rearranged fragment in addition to the normal chromosome 22 germ line *bcr* fragment. Moreover, there is a variation in the location of the chromosome 9 and 22 break points in individual Ph chromosomes such that the size of the rearranged *bcr* fragment will vary from sample to sample. Thus a restriction endonuclease analysis of CML genomic DNA utilizing *bcr* probes provides a means of "fingerprinting" individual translocation events representing the Ph chromosomes. In samples with more than one Ph chromosome, a second rearranged *bcr* fragment will be noted on a Southern blot analysis if this second Ph arose from a second independent 9:22 translocation. However, in all six CML samples harboring multiple Ph chromosomes that we have analyzed, we have noted only a single rearranged *bcr* fragment, suggesting that the new Ph chromosomes arose from a duplication (or triplication) of the initial Ph, rather than from an independent translocation event [14]. This analysis, however, was limited to analyzing break points in the break point cluster region itself and does not completely rule out a second, independent Ph translocation. For example, in a significant number of cases of Ph-positive acute lymphoblastic leukemia, the chromosome 22 break point lies within the *bcr* gene further upstream from the break point cluster region itself. Chromosome breaks in this region would not be detected with the usual break point cluster region probes. Conceivably, a second Ph chromosome involving a chromosome 22 break further upstream in the *bcr* could arise during the progression of CML from chronic phase to blast crisis, and probes further upstream in the *bcr* gene would be necessary to detect such breaks.

What is the relationship between the acquisition of multiple Ph chromosomes and the evolution of CML from chronic phase to blast crisis? Studies we have performed on a patient with CML who presented in lymphoid blast crisis with three Ph chromosomes/ metaphase provide some insight into the relationship between the acquisition of multiple Ph chromosomes and the evolution of CML to blast crisis [15]. At clinical presentation, patient's peripheral blood contained a significant number of terminally differentiating granulocytes in addition to the lymphoid blast crisis cells, and this allowed for ready separation of chronic phase granulocytes from the lymphoid blasts. A Southern blot analysis readily distinguished these lymphoid blast crisis cells from the terminally differentiating granulocytes, since only the blasts exhibited a rearrangement of the immunoglobulin heavy-chain J_H locus. In addition, Southern blot analysis of the patient blasts utilizing a 5′ c-*abl* probe revealed a rearranged c-*abl* fragment as well as the expected germ line c-*abl* fragment. Southern blot analysis indicated that this rearranged c-*abl* fragment was amplified in the blasts, most likely reflecting the presence of the multiple Ph chromosomes/ metaphase. Thus by comparing the intensity of hybridization of the amplified, rearranged c-*abl* fragment to the germ line c-*abl* fragment, the number of Ph chromosomes harbored by each isolated cell population could be estimated. We isolated blast crisis cells from this patient at the time of initial presentation and at the time of relapse. We also isolated terminally differentiated granulocytes at initial presentation and during a chemotherapy-induced partial "remission" when the peripheral blood blast population fell below 5%. We then performed a Southern blot analysis on each of these samples to see which exhibited amplification of the rearranged c-*abl* fragment. If one of the major genetic events that distinguishes the lymphoid blast crisis cells from the terminally differentiated chronic phase granulocytes is the development of multiple Ph chromosomes with concomitant amplification of the rearranged c-*abl* fragment, then one would predict that the terminally differentiated granulocytes would not display amplification of the rearranged c-*abl* fragment (i.e., precursors of these granulocytes display a single Ph). However, we were somewhat surprised to note that this was clearly not the case; rather, the Southern blot analysis indicated that the c-*abl* fragment was amplified to a similar degree in both the lymphoid blast crisis and granulocyte cell populations. In this particular patient it can be concluded that the development of multiple Ph chromosomes per se was not the sole genetic event leading to blast crisis, suggesting that there are genetic factors besides multiple Ph chromosomes that distinguish CML blast crisis from chronic phase cells.

IV. *bcr-abl* EXPRESSION IN CHRONIC PHASE VERSUS BLAST CRISIS

We had noted earlier by Northern blot and dot blot analysis that CML blast crisis cell lines including K-562, EM-2, and KCL-22 display enhanced expression of the *bcr-abl* fusion transcript when compared with chronic phase CML cells [6]. This enhanced RNA expression is present in the EM-2 and KCL-22 cell lines in the absence of genomic amplification of the *bcr-abl* fusion gene [6]. These observations suggest that increased expression of the aberrant *bcr-abl* fusion transcript might be critical to the progression of CML from chronic phase to blast crisis. Moreover this enhanced expression may not be absolutely dependent on the presence of genomic amplification of *bcr-abl* or of the acquisition of multiple Ph chromosomes. To determine the relationship between enhanced expression of the *bcr-abl* fusion transcript and the progression of CML, we quantitated the relative levels of this hybrid transcript in blast crisis versus chronic phase CML cells [16].

For this study we utilized poly(A^+) RNA extracted from the peripheral blood leukocytes of patients with CML in both chronic phase and blast crisis. Patients were considered to be in chronic phase when peripheral blood indicated a leukocytosis of mostly mature polymorphonuclear leukocytes with fewer than 5% blasts. Patients were considered to be in blast crisis if the peripheral blood smear revealed greater than 15% blasts. Most blast crisis patients in this study had greater than 50% blasts in the peripheral blood. The blast cell population was further enriched on a Ficoll gradient, and in all cases the leukocyte fraction from blast crisis patients consisted of greater than 70% blasts prior to RNA and DNA extraction.

We probed Northern blots of poly(A^+) RNA extracted from these patient samples with a v-*abl* probe, which readily identified the aberrant 8 kb *bcr-abl* fusion transcript originating from the Ph chromosome as well as the normal 7 and 6 kb c-*abl* mRNA transcripts, which are transcribed from the untranslocated c-*abl* gene on chromosome 9. All these CML samples, both chronic phase and blast crisis, exhibited detectable levels of the *bcr-abl* fusion transcript. In virtually all these individual CML blast crisis samples, a comparison of hybridization intensity of these three *abl* transcripts indicated that the aberrant 8 kb *bcr-abl* fusion transcript is the predominant *abl*-related message. Nevertheless we found considerable variation in the amount of this aberrant transcript from sample to sample, with approximately 50% of the blast crisis patients exhibited relatively high amounts of this aberrant transcript, while in the remaining samples the level of the *bcr-abl* transcript was comparable to the relatively low levels exhibited by chronic phase CML samples [16]. Relatively high levels of the *bcr-abl* fusion transcript were noted only in blast crisis samples. We have not to date observed a chronic phase patient who exhibits relatively high levels of *bcr-abl* transcript. Our observations indicate no consistent relationship between enhanced expression of the *bcr-abl* hybrid transcript and the blast crisis phase of CML.

Some of the patient blast crisis CML cells exhibit levels of the aberrant *bcr-abl* fusion transcript comparable to the K-562 blast crisis cell line, which has amplified the c-*abl* oncogene four- to eightfold. However while some of these high-expressing blast crisis samples displayed multiple Ph chromosomes/ metaphase, others harbored a single Ph chromosome and did not demonstrate any evidence of genomic amplification of c-*abl* by Southern blot analysis [16]. Therefore, as is the case with the EM-2 and KCL-22 CML blast crisis cell lines noted above, mechanisms other than increased copy number of the *bcr-abl* fusion gene must be involved in the enhanced expression of the *bcr-abl* transcript in certain fresh patient blast crisis samples.

V. SUMMARY

The aberrant *bcr-abl* fusion gene is noted in more than 90% of cases of chronic myelogenous leukemia, and thus aberrant expression of an altered c-*abl* cellular oncogene appears to be critical to the pathogenesis of CML. Interestingly, with the exception of Ph chromosome positive acute lymphatic leukemia, abnormalities in the structure and expression of c-*abl* have not been noted in any other human malignancy. The normal cell function of c-*abl* and the physiological role of the altered *bcr-abl* fusion gene in the development of CML remain unclear. Our studies indicate that genomic amplification of the *bcr-abl* fusion gene is relatively rare in both CML chronic phase and blast crisis. Relatively low level expression of *bcr-abl* appears to sufficiently alter hematopoietic stem cell function, with the consequent development of chronic phase CML. Although increased *bcr-abl* expression is noted in some cases of CML blast crisis, this is by no means a universal development. It appears that genetic events other than alteration of *bcr-abl* structure and expression, perhaps involving "tumor suppressor" genes such as p53 [17], may play an important role in the evolution of CML from chronic phase to blast crisis.

REFERENCES

1. Abelson, H., and Rabstein L. Lymphosarcoma: Virus-induced thymic independent disease in mice. *Cancer Res.* *30*:2213–2222 (1970).
2. Wang, J., and Baltimore, D. Cellular RNA homologous to the Abelson murine leukemia virus transforming gene: Expression and relationship to the viral sequence. *Mol. Cell. Biol.* *3*:773–779 (1983).
3. Fialkow, P., Jacobson, R., and Papayannopoulou, T. Chronic myelocytic leukemia: Clonal origin in a stem cell common to the granulocytic, erythrocyte, platelet and monocyte/ macrophage. *Am. J. Med.* *63*:125–131 (1977).
4. Groffen, J., Stephenson, J., Heisterkamp, N., deKlein, A., Bartram, C., and Grosveld, G. Philadelphia chromosomal breakpoints are clustered within a limited region, *bcr*, on chromosome 22. *Cell*, *36*:93–99 (1984).
5. Heisterkamp, N., Stam, J., Groffen, J., deKlein, A., and Grosveld, G. Structural organization of the *bcr* gene and its role in the Ph translocation. *Nature*, *315*:758–761 (1985).
6. Collins, S., Kubonishi, I., Miyoshi, I., and Groudine, M. Altered transcription of the c-*abl* oncogene in K-562 and other chronic myelogenous leukemia cells. *Science*, *225*: 72–75 (1984).
7. Shtivelman, E., Lifshitz, B., Gale, R., and Canaani, E. Fused transcripts of *abl* and *bcr* genes in chronic myelogenous leukemia. *Nature*, *315*:758–761 (1985).
8. Konopka, J., Watanabe, S., and Witte, O. An alteration of the human c-*abl* protein in K-562 leukemia cells unmasks associated tyrosine kinase activity. *Cell*, *37*:1035–1042 (1984).
9. Collins, S., and Groudine, M. Rearrangement and amplification of c-*abl* sequences in the human chronic myelogenous leukemia cell line K-562. *Proc. Natl. Acad. Sci. USA*, *80*:4813–4817 (1983).
10. Selden, J., Emanuel, B., Wang, E., Cannizzaro, L., Palumbo, A., Erikson, J., Nowell, P., Rovera, G., and Croce, C. Amplified C lambda and c-*abl* gene are on the same marker chromosome in K-562 cells. *Proc. Natl. Acad. Sci. USA*, *80*:7289–7292 (1983).

11. Westbrook, C., Rubin, C., Carrino, J., LeBeau, M., Bernards, A., and Rowley, J. Long-range mapping of the Philadelphia chromosome by pulsed-field gel electrophoresis. *Blood*, *71*:697–702 (1988).
12. Brodeur, G., Seeger, R., Schwab, M., Varmus, H., and Bishop, J. Amplification of N-*myc* in untreated human neuroblastoma correlates with advanced disease stage. *Science*, *224*:1121–1124 (1984).
13. Slamon, D., Clark, G., Wong, S., Levin, W., Ullrich, A., and McGuire, W. Human breast cancer: Correlation of relapse and survival with amplification of the *HER-2/neu* oncogene. *Science*, *235*:177–182 (1987).
14. Collins, S. Breakpoints on chromosomes 9 and 22 in Philadelphia chromosome positive CML: Amplification of rearranged c-*abl* oncogenes in CML blast crisis. *J. Clin. Invest*. *78*:1392–1396 (1986).
15. Collins, S., and Groudine, M. Chronic myelogenous leukemia: Amplification of a rearranged c-*abl* oncogene in both chronic phase and blast crisis. *Blood* *69*:893–898 (1987).
16. Andrews, D., and Collins, S. Heterogeneity in expression of the *bcr-abl* fusion transcript in CML blast crisis. *Leukemia* *1*:718–724 (1987).
17. Ahuja, H., Bar-Eli, M., Advani, S., Benchimol, S., and Cline, M. Alterations in the p53 gene and the clonal evolution of the blast crisis of chronic myelocytic leukemia. *Proc. Natl. Acad. Sci. USA*, *86*:6783–6787 (1989).

31

Gene Amplification in Small Cell Carcinoma of the Lung

Edward A. Sausville *National Cancer Institute, National Institutes of Health, Bethesda, Maryland*

I. INTRODUCTION

Epithelial bronchogenic neoplasms are of four histological types. Epidermoid (squamous) carcinoma, adenocarcinoma, and large cell undifferentiated carcinoma collectively are called "non-small-cell" carcinoma and are initially approached by consideration of surgical resection. In contrast, small cell carcinoma (SCLC) of the lung, expected to account for approximately 25% of the 150,000 cases of lung cancer diagnosed in the United States in 1990, has a distinctive clinical and biological presentation. There is an early propensity for metastatic spread, rendering attempts at resection usually not possible. Frequently, SCLC expresses markers of neuroendocrine differentiation, including the elaboration of peptide hormones, the production of neuronal isoenzymes such as neuron-specific enolase, the B isoenzyme of creatine phosphokinase, and enzymes of catecholamine synthesis such as dopamine decarboxylase. A detailed cell biological and molecular analysis of several dozen cell lines derived from patients with SCLC as well as examination of tumor tissue taken directly from patients reveals the frequent occurrence of nonrandom chromosomal deletions, autocrine growth factor elaboration, and gene amplification [1]. This chapter focuses principally on the amplification of the *myc* proto-oncogene family in small cell lung carcinoma cell lines. For a systematic discussion of the *myc* gene family products, their occurrence in SCLC, as well as their relation to other types of lung carcinoma, the review by Saksela is very useful [2].

II. AMPLIFICATION OF c-*myc*

A. In SCLC Cell Lines or Tumors

Little and colleagues [3] demonstrated that 8 of 18 lung cancer cell lines had detectable (approximately 20–80-fold) amplification of c-*myc* (Fig. 1). In all cases the cell lines amplified for c-*myc* had greatly increased levels of c-*myc* mRNA, with three of five amplified cell lines also contained double minute chromosomes or homogeneously staining regions.

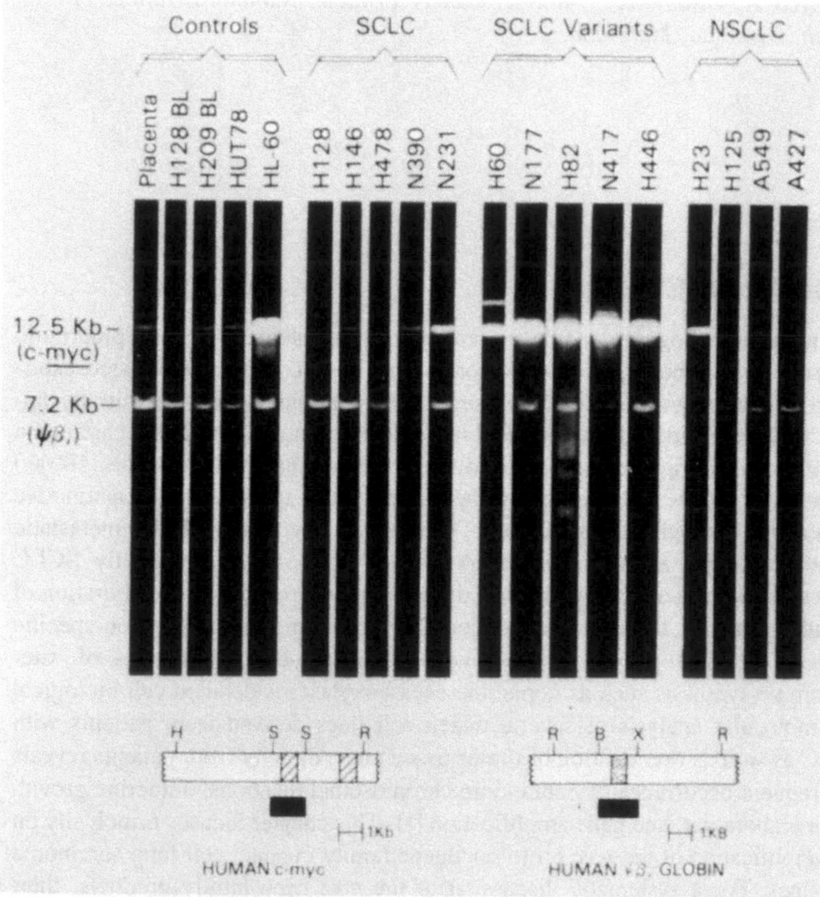

Figure 1 Amplification of c-*myc* in lung cancer cell lines. *Eco*RI-digested DNA from the indicated cell lines was electrophoresed, blotted, and hybridized to the indicated probes from c-*myc* or a pseudo-β-globin gene. (Adapted from Ref. 3.)

Two of six SCLC cell lines examined by Saksela et al. [4] also had approximately 20-fold amplification of c-*myc*, as well as four of the SCLC cell lines studied by Buys et al. [5]. Interestingly, cells from the pleural effusion that gave rise to one of the latter cell lines had only about five-fold amplification of c-*myc*, compared to greater than 30-fold amplification (at least) observed in the patient's derived cell line. In fact, Saksela [2] has noted that the frequency of c-*myc* gene amplification in clinically derived specimens of SCLC appears to be less than the readily documented frequency of this gene's amplification in SCLC cell lines in vitro.

The gross structural features of the c-*myc* gene in amplified cell lines as compared to an unamplified SCLC cell line has been studied by Seifter et al. [6]. An SCLC cell line that had a 40–50-fold amplification of c-*myc* had at least a 250-fold increase in steady state c-*myc* mRNA. As shown in Figure 2, the DNAse I hypersensitivity sites present in chromatin were identical in the amplified as compared to the unamplified cell line, and were similar to those observed in a B lymphoblastoid cell line. This result argues against a determining role of such DNAse I sensitive sites in tissue-specific or relative steady state expression of the gene. Also, the pattern of DNA methylation was very similar in both the c-*myc* amplified and unamplified cell lines.

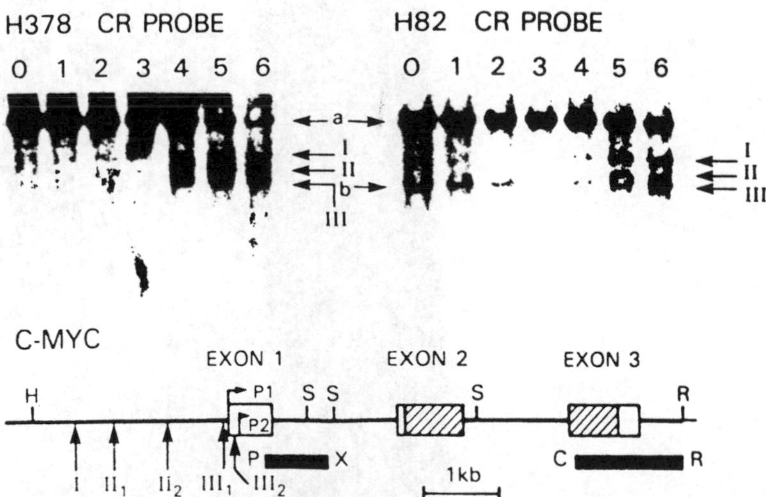

Figure 2 DNAse hypersensitivity sites in c-*myc*-amplified and single copy cell lines. Isolated nuclei from NCI-H378 (not amplified for c-*myc*) and from NCI-H82 (amplified for c-*myc*) were digested with increasing amounts of DNAse I, electrophoresed, and blotted. Hybridization was then with the CR exon 3 probe. (Adapted from Ref. 6.)

B. Correlation with SCLC Cell Biology

Gazdar et al. [7] clearly described the correlation of c-*myc* amplification with morphological and growth-related features of SCLC cell lines in vitro. "Classic" SCLC cell lines display in nude mouse xenografts a histology highly reminiscent of the intermediate subtype of SCLCs derived from biopsies of patients. These cell lines have a number of common features: growth in tightly packed floating cellular aggregates, long doubling times, low (~2%) colony-forming efficiency, and retention of neuroendocrine marker expression. In contrast, "variant" SCLC cell lines have discordant expression of these features. As outlined by Gazdar et al. [7], variant cell lines retain expression of the B isoenzyme of creatine phosphokinase, but they have less neuron-specific enolase, have lost dopamine decarboxylase activity, and also have undergone a decrease in bombesin-like immunoreactivity. Two types of variant cell line may be distinguished. "Morphologic variants" have a strikingly different appearance in cell culture (Fig. 3), with growth in loose,

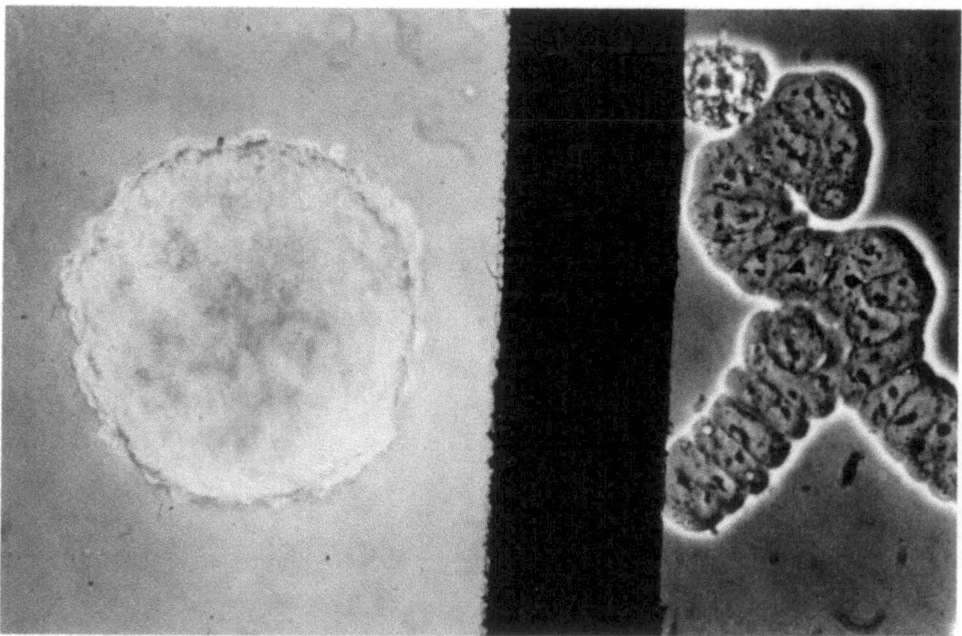

Figure 3 Morphology of SCLC cell lines: (left) the densely packed, aggregated appearance of a "classic" SCLC cell line and (right) the loose, open, "chainlike" appearance of a "variant" SCLC cell line, with large nuclei and prominent nucleoli. (Photograph courtesy of Dr. A. Gadzar.)

open aggregates, markedly decreased doubling time and high (up to 30%) cloning efficiency. In nude mouse xenografts, SCLC morphological variant cell lines have an appearance more reminiscent of large cell carcinoma of the lung, despite having arisen from patients with small cell histology. "Biochemical variants" of SCLC cell lines have lost the typical neuroendocrine markers but retain the morphological appearance of classic SCLC cell lines.

A notable correlation is that the morphologic variant SCLC cell lines in seven of nine cases had amplification of c-*myc* (Fig. 1) ranging from 5- to 76-fold. In contrast, 1 of 23 classic SCLC cases and none of 3 biochemical variant SCLC cases had evidence of c-*myc* amplification. One model that could account for the loss of neuropeptide hormone expression in the morphological variant SCLC cell lines with c-*myc* amplification relates to the observation that in model systems such as Swiss 3T3 cells, neuropeptide agonists such as bombesin can stimulate c-*myc* transcription. This stimulation is followed by DNA synthesis [8]. Thus, the SCLC cell lines that have achieved elevated c-*myc* expression by amplification have no selection pressure to retain autocrine peptide hormone expression. Indeed, as outlined by Sausville et al. [9], variant SCLC cell lines with increased c-*myc* expression are among those that do not increase intracellular Ca^{2+} in response to bombesin agonists.

A more direct test of this hypothesis was pursued by Johnson et al. [10], who transfected the c-*myc* gene into a classic SCLC cell line. These workers achieved integration in a number of clones of one to six copies of the plasmid-derived c-*myc*, with clear increase in the expression of c-*myc* mRNA. Expression of c-*myc* in the transfected clones was associated with a morphological appearance similar to the variant SCLC cell lines, with the acquisition by the transfected cell lines of shorter doubling times and increased cloning efficiency. For several generations after transfection, however, the c-*myc* transfectants did not decrease expression of DOPA decarboxylase or bombesin-like immunoreactivity. Nonetheless, these findings reinforced the correlation of c-*myc* expression with some of the phenotypic differences between classic and variant SCLC cell lines, most notably morphological appearance, increase in growth rate, and cloning efficiency.

C. Correlation with Clinical Biology

Radice et al. [11] have described samples from a small fraction (12%) of patients at initial diagnosis that had the histological appearance of SCLC admixed with elements of a more anaplastic, large cell histology. The latter component is reminiscent of the morphology of SCLC variant cell lines in nude mouse xenografts. These patients with mixed "small cell/large cell" histology have an inferior response to therapy and inferior prognosis compared to patients who do not have such "mixed" histology. While it is tempting to speculate that such patients may represent the in vivo correlate to the in vitro phenomenon of SCLC variant cell lines, no study has directly addressed c-*myc* amplification or expression in such "mixed" tumors.

Johnson and colleagues [12] instead demonstrated that SCLC cell lines from previously untreated patients generally do not have c-*myc* amplification; however, cell lines derived from patients who have relapsed after initial response to chemotherapy had in 5 of 25 cases amplification of c-*myc*. These patients also had a strong trend toward an inferior survival compared to patients whose derived cell lines did not have c-*myc* amplification (Fig. 4). This experience parallels that of Buys et al. [5], who noted in a smaller series of patients that their variant cell lines all had c-*myc* amplification and came from patients with poor response to chemotherapy and inferior survival.

Thus, while the details of the mechanism are certainly not clear, especially whether c-*myc* amplification is causative or correlative with an inferior prognosis, it appears that the capacity of an SCLC cell line to amplify c-*myc* in vitro may be

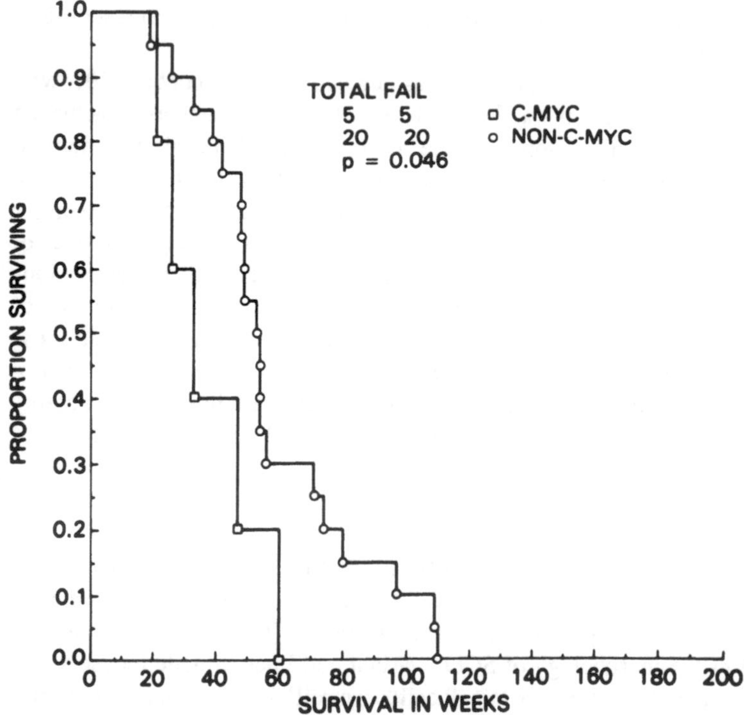

Figure 4 Survival of patients according to amplification of c-*myc* in derived cell lines. Patients with amplification of c-*myc* in their derived cell lines had a trend toward inferior survival compared to those without c-*myc* amplification. (Adapted from Ref. 12.)

an expression of an underlying more virulent SCLC tumor biology. A detailed analysis of amplification units containing no c-*myc* in the variant SCLC cell lines would be of interest in this regard, as well as an examination of what amplified segments may be present in "classic" SCLC.

III. AMPLIFICATION OF OTHER *myc* GENES IN SCLC

A. N-*myc*

As shown in Figure 5, while looking for evidence of amplification of the c-*myc* gene using an exon 2 c-*myc* probe (such as in SCLC cell line H82, with a 12.7 kb species), examples were encountered of other hybridizing fragments with a size distinct from that of c-*myc*. Since one of these amplified species migrated as a 2 kb *Eco*RI fragment (as in NCI-H249), analogous to the germ line position of the human N-*myc* gene, direct evidence for amplification of N-*myc* in SCLC was obtained using an N-*myc* probe. Nau et al. [13] found that 5 of 10 SCLC cell lines examined had amplified N-*myc*. Moreover, as shown in Figure 6, tumor tissue from the patient in addition to the cell lines to NCI-H526 and NCI-H689 showed evidence of N-*myc* amplification. Each case of N-*myc* amplification in SCLC cell lines was accompanied by evidence of increased steady state expression of the gene. N-*myc* amplification as found in one of 19 cell lines derived from untreated patients, and in 3 of 25 cell lines obtained after treatment [12]. Four of 38 tumor fresh SCLC tumor specimens had evidence of N-*myc* amplification [14].

Kiefer et al. [15] demonstrated 7 of 12 SCLC cell lines showed 20–35-fold amplification of c-*myc*, while 3 of the 12 showed 80–130-fold amplification of N-*myc*. It is of interest in this study that consistent high level expression without DNA amplification of N-*ras*, Ki-*ras*, Ha-*ras*, and *raf*l was observed in SCLC cell lines. Furthermore, in a subset of the cell lines, c-*myb* expression and a v-*fms*-related transcript were documented again without DNA amplification. Thus,

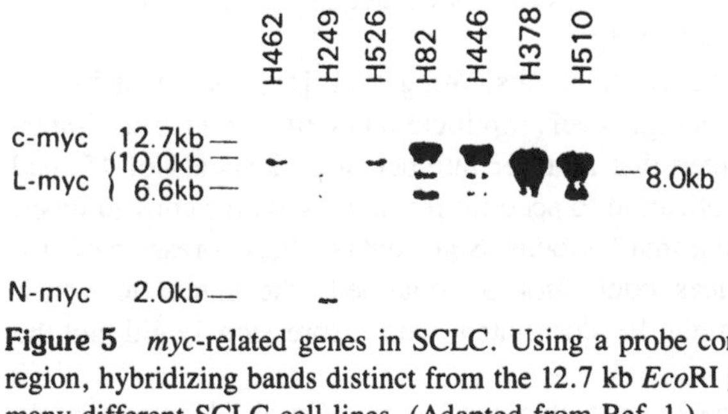

Figure 5 *myc*-related genes in SCLC. Using a probe containing the exon 2 homology region, hybridizing bands distinct from the 12.7 kb *Eco*RI germ line band can be seen in many different SCLC cell lines. (Adapted from Ref. 1.)

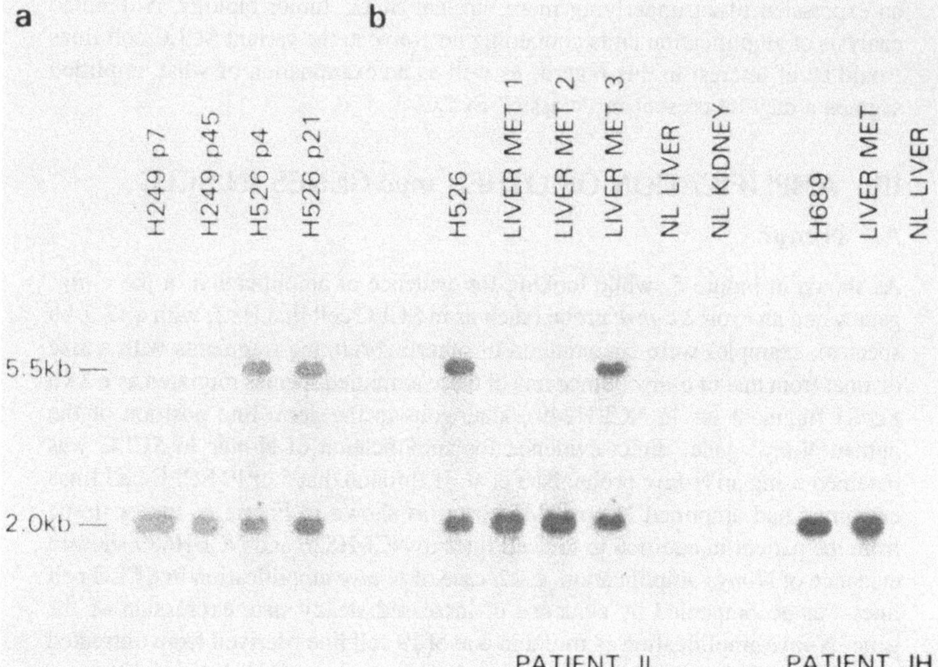

Figure 6 (a) N-*myc* amplification in tumor tissue and cell lines. Using an N-*myc* probe, evidence of amplification in both early and late passage NCI-H249 and H-526 is observed. (b) All three liver metastases have amplification of a 2.0 kb *Eco*RI N-*myc* hybridizing band, as does the patient's derived cell line. Liver metastasis 3 shows in addition a rearranged 5.5 kb fragment, also evident in the patient's derived cell line. No amplification is evident in normal liver or kidney. In NCI-H689, amplification of N-*myc* in both a liver metastasis and the cell line is evident. (Adapted from Ref. 13.)

although the incidence of *myc* family gene amplification was apparently more evident in SCLC cell lines than fresh SCLC tumor tissue, amplification of a variety of other oncogenes was not detected.

Using DNA extracted from paraffin blocks, Wong et al. [16] found that 5 of 45 patients had SCLC tumors with high level amplification of either c-*myc* or N-*myc*. These workers also documented that a larger number of patients (14 of 45) had evidence of chromosomal duplication to account for increased *myc* copy number, as opposed to amplification of a smaller locus. Since data on the expression of *myc* genes in these latter instances could not be obtained, the extent to which chromosomal duplication might be correlated with expression could not be assessed.

Formalin-fixed paraffin blocks were also used by Funa et al. [17] to correlate

N-*myc* expression with clinical course. Patients expressing N-*myc* had a median survival of 13 months, whereas patients who did not express N-*myc* had a median survival of 22 months. Also, patients without N-*myc* expression tended to have a greater incidence of complete response to chemotherapy.

B. L-*myc*

In addition to the signal observed with an exon 2 c-*myc* probe at the position of N-*myc* (Fig. 5), hybridization signals at 10 and 6 kb could be observed (e.g., NCI-H510 and NCI-H378). The novel 10.0 kb *Eco*RI fragment was cloned from a size-fractionated genomic library [18]. These experiments revealed a novel *myc* family member, L-*myc*, so named since it was initially derived from a lung neoplasm. This gene is found at chromosome 1p32, and possesses an *Eco*RI polymorphism to account for the exon 2 c-*myc*-related 6.0 kb hybridization signal (Fig. 5).

Detailed analysis of L-*myc* structure [19] and expression revealed that in contrast to other *myc* family members, L-*myc* demonstrates extensive alternative splicing as well as the use of polyadenylation sites within an intron. Consequently, "long form" L-*myc* transcripts have been described encoding an L-*myc* protein analogous in form and structure to other *myc* proteins described, as well as "short form" transcripts, which would if expressed possess a unique carboxyl-terminus after termination in the second intron. Extensive conservation of homology by the L-*myc* long form protein with c-*myc* suggests retention of regions of the protein expected to be important in transformation and nuclear localization. DeGreve et al. [20] demonstrated that in an SCLC cell line amplified for L-*myc*, two phosphorylated nuclear proteins are encoded from "long form" transcripts. Although "short form" L-*myc* proteins can be synthesized in vitro from the forms of the mRNA that prematurely terminate in intron 2, no short form proteins have yet been detected in vivo. This result leaves open the question of what role might be played by the "short form" L-*myc* mRNAs.

L-*myc* amplification was detected in 2 of 38 primary SCLC tumor specimens [14]. Kawashima et al [21] demonstrated that patients with only the 6 kb *Eco*RI polymorphic band or who possessed the 6 and 10 kb bands had an increased incidence of lymph node metastasis from a primary lung carcinoma, in most cases actually a non-small-cell lung carcinoma. In contrast, patients who had only the 10 kb band had a lower incidence of metastatic spread.

In general, SCLC cell lines that express L-*myc* tend to be relatively slow growing, to exhibit "classic" small cell morphology, to express the entire repertoire of neuroendocrine markers, and to respond to bombesin congeners with increases in intracellular Ca^{2+} [11]. Birrer et al. [22] demonstrated that the L-*myc* long form mRNA under the control of a Moloney long terminal repeat complemented an activated *ras* gene in the transformation of primary rat embryo fibroblasts. However, the efficiency of transformation was considerably less than c-*myc* or SV40 controls. This finding raises the possibility that although L-*myc* expression can promote neoplasia, it is not as potent a growth-promoting influence as, for example, c-*myc* and, in fact, may actually be a "partial agonist" for the function(s) subserved by c-*myc*.

IV. CONSEQUENCES OF GENE AMPLIFICATION IN SCLC

A. *myc* Family Transcriptional Regulation

Krystal et al. [23] demonstrated that distinctive patterns of transcriptional regulation are retained after amplification of *myc* family genes in SCLC cell lines. c-*myc* mRNA had a half-life of 36 of 50 min in unamplified or amplified SCLC, respectively. Transcripts of c-*myc* in amplified cell lines showed attenuation of transcription in the first intron: there was a fivefold excess of sense exon 1 compared to sense exon 2 transcription in an amplified cell line. In contrast, a non-c-*myc*-amplified SCLC cell line was encountered that had abundant steady state expression of c-*myc* mRNA but had lost attenuation of transcription through intron 1. This result therefore suggests that loss of transcriptional attenuation is a mechanism in addition to gene amplification which can be adopted by a tumor cell to increase steady state c-*myc* mRNA levels. L-*myc* mRNA had a half-life of 90–153 mins. As with c-*myc*, there was evidence of attenuation of transcription of exon 2 compared to exon 1 in an amplified cell line, whereas again there was an example of a nonamplified cell line achieving notably increased levels of L-*myc* mRNA by loss of attenuation. N-*myc* transcriptional regulation was distinctive in that no interexon attenuation was ever observed, but clear-cut antisense exon 1, intron 1, and exon 2 transcripts were evident and could conceivably play a regulatory role.

B. Drug Resistance

Although most work on gene amplification in SCLC has focused on *myc* family genes, there is good evidence that amplification of dihydrofolate reductase occurred in an SCLC tumor in a patient after treatment with methotrexate [24] and was evident in the cell line derived from this patient. Interestingly, the same cell line also had amplification of N-*myc*. Loss of drug resistance during culture of the cell line without methotrexate, along with loss of double minute chromosomes and decline of dihydrofolate reductase levels, strengthened the association between gene amplification and clinically manifest drug resistance in this case.

V. SUMMARY AND OVERVIEW

The experiments reviewed here emphasize that small cell carcinoma of the lung is a disease in which the capacity to amplify genes whose expression conveys a proliferative advantage may be important in the clinically relevant biology of the neoplasm. A further understanding of the biochemical means by which amplification is accomplished will be of great interest. Since this process is not evident in normal tissue, it is conceivable that the design of agents selective for the biochemical processes involved in gene amplification will lead to useful maneuvers in blocking the malignant progression of SCLC. Such treatments would not be extremely valuable in established bulky disease; they might, however, be most useful as an adjuvant after primary cytoreduction with conventional agents.

REFERENCES

1. Minna, J. D., Battey, J. F., Brooks, B. J., Cuttitta, F., Gazdar, A. F., Johnson, B. E., Ihde, D. C., Lebacq-Verheyden, A. M., Mulshine, J., Nau, M. M., Oie, H. K., Sausville, E. A., Seifter, E., and Vinocour, M. Molecular genetic analysis reveals chromosomal deletion, gene amplification, and autocrine growth factor production in the pathogenesis of human lung cancer. *Cold Spring Harbor Symp. Quant. Biol. 51*: 843–853 (1986).
2. Saksela, K. *myc* genes and their deregulation in lung cancer. *J. Cell. Biochem. 42*: 153–180 (1990).
3. Little, C. D., Nau, M. M., Carney, D. N., Gazdar, A. F., and Minna, J. D. Amplification and expression of the c-*myc* oncogene in human lung cancer cell lines. *Nature*, *306*:194–196 (1983).
4. Saksela, K., Bergh, J., Lehto, V., Nilsson, K., and Alitalo, K. Amplification of the c-*myc* oncogene in a subpopulation of human small cell lung cancer. *Cancer Res. 45*: 1823–1827 (1985).
5. Buys, C. H., Osinga, J., VanderVeen, A. Y., Mooibroek, H., VanderHout, A. H., DeLeij, L. D., Postmus, P. E., and Carritt, B. Genome analysis of small cell lung cancer (SCLC) and clinical significance. *Eur. J. Respir. Dis. 70*:29–36 (1987) [Suppl. 140].
6. Seifter, E. J., Sausville, E. A., and Battey, J. Comparison of amplified and unamplified c-*myc* gene structure and expression in human small cell lung carcinoma cell lines. *Cancer Res. 46*:2050–2055 (1986).
7. Gazdar, A. F., Carney, D. N., Nau, M. M., and Minna, J. D. Characterization of variant subclasses of cell lines derived from small cell lung cancer having distinctive biochemical, morphological, and growth properties. *Cancer Res. 45*:2924–2930 (1985).
8. Letterio, J. J., Coughlin, S. R., and Williams, L. T. Pertussis toxin-sensitive pathway in the stimulation of c-*myc* expression and DNA synthesis by bombesin. *Science*, *234*: 1117–1119 (1986).
9. Sausville, E. A., Moyer, J. D., Heikkila, R., Neckers, L. M., and Trepel, J. B. A correlation of bombesin-responsiveness with *myc*-family gene expression in small cell lung carcinoma cell lines. *Ann. NY Acad. Sci. 547*:310–321 (1988).
10. Johnson, B. E., Battey, J., Linnoila, I., Becker, K. L., Makuch, R. W., Snider, R. H., Carney, D. N., and Minna, J. D. Changes in the phenotype of human small cell lung cancer cell lines after transfection and expression of the c-*myc* proto-oncogene. *J. Clin. Invest. 78*:525–532 (1986).
11. Radice, P. A., Matthews, M. J,. Ihde, D. C., Gazdar, A. F., Carney, D. N., Bunn, P. A., Cohen, M. H., Fossieck, B. E., Makuch, R. W., and Minna, J. D. The clinical

behavior of "mixed" small cell/large cell bronchogenic carcinoma compared to "pure" small cell subtypes. *Cancer*, *50*:2894–2902 (1982).

12. Johnson, B. E., Ihde, D. C., Makuch, R. W., Gazdar, A. F., Carney, D. N., Oie, H., Russell, E., Nau, M. M., and Minna, J. D., *myc* family oncogene amplification in tumor cell lines established from small cell lung cancer patients and its relationship to clinical status and course. *J. Clin. Invest.* *79*:1629–1634 (1987).
13. Nau, M. M., Brooks, B. J., Carney, D. N., Gazdar, A. F., Battey, J. F., Sausville, E. A., and Minna, J. D. Human small-cell lung cancers show amplification and expression of the N-*myc* gene. *Proc. Natl. Acad. Sci. USA* *83*:1092–1096 (1986).
14. Johnson, B. E., Makuch, R. W., Simmons, A. F., Gazdar, A. F., Burch, D., and Cashell, A. W. *myc* family DNA amplification in small cell lung cancer patients' tumors and corresponding cell lines. *Cancer Res.* *48*:5163–5166 (1988).
15. Kiefer, P. E., Bepler, G., Kabasch, M., and Havemann, K. Amplification and expression of protooncogenes in human small cell lung cancer cell lines. *Cancer Res.* *47*:6236–6242 (1987).
16. Wong, A. J., Ruppert, J. M., Eggleston, J., Hamilton, S. R., Bayline, S. B., and Vogelstein, B. Gene amplification of c-*myc* and N-*myc* in small cell carcinoma of the lung. *Science*, *233*:461–464 (1986).
17. Funa, K., Steinholtz, L., Nou, E., and Bergh, J. Increased expression of N-*myc* in human small cell lung cancer biopsies predicts lack of response to chemotherapy and poor prognosis. *Am. J. Clin. Pathol.* *88*:216–220 (1987).
18. Nau, M. M., Brooks, B. J., Battey, J., Sausville, E., Gazdar, A. F., Kirsch, I. R., McBride, O. W., Bertness, V., Hollis, G. F., and Minna, J. D. L-*myc*, a new *myc*-related gene amplified and expressed in human small cell lung cancer. *Nature*, *318*: 69–73 (1985).
19. Kaye, F., Battey, J., Nau, M., Brooks, B., Seifter, E., Degreve, J., Birrer, M., Sausville, E., and Minna, J. Structure and expression of the human L-*myc* gene reveal a complex pattern of alternative mRNA processing. *Mol. Cell. Biol.* *8*:186–195 (1988).
20. DeGreve, J., Battey, J., Fedorko, J., Birrer, M., Evan, G., Kaye, F., Sausville, E., and Minna, J. The human L-*myc* gene encodes multiple nuclear phosphoproteins from alternatively processed mRNAs. *Mol. Cell. Biol.* *8*:4381–4388 (1988).
21. Kawashima, K., Shikama, H., Imoto, K., Izawa, M., Naruke, T., Okabayashi, K., and Nishimura, S. Close correlation between restriction fragment length polymorphism of the L-*myc* gene and metastasis of human lung cancer to the lymph nodes and other organs. *Proc. Natl. Acad. Sci. USA*, *85*:2353–2356 (1988).
22. Birrer, M. J., Segal, S., DeGreve, J S., Kaye, F., Sausville, E. A., and Minna, J. D. L-*myc* cooperates with *ras* to transform primary rat embryo fibroblasts. *Mol. Cell. Biol.* *8*:2668–2673 (1988).
23. Krystal, G., Birrer, M., Way, J., Nau, M., Sausville, E., Thompson, C., Minna, J., and Battey, J. Multiple mechanisms for transcriptional regulation of the *myc* gene family in small-cell lung cancer. *Mol. Cell. Biol.* *8*:3373–3381 (1988).
24. Curt, G. A., Carney, D. N., Cowan, K. H., Jolivet, J., Bailey, B. D., Drake, J. C., Kao-Shan, C. S., Minna, J. D., and Chabner, B. A. Unstable methotrexate resistance in human small-cell carcinoma associated with double minute chromosomes. *New Engl. J. Med.* *308*:199–202 (1983).

32

Amplification and Mutagenesis of the Acetylcholinesterase and Butyrylcholinesterase Genes in Primary Human Tumors

Hermona Soreq, Gal Ehrlich, Averell Gnatt, Lewis Neville, Revital Ben-Aziz-Aloya, Shlomo Seidman, and Dalia Ginzberg *The Hebrew University of Jerusalem, Jerusalem, Israel*

Yaron Lapidot-Lifson* and Haim Zakut *Tel Aviv University, Tel Aviv, Israel*

I. INTRODUCTION

The genes encoding cholinesterase (*CHE* genes) have only recently joined the family of known amplifiable sequences [1]. CHEs are carboxylesterase type B enzymes capable of hydrolyzing the neurotransmitter acetylcholine (ACh). They are classified primarily according to their substrate specificity and sensitivity to selective inhibitors into acetylcholinesterase (acetylcholine acetylhydrolase, ACHE, EC 3.1.1.7), which is the primary CHE in brain and muscle, and butyryl-cholinesterase (acylcholine acyl hydrolase, BCHE, EC 3.1.1.8), the major CHE in liver and plasma. Molecular cloning revealed that two distinct genes encode these two enzymes, and that the *BCHE* gene is rich in A,T residues whereas the *ACHE* gene is rich in G,C residues [1]. CHEs are present in many tissue and cell types, including "noncholinergic" embryonic or cancerous ones. Although this distribution probably indicates a growth-related role for mammalian CHEs, no specific

*Also affiliated with the Department of Biological Chemistry, The Life Sciences Institute, The Hebrew University of Jerusalem, Jerusalem, Israel.

function other than acetylcholine hydrolysis has so far been attributed to these enzymes. Since ACh cannot be defined as metabolically essential in the general sense, it was difficult to envisage a selection advantage that would maintain amplified *CHE* sequences in the genome of affected cells. Nonetheless, accumulating evidence correlating abnormal *CHE* gene expression, *CHE* gene amplification, and neoplastic growth decisively implicates the *CHE* genes in tumorigenic processes and reinforces the notion of a growth-related function for their gene products.

II. *CHE* GENE AMPLIFICATIONS IN MALIGNANT OVARIAN CARCINOMAS COINCIDE WITH ONCOGENE AMPLIFICATIONS

Unusually high levels of CHE activities have been noted in a number of primary tumors, including glioblastoma multiform, meningiomas, ovarian adenocarci-

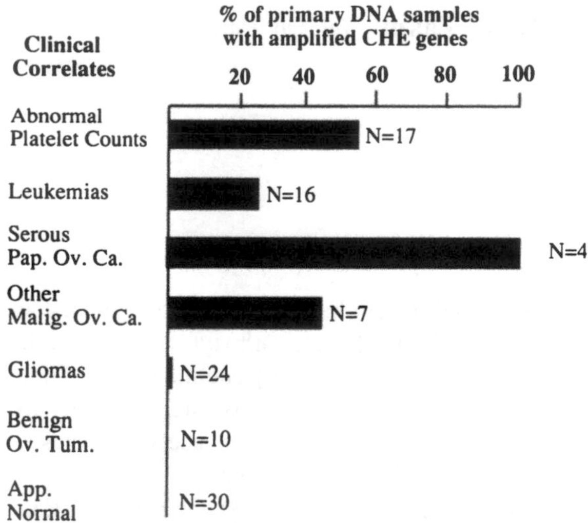

Figure 1 *CHE* gene amplifications in primary human DNA samples of various cell type origins. *CHE* gene amplifications were searched for by DNA blot hybridizations in peripheral blood cells from patients with abnormal platelet counts and/or leukemias [5] as well as in primary tumor DNAs from serous papillary ovarian carcinomas (Serous Pap.Ov.Ca.), other malignant adenocarcinomas of the ovary (other Malig.Ov.Ca. [3]), glioblastoma multiform (Gliomas), benign ovarian tumors (Benign Ov. Tum.), and peripheral blood cell DNA from apparently normal, healthy individuals (App.Normal). The number of samples examined in each group is noted; columns represent percentage of primary DNA samples with amplified *ACHE* and *BCHE* genes in each group.

nomas, and pheochromocytomas (reviewed in Ref. 1). Furthermore, a novel dimeric CHE component, sensitive to selective inhibitors of both ACHE and BCHE, appeared in the serum of patients with various carcinomas (reviewed in Ref. 1). More recently, abnormal termination resulting in 3′-extended *BCHE* mRNA transcripts was observed in glioblastoma and neuroblastoma cells [2]. To examine the possibility that gene amplification may be etiologically involved in altered CHE production in these tissues, DNA from primary ovarian carcinomas was selected as a model.

Amplification of both the *BCHE* and the *ACHE* genes was frequently observed in serous papillary adenocarcinomas as well as in other malignant tumors of the ovary, but not in benign ones (Fig. 1). In DNA blot hybridization, the amplified genes displayed intensively labeled DNA fragments comigrating with some of those of the normal, nonamplified genes (Fig. 2). Similar levels of amplification were found in these tumors for the oncogenes c-*RAF1*, v-*Sis*, and c-*FES*, known to amplify in ovarian carcinomas. N-*MYC*, however, which amplifies in nervous system tumors, remained in low copy numbers (see Ref. 3 and Fig. 3).

To determine whether the amplified *CHE* genes in ovarian carcinomas are expressed, mRNA from these primary tumors was analyzed by *Xenopus* oocyte microinjection. In addition, blot and in situ hybridization techniques as well as immuno- and activity-based methods of cytochemical staining were employed. These analyses revealed translatable *ACHE* mRNA and *BCHE* mRNA and their active enzyme products in discrete tumor foci rich in small, rapidly dividing cells [3]. Therefore, the overexpression of CHEs in ovarian carcinomas may be accounted for, at least in part, by gene amplification, and it differs from thymidylate synthetase overexpression in these cells, which is induced by enhanced transcription [4]. The correlation between CHE and oncogene amplifications implicated the concept of growth-related role(s) for these enzymes in tumors and perhaps other rapidly dividing cell types.

III. *ACHE* AND *BCHE* GENE AMPLIFICATIONS IN BLOOD CELL PROGENITORS MAY PROVIDE ADVANTAGES FOR CELL GROWTH

Acetylcholine analogues and CHE inhibitors have been shown to induce megakaryocytopoiesis, leading to abnormal platelet production in the mouse (reviewed in Ref. 5). Furthermore, chromosomal breakage on the long arm of chromosome 3, where the *BCHE* gene maps, has been observed in leukemic patients with abnormal platelet counts (reviewed in Ref. 1). These observations implicated CHEs in hematocytopoiesis and led to a search for *CHE* gene aberrations in leukemias and platelet disorders. Both the *BCHE* and the *ACHE* genes were found to be amplified 10–200-fold in 25% and 45% of the examined peripheral blood DNA samples originating from patients with leukemias or with platelet disorders,

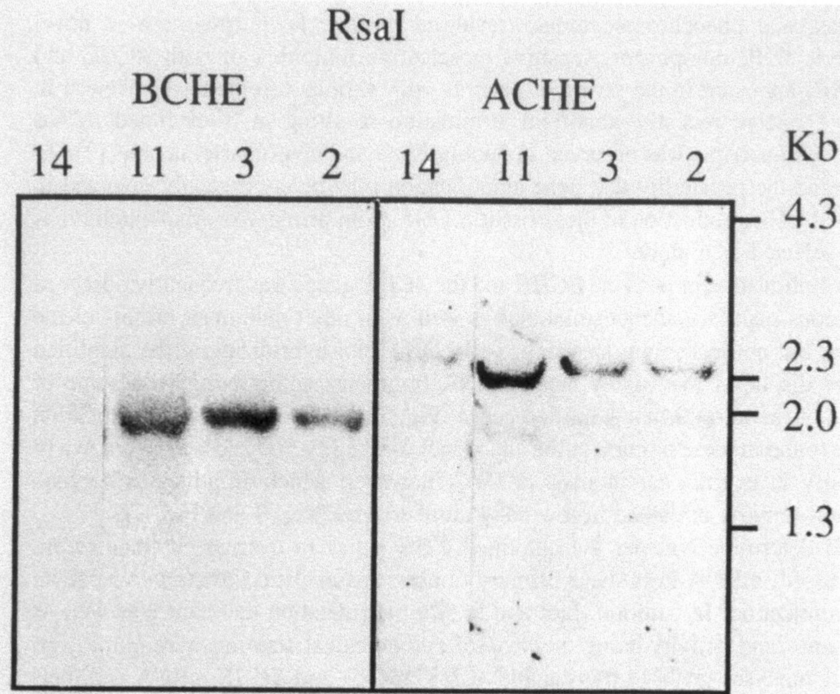

Figure 2 DNA blot hybridization reveals *ACHE* and *BCHE* gene amplifications in malignant ovarian tumors. DNA samples from three different primary adenocarcinomas of the ovary (2, 3, and 11; see Ref. 3 for details) were subjected to enzymatic digestion with *Rsa* I followed by DNA blot hybridization with ^{32}P-labeled probes for the *ACHE* and *BCHE* genes and autoradiography. DNA from a healthy tissue (14, see Ref. 3) served for control. Note the various extents of *CHE* gene amplifications reflected by labeling intensities of the relevant DNA fragments.

respectively (Ref. 6 and Fig. 1). *CHE* gene amplification in blood cells demonstrating uncontrolled growth suggested that overproduction of these enzymes may impart a growth advantage to cells carrying the amplified gene(s). We were thus led to postulate a role for these genes in normal hematocytopoietic processes as well.

One approach to examining this hypothesis was to arrest *BCHE* gene expression in developing hemopoietic cells by a selective "antisense" oligodeoxynucleotide. A 50-mer phosphorothioate anti-*BCHE* sequence was thus shown to selectively inhibit megakaryocytopoiesis in cultured murine bone marrow cells [5]. These results directly implicated BCHE in normal megakaryocytopoiesis and indicated that amplification of the gene encoding this enzyme may indeed be

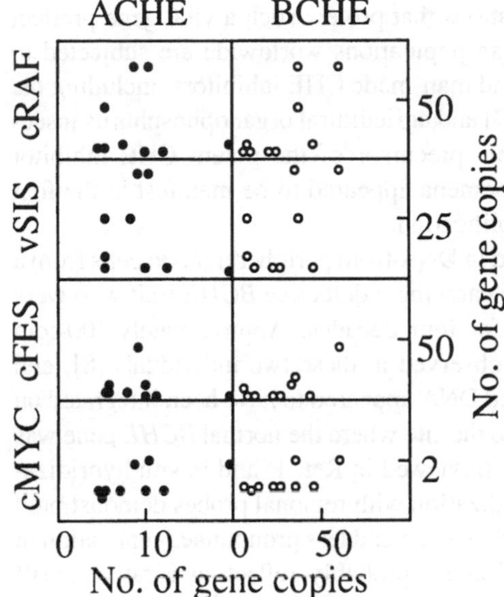

Figure 3 *CHE* gene amplifications in ovarian tumors coincide with oncogene amplifications. Number of gene copies for the *ACHE* and *BCHE* genes as well as for the c-*RAF*, v-*SIS*, c-*FES*, and c-*MYC* genes were evaluated by DNA blot hybridizations in nine different ovarian tumors [3]. Note similarities in the relationship between gene copy number of each of the two *CHE* genes and the four oncogenes.

capable of stimulating pathologically rapid cell growth within the megakaryocytopoietic cell lineage. A similar effect may be postulated for the *ACHE* gene in erythroid and/or myeloid progenitor cells where its protein product is present but unexplained.

IV. HERITABLE AMPLIFICATION OF THE *BCHE* GENE IN ORGANOPHOSPHORUS-EXPOSED INDIVIDUALS WITH A DEFECTIVE *BCHE* GENOTYPE

The association of *CHE* gene amplification with pathological growth led to the demonstration of *BCHE* gene expression in a normal growth state in differentiating blood cells, where the presence of BCHE itself had been previously unsuspected. If, in fact, it is true that (1) the *CHE* genes play a role in the growth and proliferation of specific cell types, and (2) *CHE* gene amplification confers a growth advantage for neoplastic cells, then one would expect to find that *CHE* gene amplification confers a survival advantage to certain healthy cell types under environmental stress directed toward these gene products. Moreover, one might expect to discover advantageous mutations that protect such a vital gene product from intoxication. Interestingly, human populations worldwide are subjected to subacute exposure to several natural and man-made CHE inhibitors, including the poisonous potato alkaloid solanidine [7] and agricultural organophosphorus insecticides like parathion, a biodegradable precursor of the potent CHE inhibitor paraoxon [8]. Indeed, both these phenomena appeared to be manifest in the first documented case of *BCHE* gene amplification.

BCHE gene amplification was found in DNA from peripheral blood cells from a father and son of an agronomic family carrying a defective *BCHE* trait who were occupationally exposed to parathion for four decades. Approximately 100-fold amplification of the *BCHE* gene was observed in these two individuals [8], and upon in situ hybridization, the amplified DNA appeared to have been integrated on the long arm of chromosome 3, close to the site where the normal *BCHE* gene was previously mapped by genetic linkage (reviewed in Ref. 1) and in situ hybridization [2,8] (see Fig. 4). DNA blot hybridization with regional probes demonstrated intensive amplification in the major coding exon and less pronounced replication in the 3′ and 5′ domains of the *BCHE* gene, probably reflecting a rather small amplification unit [8]. The most likely explanation for this unprecedented de novo amplification of the *BCHE* gene in apparently healthy individuals is that it is related on the one hand to their exposure to parathion and on the other hand to their defective BCHE phenotype. How this might be mediated is discussed in Section V. Nonetheless, the implied germ cell origin for this inheritable phenomenon suggested that BCHE is essential for the survival and well-being of sperm cells [1,8], further reinforcing the claim for CHE involvement in normal cell growth or differentiation, and emphasizing the potential involvement of environmental exposure to CHE inhibitors in *CHE* gene amplifications.

V. ALLELIC POLYMORPHISM AS POTENTIAL FACTOR IN *BCHE* GENE AMPLIFICATION

Having been led from abnormal *CHE* gene expression to *CHE* gene amplification and finally to a previously undemonstrated role for BCHE in normal cellular processes, we may now suggest a correlation between these phenomena and the long noted allelic polymorphism of the *BCHE* gene in man. Ten different point mutations have been discovered in the major coding exon of the *BCHE* gene within the past 3 years [2,7,9]. These imply amino acid changes at various positions within the catalytic subunit, at least eight of which confer measurable biochemical alterations in the mature BCHE protein. As many as six of these mutations occurred in G residues, which constitute only 15% of the nucleotide composition in this gene. Figure 5 summarizes the currently available data on *BCHE* gene polymorphism. Several of these mutations appear to be genetically linked to each other [2,9] and have been shown to confer resistance to various CHE inhibitors

[7]. However, at least one of these mutants appears to give rise to an unstable mRNA, resulting in deficient production of active enzyme [2]. If indeed CHEs are metabolically or developmentally essential, the combination of variant *BCHE* genotypes and sustained exposure to CHE inhibitors could create conditions favoring *CHE* gene amplification. Considering, however, factors such as the developmental regulation of these genes, the abundance of allelic polymorphs, haploid versus diploid genomes, and the infinite possibilities for exposure to CHE inhibitors, dissecting a singular pathway leading to *CHE* gene amplification is complex indeed. In fact, there are likely multiple independent or interrelated factors capable of generating a stable amplification event, resulting in beneficial or deleterious effects depending on the circumstances.

VI. SUMMARY

Available examples of gene amplification events in mammals are generally classified into those induced under exposure to toxic compounds [10,11] and those providing ample gene products during cellular differentiation stages when such products improve cell growth or division rate [12]. In contrast to findings in insects and amphibia, there are no examples in mammals for the third type of gene amplification, which is developmentally induced during normal embryogenesis or germ cell formation. Since, however, this could be due to technical difficulties, the possibility should not be excluded. Interestingly, CHEs could fit any one of these classifications because they are targets of toxic inhibitors and also transiently expressed during embryogenesis and germ cell development [1]. The already available examples of *CHE* gene amplifications in normal and tumor cells support both notions.

There is no oncogene in the vicinity of the *CHE* gene [1,8], suggesting that it amplifies on its own merit. Also, both the *ACHE* and the *BCHE* genes appear to include the inverted repeats and hairpin structures [1] assumed to be prerequisite to the initiation of gene amplification [13]. In addition, their expression in embryogenesis is concomitant with, or immediately follows, phases of DNA synthesis (reviewed in Ref. 1), a period during which overreplication of DNA has been noted [14]. This is particularly intriguing in view of the drastically different base composition in these two genes, which probably reflects a major evolutionary distance between them, and predicts for the *ACHE* and the *BCHE* genes distinct localization on different chromosomal bands and various replication and transcription times during the cell cycle [15]. This could, in essence, suggest that either one of these genes, or both of them together, might be prone to amplify in different cells and cell cycle phases.

Further search for *CHE* gene amplifications and their mechanism and structure will be required to reveal the contributions of environmental and developmental pressures toward this phenomenon.

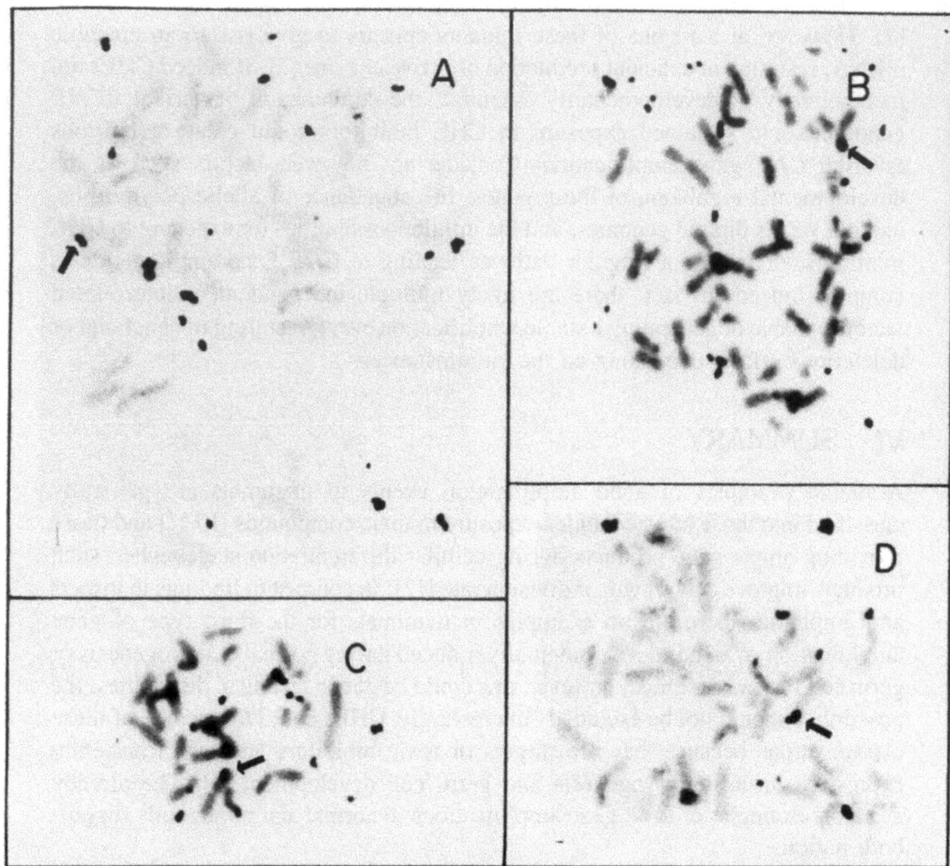

Figure 4 The *BCHE* gene maps to the long arm of chromosome 3 in its normal and amplified states. Chromosomal *BCHE* gene mapping was performed by in situ hybridization with spread chromosomes from peripheral blood cells from individuals with the normal, nonamplified *BCHE* gene (cont) and from an individual with 100 copies of the *BCHE* gene (amplified) [8]. (A)–(D) Individual spreads of chromosomes from the peripheral blood cells of an individual with amplified *BCHE* gene [8]. (E) Percentage of total grains associated with specific chromosomal bands is presented by columns. Insets: two representative pairs of normal human chromosomes 3 following in situ hybridization, with silver grains labeling the indicated 3q26-ter domain.

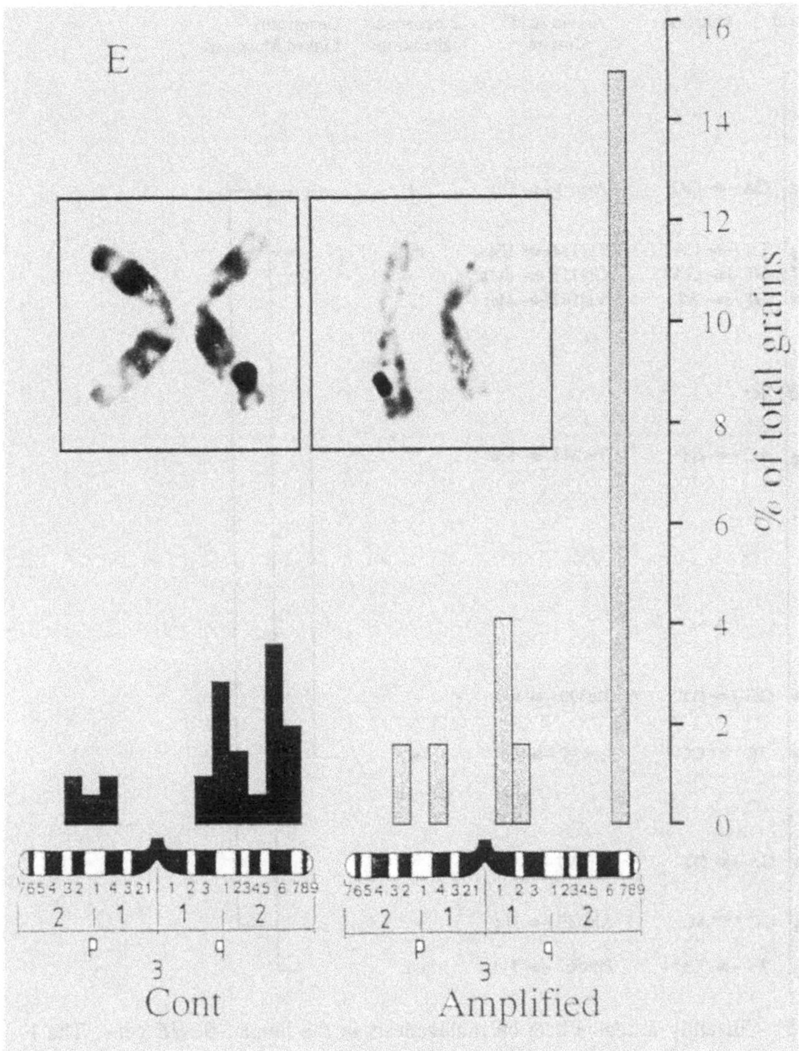
E
% of total grains
0
2
4
6
8
10
12
14
16
7654 32 1 4 321 1 2 3 12345 6 789
2 1 1 2
p q
3
Cont
7654 32 1 4 321 1 2 3 12345 6 789
2 1 1 2
p q
3
Amplified

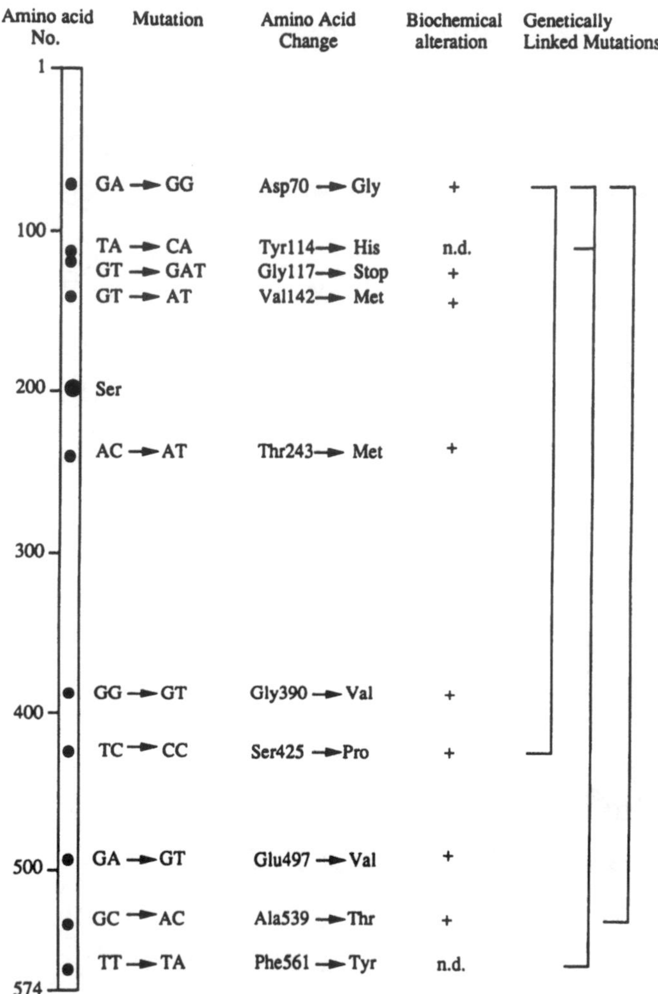

Figure 5 Currently available data on mutagenesis in the human *BCHE* gene. The 10 different point mutations recently discovered in the human *BCHE* gene (Refs. 2, 9, and unpublished observations) are presented schematically on a scale of the primary amino acid sequence in the BCHE protein; active site serine 200 noted by a larger dot. The nucleotide substitutions and the amino acid changes conferred in each mutation are displayed. Evidence for alterations in the biochemical properties of the resultant variant proteins is noted by +; n.d., not determined. Vertical lines bracket cases in which more than one mutation appeared in a single gene (Refs. 2, 9, and unpublished information).

ACKNOWLEDGMENTS

This work was supported by the Medical Research and Development Command, the U.S. Army (contract DAMD 17-C-7169, to H.S.), and the Research Fund at the Edith Wolfson Medical Center (to H.Z.). H.S. holds the Charlotte Slesinger Chair in Cancer Research at the Hebrew University of Jerusalem. G. E. and L. F. N. hold Golda Meir Fellowships (for Ph.D. and postdoctoral studies, respectively), and Y. L.-L. is a Levi Eshkol postdoctoral fellow.

REFERENCES

1. Soreq, H., and Zakut, H. Cholinesterase genes: Multileveled regulation, in *Monographs in Human Genetics*, Vol. 13 (R. S. Sparkes ed.), Karger, Basel, 1990.
2. Gnatt, A., Prody, C. A., Zamir, R., Lieman-Hurwitz, J., Zakut, H., and Soreq, H. Expression of alternatively terminated unusual cholinesterase messenger RNA transcripts, mapping to chromosome 3q26-ter, in nervous system tumors. *Cancer Res. 50*:1983–1987 (1990).
3. Zakut, H., Ehrlich, G., Ayalon, A., Prody, C. A., Malinger, G., Seidman, S., Ginzberg, D., Kehlenbach, R., and Soreq, H. Acetylcholinesterase and butyrylcholinesterase genes co-amplify in primary ovarian carcinomas. *J. Clin. Invest.* (1990).
4. Scanlon, K. J., and Kashami-Sabet, M. Elevated expression of thymidylate synthase cycle genes in cisplatin-resistant human ovarian carcinoma A2780 cells. *Proc. Natl. Acad. Sci. USA, 85*:650–653 (1988).
5. Patinkin, D., Seidman, S., Eckstein, F., Benseler, F., Zakut, H., and Soreq, H. Manipulations of cholinesterase gene expression modulate murine megakaryocytopoiesis in vitro. *Mol. Cell. Biol.* (1990).
6. Lapidot-Lifson, Y., Prody, C. A., Ginzberg, D., Meytes, D., Zakut, H., and Soreq, H. Coamplification of human acetylcholinesterase and butyrylcholinesterase genes in blood cells: Correlation with various leukemias and abnormal megakaryocytopoiesis. *Proc. Natl. Acad. Sci. USA, 86*:4715–4719 (1989).
7. Neville, L. F., Gnatt, A., Lowenstein, Y., and Soreq, H. Aspartate 70 to glycine substitution confers resistance to naturally occurring and synthetic anionic-site ligands on in ovo produced human butyrylcholinesterase *J. Neurosci. Res.* (1990).
8. Prody, C. A., Dreyfus, P., Zamir, R., Zakut, H., and Soreq, H. De novo amplification within a "silent" human cholinesterase gene in a family subjected to prolonged exposure to organophosphorous insecticides. *Proc. Natl. Acad. Sci. USA 86*:690–694 (1989).
9. LaDu, B. N. Identification of human serum cholinesterase variants using the polymerase chain reaction technique. *Trends Pharmacol. Sci. 10*:309–313 (1989).
10. Lee, T.-C., Tanaka, N., Lamb, P. W., Gilner, T. M., and Barrett, J. C. Induction of gene amplification by arsenic. *Science 241*:79–81 (1988).
11. Scott, K. W., Biedler, J. L., and Melera, P. W. Amplification and expression of genes associated with multidrug resistance in mammalian cells. *Science 232*:751–755 (1986).

12. Delidakis, C., Swimmer, C., and Kafatos, F. C. Gene amplification: An example of genome rearrangement. *Curr. Opinion Biol. 1*:488–496 (1989).
13. Ruiz, J. C., and Wahl, G. M. Formation of an inverted duplication can be an initial step in gene amplification. *Mol. Cell. Biol. 8*:4302–4313 (1988).
14. Hoy, C. A., Rice, G. C., Kovacs, M., and Schimke, R. T. Over-replication of DNA in S phase Chinese hamster ovary cells after DNA synthesis inhibition. *J. Biol. Chem. 262*:11927–11934 (1987).
15. Holmquist, G. DNA sequences in G-bands and R-bands, in *Chromosomes and Chromatin*, Vol. II (K. W. Adolph, ed.), CRC Press, Boca-Raton, Florida, 1988, pp. 75–121.

33

Gene Amplification in Nontumorigenic and Tumorigenic Mammalian Cells: Relationship to Carcinogenesis

Thea D. Tlsty *University of North Carolina, Chapel Hill, North Carolina*

I. INTRODUCTION

Amplification of many different genes has been described in the literature [1,2]. The genes that confer drug resistance are particularly amenable to study, and as such they have been used extensively to attempt to unravel the molecular basis of gene amplification. Our laboratory has been particularly interested in the regulation of gene amplification as first discovered in tumor cells. Previous work has identified conditions that enhance the frequency of this event [3–5]. Currently, we were interested in knowing whether all cells have the ability to amplify genomic loci. To this purpose we have conducted studies that are designed to measure amplification incidence and rate in a multitude of cell populations. The data should provide information on the distribution of this basic biological phenomenon.

II. MEASUREMENT OF GENE AMPLIFICATION

Amplification of endogenous genes has been measured using several different methods. Not until just recently have these methods been used quantitatively.

Perhaps the most common method used is the clonogenic assay [4,5]. Cells are plated and exposed to the selective agent, and the surviving cells that grow to form colonies are counted. This relative plating efficiency (normalized to the number of surviving cells in the nonselected population) estimates the incidence of drug resistance. If the selective agent is a drug such as *N*-(phosphonoacetyl)-L-aspartate (PALA), where the only known mechanism of resistance is the amplification of the *CAD* gene, then the determination of drug resistance and the determination of amplification incidence are one and the same. However, if the selective agent is a drug such as methotrexate, to which multiple mechanisms of resistance have been documented, then a molecular analysis of gene copy number is required before any conclusions concerning amplification incidence can be made. The incidence of amplification for any gene is dependent on the stringency of selection (i.e., the concentration of the drug). At low concentrations of selective agent, a greater number of cells survive and the incidence of drug resistance is high. At high concentrations of selective agent, fewer cells survive, and the incidence of drug resistance is correspondingly lower. In our studies we have chosen to identify a drug concentration relative to the LD_{50} of the individual cell line for the comparison of multiple cell lines.

An alternate method for measuring any mutation is to calculate its rate (incidence per cell per generation) using the Luria–Delbrück fluctuation analysis. The Luria–Delbrück fluctuation analysis is a combined experimental and statistical method that allows one to distinguish between variant cells arising by rare spontaneous mutations and variant cells arising through adaptation to an environmental selection [6]. The analysis is based on the variation that is seen in the emergence of colonies from parallel cultures [7,8]. First used in the analysis of bacterial mutation rates, it is now widely used in determinations of mammalian mutation rates [9].

Figure 1, a schematic diagram of a fluctuation experiment, indicates the hypothetical results one could obtain when analyzing a spontaneous mutation. A large parent population of cells is procured and the experimental samples are divided into two categories. The first category (indicated by the left arrow), contains replicate samples in which aliquots of cells from the parent population are plated directly into selection medium and analyzed for the number of resistant colonies that emerge. The colonies on these plates will represent rare resistant, preexisting mutants. The number of colonies on each plate should exhibit a Poisson distribution, and the mean number of colonies per plate will reflect the prevalence of resistant mutants in the parental population. Replicate platings from the same parent population should show variation due *only* to random sampling; the variance from these replicate samples should equal the mean. This category of samples is used to demonstrate that the method of sampling and plating has not introduced any fluctuations into the results besides random sampling error.

For the second category, a small number of cells (small enough to assure that no preexisting mutants are present) is plated and allowed to propagate under nonselective conditions for a given amount of time. When the individual populations have reached the same cell density plated in category 1, they are transferred to fresh plates for an even distribution of cells and placed under selection (in this

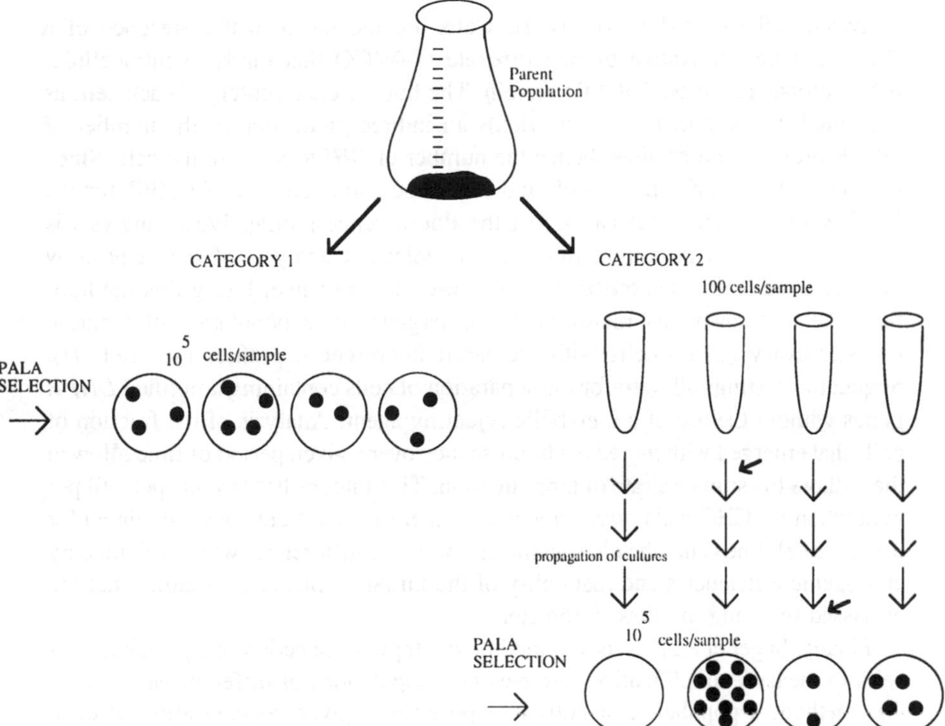

Figure 1 Luria-Delbrück fluctuation analysis.

case, PALA). If the drug-resistant cells are the result of exposure to PALA (i.e., adaptive), all cells should have similar probability for survival in selective media and the appearance of resistant mutants should be similar in all plates. The variation from plate to plate would be consistent with the Poisson model. If the events that lead to resistance are spontaneous, as each parallel culture expands it will have a given probability for generating resistant mutants with each cell division. In some cultures the event (amplification) will occur early, and many of the progeny of the resistant cell will be present to form drug resistant colonies. In others, the event will occur during one of the last cell divisions, and few drug-resistant progeny will result. The appearance of mutants will be random, and the contribution to the surviving colonies when the cells are placed in selection will vary greatly depending on when in the propagation of the population the mutation occurred. Statistical analysis of this variation allows one to calculate the rate of appearance of spontaneous mutants.

A third method that quantitates gene amplification utilized the fluorescence-activated cell sorter (FACS) [10,11]. Cells are incubated in the presence of a fluoresceinated derivative of methotrexate (F-MTX) that binds to intracellular dihydrofolate reductase (DHFR) protein. The fluorescence content of each cell, as measured by the flow cytometer, yields an indirect indication of the number of DHFR protein binding sites, hence the number of *DHFR* genes in the cell. Since transport of F-MTX into the cell and the association constant of DHFR for the F-MTX are variables that can affect the fluorescence profile, FACS analysis is usually performed in conjunction with a molecular analysis of the gene copy number in cells that demonstrate an increased fluorescence. Using this method, Schimke and coworkers measured the emergence of a population of Chinese hamster ovary (CHO) cells with increased fluorescence profiles (see Ref. 11). Sequential sortings allowed for the separation of cells containing amplified *DHFR* genes without the use of a metabolic selecting agent. Analysis of the fraction of cells that emerged with increased fluorescence over a given period of time allowed the authors to estimate a rate of amplification. This rate, $\sim 10^{-3}$ events per cell per generation for CHO cells, was somewhat higher than the frequency calculated for the same cell line using the clonogenic assay. These differences were explained by clonogenic efficiencies and instability of the initial events, two variables that are bypassed by using the flow cytometer.

Finally, Sager et al. [12] used a method of stepwise selection of a population of cells to measure amplification rate between populations of different variants. In this method, a population of cells is exposed to a given concentration of drug (in this study, MTX), and the remaining viable cells are expanded until they reach a given cell density. At this point, the selection is repeated on the entire, expanded population. Resistant populations emerge at a given rate that was then plotted as the amount of MTX resistance attained per unit of time. Potential problems with this method arise when the growth rates of the initial or emerging populations differ and when the cells have different starting sensitivities to the selecting agent.

In our studies we utilized the first two methods (clonogenic assay for incidence and the Luria–Delbrück fluctuation analysis for rate) to estimate amplification events in various sets of nontumorigenic and tumorigenic cells. The Luria–Delbrück fluctuation analysis has the added advantage of allowing one to determine whether spontaneous events are occurring.

The endogenous gene we chose to use to study amplification potential in different cell lines was the gene coding for the multifunctional CAD protein. The *CAD* gene codes for a protein that catalyzes the first three steps of pyrimidine biosynthesis. Amplification of the *CAD* gene is detected when cells are exposed to the cytotoxic drug (PALA), which inhibits aspartate transcarbamylase activity. Unlike resistance to methotrexate, which may occur through multiple mechanisms, resistance to PALA has been reported to only occur through amplification of the *CAD* gene [13].

III. INCIDENCE AND RATE OF AMPLIFICATION IN TUMORIGENIC CELLS

Our initial studies utilized a panel of rat liver epithelial cell lines that differed widely in their tumorigenic potential [14,15]. Using the clonogenic assay described above, we measured amplification incidence at the *CAD* locus in each of these cell lines [16]. We found that highly tumorigenic cells amplified the endogenous gene at a very high incidence (10^{-4}–10^{-3}). Molecular analysis confirmed that *CAD* gene amplification was the basis of drug resistance, and quantitation of the gene copy number indicated that the gene was increased two- to fourfold compared to the level in the corresponding sensitive cell population. This incidence of mutation is some 100-fold higher than the incidence of point mutations usually measured in mammalian cells. Using the Luria–Delbrück fluctuation analysis on the same tumorigenic rat liver epithelial cells, we made two important observations. First, analysis of the variation in colony number on the various plates indicated that *CAD* gene amplification was occurring spontaneously in these cells (discussed below). Second, we found that the rate of *CAD* gene amplification was high, again about 100-fold higher than the point mutation rate measured in mammalian cells.

We have also used the clonogenic assay to examine *CAD* gene amplification incidence in human tumor cell lines and hamster tumor cell lines. In all cases, the amplification event occurs frequently. Interestingly, highly tumorigenic human cells had a similar incidence of *CAD* gene amplification to that seen in the highly tumorigenic rodent cell lines, both rat and hamster. This is notable because the rodent genome has been observed to be more genetically unstable than the human genome. At least for one type of genetic instability, gene amplification, this is not so.

IV. INCIDENCE AND RATE OF AMPLIFICATION IN NONTUMORIGENIC CELLS

The clonogenic assay and Luria–Delbrück fluctuation analysis were also used to measure amplification in nontumorigenic cells. The tumorigenic phenotype is believed to arise through a multistep accumulation of genetic changes [17]. Normal diploid cells have been reported to acquire several phenotypic and genotypic changes in their progression to tumorigenicity. These include the acquisition of immortality, the transformed phenotype, and inactivation of tumor suppressor genes, as well as activation of growth-stimulating gene products. Only after the appropriate genetic changes have been accumulated does the cell display tumorigenicity [18,19].

We have begun to look at cells that acquired one or several of these alterations to see how they effect the incidence of gene amplification. Our studies have shown that a significant difference in amplification potential exists along this continuum of cell lines as they progress to tumorigenicity. Our initial studies measured *CAD* gene amplification in rat liver epithelial cell lines that exhibited no tumorigenicity upon injection into isogenic newborn rats but had acquired one or several of the genetic changes needed for the end result of tumorigenicity [16]. We found that nontumorigenic cells did have detectable incidences of gene amplification but that they were significantly lower than that measured in highly tumorigenic lines. For example, a rat liver epithelial cell line that demonstrated no tumorigenicity amplifies the *CAD* gene at an incidence that is almost 100-fold lower than the highly tumorigenic cell line described above. Using the same model system, we compared the rates of amplification in nontumorigenic and tumorigenic cell lines [20], using the Luria–Delbrück fluctuation analysis. These results mirrored those obtained measuring the incidence. The highly tumorigenic cells had a *rate* of gene amplification that was almost 100-fold higher than the rate measured in nontumorigenic cells. The power of the second type of analysis, the Luria–Delbrück fluctuation analysis, is that it allows us to draw several conclusions and provides insights into the nature of the cells that amplify (see Section V).

Perhaps the most striking finding in this series of studies was that normal, diploid cell populations lacked detectable gene amplification completely. We selected more than 10^9 viable cells of each population and failed to detect PALA-resistant colonies or amplification of the *CAD* gene in both rodent and human cell populations. These same cells generated ouabain-resistant mutants (due to point mutations) at previously reported rates (10^{-7}), indicating that our lack of PALA-resistant mutants was not due to problems with clonigenicity of the mutants [21]. These results, obtained with human, hamster, and rat diploid cell populations, demonstrate that a major difference exists between diploid and tumorigenic cells in their generation of cells containing amplified genes.

V. INSIGHTS INTO THE GENERATION OF PALA-RESISTANT COLONIES IN NONTUMORIGENIC AND TUMORIGENIC CELL LINES

The generation of a PALA-resistant colony may be viewed according to Figure 2. To understand the appearance of the colonies, individual steps in the process must be identified and analyzed. In Figure 2, the appearance of a PALA-resistant colony in plate C must result from the spontaneous rate at which it arises and is eliminated (kinetics of amplification, k_{amp} and the kinetics of deamplification, k_{deamp}), its time of appearance (early or late in the expansion of the culture), and the reproductive/survival capacity of the individual colony (kinetics of proliferation, ,k_p and the kinetics of depletion, k_{dp}). We dissected these steps in our previously published study [20] and discuss their ramifications here. Our previous studies reported an increased number of PALA-resistant colonies for GP_9 cells as compared to WB_{20}

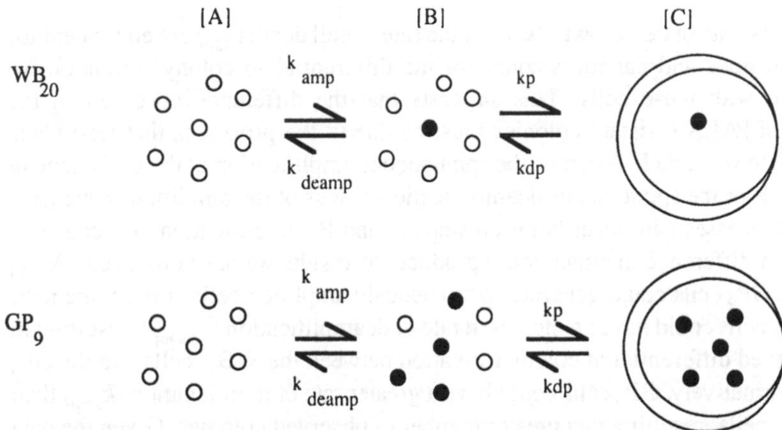

Figure 2 The kinetics of amplification.

cells (represented by the solid circles in plate C of Fig. 2). In this section we analyze the cellular kinetics of producing those PALA-resistant colonies, i.e., the events that occur between plates A and C in Figure 2.

In the most elementary situation, both WB_{20} and GP_9 cell populations could contain the same number of preexisting variants, but because of a greater viability or robustness of GP_9 cells (for any number of reasons), plating of this cell line in PALA would result in a greater number of observed colonies. We know this is not the case because the Luria–Delbrück fluctuation analysis tells us that the variants are arising from nonamplified cells in each cell line. In our experiments, we started with a single cell or a small number of cells that precluded the existence of an amplified variant. (For example: if one cell of 100 initial cells were amplified, the expansion of the population for the fluctuation analysis would yield ~ 1000 resistant progeny as the population went from 10^2 cells to 10^5 cells. This result was not detected.) Since we start with small populations (or even single cells) that have not previously amplified their *CAD* genes, we know that the number of preexisting variants cannot explain the observed difference in amplification rates.

An alternative explanation could be that although both WB_{20} and GP_9 cells spontaneously amplify at the same rate, the viability of the drug-resistant colonies is different. In terms of Figure 2, k_{amp} and k_{deamp} for both cell lines would be the same but k_p and k_{dp} would be different. The differential we see in PALA-resistant colonies in plate C could be due to a greater robustness of GP_9 PALA-resistant colonies relative to WB_{20} PALA-resistant colonies. We measured the reproductive/survival capacity of several subclones from each of these PALA-resistant colonies and found that they are comparable; both plating efficiency and cell doubling time were measured for individual drug-sensitive and drug-resistant subclones [20]. In

Figure 2 the rate of cell growth (k_p) and the rate of cell death (k_{dp}) are equivalent for these cell lines and cannot account for the differential in colony formation we measured with these cells. This suggests that the difference observed in the number of PALA-resistant colonies must be due to the processes that take place between steps A and B—that is, the spontaneous amplification of the *CAD* gene in single cells or the spontaneous deamplification or loss of the amplified sequences.

Two processes can occur between steps A and B, amplification and deamplification. A difference in either could produce the results we have observed. WB_{20} cells and GP_9 cells could generate spontaneously amplified cells at the same rate, but WB_{20} cells could have an increased rate of deamplification (k_{deamp}), resulting in the observed differential in colony formation between the WB_{20} cells and the GP_9 cells. Alternatively, GP_9 cells could have a greater rate of amplification (k_{amp}) than the WB_{20} cells, resulting in a greater number of observed colonies. Given the data presented in this chapter and given the techniques used in this study, we cannot definitively distinguish between these two explanations. Both are cases of increased genetic fluidity; in one situation WB_{20} cells have an increased rate of loss of the amplified sequences, in the alternative situation, GP_9 cells could have an increased rate of generation of amplified sequences. At the present time the enzymes involved in these processes are unknown. We have attempted to distinguish between these two possibilities by measuring the stability of newly emerged PALA-resistant subclones from each cell line. Both the nontumorigenic and highly tumorigenic drug resistant subclones have demonstrated stability of the amplified *CAD* genes and drug resistance at early times after emergence, which suggests that it is the rate of amplification (k_{amp}) that differs between the two cell lines.

It has been argued that it is difficult to use the Luria–Delbrück fluctuation analysis to determine true rates of mutation in mammalian cells because of the assumptions outlined above concerning culture conditions [22]. In this study we controlled for many of these assumptions, optimized the accuracy of our determinations, and obtained a minimum estimation of the rate of genetic fluidity in tumorigenic and nontumorigenic cell lines. We find that the rates in these cell lines can differ by almost 2 orders of magnitude and that in confirmation of the results obtained with the incidence determination, tumorigenicity is correlated with genetic fluidity as measured by gene amplification.

VI. CONCLUSIONS

The *rate* of gene amplification reported for the tumorigenic cell lines in this study ($1–8 \times 10^{-5}$ event/cell/generation) is quite high compared to the rate of point mutations usually observed in mammalian cells ($10^{-7}–10^{-8}$ event/cell/generation) [9]. Similar rates of *CAD* gene amplification have been reported for a Syrian hamster line, $\sim 2 \times 10^{-5}$ [23]. If this rate is indicative of amplification rates in tumorigenic cells in general, it is expected that the types of phenotypic change that result from gene amplification would make a significantly greater contribution to the mutagenic burden of tumorigenic cells than that produced by point mutations. Recent studies on the frequency of large spontaneous deletions suggest that rearrangements of these types are also occurring quite often [24]. The observation of deletions and amplifications in tumorigenic cells is well documented and may provide the mechanisms underlying the phenotypic and genotypic heterogeneity seen in neoplastic cells.

Several studies have linked the frequency and extent of amplification of specific oncogenes with the malignancy of the disease [18,25,26]. Several of these examples are covered in this volume. In neuroblastomas, the frequency and extent of N-*myc* amplification is reported to be a better prognostic indicator than any other determinant [25, and Chapter 29, this volume]. In breast and ovarian cancers, *HER*-2/*neu* gene amplification correlates with malignancy. While in some cases amplification of specific oncogenes is correlated with disease progression, amplification of other oncogenes, often in the same cells, is not [18]. Our results provide a framework for understanding these clinical data. The more tumorigenic a cell, the higher the probability that it will amplify a segment of its genome. This is consistent with the multiple reports of amplification in neoplastic tissue. Perhaps even more intriguing, however, is the observation that normal diploid cells lack a detectable frequency of gene amplification. Studies on gene amplification in diploid cells and tumors could provide us with insights into fundamental biological differences that lead to malignancy.

REFERENCES

1. Schimke, R. T. *Cell*, *37*:705–713 (1989).
2. Stark, G. R., and Wahl, G. M. *Annu. Rev. Biochem*. *53*:447–503 (1984).
3. Tlsty, T. D., Brown, P., Johnston, R., and Schimke, R. T. In *Gene Amplification* (R. T. Schimke, ed.), Cold Spring Harbor Laboratory, Cold Spring Harbor, New York, 1982, pp. 231–238.
4. Brown, P. C., Tlsty, T. D., and Schimke, R. T. *Mol. Cell. Biol. 3*:1097–1107 (1983).
5. Tlsty, T. D., Brown, P., and Schimke, R. T. *Mol. Cell. Biol. 4*:1050–1056 (1984).
6. Luria, S. E., and Delbrück, *Genetics*, *28*:491–511 (1943).
7. Lea, D. A., and Coulson, C. A. *J. Genet*. *49*:264–285 (1949).
8. Capizzi, R. L., and Jameson, J. W. *Mutat. Res*. *17*:147–148 (1973).
9. Baker, R. M., Burnett, D. M., Mankovitz, R., Thompson, L. H., Whitmore, G. F., Siminovitch, L., and Till, J. E., *Cell*, *1*:9–21 (1974).
10. Johnston, R. N., Beverley, S. M., and Schimke, R. T. *Proc. Natl. Acad. Sci. USA*, *80*:3711–3715 (1983).
11. Kavathus, P., and Herzenberg, L. *Nature*, *306*:385–387 (1983).
12. Sager, R., Gadi, I., Stephens, L., and Grabowy, C. T. *Proc. Natl. Acad. Sci. USA*, *82*:7015–7019 (1985).
13. Wahl, G. M., Padgett, R. A., and Stark, G. R. *J. Biol. Chem*. *254*:8679–8689 (1979).

14. Tsao, M., Grisham, J. W., Chou, B. B., and Smith, J. D. *Cancer Res. 45*:5134–5138 (1985).
15. Tsao, M., Grisham, J. W., and Nelson, K. G. *Cancer Res. 45*:5139–5144 (1985).
16. Otto, E., McCord, S., and Tlsty, T. D. *J. Biol. Chem. 264*:3390–3396 (1989).
17. Nowell, P. C. *Science, 194*:23–28 (1976).
18. Wong, A. J., Ruppert, J. M., Eggleston, J., Hamilton, S. R., Baylin, S. B., and Vogelstein, B. *Science, 233*:461–464 (1986).
19. Fearon, E., and Vogelstein, B. *Cell, 61*:567–579 (1990).
20. Tlsty, T. D., Margolin, B., and Lum, K. *Proc. Natl. Acad. Sci. USA, 86*:9441–9445 (1989).
21. Elmore, E., Kakunaga, T., and Barrett, J. C. *Cancer Res. 43*:1650–1655 (1983).
22. Kendal, W. S., and Frost, P. *Cancer Res. 48*:11060–11065 (1988).
23. Kempe, T., Swyryd, E. A., Bruist, M., and Stark, G. A. *Cell, 9*:541–550 (1976).
24. Kaden, D. A., Bardwell, L., Newmark, P., Anisowicz, A., Skopek, T., and Sager, R. *Proc. Natl. Acad. Sci. USA, 86*:2306–2310 (1989).
25. Seeger, R. C., Brodeur, G. M., Sather, H., Dalton, A., Siegel, S. E., Wong, K. Y., and Hammond, D. *New Engl. J. Med. 313*:1111–1116 (1985).
26. Slamon, D. J., Clark, G. M., Wong, S. G., Levin, W. J., Ullrich, A., and McGuire, W. L. *Science, 235*:177–182 (1987).

V

The Mechanism of Gene Amplification

34

Mechanisms of Arsenic-Induced Gene Amplification

J. Carl Barrett *National Institute of Environmental Health Sciences, National Institutes of Health, Research Triangle Park, North Carolina*

Te-Chang Lee *Academia Sinica, Taipei, Taiwan, Republic of China*

I. INTRODUCTION

Arsenic is a very interesting environmental carcinogen because it is one of the few chemicals, possibly the only one, for which there is sufficient evidence for carcinogenicity in humans but inadequate evidence in animals [1,2]. Exposure of humans to inorganic arsenic compounds in drugs, drinking water, and occupational environments is associated with increased risks of skin cancer, lung cancer and possibly liver cancer [2]. However, little evidence exists for the carcinogenicity of arsenic to animals [1,2]. Yet, sodium arsenite and sodium arsenate reproducibly induce morphological transformation of rodent cells in culture [3]. A distinction does exist, however, between arsenic and most chemicals that induce cell transformation in that arsenic does not induce gene mutations at specific genetic loci [3].

We previously reported that arsenic induces amplification of the gene for dihydrofolate reductase (DHFR) and proposed that arsenic-induced amplification of proto-oncogenes may play a role in neoplastic progression of arsenic-related cancers in humans [1]. This chapter describes our results on arsenic-induced gene amplification and discusses the possible mechanisms for this effect.

II. INDUCTION OF GENE AMPLIFICATION BY ARSENIC

Amplification of the *DHFR* gene can be measured in mouse 3T6 cells by selecting cells that form colonies in the presence of methotrexate. Cells with an increased copy number of the *DHFR* gene have increased levels of DHFR enzyme and are resistant to methotrexate (MTX^R) [1]. Treatment of mouse 3T6 cells with sodium arsenite or sodium arsenate induced dose-dependent increases in the number of MTX^R colonies (Fig. 1). Sodium arsenite was active at a lower concentration than sodium arsenate. This is the same relative potency of the two compounds in cell transformation assays [3], and the dose–response results for induction of gene amplification and cell transformation are similar. Sodium arsenite has been noted to be more active than sodium arsenate in other biological assays, and this has been attributed to differences in the uptake of trivalent versus pentavalent arsenic [1].

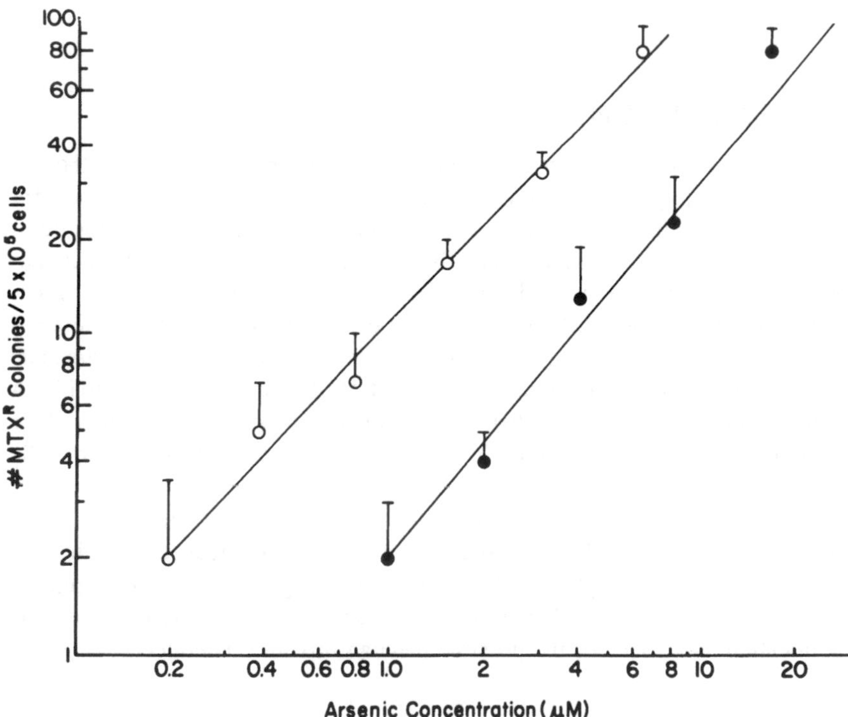

Figure 1 Induction of MTX^R colonies of mouse 3T6 cells by sodium arsenite (open circles) or sodium arsenate (solid circles). In the absence of arsenic, an average of 0.67 ± 0.58 MTX^R colony/5×10^5 cells was observed. (Data from Ref. 1.)

Sodium arsenite and sodium arsenate induced MTX^R colonies at MTX concentrations of 150–300 nm (Fig. 2).

The surviving MTX^R 3T6 cells, when isolated, were resistant to MTX and had amplified *DHFR* genes. The copy number of the *DHFR* gene was estimated by comparing the extent of hybridization by dot blot analysis of radiolabeled *DHFR* or actin cDNAs to DNA isolated from parental 3T6 cells and MTX^R clones. Approximately 50% (9 of 17) of the MTX^R clones induced by arsenic (0.2–0.8 μg/ml) had amplified copy numbers of the *DHFR* gene, ranging from 2- to 11-fold, which is consistent with the findings of others [4].

The data in Figures 1 and 2 are expressed as MTX^R colonies per cell treated and show that arsenic induces a large increase in the absolute number of MTX^R cells. The toxicity of arsenic to mammalian cells was measured by a reduction in the colony-forming efficiency of the treated cells. Arsenic induced MTX^R cells at doses that were nontoxic as well as at toxic doses. When the number of MTX^R cells per surviving cell was calculated, the frequency of MTX^R cells was observed to be greater than 10^{-2} after treatment with the highest doses of sodium arsenite or sodium arsenate [1].

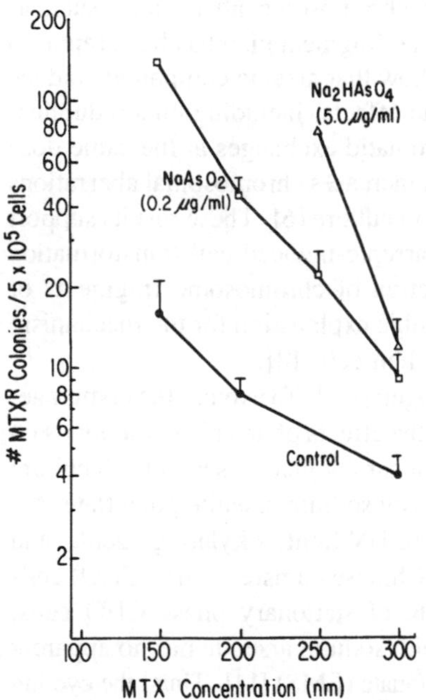

Figure 2 Effect of varying MTX concentration of MTX^R cells induced by sodium arsenate or sodium arsenite. The cells were either untreated, exposed to 16 μM sodium arsenate, or exposed to 1.6 μM sodium arsenite. (Data from Ref. 1.)

To determine whether the ability to induce gene amplification was a general property of carcinogenic metals, nickel chloride and nickel sulfate were tested over the concentration range (1–20 μg/ml) that induces morphological cell transformation of Syrian hamster embryo (SHE) cells. Neither compound increased the number of MTX^R colonies at any of the doses tested in several assays.

III. MECHANISM OF ARSENIC-INDUCED GENE AMPLIFICATION

The results of our study clearly indicate that arsenic is an effective enhancer of *DHFR* gene amplification. The mechanism by which arsenic induces gene amplification remains unknown, although several possibilities can be considered. Mutagens and carcinogens, such as UV light and alkylating agents, have been shown to induce gene amplification in rodent cells [4,5]. However, arsenic by itself is not a mutagen, or only a weak one, in both prokaryotic and mammalian cell systems. Therefore, arsenic-induced gene amplification is unlikely to be due to direct DNA base damage [1].

Gene amplification has been hypothesized to arise either by a recombinational mechanism or by inhibition of DNA synthesis and overreplication [5]. An association between gene amplification and chromosome aberrations such as polyploidy, endoreduplication, and chromosome fragmentation has been reported in several cell systems [1,5–7]. Our studies show that arsenic compounds induce morphological transformation and cytogenetic effects, including endoreduplication, chromosome aberrations, and sister chromatid exchanges at the same dose range in SHE cells [3]. Similarly, hydroxyurea increases chromosomal aberrations and induces *DFHR* gene amplification in cells in culture [6]. These results support a possible role for chromosomal changes in arsenic-induced cell transformation and gene amplification. Asymmetric segregation of chromosome fragments or unequal sister chromatid exchange is one possible explanation for the mechanism of gene amplification in drug-treated mammalian cells [8].

Arsenic is also a comutagen and a cocarcinogen [1,9]. To explore the comutagenicity of arsenic compounds, we have studied the effects of arsenic compounds on the genotoxicity of established mutagens and/or carcinogens in several cell culture systems. Our results show that posttreatment with sodium arsenite potentiates the cytotoxicity, clastogenicity, and mutagenicity of UV light, alkylating agents, and DNA crosslinking agents in actively dividing Chinese hamster ovary (CHO) cells and human fibroblasts [10–13]. Posttreatment of stationary phase CHO cells, human fibroblasts, and human lymphocytes with sodium arsenite has no apparent effect on the clastogenicity of ethyl methanesulfonate (EMS) [14]. Thus, the cycling state of the cell may play a crucial role in the action of arsenite coclastogenicity.

A number of agents, including chemical carcinogens [5], UV irradiation, hypoxia, tumor promoters, and DNA synthesis inhibitors such as hydroxyurea, methotrexate, and 5-fluorodeoxyuridine, increase the frequency of drug-resistant cell variants if they are applied before drug selection in cultured mammalian cells [5,15,16]. Inhibition of DNA synthesis per se is a common effect of agents that enhance gene amplification [15,17]. One model of gene amplification, the onion-skin model, suggests that gene amplification is caused by "misfiring" of DNA replication due to the stalling of DNA growing points at replicons following inhibition of DNA synthesis. This stalling may lead to abnormal reinitiation at the replication origin(s), resulting in a twofold or more amplification of that region of the genome. Arsenic compounds inhibit DNA synthesis [18] and prolong the cell cycle [10] of cultured cells. Our unpublished data also demonstrate that sodium arsenite prolongs the S phase in CHO cells. Prolonged S phase is often observed in cells treated with carcinogens.

Carcinogen-induced inhibition of DNA synthesis might represent an adaptive response to facilitate cell survival under adverse growth conditions, which is analogous to the pleiotropic SOS system documented in bacteria [19]. Induction of amplification of viral genes has been reported to be mediated by one or more carcinogen-induced transacting factors [20]. This is similar to the SOS system, which is also initiated by inducible cellular factors [19]. Treatment of cells with arsenic has been shown to induce a set of stress proteins [21], which are encoded from a family of genes that respond to diverse types of environmental stress including heat shock, heavy metals, amino acid analogues, and oxidizing agents [22] in both eukaryotes and prokaryotes. However, the functions of stress proteins remain to be fully elucidated. Whether arsenic-induced stress proteins play a certain role in induction of gene amplification is an interesting question for further investigation.

REFERENCES

1. Lee, T. C., Tanaka, N., Lamb, P. W., Gilmer, T. M., and Barrett, J. C. Induction of gene amplification by arsenic. *Science*, *241*:79–81 (1988).
2. International Agency for Research on Cancer. Monograph on the evaluation of the carcinogenic risk of chemicals to humans, Suppl. 4, 50–51 (1982).
3. Lee, T. C., Oshimura, M., and Barrett, J. C. Comparison of arsenic-induced cell transformation, cytotoxicity, mutation, and cytogenetic effects in Syrian hamster embryo cells in culture. *Carcinogenesis*, *6*:1421–1426 (1985).
4. Tlsty, T. D., Brown, P. C., and Schimke, R. T. UV irradiation facilitates methotrexate resistance and amplification of the dihydrofolate reductase gene in cultured 3T6 mouse cells. *Mol. Cell. Biol.* *4*:1050–1056 (1984).
5. Schimke, R. T. Gene amplification, drug resistance, and cancer. *Cancer Res.* *44*: 1735–1742 (1984).
6. Hill, A. B., and Schimke, R. T. Increase gene amplification in L5178Y mouse

lymphoma cells with hydroxyurea-induced chromosomal aberrations. *Cancer Res.* *45*:5050–5057 (1985).

7. Ottaggio, L., Agnese, C., Bonatti, S., Cavolina, P., De Ambrosis, A., Degan, P., Di Leonardo, A., Miele, M., Randazzo, R., and Abbondandolo, A. Chromosome aberrations associated with *CAD* gene amplification in Chinese hamster cultured cells. *Mutat. Res.* *199*:111–121 (1988).
8. Hahn, P., Kapp, L. N., Morgan, W. F., and Painter, R. B. Chromosomal changes without DNA overproduction in hydroxyurea-treated mammalian cells: Implications for gene amplification. *Cancer Res.* *46*:4607–4612 (1986).
9. Rossman, T. G. Enhancement of UV-mutagenesis by low concentrations of arsenite in *E. coli. Mutat. Res.* *91*:207–211 (1981).
10. Lee, T. C., Huang, T. Y., and Jan, K. Y. Sodium arsenite enhances the cytotoxicity, clastogenicity, and 6-thioguanine-resistant mutagenicity of ultraviolet light in Chinese hamster ovary cells. *Mutat. Res.* *148*:83–89 (1985).
11. Lee, T. C., Lee, K. C., Tzeng, Y. J., Huang, R. Y., and Jan, K. Y. Sodium arsenite potentiates the clastogenicity and mutagenicity of DNA crosslinking agents. *Environ. Mutagens.* *8*:119–128 (1986).
12. Lee, T. C., Wang-Wuu, S., Huang, R. Y., Lee, K. C., and Jan, K. Y. Differential effects of pre- and post-treatment of sodium arsenite on the genotoxicity of methyl methanesulfonate in Chinese hamster ovary cells. *Cancer Res.* *46*:1854–1857 (1986).
13. Jan, K. Y., Huang, R. Y., and Lee, T. C. Different modes of action of sodium arsenite, 3-aminobenzamide and caffeine on the enhancement of ethyl methanesulfonate clastogenicity. *Cytogenet. Cell Genet.* *41*:202–208 (1986).
14. Huang, R. Y., Jan, K. Y., and Lee, T. C. Posttreatment with sodium arsenite is coclastogenic in log phase but not in stationary phase. *Hum. Genet.* *75*:159–162 (1987).
15. Schimke, R. T. Gene amplification in cultured cells. *J. Biol. Chem.* *263*:5989–5992 (1988).
16. Sharma, R. C., and Schimke, R. T. Enhancement of the frequency of methotrexate resistance by γ-radiation in Chinese hamster ovary and mouse 3T6 cells. *Cancer Res.* *49*:3861–3866 (1989).
17. Schimke, R. T., Sherwood, S. W., Hill, A. B., and Johnston, R. N. Overreplication and recombination of DNA in higher eucaryotes: Potential consequences and biological implications. *Proc. Natl. Acad. Sci. USA*, *83*:2157–2161 (1986).
18. Change, M. J. W. Toxic effects of arsenic compounds on culture human fibroblasts. *In Vitro Toxicol.* *1*:103–110 (1987).
19. Elespuru, R. K. Inducible responses to DNA damage in bacteria and mammalian cells. *Environ. Mol. Mutagens.* *10*:97–116 (1987).
20. Lambert, M. E., Pellegrini, S., Gattoni-Celli, S., and Weinstein, I. B. Carcinogen induced asynchronous replication of polyoma DNA is mediated by a trans-acting factor. *Carcinogenesis*, *7*:1011–1017 (1986).
21. Li, G. C. Induction of thermotolerance and enhanced heat shock protein synthesis in Chinese hamster fibroblasts by sodium arsenite and ethanol. *J. Cell. Physiol.* *115*:116–122 (1983).
22. Lindquist, S. The heat-shock response. *Annu. Rev. Biochem.* *55*:1151–1191 (1986).

35

Involvement of Inverted Duplications in the Generation of Gene Amplification in Mammalian Cells

Mike Fried, Salvatore Feo,* and Edith Heard[†] *Imperial Cancer Research Fund, London, England*

I. INTRODUCTION

To understand the mechanism(s) responsible for the generation of amplified DNA in mammalian cells, it is important to determine the organization and structure of amplicons. Initially it was thought that amplicons were all in arrays of head-to-tail tandem repeats and that the join between different amplicons was heterogeneous. Early amplification models incorporated these features. More recently it has been discovered that amplicons can exist in arrays of head-to-head and tail-to-tail inverted repeats, and it has been suggested that the formation of inverted duplications is involved in the generation of amplified DNA [1–3]. The features of amplified inverted duplications and the findings related to the major new models incorporating inverted duplications in the generation of amplified DNA are reviewed in this chapter.

II. DISCOVERY OF THE ASSOCIATION OF INVERTED DUPLICATIONS WITH GENE AMPLIFICATION

Some years ago in a quest to isolate transcriptional enhancers we discovered that amplicons were in arrays of inverted duplications, as opposed to arrays of tandem

---

Current affiliations:
*Institute of Developmental Biology of the Italian National Research Council, Palermo, Italy.
[†]Unité de Genetique Moléculaire Morine, Institut Pasteur, Paris, France.

duplications, in the 3B rat cell line that contained an amplified polyoma virus (Py) oncogene [1]. The 3B cell line was generated after transfection of Rat-2 cells with a ligated mixture of enhancerless Py transforming region and mouse cellular DNA followed by selection for morphologically transformed colonies, which resulted from the activation of the *Py* oncogene. In 3B cells transformation resulted from a 20–40-fold amplification of the DNA region containing the *Py* oncogene, which had been only weakly activated by a cellular transcriptional enhancer. Initially we demonstrated the presence of an inverted duplication in a cloned fragment of amplified 3B DNA. We were able to show that the amplified DNA in 3B cellular DNA was also in the form of inverted duplications by its ability to "snap back," after denaturation and rapid renaturation, into a double-stranded form and thus resist hydrolysis with the single strand specific nuclease S1 [1]. We also directly demonstrated the presence of inverted duplications by DNA sequencing of part of the 3B cloned insert [2]. The inverted repeat was shown not to preexist in the chromosomal DNA of the parental Rat-2 cells. Since the 3B inversion involved not only the transfected mouse and *Py* sequences but also rat chromosomal DNA, the inversion must have taken place at the time of or after integration of the transfected DNA into the rat chromosome and probably was formed as a primary event in the amplification process [2].

At the time of this discovery, the available models for gene amplification mainly predicted that amplicons would be arranged as arrays of direct repeats rather than inverted repeats. To determine whether the structure of the 3B amplified inverted repeats was an isolated example or represented a more common feature of gene amplification, we analyzed other gene amplification events. Using the "snap back" method [1], we have shown that the amplification of the *myc* gene in a number of human tumor cell lines and the *CAD* gene in a number of Syrian hamster cell lines resistant to *N*-(phosphonacetyl)-L-aspartate (PALA) also contained at least part of their amplified DNA in the form of inverted duplications [3]. The cell lines we analyzed in this study contained between 16 and 200 amplified copies, in the form of either extrachromosomal double minutes or chromosomally located homogeneous staining regions. The presence of the *CAD* amplicon in the form of an inverted duplication in the 165-28 cell line predicted by our "snap back" analysis [3] was later confirmed by cloning and sequence analysis of the inversion joint [4].

III. FEATURES OF AMPLIFIED INVERTED DUPLICATIONS

Our initial findings of the presence of inverted repeats as a common feature of gene amplification of oncogenes and genes involved in drug resistance [1–3] were confirmed by others for the *AMPD* [5], *APRT* [6], *CAD* [4,7], and *DHFR* [8–10] amplified genes. Three amplifications [1,9,10] involved transfected genes, while the others involved amplifications of endogenous genes. In four cases [2,4–6], the sequence around one of the inversion joints was determined and found to be relatively A+T-rich and to have the potential to form hairpin structures, which may be involved in the recombination events at the sites of the inversions. No unusual sequence elements, such as transposition elements or other special features, were detected directly at the inversion joints, and the joints appear to have been formed by simple illegitimate recombination events involving 1–3 bases of homology. In one case [4], a 16 bp insertion found at the joint was an exact duplication of a sequence found further upstream in the cellular DNA. The palindromic arms of the inverted duplication are quite large, being greater than 100 kb [8,9,11]. In all the joints studied by sequence analysis, the palindromic arms of the inverted duplication were found to be separated by a short stretch of noninverted DNA of 150–1000 bp. Furthermore it was determined that this noninverted DNA was contiguous with the sequence of one of the palindromic arms in the unamplified cellular DNA. In studies where the level of amplification is relatively low, having presumably arisen from a single amplification event, it was determined that the same inversion joint was contained in all the amplicons [1,6,9], showing that there was no heterogeneity at least at one end of the amplicons.

IV. GENE AMPLIFICATION MODELS BASED ON INVERTED DUPLICATIONS

Two models were formerly favored to explain gene amplification. One involves unequal sister chromatid exchange in which there is a twofold increase in the number of amplicons with each cell division, and multiple cell cycles are required to reach a modest (10–40 copies) amplification level (see Ref. 12). The other involves disproportionate DNA replication with multiple firing of a specific origin of DNA replication during a single S phase, resulting in the form of an "onionskin" of overreplication of the same DNA region. The resolution (mechanism unclear) of this "onionskin" can result in amplification (see Ref. 12 and references therein). Although in this model amplification to a modest copy number can take place in a single cell cycle, the firing of the same origin of DNA replication many times during the same S phase cannot be beneficial to a cell that will continue to divide. Neither of these models predicts the presence of amplified inverted duplications containing homogeneous ends as an obligatory feature. The presence of inverted duplications as a general feature of gene amplification could not be easily explained by these models. This has led to the postulation of new models in which the formation of inverted duplications is a prerequisite for the generation of multiple gene copies, and which involve only a *single* initiation of DNA replication at a particular origin in a *single* S phase.

The gene amplification model postulated by Passananti et al. [2] (extrachromosomal double rolling circle model) involving inverted duplications is presented in Figure 1. In this model the DNA to be amplified is excised from the chromosome by two assymmetric nonhomologous recombination events between newly replicated chromosomal DNA strands to generate a circle containing an

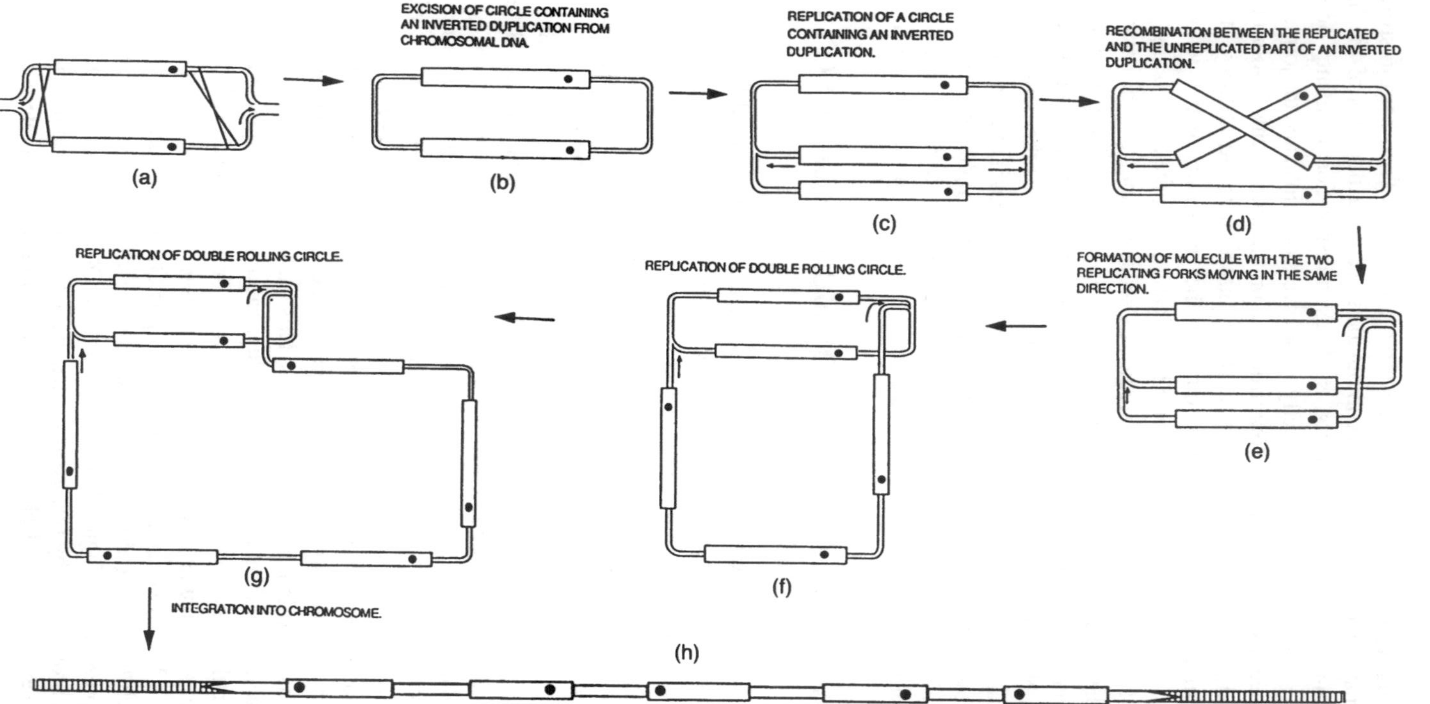
EXCISION OF CIRCLE CONTAINING AN INVERTED DUPLICATION FROM CHROMOSOMAL DNA.
REPLICATION OF A CIRCLE CONTAINING AN INVERTED DUPLICATION.
RECOMBINATION BETWEEN THE REPLICATED AND THE UNREPLICATED PART OF AN INVERTED DUPLICATION.
(a)
(b)
(c)
(d)
FORMATION OF MOLECULE WITH THE TWO REPLICATING FORKS MOVING IN THE SAME DIRECTION.
REPLICATION OF DOUBLE ROLLING CIRCLE.
REPLICATION OF DOUBLE ROLLING CIRCLE.
(e)
(f)
(g)
INTEGRATION INTO CHROMOSOME.
(h)

Figure 1 The extrachromosomal double rolling circle model. A model for inverted duplication formation via extrachromosomal amplification. A circular molecule (b) containing an inverted duplication (rectangle) harboring the gene to be amplified (dot) is excised from a replicating chromosome (a) (small arrows indicate replication forks), possibly by a nonhomologous recombination event between the two newly replicated DNA strands. This results in chromosome breakage at the site of the excision, which can lead to an interstitial deletion or loss of a part or all of the chromosome. In a subsequent S phase, replication is initiated from within the extrachromosomal circle from a sequence that can act as an origin of DNA replication. Following a single replication of an inverted sequence containing a target gene (c), a homologous recombination event takes place between a copy of the newly replicated inverted sequence and the copy of the inverted duplication that has yet to be replicated (d). Resolution of this structure results in a flip of one of the replicating forks (small arrows) so that both replication forks now proceed in the same direction (chase each other) in the form of a double rolling circle molecule (e). Further replication results in the generation of an amplified array of inverted duplications (f,g). The amplified array could either remain extrachromosomal (in the form of an episome and/or double minute) (g) or become stabilized by integration into chromosomal DNA at a new chromosomal location (h). The consequences of this model, in which a modest amount of amplification can occur in a *single* S phase from a *single* initiation of DNA replication, is a rearrangement at the original chromosomal site (resulting from the excision of the DNA that is amplified extrachromosomally), the location of the amplified DNA at a new chromosomal position, and homogeneity of the joints between the inverted repeats. (Adapted from Ref. 2.)

inverted duplication, which can replicate extrachromosomally. In a subsequent S phase, a homologous recombination event takes place between a replicated copy and an unreplicated copy of the duplicated sequence, resulting in the flipping of a replication fork so that both forks now proceed in the same direction and will not converge. The consequence of such a replicating structure (double rolling circle) is the production of an amplified array of inverted duplications that may either remain extrachromosomal (double minute) in an unstable form or become stabilized by integration into the chromosomal DNA (homogeneously staining regions). The extrachromosomal forms (Fig. 1) could be the source of the circular DNA molecules detected in many cells containing amplified DNA without evidence of double minute or homogeneously staining regions (see Ref. 13 and references therein).

An alternative model to explain the role of inverted duplications in the amplification process proposed by Hyrien et al. [5] (chromosomal spiral model) is presented in Figure 2. In this model an inverted duplication is generated by strand switching across a replicating fork [6]. To produce a noninverted central region between the two palindromic arms, the switch should occur at different regions on the two strands. This can result in formation of single-stranded loop of one of the DNA strands at the switch point, which is excised to create a gap. The resulting structure can prime replication (repair synthesis), which now proceeds in the direction of the previously replicated DNA. The outcome of a second such event on the other side of the replication bubble results in an intrachromosomal double rolling circle, which will generate an array of inverted duplications. This structure can be resolved by an excisional homologous recombination event encompassing the replication forks into an intrachromosomal inverted duplication array and an extrachromosomal replicating double rolling circle as postulated in Passananti et al. [2] (Fig. 1).

These models are attractive because they provide a means of achieving an increase in gene copy number within a *single* S phase without evoking multiple firings of the same origin of DNA replication. The two models predict different outcomes for the original chromosome locus involved in the amplification as well

Figure 2 The chromosome spiral model for inverted duplication formation and intrachromosomal amplification. (a)–(d): Formation of a head-to-head C′/BC) joint by copy choice recombination via the formation and excision of a hairpin loop. (d′): Structure obtained if a head-to-head (C′/BC) joint and a tail-to-tail (G/H′G′) joint are formed by similar events occurring at both ends of the replication bubble. (d′)–(f): Amplification of the resulting inverted duplication by intrachromosomal double rolling circle replication in the form of an indefinite spiral, which results in an intrachromosomal amplified array of inverted duplications without relocation. (g): Resolution into a chromosomal tandem array of inverted duplications with possible release of an exchromosomal circle. The consequences of this model are intrachromosomal amplification in loco at the original locus and homogeneity of the joints between the inverted repeats. (From Ref. 5.)

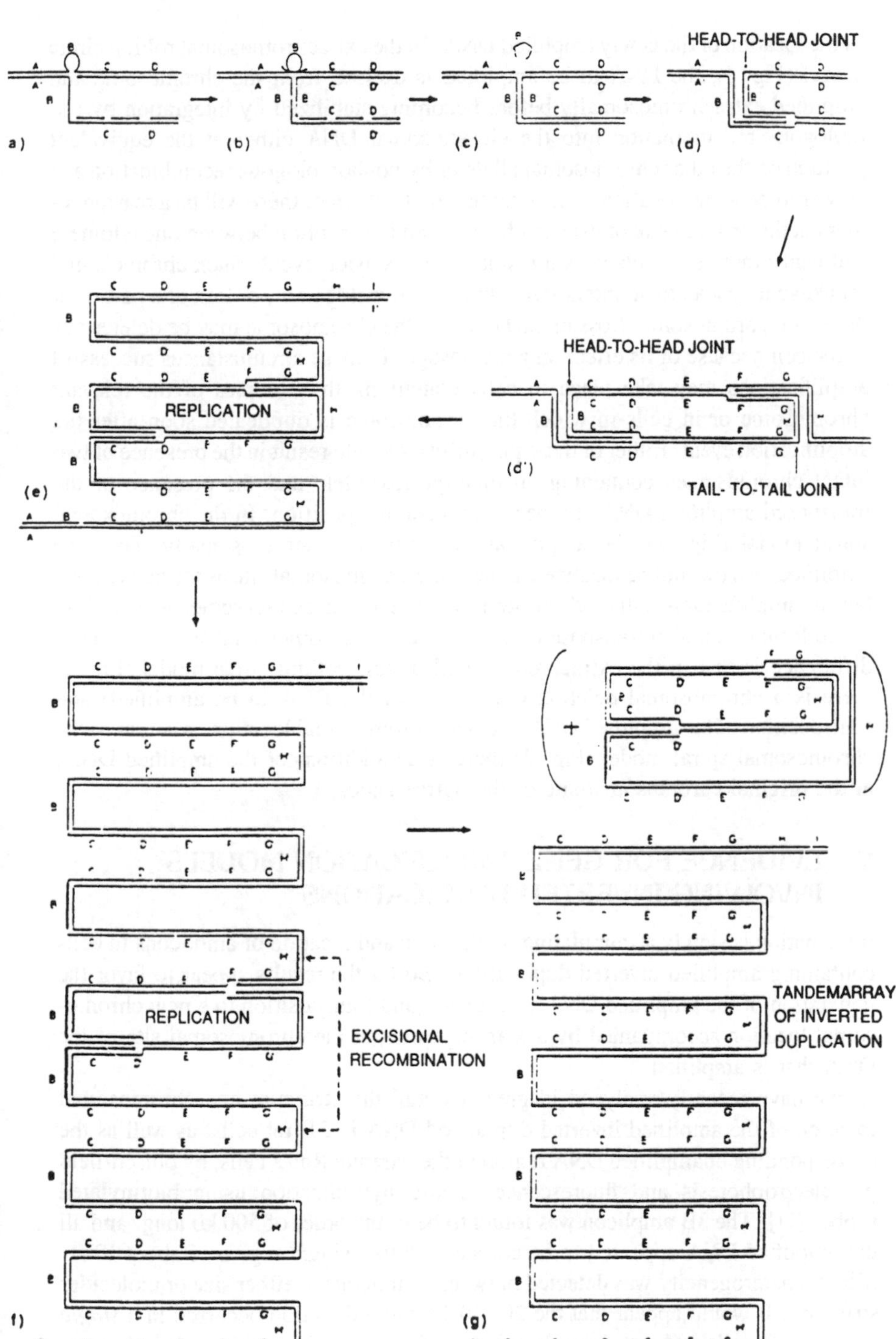
HEAD-TO-HEAD JOINT
a)
(b)
(c)
(d)
HEAD-TO-HEAD JOINT
REPLICATION
(d')
TAIL- TO-TAIL JOINT
(e)
REPLICATION
EXCISIONAL
RECOMBINATION
TANDEMARRAY
OF INVERTED
DUPLICATION
f)
(g)

as the location of the newly amplified DNA. In the extrachromosomal rolling circle model (Fig. 1), the DNA to be amplified is excised from the chromosome and amplified extrachromosomally before becoming stabilized by integration by homologous recombination into the chromosomal DNA either at the equivalent position on the other chromosomal allele or by nonhomologous recombination at a new chromosomal location. Whatever the final outcome, there will be a rearrangement at the original site of excision because an interruption between one telomere and centromere takes place as a result of the excision event. Such chromosomal breakage may lead to an interstital deletion, a translocation, or loss of a part of or the entire chromosome. Loss of part or all of the chromosome may be deleterious to the cell because of its effect on gene dosage. In these circumstances successful amplifications may take place in cells containing three copies of the relevant chromosome or in cells in which the chromosome is duplicated soon after the amplification event. Either of these possibilities would result in the presence of two intact chromosomes containing an unamplified allele and the presence of the transposed amplified DNA at a new chromosomal position. In the chromosomal spiral model (Fig. 2), the amplification occurs intrachromosomally. Thus the amplified DNA would be localized at the same chromosomal site as the native gene before amplification, although in addition an excised extrachromosomal rolling circle form could also transpose to a new (second) chromosomal site. The major difference is that in the extrachromosomal double rolling circle model (Fig. 1) there is a chromosomal deletion (consisting of the DNA to be amplified) and transposition of the amplified DNA to a new chromosomal location, whereas in the chromosomal spiral model (Fig. 2) there is an addition, of the amplified DNA, at the original chromosomal site of the native gene.

V. EVIDENCE FOR GENE AMPLIFICATION MODELS INVOLVING INVERTED DUPLICATIONS

Information is slowly accumulating on the state and location of amplicons in cells containing amplified inverted duplications. So far the results appear to favor the generation of the amplified DNA by excision and transposition to a new chromosomal location accompanied by a rearrangement at the chromosomal site of the DNA that is amplified.

We have recently analyzed in greater detail the structure and chromosomal location of the amplified inverted duplicated DNA in 3B rat cells, as well as the corresponding unamplified DNA region in the parental Rat-2 cells, by pulsed-field gel electrophoresis and fluorescence in situ hybridization using biotinylated probes [11]. The 3B amplicon was found to be of the order of 500 kb long, and all the amplified DNA appeared to be colocalized to a single region of about 10–11 Mb. No heterogeneity was detected between amplicons in either size or molecular structure. It would appear that the 20–40-fold amplification occurred in a *single event* in 3B cells and not through a series of events that would result in heterogeneity among the amplicons. Using a cloned rat DNA probe (15.5 kb) from within the 3B duplicated region (20 kb from the *Py* oncogene), three unamplified chromosomal locations were detected in the parental Rat-2 cells (Fig. 3a). Two of these have been identified at the same position on the two unaltered copies of rat chromosome 2 and the third to a similar location on a marker chromosome that contains a fusion of chromosome 2 to chromosome 3. A similar analysis of 3B metaphases (Fig. 3b) revealed a single copy signal on the fusion marker chromosome, as was detected in Rat-2 cells, and on one copy of a rat chromosome 2. The other copy of rat chromosome 2 is absent in 3B cells. The amplified DNA is all localized to a single site on a new marker chromosome. Exactly the same pattern was found in an early passage of 3B cells frozen down approximately 25 cell divisions after its generation. Thus in 3B cells there was a loss of one copy of chromosome 2 containing the site of the native DNA region that eventually was amplified and the presence of the amplified DNA at a new chromosomal location (Fig. 4). Experiments are under way to verify that the new chromosomal site of the amplified DNA is the result of a transposition event.

In the CHOC400 Chinese hamster cell line, which contains amplified *DHFR* genes in inverted duplicated arrays, the size of the amplicon has been estimated to be 220 kb [8]. In situ hybridization has also revealed new chromosomal sites for the amplified DNA as well as a rearrangement at the old chromosomal site [14]. Unfortunately this cell line contains about 1000 copies of the amplified *DHFR* gene, which was generated by multiple drug selections, and the amplification present may have resulted from a composite of multiple amplification steps and recombination events.

Ruiz and Wahl [9] isolated a hamster cell line that would amplify a transfected *DHFR* gene at a 50–100-fold greater frequency than other sister transformants. They observed that the transfected *DHFR* gene as well as adjacent hamster cellular DNA was in the form of a large inverted duplication (at least 160 kb long), which appears to be located on a large extrachromosomal form (episome). The donated *DHFR* DNA could not be located to a distinct chromosomal site prior to detection of the episomes, and the amplified copies were not found to be in the form of a double minute by in situ hybridization analysis. The extrachromosomal *DHFR* element can be stably integrated into a number of different chromosomal sites [15]. Although the origin of the hamster cellular DNA sequences associated with the *DHFR* gene in the extrachromosomal inverted duplications is unclear, Ruiz and Wahl [9] suggest that they arose after excision across a replicating fork of the DNA region containing the integrated transfected DNA as proposed by Passananti et al. [2] (see Fig. 1). The presence of submicroscopic extrachromosomal molecules containing the *DHFR* gene in the form of an inverted duplication, which can amplify at a very high frequency, is also consistent with this model.

Extrachromosomal forms ranging in size from 250 to 750 kb have also been found associated with the amplification of either endogenous genes such as *DHFR*, *mdr* and c-*myc* (see Ref. 13 and references therein) or a transfected *CAD* gene [16].

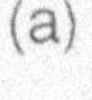

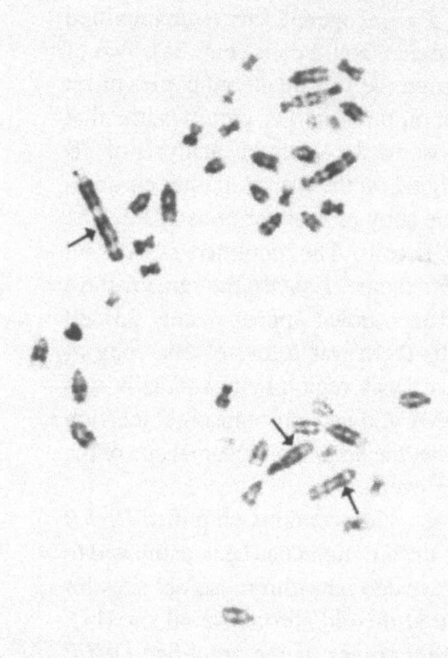

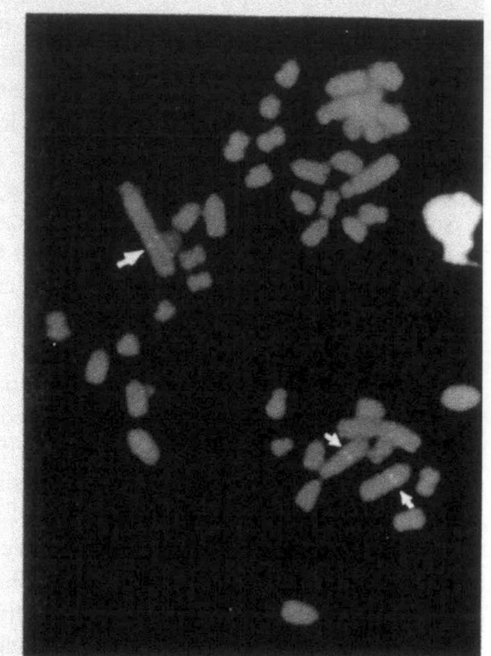

(b)

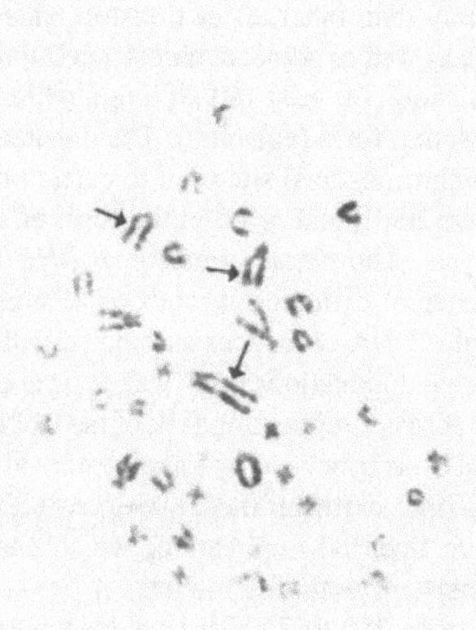

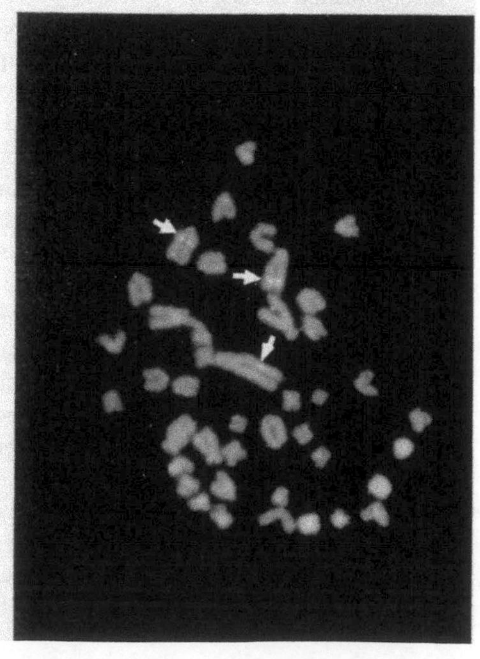

In the latter case the formation of the extrachromosomal amplified DNA was associated with deletion of the chromosomal copy of the transfected *CAD* DNA. The presence of inverted duplications in these amplified extrachromosomal forms has not yet been demonstrated.

Amplifications of oncogenes that occur in tumor cells in vivo have also been detected in the form of inverted duplications [3]. A number of amplified oncogenes have also been shown to involve a new location of the amplified DNA as well as deletion at the chromosomal site of the native gene. Analysis of a number of amplifications of c-*myc* and N-*myc* oncogenes in human tumor cells has shown that the amplified DNA has been transposed to a different chromosomal position or is in the form of double minutes (see Ref. 17 and references therein). In these amplifications it is usually difficult to examine the state of the native chromosomal site from which the amplified DNA was generated, especially in the presence of double minutes. Hunt et al. [18] were able to analyze the chromosomal site of human N-*myc* (which in normal cells is located in chromosome band 2p24) in neuroblastoma cells containing amplified N-*myc* in the form of double minutes. These investigators managed to separate and distinguish the two copies of chromosome 2 from each other and from the amplified double minute DNA in different somatic cell hybrids. In one homologue of chromosome 2 they detected a submicroscopic deletion of the DNA containing the native N-*myc* gene. It was not determined whether the amplified DNA in these cells was in the form of inverted repeats. This result provides strong evidence for a model involving excision (deletion) and extrachromosomal amplification of a target gene [2].

Figure 3 G-banding followed by fluorescence in situ hybridization of metaphase spreads of 3B cells (containing amplified DNA) and parental Rat-2 cells with a biotinylated 15.5 kb probe of rat DNA derived from within the 3B amplicon. (a) Metaphase spread of the progenitor Rat-2 cells first G-banded (left) and then fluorescence hybridized (right) with the biotinylated rat DNA probe. Arrows indicate the positions of the fluorescein signal of the unamplified alleles on the right and the corresponding positions on the G-banded chromosomes on the left. This enables the three chromosomes containing the unamplified alleles to be compared and identified as the two hologues of rat chromosome 2, and a long marker chromosome M1, which is in fact the product of a translocation between a complete rat chromosome 3 and most of rat chromosome 2, and is thus termed der(3)t(2;3). Rat-2 cells are essentially triploid for rat chromosome 2. (b) Metaphase spread of 3B cells (containing the amplified DNA) first G-banded (left) and then fluorescence hybridized (right) with the biotinylated rat DNA probe. The chromosome on which the amplified DNA is located (upper right arrowed chromosome) can be compared to the two chromosomes on which unamplified alleles are detected. The single copy, native unamplified alleles are seen on only one chromosome 2 (cf. Rat-2 spread in (a)) and on the marker chromosome M1/der(3)t(2;3). The amplified DNA lies on a chromosome whose G-banding pattern bears no obvious resemblance to chromosome 2, or other rat chromosomes, which is therefore termed marker chromosome M2. See Figure 4 for G-band pattern comparisons of chromosomes 2, M1/der(3)t(2;3), and M2. (From Ref. 11.)

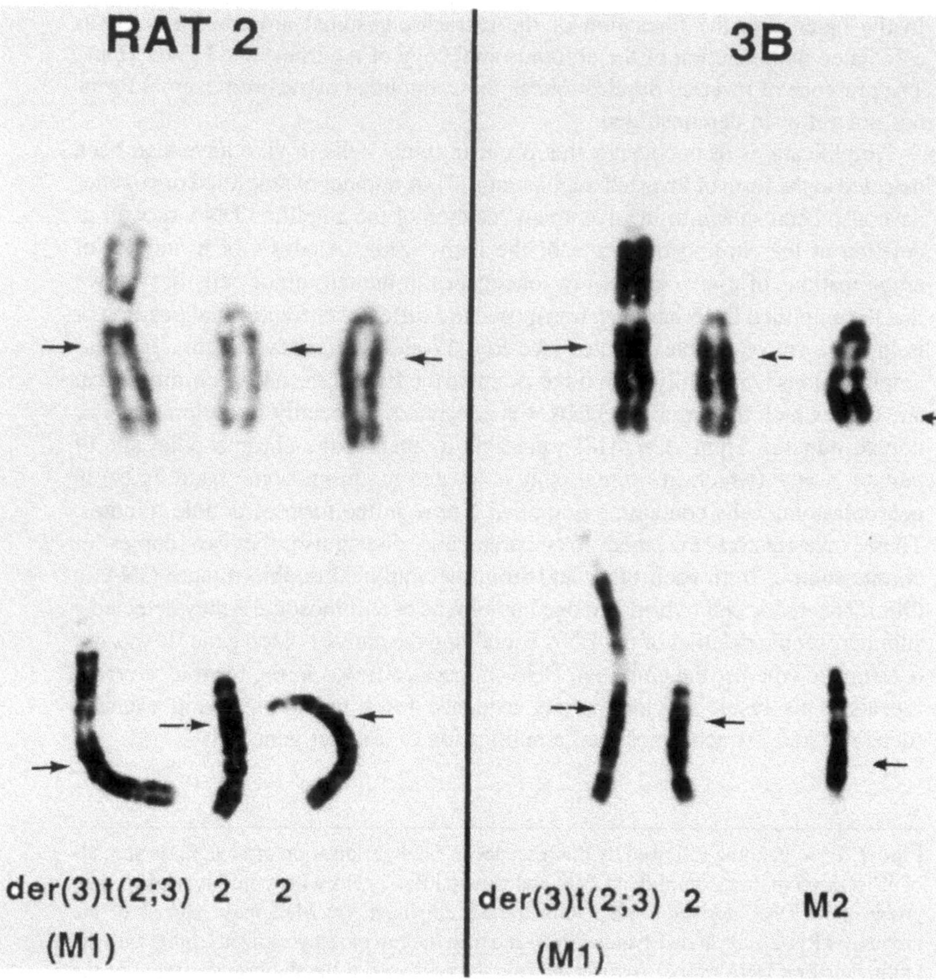

Figure 4 Partial G-banded karyotype of the chromosomes on which the biotinylated rat probe detects signal from 3B and Rat-2 cells. The G-banded chromosomes to which the biotinylated rat probe hybridizes from two Rat-2 metaphase spreads (left) and two 3B metaphase spreads (right). The arrows indicate the position of the signal detected by the probe on each chromosome (Fig. 3). The degree of chromosome condensation is similar for the upper Rat-2 and 3B spreads, as it is for the lower two spreads, thus facilitating comparison of the G-bands on the chromosomes in the two cell lines. All the amplified DNA is located on chromosome M2 in 3B cells. Unamplified alleles are located on chromosomes 2 and M1/der(3)t(2;3) in Rat-2 and 3B cells. (From Ref. 11.)

VI. SUMMARY

In recent years there has been accumulating evidence that inverted duplications are a common feature of a number of gene amplification events. There is probably more than one way to generate amplified DNA. For small increases in copy number, unequal sister chromatid exchange or an increase in the copy number of a specific chromosome number may be the method of choice. Such methods do not predict the obligatory formation of inverted duplications. Disproportionate replication, which results from the multiple firings of an origin of DNA replication during a single S phase, leading to an increase in gene copy number, would be expected to occur only in mammalian cells that do not have to divide again. Amplification may also occur after excision from the chromosome of a circular plasmid, which is able to replicate extrachromosomally [13]. To achieve a modest copy number in a single cell cycle, the episomal origin must escape the restraints, present in its normal chromosomal location, that specify that it fire only once per S phase. If the episomes can replicate only once per cell cycle, amplification to a modest copy number can be produced by unequal distribution of the episomes at each cell division, but this process will take multiple cell cycles.

The gene amplification models involving inverted duplications can account for a medium to large increase in copy number during a *single* S phase without the need for multiple firings of a single origin of DNA replication [2,5]. A number of observations that are consistent with such models have been described here. In particular there is increasing support for the extrachromosomal double rolling circle model, proposed by Passananti et al. [2], which involves excision and deletion of the native chromosomal site, followed by extrachromosomal amplification and subsequent stabilization by transposition of the amplified DNA to a new chromosomal location. Further analysis of amplicons arranged in the form of inverted duplications should contribute to our understanding of the modes of generation of amplified DNA.

ACKNOWLEDGMENTS

We thank our many colleagues at the Imperial Cancer Research Fund for their helpful discussion and interactions. S. F. is supported by a Target Project "Ingegneria Genetica" from the Italian National Research Council.

REFERENCES

1. Ford, M., Davies, B., Griffiths, M., Wilson, J., and Fried, M. Isolation of a gene enhancer within an amplified inverted duplication after "expression selection." *Proc. Natl. Acad. Sci. USA*, *82*:3370–3374 (1985).
2. Passananti, C., Davies, B., Ford, M., and Fried, M. Structure of an inverted duplication as a first step in a gene amplification event: Implications for a model of gene amplification. *EMBO J.* *6*:1697–1703 (1987).

3. Ford, M., and Fried, M. Large inverted duplications are associated with gene amplification. *Cell*, *45*:425–430 (1986).
4. Legouy, E., Fossar, N., Lhomond, G., and Brison, O. Structure of four amplified DNA novel joints. *Somatic Cell Mol. Genet.* *15*:309–320 (1989).
5. Hyrien, O., Debatisse, M., Buttin, G., and Robert de Saint Vincent, B. The multicopy appearance of a large inverted duplication and the sequence at the inversion joint suggests a new model for gene amplification. *EMBO J.* *7*:407–417 (1988).
6. Nalbantoglu, J., and Meuth, M. DNA amplification–deletion in a spontaneous mutation of the hamster *aprt* locus: Structure and sequence of a novel joint. *Nucleic Acids Res.* *14*:8361–8371 (1986).
7. Saito, I., and Stark, G. R. Charomids: Cosmid vectors for efficient cloning and mapping of large or small restriction fragments. *Proc. Natl. Acad. Sci. USA*, *83*: 8664–8668 (1986).
8. Looney, J. E., and Hamlin, J. L. Isolation of the amplified dihydrofolate reductase domain from methotrexate-resistant Chinese hamster ovary cells. *Mol. Cell. Biol.* *7*:569–577 (1987).
9. Ruiz, J. C., and Wahl, G. M. Formation of an inverted duplication can be an initial step in gene amplification. *Mol. Cell. Biol.* *8*:4302–4313 (1988).
10. Heartlein, M. W., and Latt, S. A. Amplified inverted duplication within and adjacent to heterologous selectable DNA. *Nucleic Acids Res.* *17*:1697–1716 (1989).
11. Heard, E., Williams, S. V., Sheer, D., and Fried, M. A gene amplification accompanied by the loss of a chromosome containing the native allele and the appearance of the amplified DNA at a new chromosomal location. *Proc. Natl. Acad. Sci. USA*, *88*: 8242–8246 (1991).
12. Stark, G. R., Debatisse, M., Giulotto, E., and Wahl, G. M. Recent progress in understanding mechanisms of mammalian DNA amplification. *Cell*, *57*:901–908 (1989).
13. Wahl, G. M. The importance of circular DNA in mammalian gene amplification. *Cancer Res.* *49*:1323–1340 (1989).
14. Trask, B. J., and Hamlin, J. L. Early dihydrofolate reductase gene amplification events in CHO cells usually occur on the same chromosome arm as the original locus. *Genes Dev.* *3*:1913–1925 (1989).
15. Ruiz, J. C., and Wahl, G. M. Chromosome destabilization during gene amplification. *Mol. Cell. Biol.* *10*:3056–3066 (1990).
16. Carroll, S. M., DeRose, M. L., Gaudray, P., Moore, C. M., Needham-Vandevanter, D. R., Von Hoff, D. D., and Wahl, G. M. Double minute chromosomes can be produced from precursors derived from a chromosomal deletion. *Mol. Cell. Biol.* *8*:1525–1533 (1988).
17. Alitalo, K., and Schwab, M. Oncogene amplification in tumor cells. *Adv. Cancer Res.* *47*:235–281 (1986).
18. Hunt, J. D., Valentine, M., and Tereba, A. Excision of N-*myc* from chromosome 2 in human neuroblastoma cells containing amplified N-*myc* sequences. *Mol. Cell. Biol.* *10*:823–829 (1990).

36

The Genetics of Amplification

Elena Giulotto *University of Pavia, Pavia, Italy*
Mary Ellen Perry *Imperial Cancer Research Fund, London, England*

I. THE ABILITY TO AMPLIFY IS DETERMINED GENETICALLY

Cultured cells vary enormously in their rates and frequencies of amplification. Normal, senescing cells have an undetectably low ($< 10^{-8}$) frequency of amplification in culture [1,2], whereas many cell lines have frequencies of 10^{-4}–10^{-6} [3]. Frequent detection of amplified oncogenes in tumors indicates that amplification occurs readily in tumor cells [4]. DNA amplification is one manifestation of the genetic instability that is typical of cell lines and tumors. Amplification of a defective butyrylcholinesterase gene has been demonstrated in blood cells from a man and his son exposed chronically to organophosphorus insecticides [5]. The fact that the parents of the older subject did not have extra copies of the gene in their somatic cells suggested that amplification may have occurred in the germ line. It has been recently shown that the ability to amplify is completely lost when embryonal mouse carcinoma cells are induced to differentiate by retinoic acid [6]. Thus the transformation state as well as the developmental state of a cell influences its ability to amplify.

Moreover, the rate of amplification can vary within a cell population; it was possible to isolate mutant Syrian hamster cells that had increased rates of gene amplification at different loci. These cells, described below, may carry alterations in one or more genes involved directly or indirectly in amplification [7].

These examples suggest that a wide range of genetic or epigenetic factors may influence the probability of amplification.

II. FREQUENCY OF AMPLIFICATION IN CULTURED CELLS

Drug-resistant variants of cell lines can result from mutations of various kinds, including gene amplification. In the case of resistance to *N*-(phosphonacetyl)-L-aspartate (PALA) (see Chapter 19, this volume), the only mechanism observed is chromosomal amplification of the *CAD* gene. For example, in 33 independent PALA-resistant baby hamster kidney (BHK) cell lines, the *CAD* gene was always amplified [8], and in situ hybridization experiments on several of these and other mutants showed that the amplified DNA was always chromosomal [9,10]. This is a particularly valuable system for studying gene amplification, since interpretations are not complicated by the survival of cells with mutations of other types.

Fluctuation experiments in BHK cells showed that PALA-resistant variants arise spontaneously [11]. Zieg et al. [9] measured frequencies and rates of resistance as a function of PALA concentration. As shown in Figure 1, the number of new resistant variants per cell per generation (rate) has a shallower dependence on drug concentration than the fraction of resistant cells in a population (frequency). The relatively flat appearance of the rate curve may indicate that the degree of amplification is similar in most amplification events, conferring resistance to both high and low PALA concentrations. On the other hand, the frequency curve indicates that in a steady state population the fraction of highly resistant cells is much smaller than the fraction of cells resistant to low drug concentrations.

The observations of Zieg et al. can now be explained on the basis of more recent findings. Most amplification events generate many copies of the *CAD* gene. In the absence of selection, the extra copies of *CAD*, not needed for survival, tend to be progressively lost because amplified DNA is unstable. The result would be that among all cells containing amplified DNA in a growing population, only cells that have recently amplified their DNA are resistant to high drug concentrations, while the remaining cells have lost variable numbers of copies of the gene initially amplified and can survive only in less stringent conditions. Evidence in favor of this hypothesis includes (1) the presence, in BHK cells soon after PALA selection, of many more copies of *CAD* than are needed for resistance [10] and (2) rapid and progressive loss of PALA resistance and amplified DNA in BHK cell lines selected with PALA in one step but maintained without selection [12].

III. THE AMPLIFICATOR PHENOTYPE

By analogy with mutator bacterial and cell strains, cells with increased amplification rates were called "amplificator." Since amplification of a single gene is relatively rare, the rationale for isolating amplificator mutants was based on the

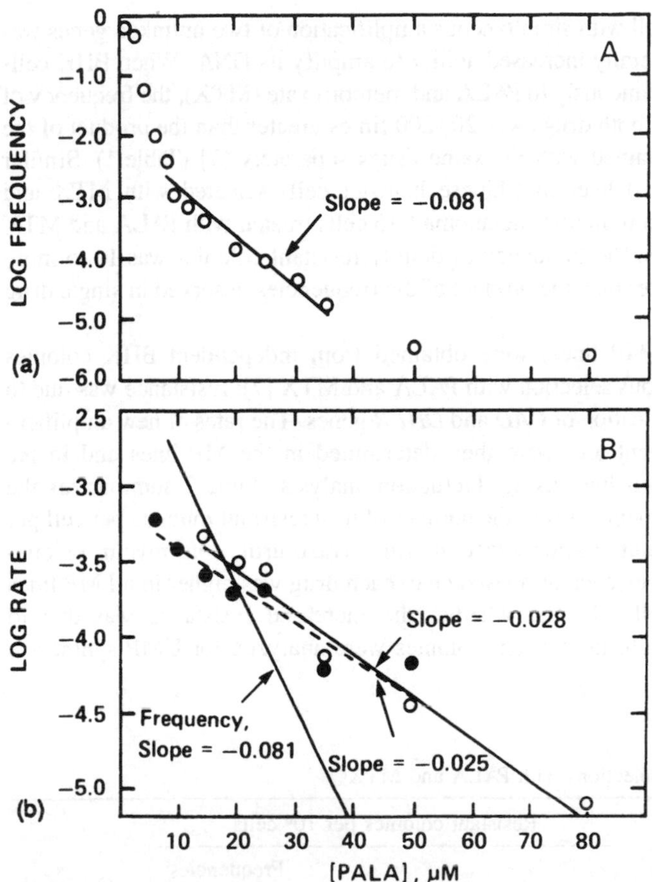

Figure 1 Frequencies and rates of resistance as a function of PALA concentration in BHK cells. To determine the frequency of resistant mutants, samples of 1×10^3–4×10^4 cells were plated directly in medium containing PALA and the number of resistant colonies was scored after 10 days. For the rate experiments, independent samples of 1×10^3 cells were plated in nonselective medium to obtain populations of 10^7 cells; 2×10^5 cells from each sample were then seeded in the presence of PALA. The number of mutants per cell per generation was determined from the median number of colonies per sample and from the number of samples without colonies. The points were fitted to straight lines by linear regression. The straight line portion of the curve in (a) is reproduced in (b), where two independent experiments are shown with different symbols. (From Ref. 9.)

assumption that a cell with simultaneous amplification of two unlinked genes was likely to have a generally increased ability to amplify its DNA. When BHK cells were exposed simultaneously to PALA and methotrexate (MTX), the frequency of colonies resistant to both drugs was 20–200 times greater than the product of the two frequencies obtained with the same drugs separately [7] (Table 1). Similar results were obtained later in Chinese hamster cells selected with MTX and doxorubicin [13] and in mouse melanoma B16 cells treated with PALA and MTX [14]. In both reports, the frequency of doubly resistant colonies was 10 to more than 300 times greater than the product of the frequencies observed in single drug selections.

Four cell lines (MP lines) were obtained from independent BHK colonies surviving simultaneous selection with PALA and MTX [7]; resistance was due to simultaneous amplification of *CAD* and *DHFR* genes. The rates of new amplifications at three different loci were then determined in the MP lines and in the BHKWT parental cell line, using fluctuation analysis. Table 2 summarizes the results of the experiments, where the number of new resistant mutants per cell per generation was measured after selection with pyrazofurin, coformycin, or ouabain. The rate of occurrence of resistance to each drug was higher in all MP lines than in parental cells. To test whether the increased resistance was due to amplification, pyrazofurin-resistant colonies were analyzed for UMP synthetase

Table 1 Combined Selections with PALA and MTX[a]

Concentration of selective drug		Resistant colonies per 10^6 cells: Frequencies observed			Frequencies calculated for independent amplifications:	Observed frequency (PALA plus MTX) ÷ calculated frequency
PALA (μM)	MTX (nM)	PALA only	MTX only	PALA plus MTX	PALA only × MTX only	
Experiment I						
15	150	85	76	1.7	0.0065	260
Experiment II						
10	150	410	−100	8.0	0.04	200
10	200	410	15	1.4	0.006	230
15	150	350	100	0.7	0.035	20

[a]In each experiment a single culture of BHK cells was split into three parts. One was treated with PALA alone, one with MTX alone, and the third with both drugs together. The number of resistant colonies was counted after 3 weeks.

Source: From Ref. 7.

Table 2 Amplificator Phenotype in MP Cell Lines[a]

	Cell lines				
	BHKWT	MP1	MP4	MP5	MP7
Selective drug	Rate	Fold increase in rate			
Pyrazofurin	1×10^{-5}	13	3	2	9
Coformycin	4×10^{-5}	>6	>7	n.d.	7
Ouabain	6×10^{-5}	27	1	3	8

[a]Independent cultures of cells, each grown from 100-cell samples in nonselective medium, were dispersed and replated in selective medium. Mutation rates were calculated from the median number of colonies per culture and from the fraction of cultures without colonies. The rates in the BHKWT parental cells are given as mutants per cell per generation. The fold increase in mutation rates compared to the BHKWT is shown for the MP cell lines. N.D., not determined.
Source: From Ref. 7.

copy number. Since all clones showed amplification of the gene, it was concluded that the MP cell lines had an amplificator phenotype.

The ability to amplify at an increased rate was a stable property of the four MP lines and was independent of resistance to PALA and MTX; the MP cells in fact retained their amplificator phenotype when resistance to the two drugs used for selection was lost upon growth in nonselective medium [7]. In the experiments with BHK cells, most if not all doubly resistant colonies were amplificators and the frequency of double resistance was of the order of 10^{-6} [7]. If we consider that very few amplificator cells would have *CAD* and *DHFR* genes amplified, we conclude that the proportion of amplificators in cell lines must be quite high.

The presence in cell populations of frequent variants with increased ability to amplify is probably one reason for failure of combination chemotherapy: the probability that tumor cells will become resistant to two drugs simultaneously cannot be assumed to equal the product of the two independent probabilities.

Cells can be stimulated to amplify their DNA at higher rates by treatment with exogenous agents that inhibit DNA synthesis or damage DNA [15]. BHK cells previously treated with 5-aza-2′-deoxycytidine have increased rates of resistance when selected in a single drug [23]. Interestingly, amplificator cells also showed higher rates of resistance after treatment with 5-aza-2′-deoxycytidine, indicating that although they were already "induced" to amplify at a high rate, this rate could be further increased. In this case, genetic and environmental factors combined to yield a very high amplification rate.

IV. ALTERED DNA METABOLISM IN AMPLIFICATOR CELLS

The mutator strains of bacteria and mammalian cells analyzed so far are altered in functions required for accurate DNA replication or repair. Analogously, amplificator mutations may be due to alterations in DNA replication, repair, or recombination. Since the genome is altered, DNA amplification can certainly be defined as a mutation. However amplification rates are not necessarily increased in mutator strains. In a mutator cell, there is a generally increased rate of mutations, which is detectable by measuring gene disruption. In an amplificator cell, the increased rate of gene amplification can be detected as occurrence of resistance to specific drugs. It is easy to imagine how a mutation that increases the probability of recombination between extended chromosomal regions may affect the frequency of gene disruption very little, while greatly favoring gene amplification. On the contrary, overaccumulation of DNA damage may or may not affect gene amplification, depending on its effect on DNA replication and recombination and on the mechanism involved in its repair, while it should certainly generate more point mutations.

A Chinese hamster ovary cell line serially selected for resistance to MTX and stably amplified for *DHFR* showed high mutation rates at two independent loci [16]. The large number of mutations arising in this mutator strain were the result of an increased number of base substitutions, while there was a lower proportion of gene rearrangements than in spontaneously arising mutations [17]. The amplification rates at two distinct loci was increased (M. Perry, M. Commane, and G. Stark, in preparation), showing that this cell line is also an amplificator. In these cells alterations of DNA replication or repair may cause increased rates of both amplification and mutation. A second mutator line, which is resistant to arabinosyl cytosine and has an expanded deoxycytidine 5′-triphosphate (dCTP) pool, creates transitions, substituting dCTP for deoxythymidine triphosphate [18]. This cell line does not have a higher amplification rate, indicating that not all mutator strains are amplificator and that the increased number of mutations in the genome of the cell is not the sole cause of amplificator status.

It is not known whether any of the BHK amplificator cell lines isolated so far is also a mutator; however, it has been recently demonstrated that the four amplificator MP lines previously described are altered in their response to DNA damage [19]. All four are hypersensitive to the killing effect of UV irradiation and to mitomycin C. In these cells recovery of DNA and RNA synthesis after UV irradiation is slower or absent (Fig. 2), and the frequency of UV-induced sister chromatid exchange is increased compared to the nonamplificator parental line. These alterations, present in all MP lines, are likely to be related directly to the amplificator phenotype. In these cells, where excision repair is normal (a slight defect was detected in one cell line only), UV hypersensitivity is probably due to defective processing of DNA damage during DNA replication and transcription.

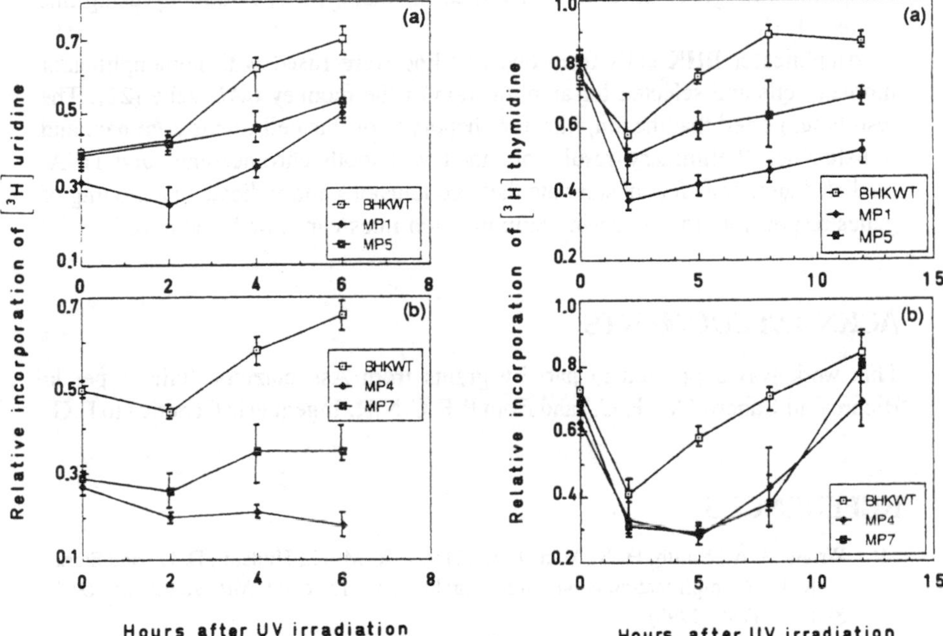

Figure 2 Inhibition of DNA and RNA synthesis by UV irradiation in amplificator MP cell lines. Cells were pulsed-labeled at different times after UV irradiation (20 J/m^2). The values represent the relative incorporation of tritiated thymidine (right) and tritiated uridine (left) in UV-irradiated cells compared to control cells. (From Ref. 19.)

The persistence of damage may then increase the probability of DNA rearrangements involved in amplification.

In their response to UV damage, MP cells show a striking similarity with cells from patients with the genetic disease Cockayne syndrome [20]. This rare disease is transmitted as an autosomal recessive trait and the affected individuals have, among other severe symptoms, an unusual sensitivity to sunlight. In Cockayne cells, as in MP cells, hypersensitivity to UV is not due to a detectable excision repair deficiency; rather, DNA and RNA synthesis does not recover normally after irradiation. It has been shown that Cockayne cells are deficient in repairing damage in actively transcribed DNA [20]. The same defect could be present in MP cells. It remains to be established whether Cockayne cells are amplificators.

Assuming that the abnormal cellular response to UV and the increased rate of amplification result from the same mutation, the defect in RNA synthesis rate after UV irradiation can now be used as a biochemical parameter to study the genetic behavior of the amplificator phenotype in complementation experiments. Three complementing genes have been identified in Cockayne syndrome by using this strategy [21].

Amplificator BHK cells from one MP line were fused with nonamplificator monkey cells and selected for amplification of the monkey *CAD* gene [22]. The results suggested that the amplificator phenotype of this cell line is dominant and transferable. Preliminary results obtained with both chromosome- and DNA-mediated gene transfer substantiate these conclusions and indicate that cloning of genes responsible for increasing amplification rates can now be achieved.

ACKNOWLEDGMENTS

This work was supported in part by grants from Associazione Italiana per la Ricerca sul Cancro (A.I.R.C.) and from P. F. C.N.R. Ingegneria Genetica to E.G.

REFERENCES

1. Wright, J. A., Smith, H. S., Watt, F. M., Hancock, M. C., Hudson, D. L., and Stark, G. R. DNA amplification is rare in normal human cells. *Proc. Natl. Acad. Sci. USA*, *87*:1791–1795 (1990).
2. Tlsty, T. D. Normal diploid human and rodent cells lack a detectable frequency of gene amplification. *Proc. Natl. Acad. Sci. USA*, *87*:3132–3136 (1990).
3. Stark, G. R., Debatisse, M., Giulotto, E., and Wahl, G. M. Recent progress in understanding mechanisms of mammalian DNA amplification. *Cell*, *57*:901–908 (1989).
4. Stark, G. R. DNA amplification in drug resistant cells and in tumors. *Cancer Surv.* *5*: 1–23 (1986).
5. Prody, C. A., Dreyfus, P., Zamir, R., Zakut, H., and Sereq, H. De novo amplification within a "silent" human cholinesterase gene in a family subjected to prolonged exposure to organophosphorous insecticides. *Proc. Natl. Acad. Sci. USA*, *86*:690–694 (1989).
6. Lucke-Huhle, C., and Herrlich, P. Retinoic acid induced differentiation prevents gene amplification in teratocarcinoma stem cells. *Int. J. Cancer*, *47*:461–465 (1991).
7. Giulotto, E., Knights, C., and Stark, G. R. Hamster cells with increased rates of DNA amplification, a new phenotype. *Cell*, *48*:837–845 (1987).
8. Giulotto, E., Saito, I., and Stark, G. R. Structure of DNA formed in the first step of *CAD* gene amplification. *EMBO J.* *5*:2115–2121 (1986).
9. Zieg, J., Clayton, C. E., Ardeshir, F., Giulotto, E., Swyryd, E. A., and Stark, G. R. Properties of single-step mutants of Syrian hamster cell lines resistant to *N*-(phosphonacetyl)-L-aspartate. *Mol. Cell. Biol.* *3*:2089–2098 (1983).
10. Smith, K. A., Gorman, P. A., Stark, M. B., Groves, R. P., and Stark, G. R. Distinctive chromosomal structures are formed very early in the amplification of *CAD* genes in Syrian hamster cells. *Cell*, *63*:1219–1227 (1990).
11. Kempe, T. D., Swyryd, E. A., Bruist, M., and Stark, G. R. Stable mutants of

mammalian cells that overproduce the first three enzymes of pyrimidine nucleotide biosynthesis. *Cell*, *9*:541–550 (1976).

12. Saito, I., Groves, R., Giulotto, E., Rolfe, M., and Stark, G. R. Evolution and stability of chromosomal DNA coamplified with the *CAD* gene. *Mol. Cell. Biol.* *9*:2445–2452 (1989).
13. Rice, G. C., Ling, V., and Schimke, R. T. Frequencies of independent and simultaneous selection of Chinese hamster cells for methotrexate and doxorubicin (adriamycin) resistance. *Proc. Natl. Acad. Sci. USA*, *84*:9261–9264 (1987).
14. McMillan, T. J., Kalebic, T., Stark, G. R., and Hart, I. R. High frequency of double drug resistance in the B16 melanoma cell line. *Eur. J. Cancer*, *26*:565–567 (1990).
15. Schimke, R. T. Gene amplification in cultured cells. *J. Biol. Chem.* *263*:5989–5992 (1988).
16. Drobetsky, E., and Meuth, M. Increased mutational rates in Chinese hamster ovary cells serially selected for drug resistance. *Mol. Cell. Biol.* *3*:1882–1885 (1983).
17. Caligo, M. A., Armstrong, W., Rossiter, B., and Meuth, M. Increased rates of base substitution in a hamster mutator strain obtained during serial selection for gene amplification. *Mol. Cell. Biol.* *10*:6805–6808 (1990).
18. Goncalves, O., Drobetsky, E., and Meuth, M. Structural alterations of the *aprt* locus induced by deoxyribonucleoside triphosphate pool imbalances in Chinese hamster ovary cells. *Mol. Cell. Biol.* *4*:1792–1799 (1984).
19. Giulotto, E., Bertoni, L., Attolini, C., Rainaldi, G., and Anglana, M. BHK cell lines with increased rates of gene amplification are hypersensitive to ultraviolet light. *Proc. Natl. Acad. Sci. USA*, *88*:3484–3488 (1991).
20. Lehmann, A. R., and Norris, P. G. DNA repair and cancer: Speculations based on studies with xeroderma pigmentosum, Cockayne's syndrome and trichothiodistrophy. *Carcinogenesis*, *10*:1353–1356 (1989).
21. Lehmann, A. R. Cockayne's syndrome and trichothiodystrophy; defective repair without cancer. *Cancer Rev.* *7*:82–103 (1987).
22. Rolfe, M., Knights, C., and Stark, G. R. Somatic cell genetic studies of amplificator cell lines. *Cancer Cells*, *6*:325–328 (1988).
23. Perry, M. E., Rolfe, M., McIntyre, P., Commane, M., and Stark, G. R. Induction of gene amplification by 5-aza-2′-deoxycytidine. *Mutat. Res.* (1992) (in press).

37

"U-Turn" Replication at Arrested Forks Leads to the Generation of Extrachromosomal Inverted Repeats

Sara Lavi, David Hassin, Mirit Aladjem, Zlil Ophir, and Shulamit Karby *Tel Aviv University, Tel Aviv, Israel*

I. INTRODUCTION

Gene amplification is commonly observed in malignant cells and in drug-resistant cell lines [1,2]. It is a dynamic process in which the structures generated early are altered during cell propagation or drug selection. Amplified sequences are often detected as paired acentric circular structures called double minute chromosomes, or as expanded chromosomal regions. Studies by several groups have demonstrated that the amplification units are often arranged as inverted repeat (IR) duplications [3,4]. It was suggested that episomal inverted repeats might represent early amplification products [3,5]. The varied structures and locations of amplified sequences are consistent with the idea that different mechanisms might be involved in their generation.

To investigate the molecular mechanisms governing the amplification process, we constructed a model system consisting of an integrated SV40 genome in virally transformed Chinese hamster cells (CO60 cells). In these cells viral DNA amplification can be induced following treatment with chemical or physical carcinogens and by drugs arresting DNA replication [6,7]. The SV40 amplification can be directly detected by hybridization without the need to select for cells harboring amplification. By using this system it is possible to examine the structure of the initial amplification intermediates that did not undergo stabilization by later processes. Our studies revealed that this phenomenon is origin and T antigen dependent; it is transient and reaches its peak at about 72 hr posttreatment. The amplified viral DNA is heterogeneous in size, is associated with both chromosomal and extrachromosomal DNA, and represents mostly the 2 kb region surrounding the origin. The amplification is bidirectional and is sensitive to aphidicolin. The involvement of carcinogen-induced cellular factors in this process was demonstrated by cell fusion experiments in which the carcinogen-induced signal directing SV40 amplification was transferred from carcinogen-treated cells that do not harbor SV40 to nontreated virally transformed cells [8].

II. SV40 AMPLIFICATION IN VITRO

To further characterize the cellular factors controlling the amplification process, we established an in vitro system in which SV40 amplification is catalyzed by extracts from carcinogen-treated Chinese hamster cells [9]. This system is based on the cell-free system for SV40 replication of Li and Kelly [10]. The in vitro amplification reaction consists of cytosolic extracts from carcinogen-treated CO60 cells, exogenous template-pSVK$_1$, a 8.23 kb plasmid containing the SV40 origin of replication [11], and immunoaffinity purified T antigen. Cytosolic extracts from CO60 cells treated with carcinogens such as *N*-methyl-*N'*-nitro-*N*-nitrosoguanidine (MNNG) or irradiated by UV, supported T-antigen-dependent SV40 replication [9]. Extracts from untreated CO60 cells did not support DNA synthesis. Analysis of the newly synthesized DNA revealed that the 2.5 kb region surrounding the origin was overreplicated, while distal sequences were not replicated [9]. Similar overreplication was observed when other template molecules of similar size but different organization were used, indicating that specific sequences are not involved in the arrested replication (D. Hassin, in preparation). In reactions catalyzed by HeLa extracts, these template molecules were evenly and completely replicated. A small plasmid (pSVK$_1$H, 3.02 kb) [11] containing the viral origin was fully replicated in reactions catalyzed by treated CO60 extracts while extracts from untreated cells did not support its replication (M. Aladjem, in preparation). Thus the size of the template molecules is an important determinant affecting the mode of SV40 DNA replication in extracts from the semipermissive Chinese hamster cells. The cellular factors involved in this restricted replication are currently being examined.

III. ACTIVATION OF INITIATION EVENTS IN EXTRACTS FROM CARCINOGEN-TREATED CELLS

To test the hypothesis that carcinogens activate specific initiation from SV40 origin, we used pulse labeling experiments. In reactions catalyzed by HeLa and MNNG-treated CO60 extracts and pSVK1 as a template, the 300 bp origin fragment is labeled after a 3 sec pulse (Fig. 1a). Prolonged incubation led to complete replication in HeLa extracts and origin-specific overreplication in treated

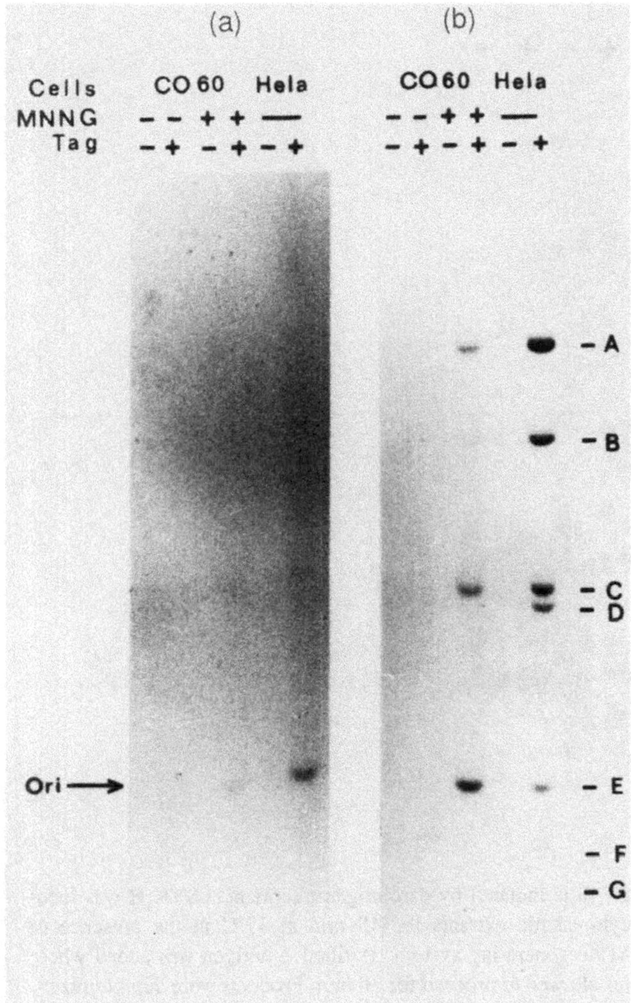

Figure 1 Initiation from the SV40 origin of replication in extracts from carcinogen-treated cells. Template DNA (pSVK$_1$) was incubated with the indicated extract, T antigen, ATP, and an energy-regenerating system for 45 min before the addition of the labeling mixture containing ^{32}P-labeled dCTP and ribo- and deoxyribonucleotides. After a 3 sec pulse (a) or after 90 min (b), the reaction was chased with nonlabeled dCTP. The purified products were digested by *Hin*dIII and *Sal* I, separated on 1.3% agarose gel, and autoradiographed [for 3 days (a) and for 1 hr (b)]. Ori, origin fragment.

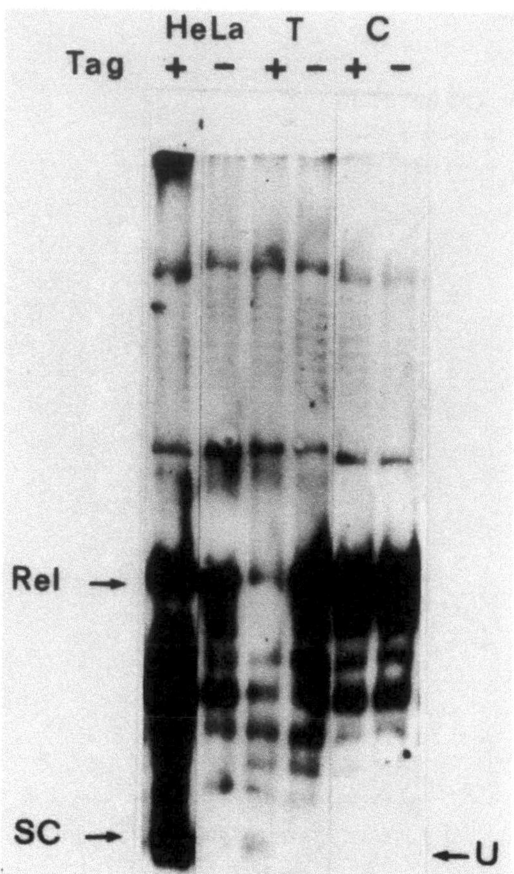

Figure 2 Unwinding of SV40 is induced by carcinogen treatment. pSVK$_1$H was incubated with the indicated cytoplasmic extracts for 45 min at 37°C in the presence of magnesium, ATP, and an ATP-regenerating system. Purified T antigen was added where indicated and the reaction was allowed to proceed for 30 min. Products were deproteinized, alcohol-precipitated, and loaded on 1.8% agarose gel containing 1.5 μg/ml chloroquine and separated at 100 V/30 cm gel for 19 hr. The DNA was transferred to nitrocellulose, hybridized with labeled pSVK$_1$, and autoradiographed. Rel, relaxed form; sc, supercoiled DNA; U, underwound.

CO60 extracts (Fig. 1b). No amplification was observed in extracts from untreated cells. Thus it seems that the acquired capability to support SV40 amplification in vitro stems from specific changes that occur in the Chinese hamster cells following carcinogen treatment and allow the activation of the viral origin in extracts from these cells. However, the inherent incapability to complete the elongation on large template molecules was not affected by the treatment.

To characterize the prereplication events leading to SV40 initiation by carcinogen-treated extracts, we analyzed the topological forms of the template DNA in the presence of various extracts. Because the separation and detection of the topoisomers in small plasmids is easier, we used the smaller plasmid pSVK$_1$H. As shown in Figure 2 in all extracts, in the presence or absence of T antigen, the supercoiled DNA is converted to topoisomers that have lower linking numbers. However, in HeLa extracts and in extracts from carcinogen-treated cells, a fast-migrating band appears, indicating the presence of underwound plasmid (U) [12]. This form is absent in extracts from nontreated CO60 cells supplemented with T antigen. It therefore can be concluded that carcinogens induce an activity that enables DNA unwinding from the SV40 origin of replication in extracts from the nonpermissive CO60 cells. The exact nature of the carcinogen-induced initiating activity is unknown. We are currently employing biochemical methods to purify and characterize the carcinogen-induced initiating factors.

IV. REREPLICATION OF THE ORIGIN REGION

The in vitro system provides us with a tool to characterize the primary amplification products and to investigate the mode of DNA amplification. Since amplification involves overreplication of specific DNA sequences, it was important to establish that some of the DNA molecules are replicated more than once. To this end the amplification reactions were carried out in a mixture containing 5-bromodeoxyuridine 5′-triphosphate as a density label, and the products were subjected to equilibrium centrifugation in a CsCl density gradient. Our results demonstrated that a fraction of the newly synthesized DNA molecules migrated in the position of heavy–heavy duplex DNA (data not shown). Such heavy–heavy DNA molecules may result from two rounds of replication. To further confirm these findings, the replication products were digested by *Mbo* I, an enzyme that cleaves only DNA molecules which underwent at least two rounds of replication and in which neither strand is methylated. Fully methylated template DNA and hemimethylated DNA molecules resulting from a single round of replication are not cleaved by this enzyme. Digestion of the amplified DNA synthesized in reactions supported by MNNG-treated CO60 extracts yielded three *Mbo* I sensitive fragments representing the origin region (Fig. 3), while replication products synthesized by HeLa extracts yielded 10 detectable fragments representing the fully replicated template molecules. Thus, though the template molecules were not fully replicated by treated CO60 extracts, the origin region underwent at least two rounds of replication.

V. EXTRACHROMOSOMAL HAIRPIN STRUCTURES ARE SYNTHESIZED IN VITRO

To analyze the structure and size of the amplified DNA, the replication products from reactions catalyzed by treated CO60 extracts were fractionated on a sucrose

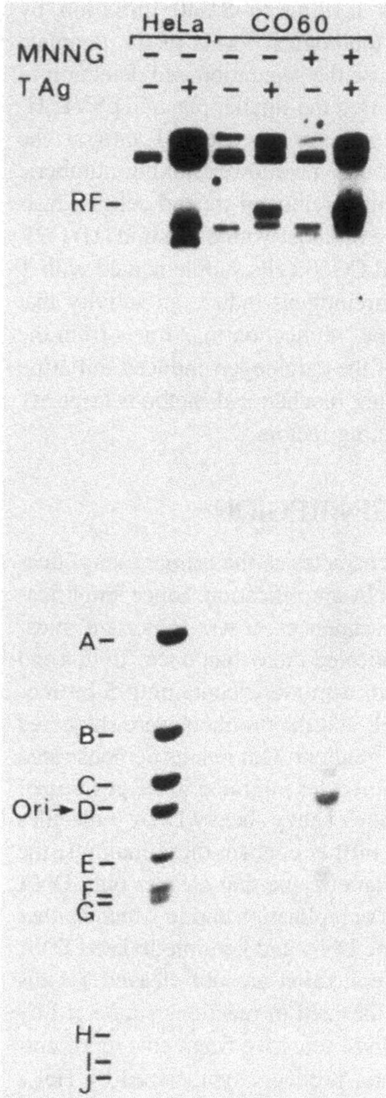

Figure 3 Rereplication of the origin region in extracts from carcinogen-treated cells. In vitro replication of $pSVK_1$ was catalyzed by extracts treated with HeLa, control, and MNNG-treated CO60 in the presence or absence of T antigen [9]. The ^{32}P-labeled DNA products were subjected to *Mbo* I digestion and then separated by 1% agarose gel electrophoresis. The gel was dried and exposed to autoradiography. RF, replicating forms; ori, origin of replication.

velocity gradient (Fig. 4). Aliquots from each fraction were either directly analyzed by gel electrophoresis for size determination (Fig. 4, uncut) or digested with *Mbo* I to determine the fractions containing the rereplicated DNA molecules (Fig. 4, *Mbo* I). The newly replicated DNA molecules consisted of two major populations. The first (fractions 5–11) contained molecules migrating in the position of replicating intermediates and supercoiled DNA molecules and were resistant to *Mbo* I digestion. The second population (fractions 13–17) consisted of molecules of heterogeneous sizes (0.3–3.0 kb). These molecules, which are not associated with the template molecules, were sensitive to *Mbo* I digestion, indicating that they resulted from two rounds of replication. These results suggested that the rereplicated DNA molecules are "peeled off" the template as double-stranded structures, in which both strands are newly replicated.

Earlier studies by several laboratories [5,13,14] suggested that the primary amplification intermediates might be extrachromosomal inverted repeat structures. To analyze the structure of the in vitro amplification products for the possible appearance of inverted repeat structures, we used the assay developed by Ford and Fried [15]. This assay involves denaturation and fast renaturation followed by the digestion with S1 nuclease, which digests single-stranded DNA. Under these experimental conditions DNA molecules containing intrastrand inverted duplications "snap back" and form double-stranded regions resistant to S1 nuclease digestion.

The replication products from reactions catalyzed by extracts from MNNG-treated CO60 cells were divided into two aliquots. One part was directly digested by *Mbo* I to determine the amplification pattern, and the second was subjected to the denaturation, renaturation, and S1 assay followed by *Mbo* I digestion. As shown in Figure 5, the pattern of *Mbo* I sensitive fragments derived from reactions catalyzed by treated CO60 extracts was not altered by the denaturation, renaturation, and S1 digestion. Hence, most of the rereplicated DNA molecules, representing the origin region, which are resistant to this treatment, contain hairpin structures. When the HeLa replication products were similarly analyzed, all the newly replicated DNA molecules were sensitive to S1 (data not shown). Thus, during SV40 replication directed by HeLa extracts, hairpin structures are not generated. Our findings suggest that the hairpin structures are exclusively produced in reactions that do not facilitate full replication of the template DNA molecules (i.e., reactions directed by the treated CO60 extracts).

VI. CHROMOSOMAL AND EXTRACHROMOSOMAL INVERTED REPEATS ARE SYNTHESIZED IN VIVO IN MNNG-TREATED CO60 CELLS

The possibility that inverted repeat duplications are generated in vivo during SV40 overreplication in the carcinogen-treated cells was investigated. Chromo-

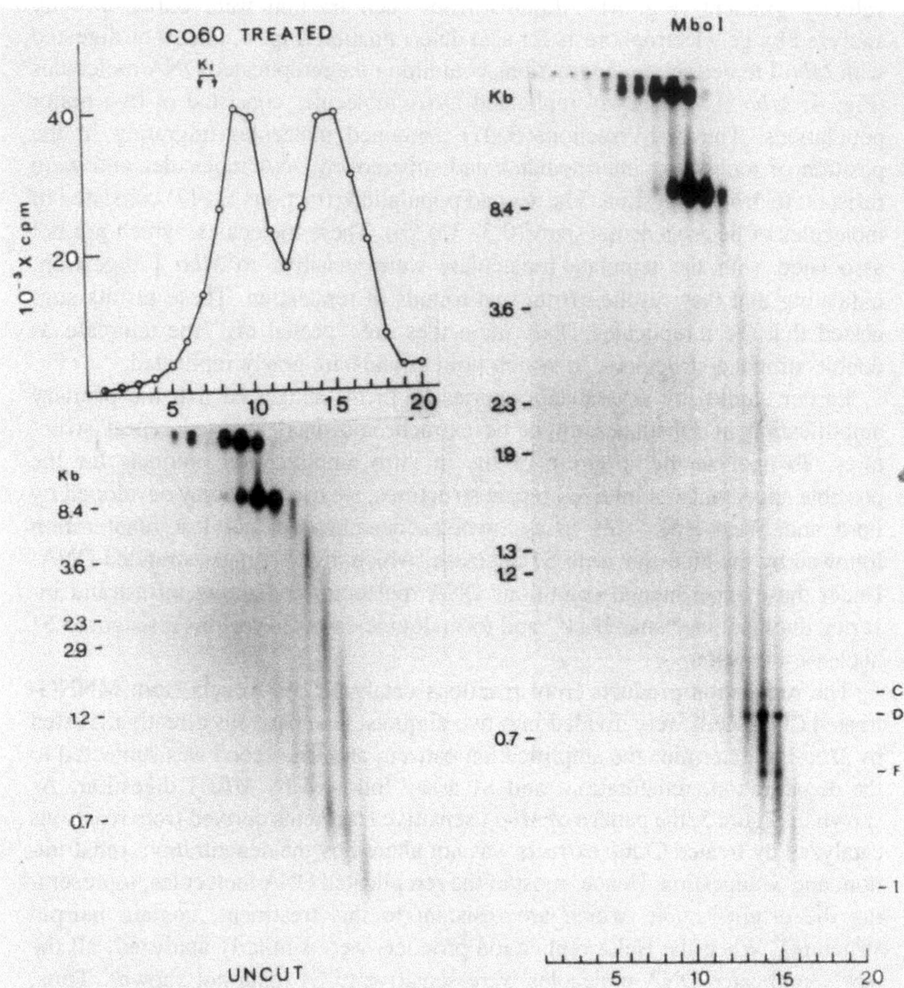

Figure 4 The replicated DNA molecules are detached from the template molecules. The in vitro replication products from reactions catalyzed by cytosolic extracts from MNNG-treated cells were fractionated by sucrose velocity gradient. Aliquots from each fraction were subjected directly to 1% agarose gel electrophoresis (uncut) or digested by *Mbo* I for the determination of the fractions containing the rereplicated DNA molecules (*Mbo* I).

Figure 5 Synthesis of inverted repeats in the in vitro replication reactions. The in vitro replication products from reactions catalyzed by extracts from HeLa (H), control (C), and treated (T) CO60 cells were cleaved by *Mbo* I (lanes a), or assayed for the presence of inverted repeats using the denaturation, renaturation, and S1 analyses [15] (lanes b). p2C is a plasmid harboring a 350 bp inverted repeat, which was used as a control for the inverted repeat assay. p2C: lane a, the plasmid cleaved by AlWN1 and end labeled; lane b, the end-labeled p2C DNA assayed for inverted repeats by denaturation, renaturation, and S1 analysis

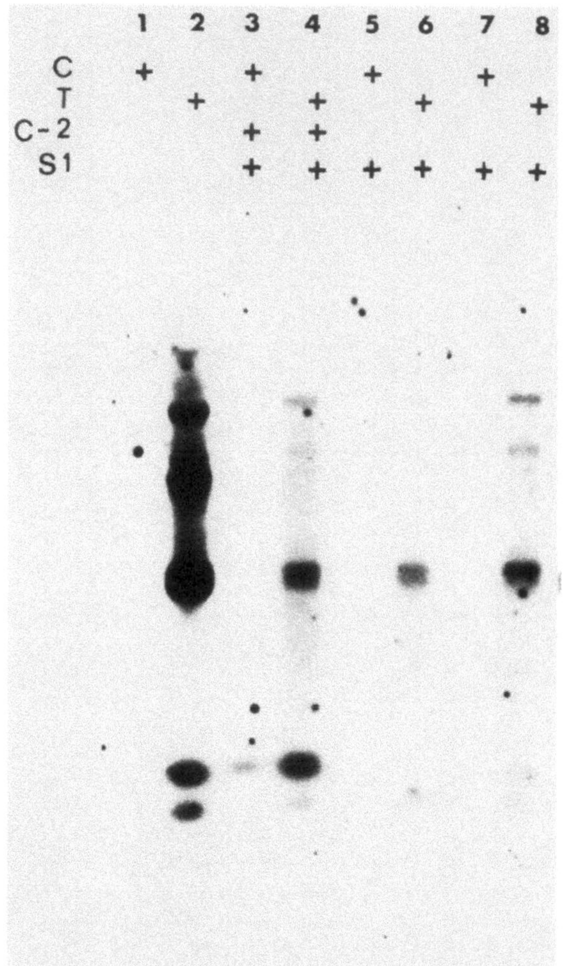

Figure 6 Chromosomal and extrachromosomal inverted repeats are generated during SV40 amplification in treated CO60 cells. Extrachromosomal (lanes 1–6) and chromosomal (lanes 7 and 8) DNA was extracted from control (lanes 1, 3, 5, and 7) and MNNG-treated (lanes 2, 4, 6, and 8) CO60 cells; 10 μg samples were directly digested by *Hin*dIII (lanes 1 and 2) or 50 μg aliquots (lanes 3–5) were processed by the denaturation, renaturation, and S1 assay for inverted repeats, then digested by *Hin*dIII. In lanes 3 and 4 p2C, a plasmid harboring an inverted repeat, was added to untreated (lane 3) and treated (lane 4) DNA prior to the assay, as an internal control for the inverted repeat assay. The strong bands appearing in lane 3 represent the inverted repeats present in p2C.

somal and extrachromosomal DNA was extracted from untreated and treated CO60 cells 72 hr after MNNG treatment. As demonstrated in Figure 6, a dramatic SV40 DNA amplification was observed in both chromosomal (data not shown) and extrachromosomal DNA (Fig. 6, lane 2) from the carcinogen-treated cells. Analysis of the in vivo amplified DNA for the presence of inverted repeat structures revealed that a substantial fraction of the amplified DNA molecules indeed contained structures that were resistant to S1 digestion following denaturation and fast renaturation. These structures appeared in both chromosomal (Fig. 6, lane 8) and extrachromosomal DNA from the treated cells (Fig. 6, lane 6). Inverted repeat duplications were not observed in the DNA from nontreated cells (Fig. 6, lanes 5 and 7). We are currently investigating the structure and organization of the inverted repeats.

VII. A MODEL FOR THE GENERATION OF EXTRACHROMOSOMAL INVERTED REPEAT DUPLICATIONS: THE "U-TURN" REPLICATION

Our results show that hairpin structures are formed during SV40 replication in extracts that are incapable of supporting the complete replication of the template molecules. We propose that these hairpin structures might be the precursors for the extrachromosomal inverted repeat duplications observed during the initial stages of gene amplification. The "U-turn" replication model describing this mechanism is presented in Figure 7. We suggest that when the advancement of the replication fork is perturbed (Fig. 7a), the DNA polymerase backtracks [16], and replication continues on a new template—the newly synthesized strand (Fig. 7b). Such U-turn replication generates hairpin structures, which are displaced from the parental DNA molecules (Fig. 7c). These hairpin structures might be converted by ligation into covalently closed, linear double-stranded DNA molecules (Fig. 7d), which may undergo a second round of replication following initiation from the viral origin of replication (Fig. 7e). This replication will lead to the formation of circles containing inverted repeat duplications (Fig. 7f). Such structures were indeed observed during gene amplification in *Leishmania* [17] and were suggested to be the precursors for the chromosomal and extrachromosomal inverted repeat duplications observed during gene amplification in higher eukaryotes [3,5]. The results regarding the generation of chromosomal and extrachromosomal inverted repeat duplications during SV40 amplification in the carcinogen-treated cells (Fig. 6) further support this hypothesis.

It was recently proposed [3,13] that strand switching of the replication fork near palindromic sequences is the mechanism responsible for the generation of inverted repeats during gene amplification. We propose that any perturbations in the advancement of the replication fork during the S phase might lead to aberrant replication and to the generation of hairpin structures in the vicinity of the arrested

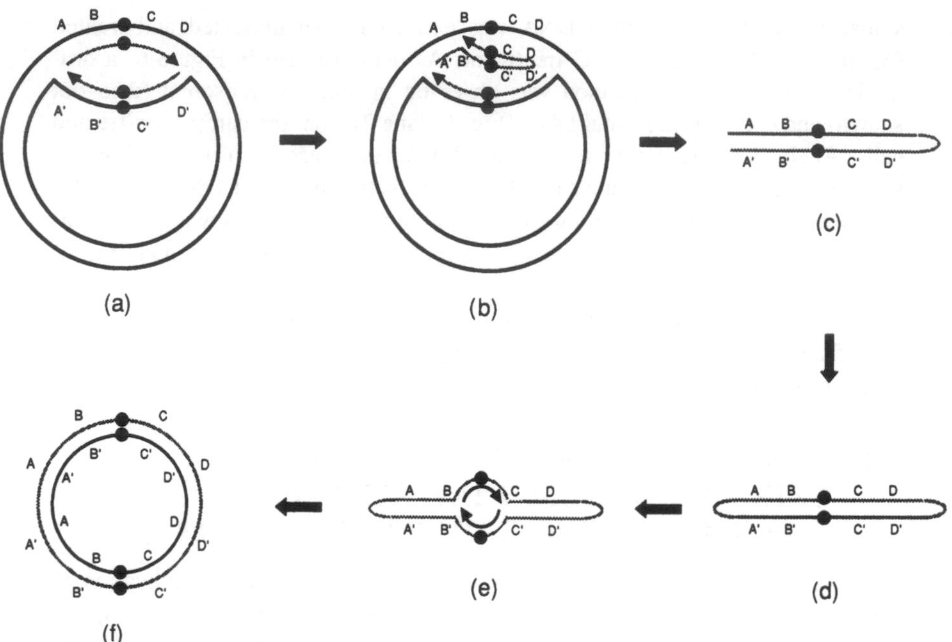

Figure 7 The U-turn replication model. Generation of inverted repeat duplications. For discussion, see text.

replication forks. Damaged template, modified DNA, nicks, triple-helix structures, sequences with specific conformation, and protein–DNA complexes might arrest the advancement of the replication fork and could lead to U-turn replication. The newly generated hairpin structures might recombine into the chromosome, yielding chromosomally associated inverted repeat duplications. Alternatively, hairpin structures harboring sequences that facilitate independent replication might replicate extrachromosomally, yielding circular inverted repeat duplications.

This mode of amplification requires a single initiation event and does not involve multiple reinitiations at the origin of replication. Almost any genomic sequence might serve as substrate for the amplification. Amplification by this mechanism is not associated with the deletion of the region that was used as a template for the amplification. The size of the amplification unit might vary from short sequences to very large genomic regions. The initial duplication might be followed by subsequent recombination events, which will gradually increase the number of the inverted repeat duplications. Recombination events might lead to insertion of the amplified DNA into the chromosome.

VIII. SUMMARY

We utilized the in vitro amplification system in an attempt to identify the carcinogen-induced trigger for amplification and to elucidate the mechanism of the initial steps in SV40 amplification. Analysis of the nature of the amplification process revealed that amplification in vitro is triggered by a carcinogen-induced activity that allows prereplicative topological changes in the template DNA molecule. Detailed analysis of the structure of the amplified DNA demonstrates that hairpin structures and inverted repeat duplications are among the earlier intermediates of amplification in vitro and in vivo. These observations led to the suggestion that the U-turn replication model for the generation of inverted duplication during SV40 amplification could apply to the generation of inverted repeats during gene amplification in general.

ACKNOWLEDGMENTS

This research was supported by grants from the Israel Academy of Sciences and from the Israel Association for Cancer Research. S. L. is a recipient of a career development award from the Israel Cancer Research Fund.

REFERENCES

1. Schimke, R. T. Gene amplification, drug resistance and cancer. *Cancer Res. 44*:1735–1742 (1984).
2. Stark, G. R., Debatisse, M., Giulotto, E., and Wahl, G. M. Recent progress in understanding mechanisms of mammalian DNA amplification. *Cell, 57*:901–913 (1989).
3. Passananti, C., Davies, B., Ford, M., and Fried M. Structure of an inverted duplication formed as a first step in a gene amplification event: Implications for a model of gene amplification. *EMBO J. 6*:1697–1703 (1987).
4. Hyrien, O., Debatisse, M., Buttin, G., and Robert de Saint Vincent, B. The multicopy appearance of a large inverted duplication and the sequence at the inversion point suggest a new model for gene amplification. *EMBO J. 7*:407–417 (1988).
5. Ruiz, J. C., and Wahl, G. M. Formation of an inverted duplication can be an initial step in gene amplification. *Mol. Cell. Biol. 8*:4303–4313 (1988).
6. Lavi, S. Carcinogen-mediated amplification of viral DNA sequences in simian virus 40 transformed Chinese hamster embryo cells. *Proc. Natl. Acad. Sci. USA, 78*:6144–6148 (1981).
7. Lavi, S., and Etkin, S. Carcinogen-mediated induction of SV40 DNA synthesis in SV40 transformed Chinese hamster embryo cells. *Carcinogenesis, 2*:417–423 (1981).
8. Berko-Flint, Y., Karby, S., and Lavi, S. Carcinogen-induced factors responsible for SV40 DNA replication and amplification in Chinese hamster cells. *Cancer Cells, 6*: 183–189 (1988).
9. Berko-Flint, Y., Karby, S., Hassin, D., and Lavi, S. Carcinogen-induced DNA

amplification in vitro: Overreplication of the simian virus 40 origin region in extracts from carcinogen-treated CO60 cells. *Mol. Cell. Biol. 10*:75-83 (1990).

10. Li, J. J., and Kelly, T. J. Simian virus 40 replication in vitro. *Proc. Natl. Acad. Sci. USA, 81*:6973–6977 (1984).
11. Dorsett, D., Deichaite, I., and Winocour, E. Circular and linear simian virus 40 DNAs differ in recombination. *Mol. Cell. Biol. 5*:869–880 (1985).
12. Dean, F. B., Bullock, P., Murakami, Y., Wobbe, C. R., Wiessbach, L., and Hurwitz, J. SV40 replication: SV40 large T antigen unwinds DNA containing the SV40 origin of replication. *Proc. Natl. Acad. Sci. USA, 84*:16–20 (1987).
13. Nalbantoglu, J., and Meuth, M. DNA amplification—Deletion in a spontaneous mutation of the hamster *aprt* locus: Structure and sequence of a novel joint. *Nucleic Acid Res. 14*:8361–8371 (1986).
14. Wahl, G. M. The importance of circular DNA in mammalian gene amplification. *Perspect. Cancer Res. 49*:1333–1340 (1989).
15. Ford, M., and Fried, M. Large inverted duplications are associated with gene amplification. *Cell, 45*:425–430 (1986).
16. Korenberg, A. *DNA Replication*. Freeman, San Francisco, 1980.
17. Beverly, S. M., Coderre, J. A., Santi, D. V., and Schimke, R. T. Unstable DNA amplification in methotrexate-resistant *Leishmania* consist of extrachromosomal circles which relocalize during stabilization. *Cell, 38*:431–439 (1984).

38

The Stimulation of Gene Amplification in Mammalian Cells by HSAG Middle Repetitive Elements

James G. McArthur,* Lenore K. Beitel,† and Clifford P. Stanners *McGill University, Montreal, Quebec, Canada*

I. INTRODUCTION

The frequency of gene amplification in mammalian cells is not constant throughout the genome [1,2] nor within cell lines derived from a common parental source [3–5]. This variability reflects the contribution of both cis- and trans-acting sequences, but the exact nature of these sequences has not been well defined. We have recently shown that several members of the HSAG family of middle repetitive human and hamster genomic elements promote the amplification of sequences *in cis* in several mammalian cell lines [6–8].

II. HSAG ELEMENTS INCREASE THE FREQUENCY OF GENE AMPLIFICATION

The effect of HSAG elements on vector amplification was first demonstrated in Chinese hamster ovary (CHO) cells transfected with the shuttle vector pSV2-*DHFR* (D) containing the 3.4 kb HSAG-1 insert (HLD) in the *Eco*RI site (Fig. 1;

Current affiliations:
*Department of Molecular and Cellular Biology, University of California at Berkeley, Berkeley, California.
†Lady Davis Institute for Medical Research, Sir Mortimer B. Davis Jewish General Hospital, Montreal, Quebec, Canada.

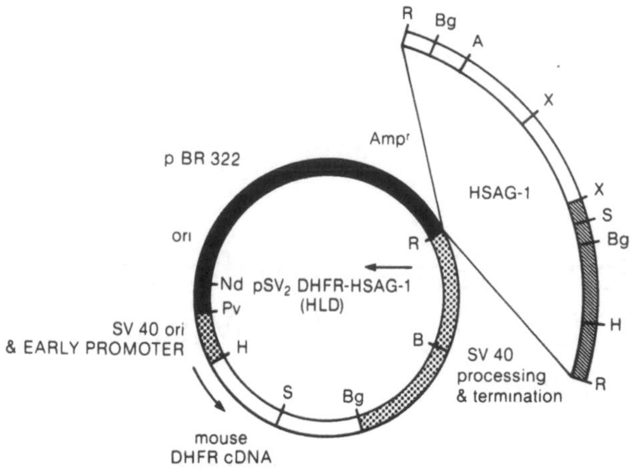

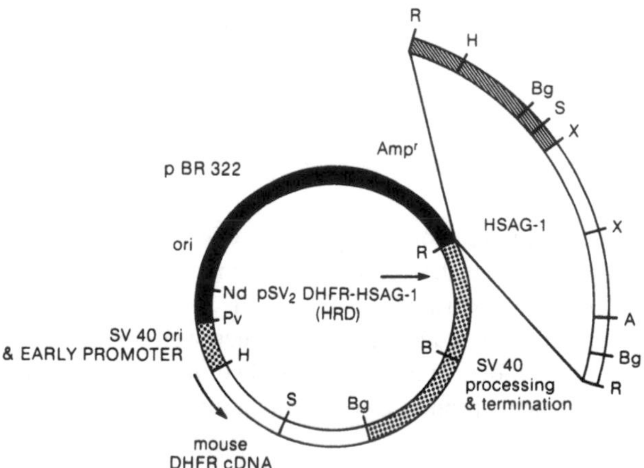

Figure 1 Map of pSV2-DHFR-HSAG-1 in the "left-hand" (HLD) and "right-hand" (HRD) orientations. Restriction enzyme sites: R, *Eco*RI; Bg, *Bgl* II; A, *Ava* I; X, *Xba* I; S, *Sst* I; H, *Hind*III; B, *Bam*HI; Pv, *Pvu* II; Nd, *Nde* I.

Ref. 6). The maximum rate of increase of resistance to methotrexate (MTX), a competitive inhibitor of dihydrofolate reductase (DHFR), was at least twofold faster in repeated experiments for HLD transfectants compared to cells transfected with the parental vector, D. The faster rate at which the drug concentration could be increased for HLD transfectants was shown to be due to a greater frequency of cells in the HLD transfectant population with amplified vector *DHFR* sequences [6]. This phenomenon was not due to a higher initial vector copy number in HLD transfectants (see below), nor was it due to any effect of HSAG-1 on vector *DHFR* transcription or translation efficiencies [6].

An in vitro assay was developed to quantify the ability of a nucleic acid sequence to increase the frequency of vector amplification. In brief, the sequence of interest was subcloned into the *Eco*RI site of pSV2-*DHFR* (Fig. 1) and the resulting construct cotransfected into a CHO-derived cell line, LR73, along with the dominant selectable marker, pSV2-AS [7]. AS^+ transfectants were selected, pooled, grown to equivalent cell densities in nonselective media, and then plated in increasing concentrations of MTX. Ten to 12 days later, the resulting colonies were stained, and the frequency of drug-resistant colonies from populations transfected with the test constructs was determined and compared to that of cells transfected with pSV2-AS and the parental vector, D.

pSV2-*DHFR* vectors containing the human chronic lymphocytic leukemia (CLL)-CHO hybrid genomic elements HSAG-1 and HSAG-2 [8], the human CLL genomic element HSAG-5 [9], and the CHO genomic HSAG-1-homologous elements LRH1 and LRH2 [10] reproducibly produced 20–200-fold more colonies resistant to 2 μM MTX than the parental vector (Fig. 2). This increase in MTX-resistant cells was due to an increased frequency of vector *DHFR* amplification in cells transfected with constructs containing HSAG elements [6].

III. STRUCTURAL FEATURES FOUND IN ACTIVE HSAG SUBFRAGMENTS

While all HSAG family members tested promoted vector amplification, they shared little sequence homology either at the nucleic acid level or at the level of predicted proteins; in fact, extensive computer analysis indicated that it was unlikely that any of these genetic elements could even encode a polypeptide. Analysis of the amplification activity of subfragments of HSAG-1, HSAG-2, and HSAG-5 suggested that the interaction of multiple positive-acting elements was required [11]. Further computer sequence analysis of these positive-acting elements detected several sequence features associated with gene amplification and/or recombination hot spots in other systems, including A+T-rich regions, purine–pyrimidine tracts, *Alu* repetitive elements, and putative stem–loop structures [5]. Deletion of several of these structural features from the highly active 1.45 kb *Eco*RI-*Xba* I fragment of HSAG-1 greatly reduced or eliminated its amplification activity. Although no individual element was shown to be sufficient or necessary for activity in all HSAG family members, combinations of these elements in an "active" configuration may be required to promote recombination events, which assist the formation of amplification-prone structures or the resolution of amplified arrays.

Computer nucleic acid sequence analysis also detected several putative yeast autonomously replicating sequence (ARS) consensus sequences in HSAG-1 and

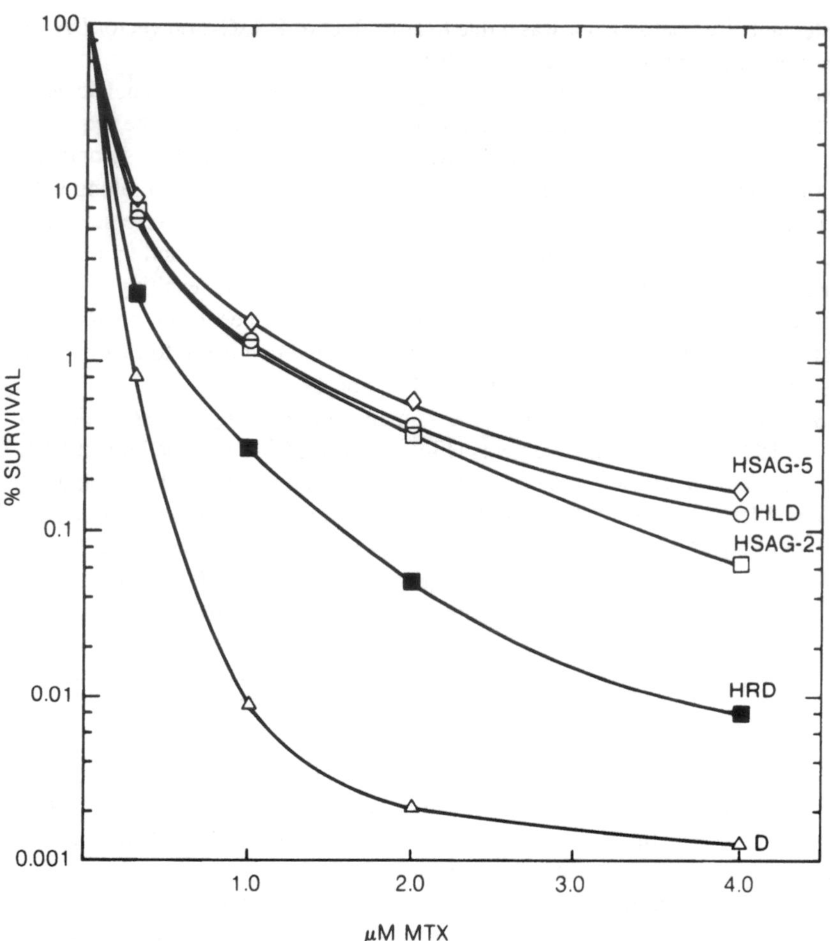

Figure 2 Semilogarithmic plot of the percentage of surviving colonies in total transfectant populations of LR-73 cells, obtained by cotransfection with the dominant selectable marker AS [7], and pSV2-DHFR-HSAG-5 (◇), pSV2-DHFR-HSAG-2 (□), HLD (○), HRD (■), and D (△), and plated in media containing different concentrations of MTX.

HSAG-2 [10]. The presence of these elements and the numerous diverged *Alu* repetitive elements, a characteristic of HSAG family members, suggested that these elements may act as origins of replication or alternatively activate the SV40 origin of replication, which is normally quiescent in LR73 cells. To examine these possibilities, COS and LR73 cells were transiently transfected with D, HLD, or a pSV2-*DHFR* construct containing the highly active 1.45 kb *Eco*RI-*Xba* I fragment of HSAG-1 (RXD). At various time points Hirt supernatants were prepared and the DNA preparations digested with restriction enzyme *Dpn* I and analyzed by the Southern blot technique to determine whether the HSAG-1 inserts affected vector autonomous replication in permissive COS or nonpermissive LR73 cells. Neither HSAG-1 nor its *Eco*RI-*Xba* I fragment increased the quantity of *Dpn* I-insensitive plasmid in either cell line compared to D transfectants, indicating that HSAG-1 does not possess a mammalian origin of replication functional in LR73 cells, nor does it further activate the SV40 origin of replication in these cell lines [10]. This does not exclude, however, an effect of HSAG-1 on cellular origins of replication adjacent to sites of vector integration.

IV. HSAG–VECTOR INTERACTIONS

The involvement of the pSV2-*DHFR* vector sequences in the amplification phenomenon was examined following experiments that demonstrated that the magnitude of the amplification activity of HSAG elements and their subfragments was affected by their orientation in the pSV2-*DHFR* vector. For example, cells transfected with the pSV2-*DHFR* construct with HSAG-1 subcloned in the "left-hand" orientation in the *Eco*RI site (HLD) showed consistently 2–10-fold more colonies resistant to 2 μM MTX than cells transfected with the vector with HSAG-1 in the "right-hand" orientation (HRD) (Fig. 2). This effect was also observed with constructs containing HSAG-2, HSAG-5, and subfragments of all three elements [10,11]. The orientation dependence was not reflected in the levels of *DHFR* transcription (Fig. 3, lower right; Ref. 6). HSAG inserts, on the other hand, were transcribed from cryptic promoters in pSV2-*DHFR*, and the level of this transcription was strongly affected by the orientation of the insert (Fig. 3, upper right) [10].

The involvement of HSAG transcription in the amplification phenomenon was further indicated by experiments in which linearization of the HLD vector at the *Bam*HI site, prior to transfection into LR73 cells, greatly decreased the amplification ability of the construct [10]. *DHFR* transcription was unaffected by this vector linearization in D and HLD transfectants, as seen in the Northern analysis of total RNA from transfectants resistant to 0.6 μM MTX hybridized with a ^{32}P-labeled DHFR probe (Fig. 3, lower left); HSAG transcription, however, was eliminated (Fig. 3, upper left). Kesler et al. [12] have reported that disrupting a similar vector, pSV2-*neo*, in the region of the SV40 polyadenylation signal, shortened the half-life of a 3.5 kb transcript that initiated from the SV40 late promoter and included the vector pBR322 sequences. Cutting the HLD vector with *Bam*HI near the polyadenylation signal may similarly greatly shorten the half-life of the HSAG-1 transcripts, resulting in their effective disappearance. Linearization of the vector per se was not responsible for the decrease in amplification activity, since cutting the vector at the *Nde* I site prior to transfection had no effect on the amplification activity of HLD or HRD [10].

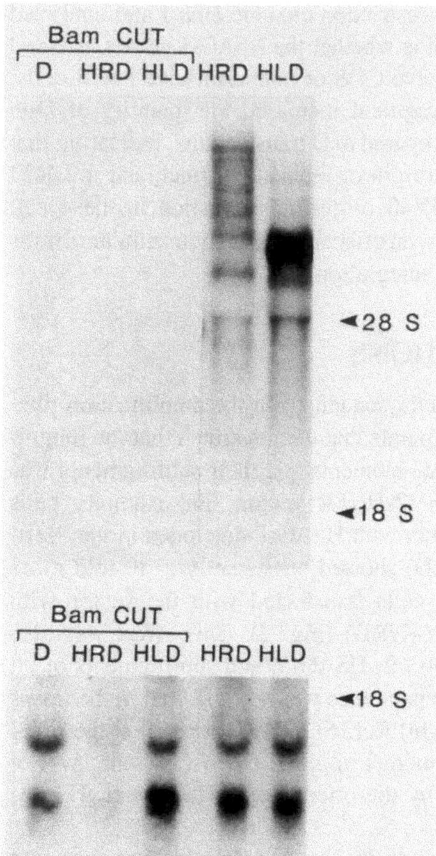

Figure 3 Effect on HSAG-1 transcription of linearizing the HRD and HLD vectors with *Bam*HI. Total RNA from LR73 cells transfected with *Bam*HI cut D, HRD, or HLD DNA or uncut HRD or HLD DNA, and resistant to 0.6 μM MTX, was analyzed by the Northern technique and probed with a ^{32}P-labeled HSAG-1-specific fragment, SS330 (13), (top) or DHFR cDNA (bottom). The positions of the 18S and 28S rRNA are indicated.

A. Effect of an Additional SV40 Promoter–Enhancer on HSAG Amplification Activity

If decreased HSAG-1 transcription of HRD was responsible for its lower amplification activity, then artificially increasing the level of this transcription might be expected to increase its activity. To test this, the 373 bp SV40 promoter–enhancer sequence from pSV2-*DHFR* (SV40 5172 bp to 273 bp) was subcloned into HRD and D at the unique *Nde* I site flanking the SV40 promoter or into the *Bam*HI site downstream of the polyadenylation signal for *DHFR* transcription, and the resulting constructs cotransfected into LR73 cells, as described above.

The addition of the SV40 promoter–enhancer sequences, in either orientation at the *Nde* I and *Bam*HI sites of pSV2-*DHFR*, decreased the frequency of MTX-resistant colonies two- to threefold and had an even greater negative effect when cloned into the *Nde* I site of HRD (Table 1). Inserting this sequence into the *Bam*HI site of HRD in either orientation had the opposite effect, increasing the frequency of MTX-resistant colonies more than 50-fold above D-*Bam*HI-SV transfectants. This increase in HRD-*Bam*HI-SV drug-resistant colonies was not due to an increase in *DHFR* transcription, as shown in the Northern analysis of Figure 4. This effect, however, was also not mediated by an increase in HSAG-1 transcription, which was unaffected by the additional SV40 promoter–enhancer sequences (data not shown).

Table 1 Effect of Additional SV40 Promoter–Enhancer Elements on the Amplification Stimulatory Activity of HSAG-Containing Vectors[a]

D	HRD	HLD
0.0076[b]	0.081[b]	0.17
1X	11X[c]	22x
D-*Bam*HI-SV (t)	HRD-*Bam*HI-SV (t)	
0.0021	0.14	
1X	68X	
D-*Bam*HI-SV (a)	HRD-*Bam*HI-SV (a)	
0.0025	0.14	
1X	55X	
D-*Nde* I-SV (t)	HRD-*Nde* I-SV (t)	
0.0054	0.0066	
1X	1X	
D-*Nde* I-SV (a)	HRD-*Nde* I-SV (a)	
0.0022	0.0024	
1X	1X	

[a](t), SV40 early promoter drives transciption toward HSAG insert; (a), SV40 early promoter drives transcription away from HSAG insert.

[b]Percentage of surviving colonies in 2 μM MTX, determined as indicated in the text and Ref 6, after transfection with the indicated vectors.

[c]Increase in percentage of surviving colonies relative to parental vector, D.

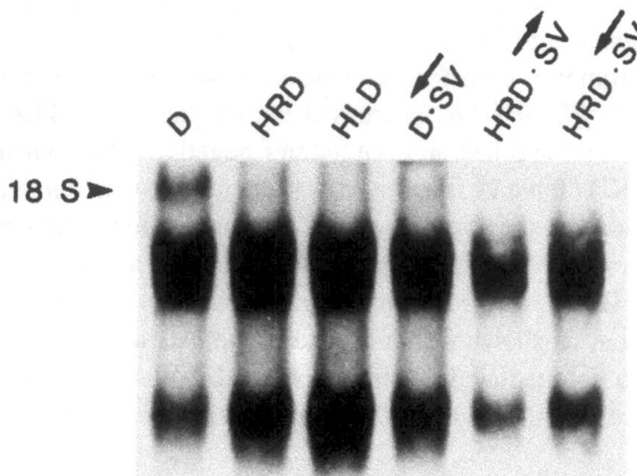

Figure 4 Effect of the SV40 promoter–enhancer on DHFR transcription of D and HRD. Total RNA from LR73 cells transfected with D, HRD, or HLD, or the D or HRD vectors with an additional SV40 promoter–enhancer (SV40 5172-273 bp) in the *Bam*HI site between HSAG-1 and the SV40 polyadenylation signal (or the *Eco*RI site and the SV40 polyadenylation signal in the case of D), which were resistant to 0.6 μM MTX, was analyzed by the Northern blot technique and probed with a ^{32}P-labeled DHFR cDNA probe. The direction of the SV40 early promoter initiated transcripts is indicated by the arrows above the letters SV: arrow pointing toward D or HRD, transcription is toward the HSAG-1 insert or *Eco*RI site; arrow pointing away from HRD or D, transcription would proceed away from the HSAG-1 insert or *Eco*RI site. The upper band is the vector DHFR transcript utilizing SV40 early promoter and polyadenylation signals. The position of the 18S rRNA is indicated.

Table 2 Position Effects of the Amplification Activity of HSAG in pSV2-*DHFR*

	Site in pSV2-DHFR		
Element/orientation	*Eco*RI	*Pvu*II	*Bam*HI
HSAG-1/left-hand (HLD)	50[a]	59	24
HSAG-1/right-hand (HRD)	18	25	4

[a]The amplification activity representing the ratio of the number of MTX resistant colonies produced by transfection with pSV2-*DHFR* containing a test genetic element relative to pSV2-*DHFR* alone [9].

B. Activity of HSAG-1 in Alternative Sites in pSV2-*DHFR*

The dependence of the amplification phenomenon on the orientation of HSAG-1 in the vector was also observed when HSAG-1 was inserted in the *Bam*HI or *Pvu* II sites of pSV2-*DHFR* (Fig. 1). The activity of the *Pvu* II constructs was similar to that of the *Eco*RI constructs HLD and HRD, while the activity of the *Bam*HI constructs was considerably lower (Table 2). It seems unlikely that interrupting the vector at this site with HSAG-1 interfered with *DHFR* transcription of the resulting construct, since linearizing it at the *Bam*HI site prior to transfection left *DHFR* transcription unaffected (Fig. 3, lower left). The explanation of this effect awaits an examination of *DHFR* and HSAG-1 transcription in these transfectants.

V. RECOMBINATION AND HSAG-1

A. Southern Analysis of the Configuration of Amplified Vector DHFR Sequences

To determine the nature of the structure of amplified D and HLD vector sequences in MTX-resistant cells, clones were isolated from transfectants prior to exposure to MTX and, following selection in 2 μM MTX, genomic DNA was prepared, digested with *Bam*HI, and subjected to Southern blot analysis using a ^{32}P-labeled *DHFR* probe. The most intense bands were the 5.0 kb unit length D vector and the 8.4 kb unit length HLD vector, presumably produced by cutting at the single *Bam*HI sites in adjacent integrated copies of the vectors (Fig. 5). Because of the larger size of the HLD vector and the transfection of equivalent weights of each vector, rather than an equal number of moles, cells were transfected with approximately 70% more copies of the D vector than the HLD vector in each experiment. This is reflected in the lower initial copy number in HLD-transfected cells prior to MTX exposure (Fig. 1; Fig. 5, HLD(0); Ref. 6). Despite this disparity, both the HLD and D vectors were amplified 30–100-fold in individual clones resistant to 2 μM MTX (cf. D(0) to D(2) clones and HLD(0) to HLD(2) clones); it should be emphasized that in this experiment 40-fold more HLD(2) colonies were obtained than D(2) colonies. It is also of interest that a large portion of the HLD(2) vector *DHFR* amplification was manifested as amplified side bands, the significance of which is discussed below.

While D(2) clone 7 had the greatest vector copy number of the clones analyzed, Northern analysis of total RNA from the D(2) and HLD(2) clones of Figure 5 demonstrated that the observed variability in vector copy number was not generally reflected in the level of *DHFR* transcripts (data not shown). Integration of the D vector into a transcriptionally inactive region of a chromosome or inactivation of copies of the vector in the early stages of gene amplification may have been responsible for this poor correlation in vector copy number and *DHFR* transcript level.

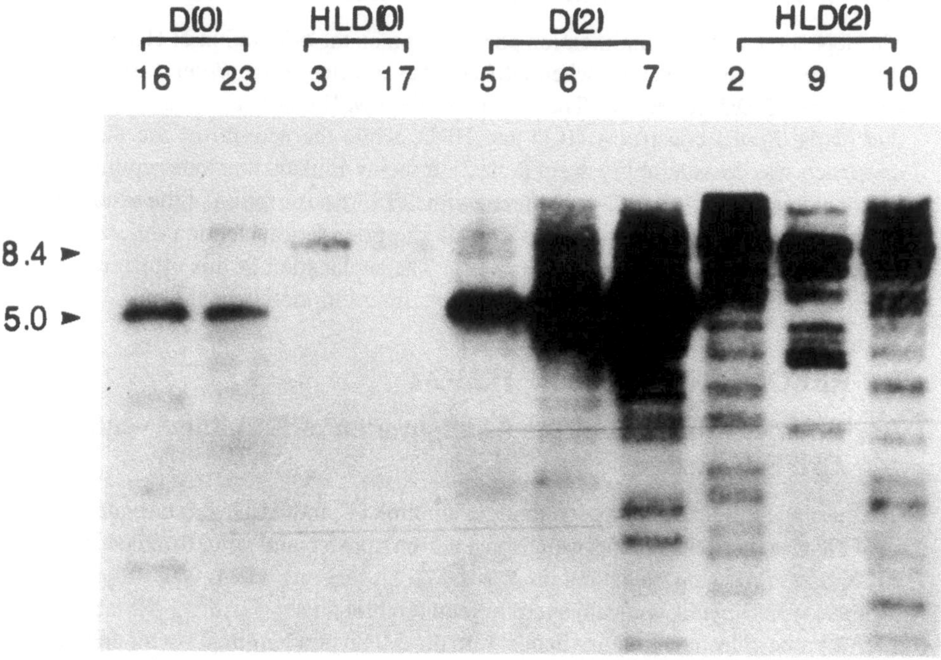

Figure 5 Southern analysis of *Bam*HI-digested genomic DNA from transfectant clones of D or HLD isolated prior to exposure to MTX (D(0) and HLD(0), respectively) or resistant to 2 μM MTX (D(2) and HLD(2), respectively). The identification number of the clone is indicated at the top. The predominant bands detected by the ^{32}P-labeled DHFR cDNA probe are the 5.0 kb D and 8.4 kb HLD unit length vector bands.

The intense unit length D and HLD bands indicate that head-to-tail concatamers of the vectors represented the predominant configuration of the amplified sequences. While head-to-head concatamers have been more frequently reported, both structures are apparently easily generated and may be a favored substrate for amplification because of their very compact size, which allows for amplification of multiple copies of the vector with each round of amplification. Along with the amplification of concatamers, a large proportion of vector amplification in HLD(2) clones was present in amplified side bands, while the overwhelming majority of amplification in D(2) clones was present in the unit length 5.0 kb band. Side bands are likely the product of vector–vector and vector–genome recombination events during the amplification process, which tends to generate *Bam*HI digestion products of sizes other than the unit length bands. The increased frequency of the HLD(2) side bands was not produced as a consequence of initial vector integration at more genomic sites: Southern analysis showed that HLD(0) clones, isolated prior to exposure to MTX, possessed fewer side bands compared to D(0) clones; only 30% of HLD(0) clones, compared with 75% of D(0) clones, had 10 or more side bands [6]. Many of the HLD(2) side bands, observed after amplification, were as intense as the 8.4 kb unit length band, indicating that the recombination events that generated these side bands probably occurred early in the amplification process.

HLD transfectants prior to exposure to MTX possessed fewer side bands and a lower vector copy number compared to D transfectants, but the HLD vector was amplified 30–100-fold in half the time relative to the D vector. From the large number of HLD(2) side bands observed after amplification, we suggest that the accelerated vector amplification in HLD transfectants is due to the stimulation of local recombination events by HSAG-1, which may assist the formation and resolution of amplified arrays at a greater number of primary integration sites.

B. Southern Analysis of the Configuration of Amplified Vector HSAG-1 Sequences

Genomic DNA from the same D(2) and HLD(2) clones investigated above was digested with the enzyme *Kpn* I and analyzed by a Southern blot hybridized with either a ^{32}P-labeled HSAG-1 specific probe, SS330 (Fig. 6, left), or a ^{32}P-labeled pBR322 probe (Fig. 6, right). *Kpn* I does not cut either the D or HLD vectors; hence any labeled bands represent amplified vector–genome recombination products or autonomously replicated plasmid. Uncut autonomously replicated plasmids should run at the same position in the Southern analysis of HLD(2) clone DNA, whether hybridized with the SS330 or the pBR322 probes. The absence of a common band smaller than 10 kb in either the D or HLD clones (D and HLD whole-plasmid DNA migrates at apparent sizes of $<$ 10 kb) suggests that these vectors do not replicate autonomously, in agreement with the experiments examining the *Dpn* I sensitivity of Hirt supernatants of D, RXD, and HLD transfectants (above; Ref. 10). These results, however, do not exclude the possibility that, following vector integration, recombination events gave rise to large episomes of various sizes in which HSAG-1 encouraged replication. The completeness of the *Kpn* I digestion is indicated by the single SS330 homologous band in D(2) clones indicated by the heavy arrow. This band represents the endogenous 20–50 copies of HSAG-1-like elements such as LRH1 and LRH2 in the CHO genome [13], since D has no HSAG-1 specific sequences.

The Southern analysis of the *Kpn* I-digested D(2) clone genomic DNA was expected to produce a greater number of smaller bands than the HLD(2) clones because of the greater initial copy number of pSV2-*DHFR* vector, but this was not the case. Both the pBR322 and SS330 probes hybridized to a greater number of bands in the HLD(2) clones than the pBR332 probe to the D(2) clones (Fig. 6). Many of the HLD(2) side bands had undergone several rounds of amplification (cf.

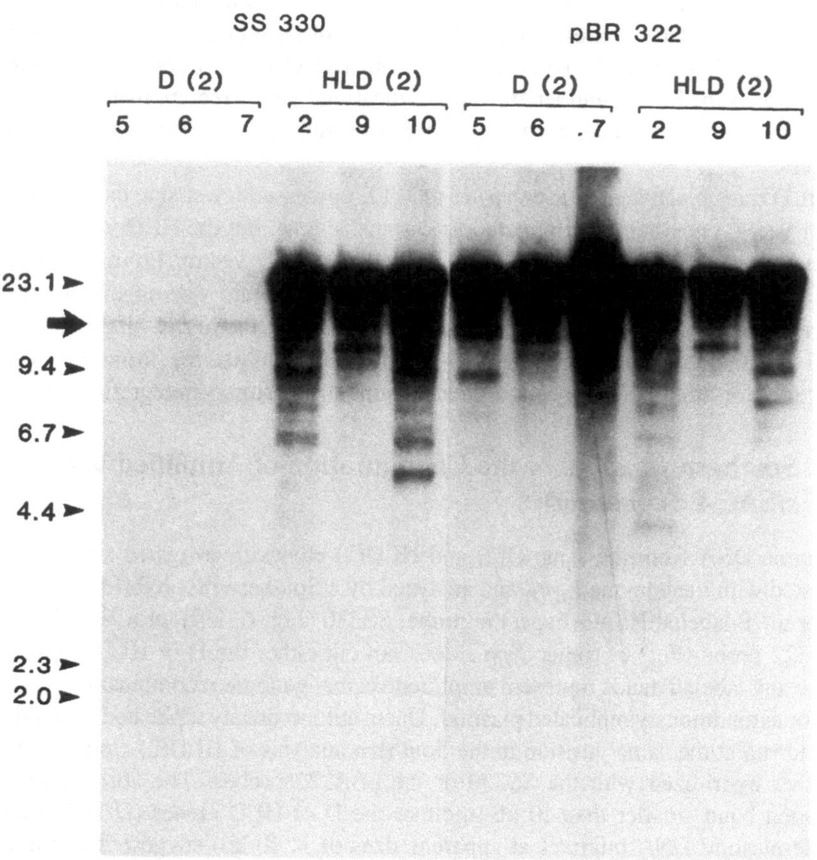

Figure 6 Southern analysis of *Kpn* I-digested genomic DNA from the D and HLD transfected clones resistant to 2 μM MTX of Figure 5, probed with a ^{32}P-labeled HSAG-1-specific SS330 fragment (left) or ^{32}P-labeled pBR322 *Pvu* II–*Eco*RI fragment (right). The band detected in D(2) clones by the SS330 probe is the endogenous 20–50 copies of HSAG-1 in the CHO genome [13]. The positions of the size markers are indicated at the left-hand margin.

the intensity of HLD(2) side bands to the 20–50 copies of endogenous CHO HSAG-1 in the left-most panel), indicating that the recombination events that produced these *Kpn* I fragments occurred early during the amplification process. Even the D(2) clone 7, with exceptionally high vector copy number, possessed fewer amplified side bands than HLD(2) clones 2 and 9. These results, along with the Southern analysis of *Bam*HI-digested amplified vector *DHFR* sequences

(Fig. 5), support the hypothesis that HSAG-1 encourages vector–chromosomal recombination events and that this is perhaps the mechanism responsible for the HSAG amplification phenomenon. An elevation in the frequency of local recombination events could generate amplification-prone structures and assist in the resolution of amplified sequences.

VI. CONCLUSION

With the analysis of an increasing number of the amplified structures in different systems, it is apparent that more than one mechanism of gene amplification exists. Each of these mechanisms, in turn, involves a series of steps, the products of which may be unstable and lost without the timely initiation of subsequent steps. A significant difference in the frequency of stable cellular *DHFR* amplification events relative to the frequency of transient increases in endogenous *DHFR* copy number have been observed in other systems [14] and may due to amplification events that did not go to completion. Similarly, the variability in the frequency of gene amplification at different sites within the genome of a cell may reflect regional differences in the ability to generate amplification-prone structures or to resolve these structures to produce stable arrays of amplified genes. Rate-limiting points for gene amplification events may include either of these intermediate steps, both of which depend on local recombination events and therefore are sensitive to sequences that promote recombination.

We have previously shown that members of the HSAG family of middle repetitive elements increase the frequency of gene amplification *in cis*. The following lines of evidence suggest that the mechanism of this phenomenon may involve the stimulation of local recombination by HSAG elements.

1. Active subfragments of HSAG-1 and HSAG-2 included sequence features associated with "hot spots" for recombination and/or gene amplification including purine–pyrimidine tracts, A+T-rich regions, and putative stem–loop structures. Deletion of several of these structures from active fragments eliminated or greatly reduced their activity.

2. An orientation dependence of HSAG elements in the vector on the magnitude of the amplification activity correlated with the level of vector-initiated HSAG transcription. Experiments in which the vector was linearized at the *Bam*HI site prior to transfection eliminated HSAG-1 transcription and the amplification stimulatory activity of HSAG-1 sequences without affecting vector *DHFR* transcription. Increased local transcription has been implicated in other systems in promoting recombination events [15].

3. The number of side bands, the product of recombination events, observed by Southern analysis was significantly higher in clones with amplified vector-HSAG sequences than in the corresponding clones with amplified-vector-alone sequences, indicating, we suggest, the facilitation of vector amplification by

HSAG-1 through an increase in the efficiency of recombination at a greater number of integration sites.

While the data indicate that HSAG elements increase the frequency of local recombination events, the mechanism of this action is not understood. HSAG elements might act as a homing signal for recombinogenic proteins or, alternatively, these sequences may be unstable and prone to nicking, thus promoting recombination. Also of interest is the question of the endogenous HSAG repetitive elements and their ability to promote gene amplification. Perhaps they are quiescent until they are placed in an active configuration or are transcribed. These questions await further investigation.

REFERENCES

1. Wahl, G., Robert de St. Vincent, B., and DeRose, M. *Nature*, *307*:516–520 (1984).
2. Giulotti, E., Knights, C., and Stark, G. *Cell*, *48*:837–845 (1988).
3. Pasion, S., Hartigan, J., and Pichler, M. *Science*, *225*:941–943 (1984).
4. Passananti, C., Davies, B., Ford, M., and Fried, M. *EMBO J. 6*:1697–1703 (1987).
5. Hyrien, O., Debatisse, M., Buttin, G., and Robert de St. Vincent, B. *EMBO J. 6*: 2401–2408 (1987).
6. McArthur, J. G., and Stanners, C. P. *J. Biol. Chem. 266*:6000–6005 (1991).
7. Cartier, M., Chang, M. W.-W., and Stanners, C. P. *Mol. Cell. Biol. 7*:1623–1628 (1987).
8. Stanners, C. P., Lam, T., Chamberlain, J. W., Stewart, S. S., and Price, G. B. *Cell*, *7*: 211–221 (1981).
9. Stanners, C. P., Chamberlain, J. W., Lam, T., and Price, G. B. In *Oncogenes: Evaluation of Basic Findings and Clinical Potential* (T. C. O'Connor, ed.), Alan R. Liss, New York, 1983, pp. 173–183.
10. McArthur, J. G., Beitel, L. K., Chamberlain, J., and Stanners, C. P. *Nucleic Acids Res.* (1991).
11. Beitel, L. K., McArthur, J. G., and Stanners, C. P. *Gene* (1991).
12. Kesler, M. M., Westhaver, M. A., Carson, D. D., and Nordstrom, J. L. *Nucleic Acids Res. 15*:631–642 (1987).
13. Beitel, L. K., Chamberlain, J. W., Benchimol, S., Lam, T., Price, G. B., and Stanners, C. P. *Nucleic Acids Res. 14*:3391–3408 (1986).
14. Johnston, R., Beverley, S., and Shimke, R. *Proc. Natl. Acad. Sci. USA*, *80*:3711–3715 (1983).
15. Nickoloff, J. A., and Reynolds, R. J. *Mol. Cell. Biol. 10*:4837–4845 (1990).

39

Amplisomes: Novel Submicroscopic Extrachromosomal Elements Containing Amplified Genes

Giovanni Pauletti,* Barry J. Maurer,† and Giuseppe Attardi
California Institute of Technology, Pasadena, California

I. INTRODUCTION

Extrachromosomal gene amplification has been conventionally associated with the karyotypic abnormality represented by double minute chromosomes (DMs) [1]. DMs generally appear as paired, spherical chromatin structures in metaphase spreads, although they may also occur as unpaired elements. They characteristically vary in number among cells of a population, and may also vary in size both among cells and within a single cell. On the contrary, developmentally regulated gene amplification, when extrachromosomal, is characterized by submicroscopic elements that are homogeneous in size. Such are the extrachromosomal rings in amphibian oocytes or the linear palindromic dimers in *Tetrahymena*, structures that carry amplified rDNA genes [2]. Elsewhere we have reported the occurrence of novel extrachromosomal submicroscopic elements in drug-resistant human cell lines [3]. Similar elements harboring the amplified *MYC* gene, the transfected *CAD* gene, or the multiple drug resistance gene have been found in other mammalian systems [4]. Furthermore, resistance to methotrexate (MTX) in several species of the lower eukaryote *Leishmania* has been associated with the presence of homogeneous extrachromosomal circles containing either the

Current affiliations:
*UCLA School of Medicine, Los Angeles, California.
†Children's Hospital of Michigan, Detroit, Michigan.

dihydrofolate reductase–thymidine synthetase gene (R circles) or an open reading frame probably coding for a P-glycoprotein (H circles) [5,6].

The extrachromosomal submicroscopic elements bearing amplified genes mentioned above have come to occupy an important role in studies of gene amplification in eukaryotic cells [1,4]. In fact, besides being a novel manifestation of this process, they may represent a useful tool for unraveling the mechanism that leads to the amplification of DNA segments in the chromosome itself. Furthermore, they may constitute the "missing link" between intrachromosomal amplification and the previously recognized extrachromosomal manifestation of gene amplification, namely DMs. We have proposed the term "amplisomes" (i.e., bodies carrying amplified genes) for these submicroscopic extrachromosomal elements [7]. The amplisomes are strikingly homogeneous in size, unlike DMs. Because of their homology to segments of the chromosomal genome and because of their chromosomal origin, they are clearly distinct from plasmids. The previously proposed term "episomes" [8] does not seem to be appropriate for these structures, because of the inherent implication that the element can alternate between integrated and free state [9], a property that has not been shown to apply to them.

II. AMPLIFIED DNA IN SLOW SEDIMENTING STRUCTURES

In other work from this laboratory [10], several variants resistant to high concentrations of MTX were isolated from the human cell lines HeLa BU25 and WI-18-VA_2-B, which exhibited a markedly pleiomorphic chromosome constitution. Several of these variants, derived from HeLa BU25, contained a small average number of minute chromosomes per cell and lacked recognizable homogeneously staining regions (HSRs). These variants exhibited the same degree of *DHFR* gene amplification (250–300-fold) that was observed in variants with a large average number of minute chromosomes and/or one or more HSRs, as judged from the amount of enzyme [10] or specific mRNA [11] that they contained, and from the increased number of *DHFR*-specific sequences in the genome [7]. These findings suggested that the amplified genes could be located in one or more chromosomal sites without obvious morphological or staining alterations. However, an alternative possibility was that the amplified genes were associated with extrachromosomal elements too small to be detectable by optical microscopy. In one of these variants, HeLa BU25-10B3 (which had been grown in 1.8×10^{-4} M DL-MTX for about one year) an analysis of Giemsa-banded metaphase spreads did not reveal any obvious HSR. In other experiments, after in situ hybridization with a tritium-labeled *DHFR* cDNA probe, no large cluster of grains was observed on any of the chromosomes of this cell line (unpublished observations).

To further examine the chromosomal distribution of the amplified *DHFR* genes in 10B3 cells, a chromosome preparation from mitotically arrested cells was fractionated in a sucrose gradient. DNA from the individual fractions was transferred onto a nitrocellulose filter, then probed for the presence of sequences homologous to a human *DHFR* cDNA clone, or to a nuclear ribosomal rDNA clone or to a human 3-hydroxy-3-methylglutaryl coenzyme A (HMG-CoA) reductase cDNA clone. As shown in Figure 1, the majority of the *DHFR*-specific sequences were associated with structures sedimenting more slowly than chromosome 5, which contains the unamplified *DHFR* gene and the HMG-CoA reductase gene, and also more slowly than chromosomes 13, 14, 15, 21, and 22, which contain the nuclear rRNA genes [12]. Although some chromosome aggregation complicated the pattern in the lower half of the gradient, it is clear that the bulk of the amplified *DHFR* genes were associated with structures smaller than normal chromosomes. When correlated with the karyotype and in situ hybridization analysis described above, this experiment provided the first indication that the amplified *DHFR* genes in 10B3 cells are in extrachromosomal structures, which are too small to be visible in the light microscope.

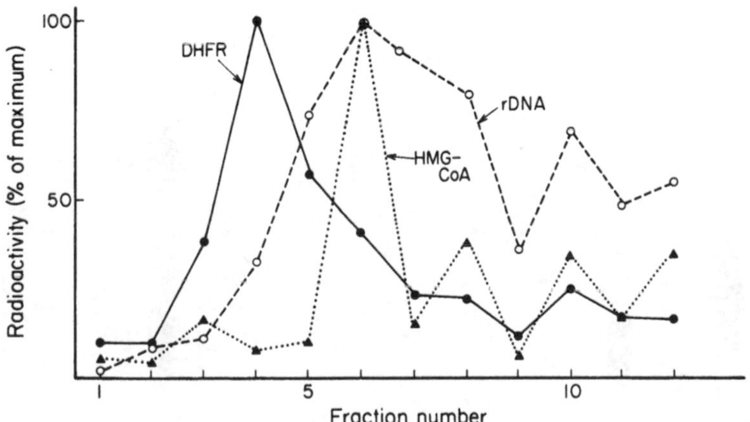

Figure 1 Distribution of sequences hybridizable with the human *DHFR* cDNA clone pHD84, a human HMG-CoA reductase cDNA probe, and a human rDNA probe among chromosomal structures isolated from HeLa BU25-10B3 cells arrested in metaphase and fractionated on a sucrose gradient. Sedimentation was from left to right. After hypotonic treatment, cells were fixed in 50% acetic acid and broken by three passages through a 23 gauge needle. The suspension was then made 1% in Tween 20 and loaded onto a 250 ml 10–40% (w/v) sucrose gradient in HEPICAT medium (12% (w/v) hexylene glycol, 10^{-4} M PIPES, 6×10^{-4} M $CaCl_2$, 0.1% Tween 20). Centrifugation at room temperature was at 700 *g* for 90 min. Next, 20 ml fractions were collected and, after addition of 10 ml of 10% (w/v) sucrose in HEPICAT medium to each one, centrifuged in a SW27 rotor at 20,000 rpm for 20 min. Pellets were then processed for slot blot analysis and hybridized with the above-indicated probes.

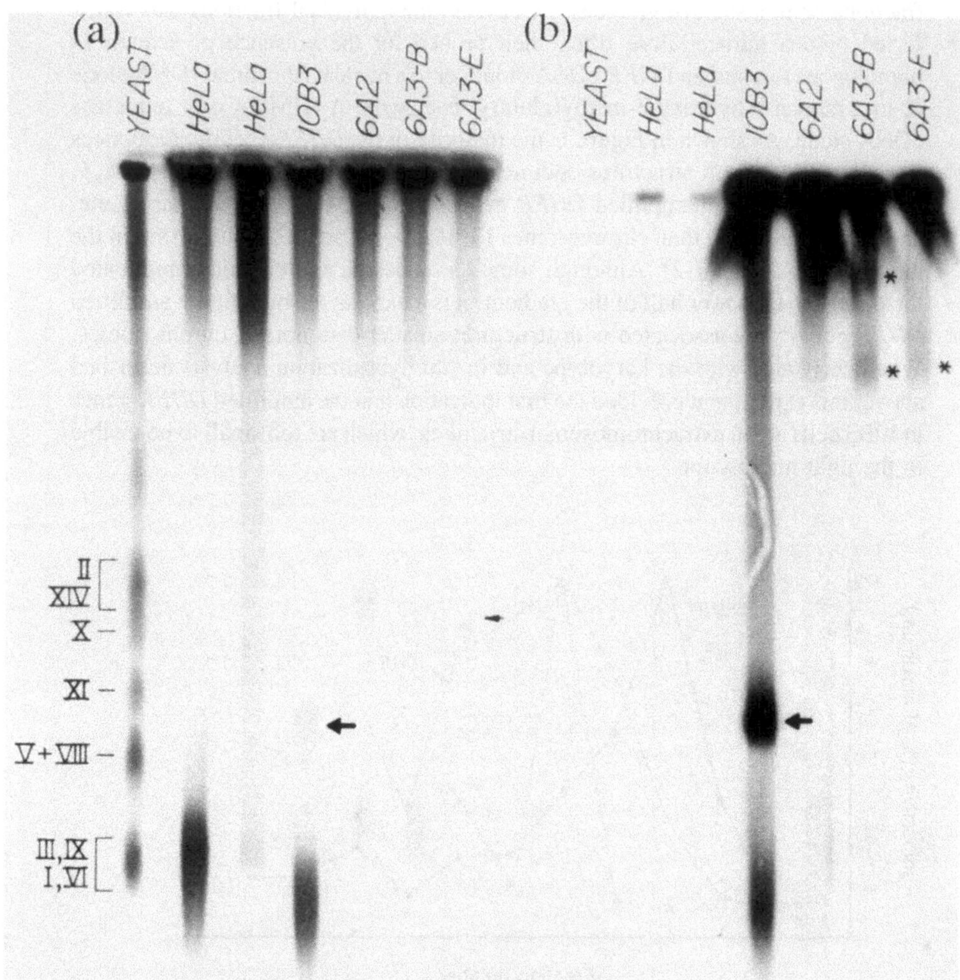

Figure 2 FIGE of genomic DNA isolated from HeLa S3 cells, or from the MTX-resistant human cell variants HeLa BU25-10B3, VA_2B-$6A_2$, VA_2B-$6A_3$-B, and VA_2B-$6A_3$-E, or from *Saccharomyces cerevisiae* (SS 327), and detection of chromosomal structures containing amplified *DHFR* genes: (a) EtBr-stained gel and (b) Southern blot analysis of the same gel with the human *DHFR* cDNA probe. Electrophoresis was run in TBE buffer at 10 V/cm for 19 hr (11°C), with a switching forward-migration interval varying linearly between 30 and 63 sec at $t = 0$ and $t = 19$ hr, and with a constant 3:1 ratio between the forward and reverse intervals.

III. PULSED-FIELD GEL ELECTROPHORESIS ANALYSIS

To further investigate the presence of extrachromosomal elements in MTX-resistant human cell lines, pulsed-field gel electrophoretic (PFGE) techniques were applied that allow the separation of large DNA molecules [13]. Figure 2 shows the results of an experiment in which field-inversion gel electrophoresis (FIGE) was used to fractionate genomic DNA samples from the MTX-resistant cell variants HeLa BU25-10B3, VA_2B-$6A_2$, VA_2B-$6A_3$-B and VA_2B-$6A_3$-E and from HeLa S3, as a control. The ethidium bromide (EtBr) staining pattern (Fig. 2a) reveals a discrete band (heavy arrow) in the 10B3 DNA lane, which is absent in the HeLa S3 DNA lane but migrates to a position between the bands represented by yeast chromosomes V and VIII and chromosome XI. This migration corresponds to a linear DNA size of ~650 kb. A band migrating approximately like yeast chromosome X (thin arrow) is present in the lanes containing $6A_2$, $6A_3$-B, and $6A_3$-E DNA.

The gel shown in Figure 2a was blotted onto a nylon membrane and probed with the human *DHFR* cDNA (pHD84). As shown in Figure 2b, the "650 kb" band of the 10B3 genomic DNA sample gave a strong signal. This experiment thus confirmed the existence of submicroscopic DNA structures containing amplified *DHFR* genes in the 10B3 cell line, which was predicted from the sucrose gradient centrifugation experiment. Moreover, FIGE analysis showed that these elements consist of a population of homogeneous molecules. Subsequent work [7] revealed that HeLa BU25-10B1 and -10B2, two MTX-resistant variants related to the 10B3 cell line, harbor the 650 kb elements and that the parental HeLa BU25 cell line lacks them. No hybridization with the *DHFR*-specific probe is exhibited by the "1000 kb" band common to the three cell lines of the VA_2-B lineage. The autoradiogram also shows two faint, slowly moving bands in $6A_3$-B DNA samples, and one such band in the $6A_3$-E DNA sample (indicated by asterisks). Further studies (unpublished) have shown that the 1000 kb band is also present in the MTX-sensitive parental VA_2-B cell line. Its significance is presently being investigated. The presence of this presumably acentromeric element in the parental cell line prior to MTX selection indicates that it is unrelated to *DHFR* gene amplification and suggests that genes essential for this cell line's growth may reside in the amplisomes.

IV. TEMPORAL DYNAMICS OF APPEARANCE AND STABILITY OF *DHFR* AMPLISOMES IN HeLa BU25-10B3 CELLS

The kinetics of appearance of submicroscopic extrachromosomal elements carrying amplified *DHFR* genes in the 10B3 cell line under increasing selective pressure (i.e., increasing concentration of MTX) and their stability after removal of MTX were investigated. Figure 3 shows diagrammatically the quantitative behavior of amplisome-associated *DHFR* sequences after FIGE fractionation and the behavior of total *DHFR* gene amplification at different times of the selection process. Densitometric data are plotted as a function of MTX exposure time and MTX concentration. The data in both curves were corrected for variation in input DNA by hybridization with a single copy probe. The results indicate a parallel increase in the total number of *DHFR* genes and the number of amplisomes during MTX selection. It is important to note that amplisomes were observed in the initial stages of the stepwise selection process, when the total *DHFR* copy number was just above that of the parental HeLa BU25 cell line and the selective pressure was 4 orders of magnitude lower than that applied at the end of the selection (1.8×10^{-4} M, DL-MTX). This observation indicates that amplisomes could be recognized as early as the first detectable amplified genes. This in turn suggests that extrachromosomal DNA segments may be involved in the primary event leading to

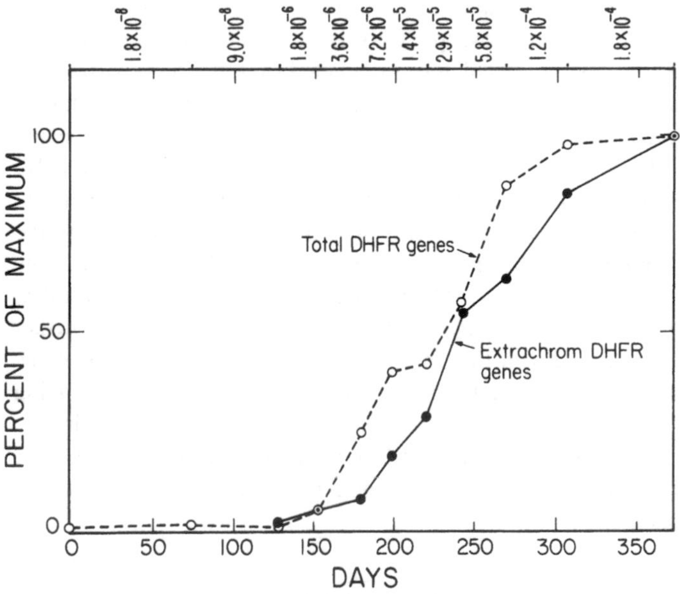

Figure 3 Time course of increase in the amount of *DHFR* genes associated with amplisomes and with total DNA in HeLa BU25-10B3 cells during stepwise selection in the presence of MTX. The densitometric data of Southern blot and dot blot analysis have been normalized for variation in input DNA content of the slices on the basis of hybridization with the pYNZ2 probe. The numbers at the top indicate the molar concentration of DL-MTX to which the cells were exposed, and the arrowheads indicate the time limits of exposure of the cells to each concentration. (From Ref. 7.)

the amplification of the *DHFR* locus in 10B3 cells. No variation in size of the 650 kb amplisomes was detected by FIGE and Southern blot analysis throughout the selection process.

The stability of these elements after removal of MTX was also investigated by FIGE. As shown in Figure 4, amplisomes are not stable in the absence of selective pressure. However, the kinetics of their loss is significantly slower than predicted by simple dilution of nonreplicating molecules. They are, in fact, still detectable after more than 60 generations of cell growth in medium without MTX. These findings suggest that they replicate autonomously, as shown earlier for DMs [14]. Again, a parallel behavior was observed between decrease in amount of amplisomes and decrease in the number of total *DHFR* genes. This finding is consistent with the evidence, discussed below, indicating that in the 10B3 cell line, the amplified *DHFR* genes are primarily harbored in these structures.

An interesting finding was that in the control cultures maintained in the presence of MTX, there was no detectable loss of amplisomes over the 240-day period in which the experiment was carried out. Furthermore, as in the case of the year-long, stepwise selection process, no size change of the amplisomes was observed. Karyotypic analysis has shown that the great majority of 10B3 metaphase spreads contain virtually no DMs. Only a few sporadic cells contain a

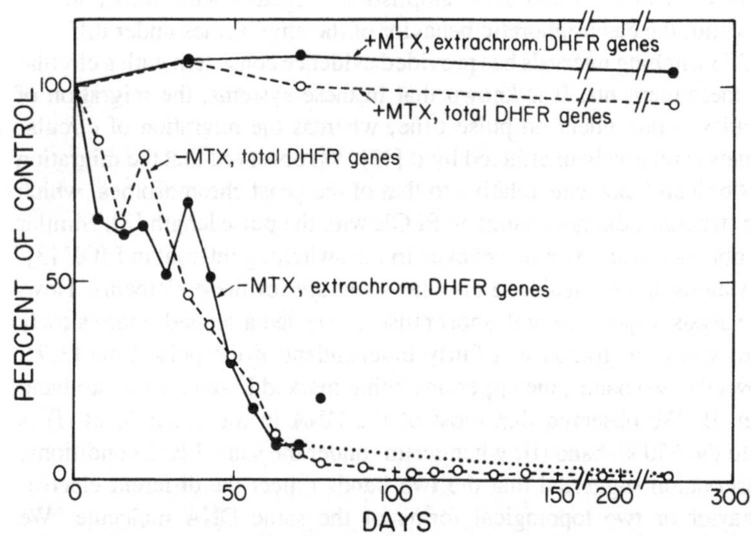

Figure 4 Quantitative behavior of the *DHFR* genes associated with amplisomes and of total *DHFR* genes in HeLa BU25-10B3 cells grown in the absence of MTX (lower curves) or in the presence of 0.18 mM DL-MTX (upper curves). Data derived and normalized as explained in Figure 3. (From Ref. 7.)

moderate to large amount of these structures. An almost total absence of DMs has also been observed in 10B3 cells adapted to grow in suspension culture in the presence of the highest concentration of MTX. Karyotypic analysis of these cells was performed after 25 months of growth, 15 of which were uninterrupted. 10B3 *DHFR* amplisomes are, therefore, clearly characterized by a long-term extrachromosomal persistence. Interestingly, a plasmid containing the 5′ region of the human c-*myc* locus and a selectable marker has been shown to persist as an extrachromosomal element in HeLa cells for more than 200 generations, when maintained under G418 selection [15]. In this system, chromosomally integrated copies of the plasmids were not detected, and removal of the selective pressure resulted in less than 5% plasmid loss per generation. By contrast, both *CAD* amplisomes derived from a transfected *CAD* gene and their larger derivatives have been shown to tend to integrate into chromosomes upon prolonged growth in selective medium of the cells carrying them [16].

V. AMPLISOME STRUCTURE

No direct evidence concerning the structure of mammalian cell amplisomes is presently available, although there is a consensus favoring a circular structure. This opinion is based on several lines of indirect evidence. First, alkaline lysis procedures developed to purify circular DNA molecules have been shown to allow a selective recovery of *CAD* and *MYC* amplisomes together with mitochondrial DNA [8]. Second, the electrophoretic behavior of the amplisomes under different PGFE or FIGE switching intervals has provided evidence consistent with a circular structure of these elements. It is known that in these systems, the migration of linear molecules is dependent on pulse time, whereas the migration of circular DNA molecules is relatively unaffected by it [13]. We observed that the migration of the 10B3 "650 kb" elements relative to that of the yeast chromosomes, which have a linear structure, did not change in PFGE with the pulse length [3]. Similar results were obtained with extreme changes in the switching interval in FIGE [3]. These observations argued against a circular structure for these elements. However, FIGE analysis, especially with short pulses, revealed a second, more slowly moving band, whose migration was fairly independent of the pulse time [3,7]. Figure 5 shows the two bands, the upper one being marked I, and the 650 kb band being marked II. We observed that most of the DNA in the upper band (I) is converted into the 650 kb band (II) when rerun (under the same FIGE conditions) [7]. This observation suggested that the two bands reflect the different electrophoretic behavior of two topological forms of the same DNA molecule. We postulate that the upper band consists of circular DNA molecules, while the lower one is the product of a linearization process of the former, probably due to mechanical rupture by the strong pulsating electric field. In *Leishmania*, electron microscopy (EM) analysis of circular extrachromosomal structures analogous to the mammalian amplisomes has provided conclusive evidence that supercoiled DNA molecules migrate into pulsed-field gels [17]. Even though the *Leishmania* supercoiled "amplisomes" (30, 60, 83, and 118 kb) are much smaller than the 650 kb 10B3 amplisomes, we are inclined to believe that the upper band observed in our studies is actually formed by supercoiled structures. Our extrapolation from the *Leishmania* findings is also supported by EM observations of chromatin preparations made from 10B3 cells arrested in metaphase, which showed the presence of "rings" consisting of chromatin fibers [3]. Figure 6 shows two examples of such structures. The calculated size of the DNA in these elements varied between 580 and 1000 kb. This variability in estimated DNA content could be due to differences in degree of compaction of the chromatin fibers or, in the case of the larger forms, to the occurrence of dimeric circles.

The results of agarose trapping experiments [8] and the observation that significant trapping of the 650 kb amplisomes occurs at the PFGE or FIGE origin [7] have provided further support for the conclusion that amplisomes are covalently closed circular molecules. Figure 5a shows the effect of increasing doses of γ-rays on the trapping of 10B3 amplisomes at FIGE origin. The autoradiogram clearly reveals a striking progressive increase in the amount of the *DHFR* amplisomes migrating into the gel as a function of the irradiation dose. On the other hand, digestion with increasing concentrations of the restriction enzyme *Sal* I, which does not cut the 10B3 amplisomes, produced, together with the expected massive release of heterogeneously sized chromosomal fragments, a different pattern of increase in the intensity of the amplisome band. In this case, a large release of amplisomes occurred at relatively low *Sal* I concentrations, whereas the intensity of the two amplisome bands remained relatively constant at higher concentrations of the enzyme. These results are also consistent with a circular structure for the 10B3 amplisomes. We believe that trapping at the gel origin is the result of two different phenomena: a strong sieving effect by the chromosomal DNA meshwork and a threading effect due to the topological properties of these elements. Both the physical (γ-rays) and the enzymatic method of DNA cleavage (*Sal* I) eliminated to a great extent the trapping due to the chromosomal network. However, only γ-irradiation by cutting the amplisomes and, thus, by relieving the topological constraints, allowed their migration into the gel in a dose-related way. The portion of amplisomes not released even by high dose γ-irradiation presumably is represented by relaxed circular molecules. Another important conclusion of these experiments was that the majority (~80%) of the amplified genes in 10B3 cells could be accounted for by amplisomes released by the physical or enzymatic treatment of the gel slice prior to electrophoresis. As shown in Figure 5a, even after extensive γ-irradiation, no smearing of *DHFR*-specific sequences was detectable in the gel. Because of the random nature of the γ-ray-induced DNA breaks, such a smearing would be expected if intrachromosomal gene amplification had taken place. The presence of massive

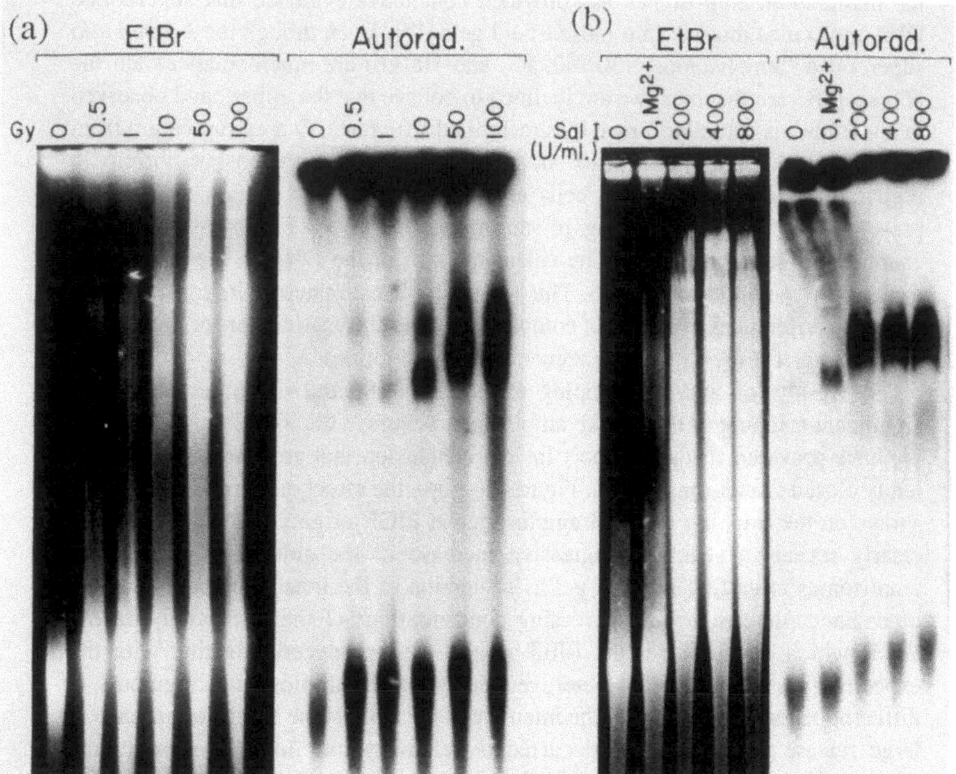

Figure 5 γ-Irradiation or *Sal* I digestion of 10B3 DNA/agarose slices reduce trapping of amplisomes at the FIGE origin. (a) FIGE of 10B3 DNA/agarose inserts from ~2×10^6 10B3 cells (grown in suspension for 10 months in 0.18 mM DL-MTX) subjected to increasing doses of γ-irradiation, as indicated in grays (Gy); EtBr staining pattern and DNA blot probed with pHD84 (Autorad). (b) FIGE of 10B3 DNA/agarose inserts (same preparation as above) subjected to *Sal* I digestion at three enzyme concentrations, as indicated in units (U) per milliliter. EtBr staining pattern and a DNA blot probed with pHD84 (Autorad) are shown. Lanes: 0, sample pretreated as the *Sal* I-digested samples; 0, Mg^{2+}, sample pretreated and incubated at 37°C as the *Sal* I-digested samples, except without the restriction enzyme. The smear in the upper portion of the first three lanes in the autoradiogram of (b) is due to diffusion of small DNA fragments from the gel slices at the origin during blotting. A ramped-pulse regime from 3 to 50 sec switching cycles, with a constant 3:1 ratio between the forward and the reverse interval, was applied for 20 hr at 10 V/cm. (From Ref. 7.)

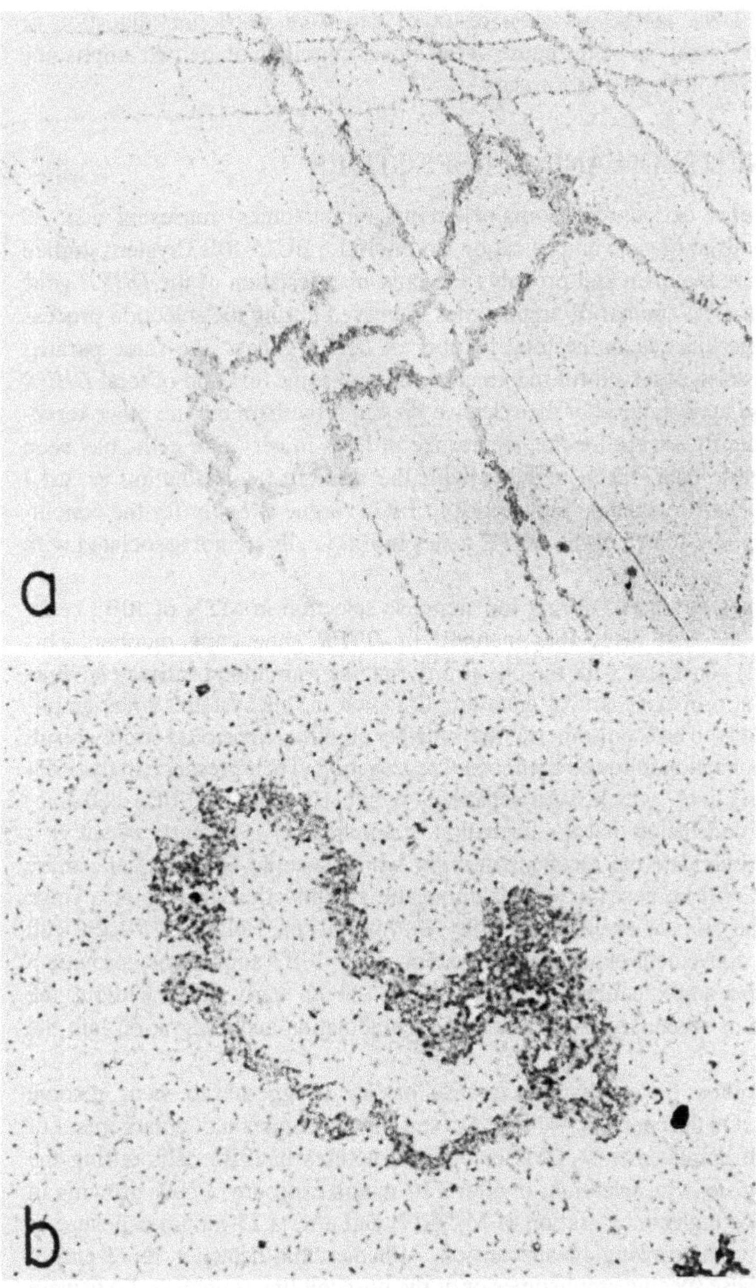

Figure 6 Electron micrographs of extrachromosomal elements in metaphase arrested 10B3 cells: (a) ~580 kb, × 30,000 and (b) ~710 kb, × 41,000. The DNA contents were estimated from the contour lengths of the molecules, assuming a 40:1 packing ratio of DNA in the 20–30 nm chromatin fiber. (From Ref. 3.)

amounts of DNA in the gel as a result of extensive restriction digestion or γ-irradiation is likely to be the cause of the closer migration of the two amplisome bands observed in these experiments.

VI. CONCLUSION AND PERSPECTIVES

Submicroscopic extrachromosomal elements (amplisomes) represent a novel structure involved in gene amplification. In the HeLa BU25-10B3 system studied by us, they are the main and probably the only manifestation of the *DHFR* gene amplification. The number of amplisomes increased during the selection process with the same kinetics as the total number of *DHFR* genes. The same parallel behavior was also observed for the kinetics of amplisome loss and of total *DHFR* gene decrease upon removal of the selective pressure. Furthermore, no other karyotypic abnormality, except for the occurrence of DMs in very rare cells, has been detected. These data, taken together with the results of γ-irradiation or *Sal* I digestion of DNA in agarose plugs prior to FIGE, argue strongly for the conclusion that very few, if any, of the *DHFR* genes in 10B3 cells are not associated with amplisomes.

Amplisomes appeared, during the stepwise selection in MTX of 10B3 cells, as early as the first detectable increase in *DHFR* gene copy number. This observation is consistent with the possibility that the formation of these elements represents the primary event of gene amplification in this system. Since amplisomes are likely to be acentromeric, their further amplification could simply result from their replication followed by unequal segregation. Their presence in the 10B1 and 10B2 cell lines, which were separated very early from the 10B3 cell line, supports the conclusion that the formation of amplisomes is an early event. It is clear that to elucidate the mechanism(s) by which specific DNA sequences are amplified, it will be necessary to examine the amplification process at a stage as close to the primary event as possible. To this end, the HeLa BU25-10B3 cell line offers a convenient experimental system, since 10B3 amplisomes provide a picture of the amplification process "frozen" at an early step, without the complication of secondary phenomena of recombination and integration into the chromosome.

A conversion of the amplisomes to DMs has been suggested to occur in some systems [16]. On the contrary, we failed to see any phenomenon of enlargement or conversion of amplisomes to DMs or other structures in 10B3 cells during the 9-month-long stepwise selection, or after a 10-month exposure of cells growing in monolayer to a high concentration of MTX [7] and after a 25-month exposure to the drug of cells growing in suspension. Whether this behavior represents a property of these specific amplisomes or of the host cell line is unclear. DMs were indeed observed in very rare cells in the population. However, it is conceivable that they arose by a different mechanism not involving amplisomes.

An interesting possibility is that the presence of 10B3 cells of one or more defects in the cellular recombination system, possibly involving one or both of the two cell topoisomerases, is responsible for the failure of the amplisomes to recombine and to be converted to larger structures. It is likewise possible that the same mechanism would account for the apparent failure of the amplisomes to integrate into chromosomes during long-term growth. This behavior is in contrast to the general tendency that has been observed for DM in cells growing in vitro [4]. It is pertinent to mention here the recent observation that, in a yeast strain carrying double mutations affecting topoisomerase I and topoisomerase II, more than half the rDNA is present as extrachromosomal rings [18]. Even more significant is the finding by the same authors that expression of plasmid-borne topoisomerase I or II genes in this mutant leads to the integration of the extrachromosomal rDNA rings back into the chromosomal rDNA cluster.

The instability of the amplisomes in the absence of selection indicates the lack of functional centromeric sequences. Studies on the replication of these elements should open the way to the identification of the origin(s) of DNA replication in the amplisomes and their isolation. Furthermore, the property of the 10B3 amplisomes to persist in a free state during long-term growth makes these elements potentially useful for the construction of mammalian vectors.

ACKNOWLEDGMENTS

This work was supported by grants from the Margaret E. Early Medical Research Trust, from the U.S. National Institutes of Health (GM11726), and from the American-Italian Foundation for Cancer Research (AIFCR 88-104).

REFERENCES

1. Stark, G. R., Debatisse, M., Giulotto, E., and Wahl, G. M. Recent progress in understanding mechanisms of mammalian DNA amplification. *Cell*, *57*:901–908 (1989).
2. Kafatos, F. C., Orr, W., and Delidakis, C. Developmentally regulated gene amplification. *Trends Genet*. *1*:301–306 (1985).
3. Maurer, B. J., Lai, E., Hamkalo, B. A., Hood, L., and Attardi, G. Novel submicroscopic extrachromosomal elements containing amplified genes in human cells. *Nature*, *327*:434–437 (1987).
4. Wahl, G. M. The importance of circular DNA in mammalian gene amplification. *Cancer Res*. *49*:1333–1340 (1989).
5. Garvey, E. P., and Santi, D. V. Stable amplified DNA in drug-resistant *Leishmania* exists as extrachromosomal circles. *Science*, *233*:535–540 (1986).
6. Ouellette, M., Fase-Fowler, F., and Borst, P. The amplified H circle of methotrexate-resistant *Leishmania tarentalae* contains a novel P-glycoprotein gene. *EMBO J. 9*: 1027–1033 (1990).

7. Pauletti, G., Lai, E., and Attardi, G. Early appearance and long-term persistence of the submicroscope extrachromosomal elements (amplisomes) containing the amplified *DHFR* genes in human cell lines. *Proc. Natl. Acad. Sci. USA*, *87*:2955–2959 (1990).
8. Carroll, S. M., Gaudray, P., De Rose, M. L., Emery, J. F., Meinkoth, J. L., Nakkim, E., Subler, M., Van Hoff, D. D., and Wahl, G. M. Characterization of an episome produced in hamster cells that amplify a transfected *CAD* gene at high frequency: Functional evidence for a mammalian replication origin. *Mol. Cell. Biol.* *7*:1740–1750 (1987).
9. Jacob, F., and Wollman, E. L. Les épisomes, éléments génétiques ajoutés. *C. R. Acad. Sci., Paris*, *247*:154–156 (1958).
10. Masters, J., Keeley, B., Gay, H., and Attardi, G. Variable content of double minute chromosomes is not correlated with degree of phenotype instability in methotrexate-resistant human cell lines. *Mol. Cell. Biol.* *2*:498–507 (1982).
11. Morandi, C., Masters, J. N., Mottes, M., and Attardi, G. Multiple forms of human dihydrofolate reductase messenger RNA. Cloning and expression in *Escherichia coli* of their DNA coding sequence. *J. Mol. Biol.* *156*:583–607 (1982).
12. Human Gene Mapping 11, *Cytogenet. Cell Genet.* *58*:1–2200 (1991).
13. Olson, M. V. Pulsed-field gel electrophoresis, in *Genetic Engineering*, Vol. 11 (J. K. Setlow and A. Hollaender, eds.), Plenum Press, New York, 1989, pp. 183–227.
14. Barker, P. E., Drwinga, H. L., Hittelman, W. N., and Maddox, A. Double minute chromosomes replicate once during S phase of the cell cycle. *Exp. Cell Res.* *130*:353–360 (1980).
15. McWhinney, C., and Leffak, M. Episomal persistence of a plasmid containing human c-*myc* DNA. *Cancer Cells, 6*:467–471 (1988).
16. Carroll, S. M., De Rose, M. L., Gaudray, P., Moore, C. M., Needham-VanDevanter, D. R., Von Hoff, D. D., and Wahl, G. M. Double minute chromosomes can be produced from precursors derived from a chromosomal deletion. *Mol. Cell. Biol.* *8*: 1525–1533 (1987).
17. Hightower, R. C., Wong, M. L., Ruiz-Perez, L., and Santi, D. V. Electron microscopy of amplified DNA forms in antifolate-resistant *Leishmania*. *J. Biol. Chem.* *262*: 14618–14624 (1987).
18. Kim, R. A., and Wang, J. C. A subthreshold level of DNA topoisomerases leads to the excision of yeast rDNA as extrachromosomal rings. *Cell*, *57*:975–985 (1989).

40

Cytogenetic and Molecular Dynamics of Mammalian Gene Amplification: Evidence Supporting Chromosome Breakage as an Initiating Event

Geoffrey M. Wahl *The Salk Institute for Biological Studies, La Jolla, California*

Susan M. Carroll *University of California San Diego School of Medicine, La Jolla, California*

Bradford E. Windle *The Cancer Therapy and Research Center, San Antonio, Texas*

I. INTRODUCTION

Normal mammalian cells rarely experience large-scale genetic rearrangements such as gene amplification, deletion, and translocation. By contrast, such genomic alterations occur frequently in neoplastic cells (see Chapter 33, this volume for a discussion), resulting in changes in gene expression that presumably confer survival advantages under constantly changing environmental conditions. The capacity to undergo large-scale genetic alterations should, therefore, be highly selected in malignant cells. Consequently, a fundamental problem in cancer biology concerns the identification of genes that can be mutated in the normal cell to enable such genetic plasticity to develop.

We have been investigating the molecular mechanisms involved in gene amplification, since this is a common genetic alteration in certain cancers. We reasoned that the elucidation of the initial molecular intermediates that generate units of amplification (amplicons) might provide insight into the types of genes whose functions could be altered to enable this type of genetic alteration, and possibly others, to occur. As described below, our studies reveal that a single type of molecular intermediate can be resolved in different ways to lead to gene amplification, large chromosomal deletions, translocations, and aneuploidy. We will propose a model in which this intermediate is a replication bubble, and the initiating event involves chromosome breakage within such a structure. While our studies do not reveal which normal genes have become altered to enable such genetic plasticity to occur, they do lead to a testable hypothesis, which predicts likely candidates.

II. EARLY AMPLIFICATION PRODUCTS ARE GENETICALLY UNSTABLE

There is general agreement that the earliest products in the amplification process are genetically unstable. A decade ago, Kaufman and Schimke (1981) reported that the products generated after a single step of methotrexate (Mtx) selection gave rise to heterogeneous dihydrofolate reductase (DHFR) gene expression and copy number, but the techniques employed did not enable a molecular or cytogenetic definition of the structures involved. The instability of the early phase products in this study rapidly gave rise to stable chromosomal products in cells subjected to multiple steps of selection. The generality of the evolution from unstable to stable products is revealed by the numerous examples of similar events in many cell types involving different amplified genes (see Wahl, 1989, for a review and Ruiz and Wahl, 1990; Von Hoff et al., 1990; Windle et al., 1991, for recent examples).

An intriguing unresolved problem is why the transition from unstable to stable products occurs at such different rates in cells of different species. To take two extremes, cells with unstable *DHFR* amplicons are replaced by cells with stable chromosomal amplicons within 30 cell doublings of the initial event in Chinese hamster ovary (CHO) cells, independent of whether the gene is amplified from its native (Windle et al., 1991) or an ectopic position (Ruiz and Wahl, 1990). By contrast, the *DHFR* gene in human cells can be amplified in unstable extrachromosomal elements, which are retained after multiple steps of selection and years of passage in cell culture (Pauletti et al., 1990).

Two different types of structure could engender the genetic instability that characterizes the early stage of most, and possibly all, amplification events. The first consists of acentric elements. These elements have the potential to generate copy number increases simply by unequal segregation at mitosis. There are two predominant size constraints on such structures: they should be (1) large enough to contain the selected gene and one or more functional DNA replication origins, and (2) not too small to be lost at unacceptable rates during the reformation of the nucleus subsequent to mitosis. Thus, such structures could range in size from entire chromosome arms to submicroscopic elements. As discussed below, there is evidence for the existence of structures comprising these two size extremes. The second type of unstable amplicon could consist of unstable intrachromosomal structures. Increased copy number could be generated from such structures by processes such as aberrant intrachromosomal replication, promiscuous intrachromosomal rearrangements, or repeated chromosome breakage over multiple cell cycles (see Stark et al., 1989 for a review). With regard to the latter, dicentric chromosomes have frequently been observed to harbor amplified genes. The ability of dicentric chromosomes to break once per cell cycle by a bridge–breakage–fusion cycle (e.g., see McClintock, 1984, for a summary of earlier work) could potentially generate copy number increase (e.g., see Kaufman et al., 1983; Ruiz and Wahl, 1990). However, we have recently found that while dicentric chromosomes are cytogenetically unstable, they do not generate copy number heterogeneity at appreciable rates (Windle et al., 1991). Also note that, in contrast to the case with acentric elements, chromosomal mechanisms for generating copy number increases generally require independent recombination events to occur over multiple cell cycles. Since cells with amplified genes arise spontaneously at rates of 10^{-4} to 10^{-6} per cell per cell doubling (see Chapter 33, this volume, and references therein), a scenario involving multiple events is unlikely a priori unless the putative chromosomal amplification precursors have structures that make them unusually recombinogenic. It is important to note that amplicon size alone is insufficient to explain recombinogenicity, since chromosomally amplified regions, including those with extremely large amplicons, are typically very stable (e.g., see Chasin et al., 1982; Cowell et al., 1982; Wahl et al., 1983). The discussion below focuses on acentric structures involved in gene amplification, since there is direct evidence for their existence as early, unstable amplification intermediates.

III. EXTRACHROMOSOMAL ELEMENTS INVOLVED IN GENE AMPLIFICATION

Many studies show that the paired, acentric chromatin structures referred to as double minute chromosomes (DMs) harbor amplified genes (for reviews and references, see Cowell, 1982; Wahl, 1989). For example, amplification of the adenosine deaminase gene in mouse cells (Yeung et al., 1983; G. Wahl, S. Carroll, and B. Windle, unpublished observations) and the *DHFR* gene in some human cells (Pauletti et al., 1990), is solely mediated by DMs and the submicroscopic precursors from which they originate (see below). DMs have been shown by density shift analyses (Carroll et al., 1987; Von Hoff et al., 1988; Ruiz et al., 1989) to replicate semiconservatively, and once per cell cycle. The initiation of replication of extrachromosomal elements has been reported to occur either throughout S phase (Tlsty and Adams, 1990), or in a discrete interval approximating that of the corresponding chromosomal gene (Fig. 1, and Carroll et al., 1991). These different results could derive from the dissimilar analytical techniques employed, or from the inclusion in some extrachromosomal amplicons of the sequences or structures that enable the correct temporal initiation of DNA synthesis to occur. The

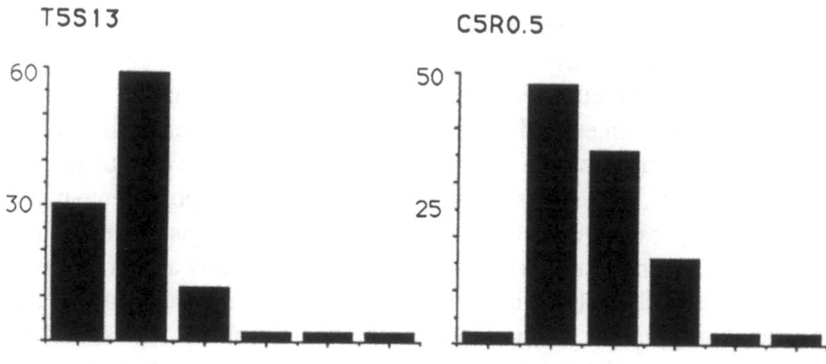
T5S13
60
30
C5R0.5
50
25

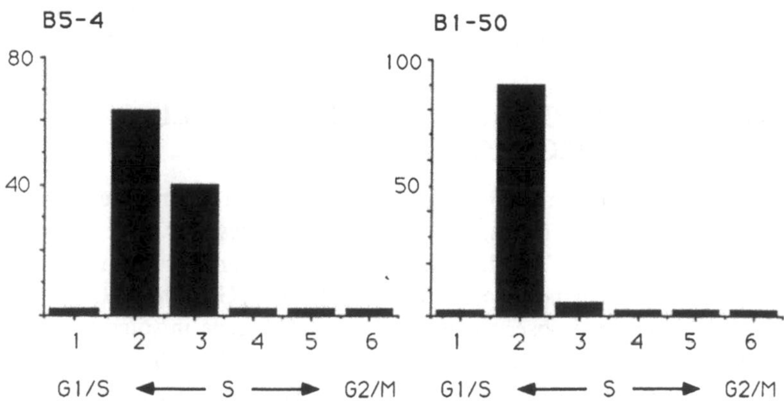
B5-4
80
40
B1-50
100
50
1 2 3 4 5 6
G1/S ← S → G2/M
1 2 3 4 5 6
G1/S ← S → G2/M

replication patterns of DMs indicate that part or all of at least one functional cellular replication origin is included in each extrachromosomal element.

Early studies suggested that a single DM could contain multiple amplicons of the same gene (see Wahl, 1989, for references). Since average-sized DMs are circular and approximately 1–3 Mbs, we reasoned that some DMs could derive from the multimerization of smaller, perhaps submicroscopic, circular precursors (Carroll et al., 1987). Many groups have now shown that submicroscopic circular molecules of varying sizes contain amplified genes in diverse systems in vitro and in vivo (Table 1).

Methods to detect circular DNA molecules and the size ranges isolated by each are shown in Figure 2. The method for mammalian cells with the highest accuracy for size determination, which is least limited by size extremes, employs γ-irradiation to introduce a solitary double strand lesion into a fraction of the circular molecules (e.g., see Ruiz et al., 1989; van der Bliek et al., 1988). The linear molecules produced by this procedure are then fractionated by size using a convenient form of field alternation gel electrophoresis, blotted, and the products detected with a cloned probe from the amplicon. This enables detection of circular molecules from ~100 kb to > 2.5 Mb at copy levels as low as one 250 kb molecule per cell in a population of 10^6 cells. Note that radiation doses that linearize small circular amplicons introduce multiple breaks in large circular and intrachromo-

---

Figure 1 Replication of extrachromosomal and intrachromosomal amplicons. The window in which replication occurs for *CAD* and *ADA* genes was determined by the retroactive synchrony method of Gilbert (1986). Exponentially growing cells were incubated in 30 μM BrdU for 90 min, and the cells were then harvested, fixed with ethanol, and stained with chromomycin A3. The cells were sorted on the basis of DNA content into six windows using the fluorescence-activated cell sorter. DNA was isolated, sheared, and fractionated according to density on CsCl gradients. The density of each gradient fraction was determined in a refractometer, and portions of each fraction were then applied to nylon membranes using a slot blotting device. The blot was hybridized with probes that detect either the *CAD* or adenosine deaminase (*ADA*) genes, washed under stringent conditions, exposed to film and the results quantitated using a densitometer. The results are expressed as the percentage of *CAD* or *ADA* DNA that had been replicated in each window of S phase (i.e., heavy-light/heavy-light + light-light). The following cell lines were employed: T5S13 is a CHO cell containing 50–100 chromosomally amplified transfected Syrian hamster *CAD* genes; C5RO.5 is the parent of T5S13, and it contains 50–100 extrachromosomally amplified *CAD* genes (see Carroll et al., 1987, 1988); B5-4 is a Syrian hamster cell with ~100 (*CAD* genes amplified near the native *CAD* gene (see Wahl et al., 1983); B1-50 is a mouse cell line with ~5000 extrachromosomally amplified *ADA* genes (see Yeung et al., 1983). The replication timing of the extrachromosomal and chromosomal *CAD* genes is the same within experimental error, as is that of the amplified and single copy *ADA* genes (S. Carroll and G. M. Wahl, in preparation).

Table 1 Episomes in Mammalian Cells

Gene	Species	Size(s)	Replication[a]	Reference
CAD	Hamster	~250	+	Carroll et al., 1987, 1988, 1991
mdr-1	Mouse	400, 800	nd	Meese et al., 1990
ADA	Mouse	~500–1400	+	Carroll et al., 1991
DHFR	Human	600–700	+	Maurer et al., 1987
DHFR	Human	200–300	nd	Von Hoff, unpublished
mdr-1	Human	600, 750	+	Ruiz et al., 1989
C-*myc*	Human	~250	+	Von Hoff et al., 1988
C-*myc*	Human	120, 160	nd	Von Hoff et al., 1988
EGFR	Human	650–2000	nd	Dolf et al., 1991
N-*myc*	Human (biopsy)	590–3400	nd	Van Devanter et al., 1991

[a]nd, not determined.

somal amplicons present in the same cell, and that these randomly sized irradiation products can obscure the signal derived from small circular amplicons.

There is now direct experimental evidence that submicroscopic circular molecules can be precursors of DMs. For example, the molecular progression from cell populations that initially contained a majority of small submicroscopic circular molecules and few if any visible DMs to populations with a majority of larger and heterogeneously sized circular molecules and DMs has been reported in two

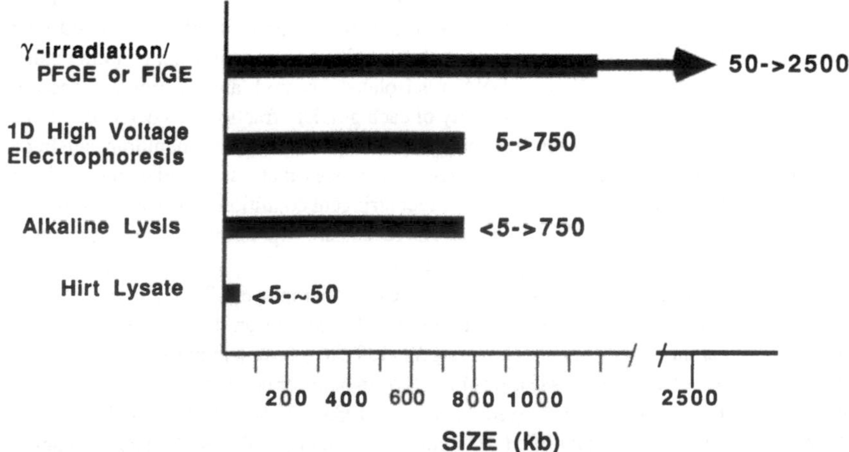

Figure 2 Methods used to detect circular DNA molecules: cumulative results of numerous experiments obtained with the indicated methods in this and other laboratories. Descriptions of experimental methods and references to original literature can be found in Carroll et al. (1987, 1988), van der Bliek et al. (1988), and Ruiz et al. (1989).

different systems (Carroll et al., 1988; Von Hoff et al., 1990). A reverse reaction involving the breakdown of microscopic DMs into submicroscopic elements has not been observed (Carroll et al., 1988; B. Windle, S. Carroll, and G. Wahl, unpublished observations).

We have used the term "episome" to distinguish submicroscopic DM precursors from DMs and to indicate that such molecules share similarities with some circular, autonomously replicating bacterial plasmids. "Amplisome" is a term that has also been used to refer to mammalian submicroscopic extrachromosomal elements involved in gene amplification (Pauletti et al., 1990; see also Chapter 39, this volume). There is one example in which a size progression has not been observed (Pauletti et al., 1990). In other cases of progression toward larger sizes, we infer that there is a selective advantage for larger extrachromosomal molecules. The reasons for the hypothesized size advantage are unknown, but several nonexclusive possibilities can be envisioned: (1) large molecules may be transmitted more effectively because they are captured more readily within the re-forming nucleus after mitosis, (2) larger molecules may adhere more tightly to chromosomes, or (3) transmission of a single large molecule confers greater selective and replicative advantages, since it contains multiple amplicons and multiple replication origins.

The data summarized above strongly indicate that DMs can be produced in many cases by enlargement of precursors substantially smaller than 1 Mb, but other evidence indicates that some extrachromosomal amplicons encompass many megabases. A single amplicon of this size could create a visible DM. One intriguing idea is that de novo formation of large amplicons (Sen et al., 1989) could derive from the fusion of a cell undergoing DNA synthesis with one in metaphase. This would lead to premature chromosome condensation and chromosome breakage. Structures with striking resemblance to DMs have been shown to be generated by such fusion events (Sen et al., 1989). Evidence in support of the occurrence of cell fusion in tumors in vivo has been presented (see Fortuna et al., 1989, for a recent example and other references). It is less likely that cell fusion plays a significant role in gene amplification under normal conditions of in vitro cell culture, since the spontaneous frequency of cell fusion is low, especially at the low cell densities employed for most drug selections. Furthermore, the initial cells manifesting gene amplification are most frequently equivalent in ploidy to the unselected parental cells, not increased in ploidy as the fusion model requires.

IV. MOLECULAR AND CYTOGENETIC EVENTS OCCURRING DURING *DHFR* GENE AMPLIFICATION IN CHO CELLS

The diversity of amplicon sizes and structures that can be generated from a single locus or from different loci are remarkable and have elicited proposals of numerous amplification mechanisms (see Stark et al., 1989). However, many of these mechanisms are based on analyses of the molecular and cytogenetic structures detected in cells subjected to multiple rounds of selection or prolonged passage in tissue culture. Since such structures may have little resemblance to the initial amplicons, or may have been generated by secondary events, it is difficult to use such information to make inferences about the events that generated the first amplicon. To gain an understanding of the initial events in the amplification process, one must analyze the chromosomal target site, as well as the amplification products, as soon as possible after the primary event occurs.

Analyses of early events in gene amplification have been difficult to perform because cells with amplification arise spontaneously at low frequencies (see Chapter 33, this volume), and there is no way of predicting which cell will undergo the process. In addition, since the initial intermediates are very short-lived in some cell lines (e.g., CHO), it is challenging to obtain sufficient material for analysis before the initial products are replaced by secondary derivatives. Further complicating the task is the fact that most of the cell lines used have at least two copies of amplification target, which prevents one from knowing unambiguously which generated the first amplicon. Such factors necessitate the use of cells that either are hemizygous for the target gene or are constructed to contain a single amplification target (e.g., by gene transfer). Furthermore, the analytical methods employed should enable cytogenetic and molecular studies to be performed on a cell-by-cell basis soon after selection is applied. The CHO *DHFR* locus has provided a valuable system for the analysis of gene amplification because appropriate cell lines, molecular strategies, and molecular probes have been developed over the past decade to enable most of these issues to be explored in considerable detail. The results of many of these studies are summarized briefly below.

A. Population Genetics

1. Cells with amplified *DHFR* genes preexist in the population (see Johnston et al., 1983, for an example and chapter 33, this volume for additional discussion); the selective agent does not induce gene amplification, but merely enables detection of those cells with elevated *DHFR* gene expression.
2. Agents that inhibit DNA replication increase the frequency of cells with *DHFR* gene amplification if they are applied transiently and near the time when the *DHFR* gene is replicated (e.g., see Tlsty et al., 1982). Contrary to initial suggestions (Mariani and Schimke, 1984), transient inhibition of DNA synthesis does not lead to measurable levels of DNA rereplication (Hahn et al., 1986).

B. Structure of the Chromosomal Target Subsequent to the Initial Event

1. Breakage of the chromosome that harbors the *DHFR* gene (chromosome 2, or a rearranged derivative called Z2) is frequently observed in cells that

have undergone *DHFR* gene amplification (Biedler 1982; Biedler et al., 1988; Flintoff et al., 1984; Windle et al., 1991). Recent studies reveal that breakage occurs as a very early and probably initial event near the amplification target (Windle et al., 1991). The acentric arm distal to the *DHFR* gene generated by breakage can be translocated to other chromosomes (Flintoff et al., 1984) or reattached to the broken derivative of the target chromosome Windle and Wahl, manuscript in preparation).
2. Sequences left at single copy level at the end of the broken chromosome can also be found in the amplicon (Windle et al., 1991). This observation indicates that the break can be asymmetric (see Fig. 3).
3. Deletion of the *DHFR* gene is observed frequently in cells that have undergone *DHFR* gene amplification. However, some sequences from the *DHFR* locus can remain at their native position at apparently single copy

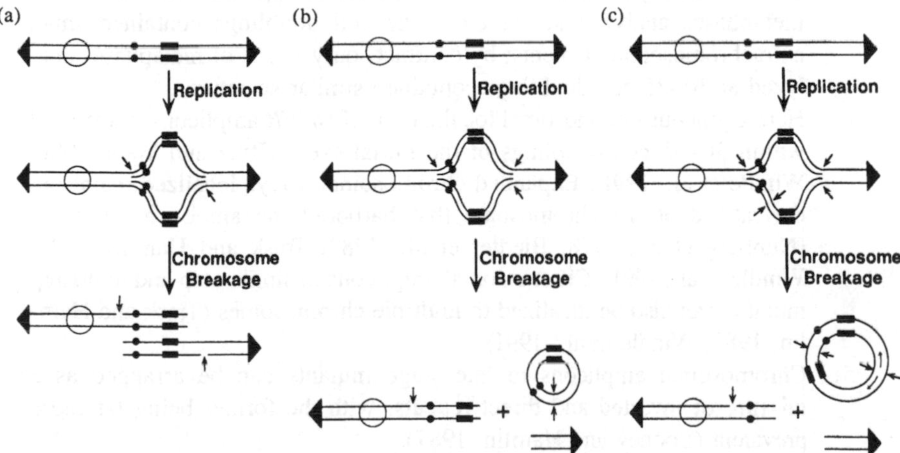

Figure 3 Production of amplicons by asymmetric chromosome breakage. As described in the text, the available data suggest that breakage occurs within a replication bubble in a manner that leaves overlapping sequences at the end of the broken chromosome and in the amplicon. Breakage at the sites indicated by the arrows will generate this result. In addition, the figure shows that changing the number of break sites produces amplicons that can vary enormously in size and molecular configuration. Breakage within a single replication bubble is shown for simplicity and is not meant to exclude the possibility that breakage could occur at sites separated by multiple replication origins to generate larger amplicons. Also, the figure is not meant to exclude the possibility that symmetrical breakage could occur in some cases. Note that with symmetrical breakage, sequences in the amplicon would not be present at the break site. This could have profound affects on the frequencies of deletion and targeted insertion of extrachromosomal amplicons. In Figures 3–5, the small solid circles represent sequences 5′ of the *DHFR* gene, and the *DHFR* gene is indicated by the solid rectangles.

level in some target chromosomes that have generated *DHFR* amplicons (Trask and Hamlin, 1989; Windle et al., 1991).

C. Characteristics of Early and Late *DHFR* Amplicons

1. Early *DHFR* amplicons are genetically unstable (Kaufman and Schimke, 1981; Windle et al., 1991). The initial amplicons generate substantial cell-to-cell copy number heterogeneity within eight or nine cell doublings of their formation.
2. Two types of amplicon are evident in independent clones analyzed eight or nine cell doublings after application of the selective agent (Windle et al., 1991). Small, extrachromosomal elements were observed in two-thirds of the clones, while one-third of the clones contained huge amplicons, which appeared to contain all the DNA between the *DHFR* gene and the telomere.
3. Extrachromosomal elements were far more abundant at earlier times. In cells containing small amplicons, an average of approximately 40% of all metaphases analyzed at eight or nine cell doublings contained small extrachromosomal elements; by contrast, only 1–2% of metaphases analyzed at 30–35 cell doublings contained similar structures.
4. Heterogeneous chromosomal localization of *DHFR* amplicons is apparent within 30–35 cell doublings of the initial event (Ruiz and Wahl, 1990; Windle et al., 1991). Expanded chromosomal arrays localized mainly to derivatives of the chromosome that harbored the amplification target (Nunberg et al., 1978; Biedler et al., 1982; Trask and Hamlin, 1989; Windle et al., 1991). Chromosomal amplicons in single-step and multistep mutants can also be localized to multiple chromosomes (Trask and Hamlin, 1989; Windle et al., 1991).
5. Chromosomal amplicons in late stage mutants can be arranged as a mixture of inverted and direct repeats, with the former being far more prevalent (Looney and Hamlin, 1987).
6. Chromosomes harboring amplified *DHFR* genes do not give rise to copy number heterogeneity or *DHFR* gene deletion at measurable frequencies. Surprisingly, this conclusion also holds for dicentric chromosomes, even though such structures do rearrange by breakage–fusion–bridge cycles in mammalian cells (Kaufman et al., 1983; Ruiz and Wahl, 1990; Windle et al., 1991).

V. A MODEL FOR GENE AMPLIFICATION AND GENE DELETION IN WHICH CHROMOSOME BREAKAGE IS THE INITIATING EVENT

Each of the currently proposed models for gene amplification accounts for some of the observations concerning CHO *DHFR* gene amplification, but none of the models readily accounts for all of them unless significant additional assumptions are made. Two of the most often cited models, DNA overreplication (the "onion-skin" model, Schimke et al., 1986) and unequal sister chromatid exchange (see Stark et al., 1989, for a review), will be considered for illustration. DNA overreplication could account for the initial genesis of unstable amplicons by "extrusion" of supernumerary acentric amplicons. At least some of the extruded molecules should contain replication origins, which would enable them to be replicated in subsequent cell cycles. On the other hand, the DNA overreplication model does not predict the formation of chromosome breaks, the frequent deletion of the target site, the formation of either terminal or interstitial deletions from a single initiating event, and the frequent localization of amplified sequences to *derivatives* of the chromosome that underwent the amplification event. Unequal sister chromatid exchange could explain the localization of the amplified sequences near or at the site of the native gene. However, this model requires multiple independent recombination events to generate the substantial copy number heterogeneity detected in early populations. Since *DHFR* gene amplification occurs spontaneously at frequencies of 10^{-5}, one would have to propose that these independent intrachromosomal recombination events occur at rates far higher than reported thus far for mammalian cells (e.g., see Bollag et al., 1990). The cytogenetic aspects of *DHFR* gene amplification are also difficult to explain by this model.

Incontrovertible evidence against any of the existing amplification models has not been presented thus far, but difficulties such as those discussed above have led us to develop the alternative model presented in Figure 3 and discussed below. The occurrence of chromosome breaks near the *DHFR* gene in cells that have undergone *DHFR* gene amplification and were analyzed within eight or nine cell doublings of the initial event suggest that breakage is an early and most likely initial event in the amplification process. A similar conclusion was reached on the basis of cytogenetic observations of multistep mutants by Biedler and colleagues (e.g., see Biedler et al., 1982, 1988). We propose that the amplicons generated by chromosome breakage are acentric molecules and that copy number increases occur by the unequal segregation of such molecules at mitosis. Thus, this model predicts that accumulation of multiple gene copies requires more than one cell cycle and that only a twofold increase in copy number should occur in a single cell cycle. Other mechanisms such as rolling circle or double rolling circle replication of early circular intermediates (e.g., Passananti et al., 1987) could potentially generate large copy number increases in a single cell cycle, but there is no direct evidence to support the occurrence of such processes in mammalian amplicons.

What is the molecular substrate for chromosome breakage? The observation that the broken chromosomes can contain terminal sequences that are also present in the amplicon suggests that two copies of such sequences were present at the time breakage occurred. Since replication generates two copies of the target sequence, and transient inhibition of replication increases amplification frequency, we

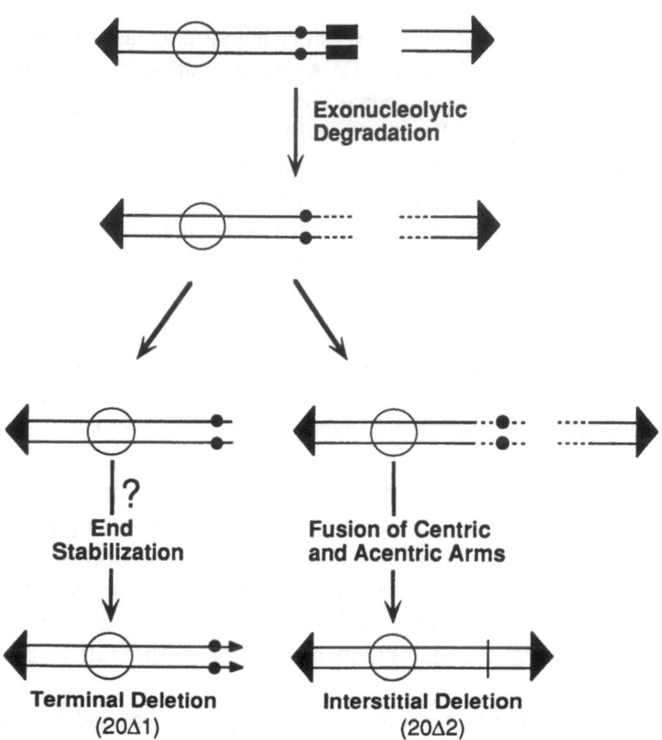

Figure 4 Generation of heterogeneous deletions from a single initiating event: random shortening of the atelomeric ends created by chromosome breakage can lead to the formation of molecularly and cytogenetically distinguishable deletions from a single initiating event. Note that breakage creates two fragments that lack telomeres (large solid arrowheads) at one end. One of the fragments also lacks a centromere (large open circle). Since the acentric fragment can replicate but cannot segregate faithfully, it will be lost from some cells (left) and retained in others (right). Processes such as de novo telomere addition or resection up to telomere-like sequences may stabilize the broken end (indicated by small solid arrowheads) so that it is no longer subject to extensive shortening, producing a chromosome with a large terminal deletion (as in the mutant 20Δ1, Windle et al., 1991; bottom left). On the other hand, the broken centric fragment may be "healed" by reattachment of the broken acentric fragment (right). This will create a chromosome with an apparent interstitial deletion of all the sequences removed prior to reattachment (as in mutant 20Δ2 described in Windle et al., 1991).

propose that stalled replication intermediates are likely to be one type of substrate in which breakage occurs.

Amplicons with the different structures and sizes detected thus far in many gene amplification systems could be produced by simply varying the number and distribution of breaks as indicated in Figure 3. For example, two breaks (Fig. 3a) generate an acentric chromosome fragment spanning the distance between the break site and the telomere. Three breaks (Fig. 3b) could produce a circle containing one or more replication origins, a broken acentric fragment distal to the *DHFR* gene, and a broken centric fragment proximal to the *DHFR* gene. If the breaks are introduced asymmetrically as shown, the broken chromosomes in all three cases will be terminated by sequences also present in the amplicons. Asymmetric breaks at four sites followed by appropriate ligation events would generate a circle with an imperfect inverted repeat of the structure present prior to amplification, as well as broken centric and acentric fragments. Circles with inverted repeats could also be generated by recombination within a replicating circle of the type produced by three breaks if recombination occurs prior to segregation of the two replication products. Note that combinations of large and small amplicons could be generated in a single cell by appropriate placement of three or more breaks (e.g., two in the top half of the bubble and one placed appropriately in the bottom half).

The asymmetric breakage–acentric element model can also explain the frequent occurrence of deletions in the *DHFR* locus, as well as their heterogeneous cytogenetic characteristics. We propose that breakage creates chromosome ends that are subject to shortening because they lack telomeres and as a consequence cannot be replicated to completion, and/or because they are substrates for nucleolytic attack (Fig. 4). Random resection of the unprotected ends will generate chromosomes with different lengths of deletion. The existence of chromosomes containing an apparent interstitial deletion can be explained by the ligation of the broken centric and acentric fragments subsequent to removal of sequences from the *DHFR* locus. Chromosomes that retain sequences from the *DHFR* locus at the broken terminus for many generations are likely to have been healed by de novo addition of telomeres, or by resection to a telomerelike sequence or to a sequence that is refractory to further resection for unknown reasons (e.g., see Zakian, 1989).

VI. ON THE GENESIS OF CHROMOSOMALLY LOCALIZED *DHFR* AMPLICONS

The frequent localization of amplified *DHFR* genes to derivatives of the broken target chromosome may also derive from the existence of atelomeric chromosome ends. It has been shown in yeast that unprotected ends generated by chromosome breakage are highly recombinogenic and dramatically increase recombination at the break site (Haber and Thorburn, 1984; Szostak et al., 1983). Additionally, the asymmetric breaks proposed in Figure 3 could leave a region as long as one amplicon length at the end of the broken chromosome. Since this region is also present in the amplicon, and the frequency of homologous recombination increases with the length of the homologous regions (Thomas and Capecchi, 1987), the frequency of targeted insertion by homologous recombination could be increased dramatically. Different extents of resection of the broken end in individual cells, followed by integration of amplicons at the atelomeric terminus by either homologous or nonhomologous recombination, could explain the localization of amplified *DHFR* genes to many different sites along the length of only one of the homologues containing the target gene (Biedler, 1982; Trask and Hamlin, 1989).

The mechanisms for generating the diverse derivatives of chromosome 2 or Z2 that harbor *DHFR* amplicons (e.g., see Trask and Hamlin, 1989; Windle et al., 1991) have not been explored in detail. We speculate that recombination events involving combinations of the broken acentric arm distal to the *DHFR* gene (which lacks a telomere on one side), extrachromosomal *DHFR* amplicons, and the broken centric chromosome could generate many of the aberrant structures detected thus far (see Fig. 5 for examples). Since the acentric arm should replicate normally, but segregate randomly, it could accumulate in some cells and be lost from others. In the latter case, recombination between *DHFR* amplicons and the broken centric fragment would produce a "metacentric" chromosome with an accumulation of

---

Figure 5 Generation of expanded chromosomal regions in derivatives of the wild-type chromosome by amplicon integration. The types of chromosome depicted have been described in many independent studies of *DHFR* gene amplification (e.g., see Biedler et al., 1982; Trask and Hamlin, 1989; Windle et al., 1991). (a) How the extensive homology shared by the broken chromosome and amplicon could lead to preferential integration of one or more amplicons at the break site. This will generate a metacentric variant of the wild-type chromosome in which the amplified *DHFR* sequences (solid rectangles) are localized terminally. Note that an atelomeric broken chromosome terminated by a duplication of the amplified region remains after the first integration, and this should provide a preferred substrate for further integration events. (b) How integration of amplicons into either the broken centric or acentric fragment, followed by fusion of the acentric fragment to the centric fragment, could generate variant wild-type chromosomes. Fusion in the correct orientation produces a chromosome with *DHFR* amplicons localized near the wild-type position (solid circles; left), while fusion in the incorrect orientation generates a chromosome with terminal *DHFR* amplicons. Note that sequences near the *DHFR* gene are retained in the cases shown. Fusion in the incorrect orientation requires ligation of the telomeric end of the chromosome to a broken end. Data to be presented elsewhere indicates that such fusions can occur in CHO cells (B. Windle and G. M. Wahl, in preparation). (c) One mechanism for producing a dicentric chromosome by ligation of the atelomeric ends of a chromosome containing terminal amplicons (for further discussion, see Kaufman et al., 1983; Ruiz and Wahl, 1990).

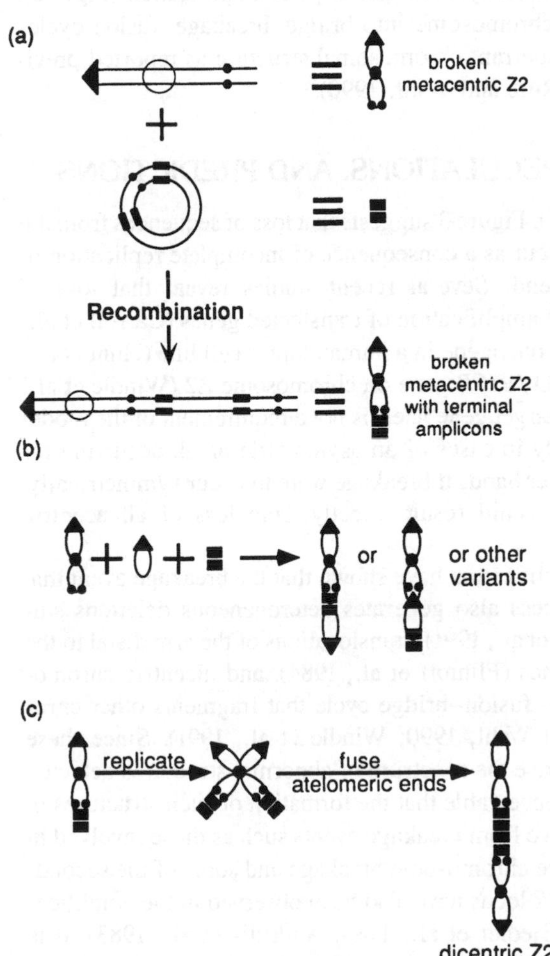
(a)
broken
metacentric Z2
Recombination
broken
metacentric Z2
with terminal
amplicons
(b)
or
or other
variants
(c)
replicate
fuse
atelomeric ends
dicentric Z2

sequences at one end (Fig. 5a). Preferential recombination between extrachromosomal amplicons and the broken acentric fragment, followed by fusion to the centric fragment, could generate other variants of the target chromosome (Fig. 5b). Finally, some of the chromosomes generated by the recombination events above should have atelomeric ends that could "heal" by fusion to generate the dicentric chromosomes found in many examples of gene amplification (Fig. 5c). The entrance of the dicentric chromosome into bridge–breakage–fusion cycles could generate other types of aberrant chromosomal structure as reported previously (Kaufman et al., 1983; Ruiz and Wahl, 1990).

VII. IMPLICATIONS, SPECULATIONS, AND PREDICTIONS

The breakage model illustrated in Figure 3 suggests that loss of sequences from the amplification target site may occur as a consequence of incomplete replication or resection of the unprotected end. Several recent studies reveal that loss of sequences is associated with the amplification of transfected genes (Carroll et al., 1988; Ruiz and Wahl, 1990), an oncogene in a human tumor cell line (Hunt et al., 1990), and the endogenous CHO *DHFR* gene on chromosome Z2 (Windle et al., 1991). Note that deletion of the target gene itself is not a requirement of the model and may occur only infrequently in cases of an asymmetric break occurring far from the target gene. On the other hand, if breakage were to occur symmetrically, cells with target site deletion would result directly from loss of all acentric amplicons.

Analyses of *DHFR* gene amplification have shown that the breakage event that we propose to initiate the process also generates heterogeneous deletions surrounding the break site (Windle et al., 1991), translocations of the arm distal to the target gene to other chromosomes (Flintoff et al., 1984), and dicentric chromosomes that establish a breakage–fusion–bridge cycle that fragments other chromosomes in the cell (Ruiz and Wahl, 1990; Windle et al., 1991). Since these chromosomal aberrations comprise the spectrum of abnormal structures detected in malignant tumor cells, it is conceivable that the formation of such structures in malignant cells in vivo may derive from breakage events such as those involved in gene amplification in vitro. Since chromosome breakage and some of the secondary events described at the *DHFR* locus have also been observed in the amplification of other genes (e.g., see Biedler et al., 1988; Andrulis et al., 1983), it is reasonable to infer that chromosome breakage will be generally important in the amplification process and that cells with chromosome breakage syndromes might undergo gene amplification at elevated rates.

New directions for research in the genesis of genetic plasticity in cancer cells are suggested by the idea that stalled replication bubbles provide the crucial intermediates for chromosome breakage. For example, it is now recognized that advanced stage cancer cells can contain numerous chromosome rearrangements.

If some of the loci involved replicate at the same time in S phase, adverse growth conditions could lead to the accumulation of stalled replication bubbles containing them. Breakage within two or more such structures could then lead to coincident amplification, deletion, and translocation of several genetic loci in a single cell. Such a process could explain the observation that cells containing an amplification of two unlinked genes can be obtained at 10–100-fold higher frequencies than would be expected for the occurrence of two independent events (Giulotto et al., 1987; Rice et al., 1986). The notion that replication intermediates could lead to genetic instability also suggests how one might search for genes that enable normal cells to undergo gene amplification and other large-scale genetic alterations. We propose that one class of candidate genes will be those which, when mutated, enable the cell to enter S phase under suboptimal growth conditions and, as a consequence, to accumulate stalled replication intermediates. Several intriguing possibilities include genes such as *ras* and *p53*, whose mutant phenotype can be created by dominant point mutations. The techniques currently available to molecular geneticists will enable a test of this hypothesis, and ultimately, a more intimate knowledge of the interplay between cell cycle regulation and the maintenance of chromosome stability.

REFERENCES

Andrulis, I. L., Duff, C., Evans-Blackler, S., Worton, R., and Siminovitch, L. Chromosomal alterations associated with overproduction of asparagine synthetase in albizziin-resistant Chinese hamster ovary cells. *Mol. Cell. Biol. 3*:391–398 (1983).

Biedler, J. L. Evidence for a transient prolonged extrachromosomal existence of amplified DNA sequences in antifolate-resistant, vincristine-resistant, and human neuroblastoma cells, in *Gene Amplification* (R. T. Schimke, ed.), Cold Spring Harbor Laboratory, Cold Spring Harbor, New York, 1982, pp. 39–45.

Biedler, J. L., Chang, T., Scotto, K. W., Melera, P. W., and Spengler, B. A. Chromosomal organization of amplified genes in multidrug-resistant Chinese hamster cells. *Cancer Res. 48*:3179–3187 (1988).

Bollag, R. J., Waldman, A. S., and Liskay, R. M. Homologous recombination in mammalian cells. *Ann. Rev. Genet. 23*:199–225 (1989).

Carroll, S. M., DeRose, M. L., Gaudray, P., Moore, C. M., Needham-Vandevanter, D., Von Hoff, D. D., and Wahl, G. M. Double minute chromosomes can be produced from precursors derived from a chromosomal deletion. *Mol. Cell. Biol. 8*:1525–1533 (1988).

Carroll, S., Gaudray, P., DeRose, M. L., Emery, J. F., Meinkoth, J. L., Nakkim, E., Subler, M., and Wahl, G. M. Characterization of an episome produced in hamster cells that amplify a transfected *CA* gene at high frequency: Functional evidence for a mammalian replication origin. *Mol. Cell Biol. 7*:1740–1750 (1987).

Carroll, S. M., Trotter, J., and Wahl, G. M. Replication timing control can be maintained in extrachromosomally amplified genes. *Mol. Cell. Biol.*, (in press).

Chasin,, L. A., Graf, L., Ellis, N., Landzberg, M., and Urlaub, G. Gene amplification in dihydrofolate reductase-deficient mutants, in *Gene Amplification* (R. T. Schimke, ed.), Cold Spring Harbor Laboratory Press, Cold Spring Harbor, New York, 1982, pp. 161–166.

Cowell, J. K. Double minutes and homogeneously staining regions: Gene amplification in mammalian cells. *Ann. Rev. Genet. 16*:21–59 (1982).

Dolf, G., Meyn, R. E., Curley, D., Prather, N., Story, M. D., Boman, B. M., Siciliano, M. J., and Hewitt, R. R. Extrachromosomal amplification of the epidermal growth factor receptor gene in a human colon carcinoma cell line. *Genes, Chromosomes Cancer, 3*, (1991).

Flintoff, W. F., Davidson, S. V., and Siminovitch, L. Isolation and partial characterization of three methotrexate-resistant phenotypes from Chinese hamster ovary cells. *Somat. Cell Genet. 2*:245–261 (1976).

Flintoff, W. E., Livingston, E., Duff, C., and Worton, R. G. Moderate-level gene amplification in methotrexate-resistant Chinese hamster ovary cells is accompanied by chromosomal translocation at or near the site of the amplified *DHFR* gene. *Mol. Cell. Biol. 4*:69–76 (1984).

Fortuna, M. B., Dewey, M. J., and Furmanski, P. Cell fusion in tumor development and progression: Occurrence of cell fusion in primary methylcholanthrene-induced tumorigenesis. *Int. J. Cancer, 44*:731–737 (1989).

Gilbert, D. Temporal order of replication of *Xenopus laevis* 5s ribosomal RNA genes in somatic cells. *Proc. Natl. Acad. Sci. USA, 83*:2924–2928 (1986).

Giulotto, E., Saito, I., and Stark, G. R. Structure of DNA formed in the first step of *CAD* amplification. *EMBO J. 5*:2115–2121 (1986).

Haber, J. E., and Thorburn, P. C. Healing of broken linear dicentric chromosomes in yeast. *Genetics, 106*:207–226 (1984).

Hahn, P., Kapp, L. N., Morgan, W. F., and Painter, R. B. Chromosomal changes without DNA overproduction in hydroxyurea-treated mammalian cells: Implications for gene amplification. *Cancer Res. 46*:4607–4612 (1986).

Hunt, J. D., Valentine, M., and Tereba, A. Excision of N-*myc* from chromosome 2 in human neuroblastoma cells containing amplified N-*myc* sequences. *Mol. Cell. Biol. 10*:823–829 (1990).

Johnston, R. N., Beverley, S. M., and Schimke, R. T. Rapid spontaneous dihydrofolate reductase gene amplification shown by fluorescence-activated cell sorting. *Proc. Natl. Acad. Sci. USA, 80*:3711–3715 (1983).

Kaufman, R. J., Sharp, P. S., and Latt, S. A. Evolution of chromosomal regions containing transfected and amplified dihydrofolate reductase sequences. *Mol. Cell. Biol. 3*:699–711 (1983).

Kaufman, R. J., and Schimke, R. T. Amplification and loss of dihydrofolate reductase genes in a Chinese hamster ovary cell line. *Mol. Cell. Biol. 1*:1069–1076 (1981).

Looney, J. E., and Hamlin, J. L. Isolation of amplified dihydrofolate reductase domain from methotrexate-resistant Chinese hamster ovary cells. *Mol. Cell. Biol. 7*:569–577 (1987).

Mariani, B. D., and Schimke, R. T. Gene amplification in a single cell cycle in Chinese hamster ovary cells. *J. Biol. Chem. 259*:1901–1910 (1984).

Maurer, B. J., Lai, E., Hamkalo, B. A., Hood, L., and Attardi, G. Novel submicroscopic extrachromosomal elements containing amplified genes in human cells. *Nature, 327*: 434–437 (1987).

McClintock, B. The significance of responses of the genome to challenge. *Science, 226*:792–801 (1984).

Meese, E., Olson, S., Horwitz, S., and Trent, J. Gene amplification is mediated by extrachromosomal elements in the multidrug-resistant J774.2 murine cell line. *Preclin. Pharm./Exp. Ther. 31*:380 (1990).

Passananti, C., Davies, B., Ford, M., and Fried, M. Structure of an inverted duplication formed as a first step in a gene amplification event: Implications for a model of gene amplification. *EMBO J. 6*:1697–1703 (1987).

Pauletti, G., Lai, E., and Attardi, G. Early appearance and long-term persistence of the submicroscopic extrachromosomal elements (amplisomes) containing the amplified *DHFR* genes in human cell lines. *Proc. Natl. Acad. Sci. USA, 87*:2955–2959 (1990).

Rice, G. C., Ling, V., and Schimke, R. T. Frequencies of independent and simultaneous selection of Chinese hamster cells for methotrexate and dioxorubicin (adriamycin) resistance. *Proc. Natl. Acad. Sci. USA, 84*:9261–9264 (1987).

Ruiz, J., Choi, K., Von Hoff, D. D., Roninson, I. B., and Wahl, G. M. Autonomously replicating episomes contain *mdr*-1 genes in a multidrug-resistant human cell line. *Mol. Cell. Biol. 9*:109–115 (1989).

Ruiz, J. C., and Wahl, G. M. Chromosomal destabilization during gene amplification. *Mol. Cell. Biol. 10*:3056–3066 (1990).

Saito, I., Groves, R., Giulotto, E., Rolfe, M., and Stark, G. R. Evolution and stability of chromosomal DNA coamplified with the *CAD* gene. *Mol. Cell. Biol. 9*:2445–2452 (1989).

Schimke, R. T., Sherwood, S. W., Hill, A. B., and Johnston, R. N. Overreplication and recombination of DNA in higher eukaryotes: Potential consequences and biological implications. *Proc. Natl. Acad. Sci. USA, 83*:2157–2161 (1986).

Sen, S., Hittleman, W. N., Teeeter, L. D., and Kuo, T. T. A model for the formation of double minutes from prematurely condensed chromosomes of replicating micronuclei in drug treated cells undergoing DNA amplifications. *Cancer Res. 49*:6731–6737 (1989).

Stark, G. R., Debatisse, M., Guilotto, E., and Wahl, G. M. Recent progress in understanding mechanisms of mammalian DNA amplification. *Cell, 57*:901–908 (1989).

Szostak, J. W., and Wu, R. Unequal crossing over in the ribosomal DNA of *Saccharomyces cerevisiae*. *Nature, 284*:426–430 (1980).

Thomas, K., and Capecchi, M. R. Site-directed mutagenesis by gene targeting in mouse embryo-derived stem cells. *Cell, 52*:503–512 (1987).

Tlsty, T. D., and Adams, P. Replication of the dihydrofolate reductase genes on double minute chromosomes in a murine cell line. *Exp. Cell Res. 188*:148–164 (1990).

Tlsty, T. D., Brown, P. C., Johnston, R., and Schimke, R. T. Enhanced frequency of generation of methotrexate resistance and gene amplification in cultured mouse and hamster cell lines, in *Gene Amplification* (R. T. Schimke, ed.), Cold Spring Harbor Laboratory Press, Cold Spring Harbor, New York, 1982, pp. 231–238.

Trask, B. J., and Hamlin, J. L. Early dihydrofolate reductase gene amplification events in CHO cells usually occur on the same chromosome arm as the original locus. *Genes Dev. 3*: 1913–1923 (1989).

van der Bliek, A. M., Lincke, C. R., and Borst, P. Circular DNA of 3T6R50 double minute chromosomes. *Nucleic Acids Res. 16*:4841–4851 (1988).

Van Devanter, D. R., Piaskowski, V. C., Casper, J. T., Douglass, E. C., and Von Hoff, D. D. Ability of circular extrachromosomal DNA molecules to carry amplified *myc*-N proto-oncogenes in human neuroblastomas in vivo. *J. Natl. Cancer Inst. 82*:1815–1821 (1990).

Von Hoff, D. D., Forseth, B., Clare, C. N., Hansen, K. L., and VanDevanter, D. Double minutes arise from circular extrachromosomal DNA intermediates which integrate into chromosomal sites in human HL-60 leukemia cells. *J. Clin. Invest. 85*:1887–1895 (1990).

Von Hoff, D. D., Needham-Vandevanter, D. R., Yucel, J., Windle, B. F., and Wahl, G. M. Amplified c-*myc* genes are contained on replicating, submicroscopic, circular DNA molecules in some HL60 and COLO320 cell lines. *Proc. Natl. Acad. Sci. USA, 85*: 4804–4808 (1988).

Wahl, G. M. The importance of circular DNA in mammalian gene amplification. *Cancer Res. 49*:1333–1340 (1989).

Wahl, G. M., Vitto, L., and Rubnitz, J. Co-amplification of rDNA genes with *CAD* genes in *N*-(phosphonacetyl)-L-aspartate-resistant Syrian hamster cells. *Mol. Cell. Biol. 3*:2066–2075 (1983).

Windle, B., Draper, B. W., Yin, Y., O'Gorman, S., and Wahl, G. M. *Genes Dev.* (1991).

Windle, B., Draper, B. W., Yin, Y., O'Gorman, S., and Wahl, G. M. A central role for chromosome breakage in gene amplification, deletion formation, and amplicon integration. *5*:160–174 (1991).

Yeung, C-Y., Ingolia, D. E., Bobonis, C., Dunbar, B. S., Riser, M. E., Siciliano, M. J., and Kellems, R. E. Selective overproduction of adenosine deaminase in cultured mouse cells. *J. Biol. Chem. 258*:8338–8345 (1983).

Zakian, V. A. Structure and function of telomeres. *Annu. Rev. Genet. 23*:579–604 (1989).

Index

Milton Keynes UK
Ingram Content Group UK Ltd.
UKHW021929071024
449327UK00022B/1736

9 780367 402662